10 - 11:40 AM
5143 CLASSROOM

6:30 - 8:20
3270 CLASSROOM

ALAN S̶H̶_____
491 DEN__
ROCHESTER, N.Y.
14616
865-7424
722-0413

TECHNICAL CALCULUS

PRENTICE-HALL SERIES IN TECHNICAL MATHEMATICS
Frank L. Juszli, *Editor*

TECHNICAL CALCULUS

Second Edition

DALE EWEN

Parkland Community College

MICHAEL A. TOPPER

Yavapai College

PRENTICE-HALL, Englewood Cliffs, New Jersey 07632

Library of Congress Cataloging-in-Publication Data

EWEN, DALE.
   Technical calculus.

   (Prentice-Hall series in technical mathematics)
   Includes index.
   1. Calculus. I. Topper, Michael A.
II. Title. III. Series.
QA303.E95 1986   515   85-12393
ISBN 0-13-898164-7

Editorial/production supervision: Mary Carnis and Colleen Brosnan
Cover design: Wanda Lubelska Design
Manufacturing buyer: John Hall

Portions of Chapters 1 and 6 were previously published in Dale Ewen and Michael A. Topper, *Mathematics for Technical Education*, 2nd ed., © 1983. Reprinted by permission of Prentice-Hall, Englewood Cliffs, N.J.

Portions of Chapter 8 were previously published in Dale Ewen and Lynn Akers, *Trigonometry with Applications*, © 1984. Reprinted by permission of Addison-Wesley, Reading, Mass.

Printed in the United States of America

10 9 8 7 6 5 4 3 2 1

ISBN 0-13-898164-7 025

PRENTICE-HALL INTERNATIONAL (UK) LIMITED, *London*
PRENTICE-HALL OF AUSTRALIA PTY. LIMITED, *Sydney*
PRENTICE-HALL CANADA INC., *Toronto*
PRENTICE-HALL HISPANOAMERICANA, S.A., *Mexico*
PRENTICE-HALL OF INDIA PRIVATE LIMITED, *New Delhi*
PRENTICE-HALL OF JAPAN, INC., *Tokyo*
PRENTICE-HALL OF SOUTHEAST ASIA PTE. LTD., *Singapore*
EDITORA PRENTICE-HALL DO BRASIL, LTDA., *Rio de Janeiro*
WHITEHALL BOOKS LIMITED, *Wellington, New Zealand*

CONTENTS

Contents   ix

PREFACE

This book is designed to provide the calculus skills for those students who are enrolled in an engineering technology program that requires a development of a practical calculus. The students are assumed to have a mathematics background of algebra and trigonometry. This text is intended for use in Associate Degree programs as well as ABET (Accrediting Board for Engineering Technology) programs and BIT (Bachelor of Industrial Technology) programs. The companion text, *Mathematics for Technical Education*, serves as a smooth transition to this book, although other equivalent texts are also feasible.

The topics are presented in an intuitive manner with technical applications being integrated wherever possible. The large number of detailed examples and the abundance and variety of exercises are features that students and instructors alike find essential.

Chapter 1 provides the basic analytic geometry needed for a study of a practical calculus. Chapters 2 and 3 present some intuitive discussions about the limit and develop basic techniques and applications of differentiation. Chapters 4 and 5 develop integration and some basic applications. Chapters 6 and 7 develop the necessary differentiation and methods of integration of the transcendental functions. Chapters 8 and 9 are new chapters that address the areas of polar coordinates, partial differentiation, and double integrals. Chapter 10 has been revised to provide a basic understanding of series. Chapters 11 and 12 provide an introduction to differential equations with technical applications.

Changes from the first edition include:

1. Condensing analytic geometry into one chapter (1).

2. Condensing the basic development of the derivative into one chapter (2).

3. Adding topics and sections in the chapter on applications of integration (5).

4. Realigning the development of the topics in the differentiation and integration of the transcendental functions (6 and 7).

5. Adding new sections on such topics as the trigonometric functions (6.1), inverse trigonometric functions (6.4), and the exponential and logarithmic functions (6.6), which may be used for review, used for more in-depth study, or may be omitted, depending on the needs and background of the class.

6. Adding a new chapter on polar coordinates and one on partial derivatives and double integrals (8 and 9).

7. Condensing some sections and adding new sections on convergence and divergence of series in the series chapter (10).

8. Condensing the tables to conform with the common and universal use of calculators.

9. Adding a variety of optional computer examples to stimulate the use of computers whenever they are available to the class and can add to the class discussion and development of certain topics. Specific uses will depend on the class's background and expertise regarding computers. They may be omitted without loss of continuity or basic development of the text.

10. Strengthening most exercise sets by providing more exercises in both number and variety.

11. Critically reviewing all sections, which has resulted in numerous changes that give better development, clearer explanations, and more or better examples.

I thank the many colleagues who assisted directly or indirectly, especially the reviewers: Joseph Jordan, John Tyler Community College; Frank L. Juszli, Prentice-Hall Series Editor; Kenneth J. Morgan, DeVry Institute of Technology; and Douglas W. Peterson, Green River Community College.

If you wish to offer suggestions or criticisms or have questions, please feel free to contact Dale Ewen directly at Parkland College, 2400 West Bradley, Champaign, Illinois 61821 or through Prentice-Hall.

I extend my sincere and special thanks to the excellent Prentice-Hall editorial-production staff under the most capable supervision of Colleen Brosnan; to my colleague, Michael Emerick, for preparing the solutions manual, checking the answers, and proofing; to Lynn Akers for his assistance with the computer examples; to George Morris for his usual outstanding artwork; and to Joyce Ewen for her excellent proofing assistance.

Finally, I thank Michael Topper for working with me on the first edition.

Dale Ewen

1

ANALYTIC GEOMETRY

1.1 RECTANGULAR COORDINATES

The **positive integers** are the counting numbers; that is, 1, 2, 3, ... Note that the positive integers form an infinite set. The **negative integers** may be defined as the set of opposites of the positive integers; that is, -1, -2, -3, ... **Zero** is the dividing point between the positive integers and the negative integers and is neither positive nor negative. The set of integers consists of the positive integers, the negative integers, and zero. The set may be represented on a number line as in Fig. 1.1.

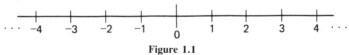

Figure 1.1

The **rational numbers** are those numbers that can be represented as the ratio of two integers, such as $\frac{3}{4}$, $\frac{-7}{5}$, and $\frac{5}{1}$. The **irrational numbers** are those numbers that cannot be represented as the ratio of two integers, such as $\sqrt{3}$, $\sqrt[3]{16}$, and π.

The set of **real numbers** is the set consisting of the rational numbers and the irrational numbers. With respect to the number line, we say there is a one-to-one correspondence between the real numbers and the points on the number line; that is, for each real number there is a corresponding point on the number line, and for each point on the number line there is a corresponding real number. As a result, we say the number line is dense, or "filled."

1

Inequalities are statements involving less than or greater than and may be used to describe various intervals on the number line, as follows:

Type of interval	Symbols	Meaning	Number line graph
Open	$x > a$	x is greater than a	(open circle at a, extending right)
	$x < b$	x is less than b	(open circle at b, extending left)
	$a < x < b$	x is between a and b	(open circles at a and b)
Half-open	$x \geq a$	x is greater than or equal to a	(closed circle at a, extending right)
	$x \leq b$	x is less than or equal to b	(closed circle at b, extending left)
	$a < x \leq b$	x is between a and b, including b but excluding a	(open circle at a, closed at b)
	$a \leq x < b$	x is between a and b, including a but excluding b	(closed circle at a, open at b)
Closed	$a \leq x \leq b$	x is between a and b, including both a and b	(closed circles at a and b)

Analytic geometry is the study of the relationships between algebra and geometry. The concepts of analytic geometry provide us with ways of algebraically analyzing a geometrical problem. Likewise, with these concepts we can often solve an algebraic problem by viewing it geometrically. We will develop several basic relations between equations and their graphs.

In mathematics we define a **relation** as a set of ordered pairs, usually in the form (x, y). Sometimes a description or rule is also included that states the relationship between x and y. In an ordered pair the first element may be represented by any letter, but x is normally used. This first element is called the **independent variable**. The letter y is normally used as the second element of an ordered pair and is called the **dependent variable**.

All those numbers that are used as a first element of an ordered pair of a given relation form a set of numbers called the **domain**. The domain is often referred to as the set of all x's.

The **range** of a relation is the set of those numbers that are used as a second element of an ordered pair. The range may be referred to as the set of all y's.

Example 1

Given the relation $A = \{(1, 2), (3, 5), (7, 9), (6, 3)\}$, find its domain and range.

The domain is the set of first elements: $\{1, 3, 6, 7\}$. The range is the set of second elements: $\{2, 3, 5, 9\}$. *Note:* Braces { } may be used to designate sets.

Example 2

Consider the relation given by the equation $y = x^2$. Find its domain and range.

Note that there are no restrictions on the substitutions for x. That is, we may replace x by any real number. This is another way of saying that the domain is the set of real numbers. After each replacement of x, there is no way that we can get a negative value for y because the square of any real number is always positive or zero. This means that the range is the set of nonnegative real numbers, or $y \geq 0$.

Example 3

Consider the relation $y = \sqrt{x - 4}$. Find its domain and range.

Note that no value of x less than 4 may be used because the square root of any negative number is not a real number. We then say that the domain is the set of real numbers greater than 4 or equal to 4, or $x \geq 4$. The square root of a real number is never negative. So the range is $y \geq 0$.

Function

A **function** is a special relation: a set of ordered pairs in which no two distinct ordered pairs have the same first element.

Example 4

Consider $B = \{(3, 2), (6, 7), (5, 3), (1, 1), (3, 7)\}$. Is B a function? Find its domain and range.

Note that B is not a function because it contains two ordered pairs which have the same first element: (3, 2) and (3, 7). The domain is $\{1, 3, 5, 6\}$ and the range is $\{1, 2, 3, 7\}$.

Is set $A = \{(1, 2), (3, 5), (7, 9), (6, 3)\}$ from Example 1 a function? Yes, because no two ordered pairs have the same first element.

Is the relation $x = y^2$ a function? Can we find two ordered pairs which have the same first element? Yes, for example (9, 3) and (9, -3), or (16, 4) and (16, -4), or (36, 6) and (36, -6), and so forth. Therefore, $x = y^2$ is not a function. The domain is the set of nonnegative real numbers; that is, $x \geq 0$ because each x-value, being the square of a y-value, can never be negative. The range is the set of real numbers because there are no restrictions on y.

Consider the relations in Examples 2 and 3. Are they functions? Note that in the relation $y = x^2$, for each value of x there is only one corresponding value of y. For example, (2, 4), (-2, 4), (3, 9), (-3, 9), (4, 16), (-4, 16), and so forth. Therefore, $y = x^2$ is a function.

Example 3 was the relation $y = \sqrt{x - 4}$. Here we find that for each x-value there corresponds only one y-value [for example, (5, 1), (8, 2), (10, $\sqrt{6}$), . . .]. Therefore, $y = \sqrt{x - 4}$ is a function.

In summary, a function is a relationship between two sets of numbers, the domain and range, which relates with each number x in the domain one and only one number y in the range.

Next, consider a plane in which two number lines intersect at right angles. Let the point of intersection be the zero point of each line and call it the **origin**.

Each line is called an **axis**. The horizontal number line is usually called the **x-axis** and the vertical line is usually called the **y-axis**. Normally, the units are the same for each line. Such a system is called the **rectangular coordinate system**, or the **Cartesian coordinate system**. (The name *Cartesian* is after Descartes, a seventeenth-century French mathematician. He first conceived this idea of combining algebra and geometry together in such a way that each could aid the study of the other.) The plane is divided by the axes into four regions called **quadrants**. The quadrants are numbered as shown in Fig. 1.2.

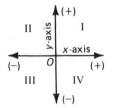

Figure 1.2

In the plane there is a point which corresponds to each ordered pair of real numbers (a, b). Likewise, there is an ordered pair (a, b) which corresponds to each point in the plane. Together a and b are called the **coordinates** of the point; a is called the **abscissa** and b is called the **ordinate**. This relationship is called a **one-to-one correspondence**. The location, or position, of a point in the plane corresponding to a given ordered pair is found by first counting right or left from O (origin) the number of units along the x-axis indicated by the first number of the ordered pair (right if positive, left if negative). Then from this point reached on the x-axis count up or down the number of units indicated by the second number of the ordered pair (up if positive, down if negative).

Example 5

Plot the point corresponding to each ordered pair in the number plane. (See Fig. 1.3.)

$A (3, 1)$
$B (2, -3)$
$C (-4, -2)$
$D (-3, 0)$
$E (-6, 2)$
$F (0, 2)$

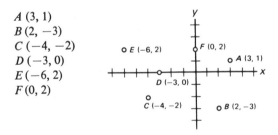

Figure 1.3

To graph equations we plot a sample of ordered pairs and connect them with a smooth curve. To obtain the sample, we need to generate ordered pairs from a given equation. One way to generate these ordered pairs is by randomly choosing a value for x, replacing this value for x in the equation, and solving for y.

4   Analytic Geometry   Chap. 1

Example 6

Graph $y = 2x - 3$.

Plot the ordered pairs and connect with a smooth line as in Fig. 1.4.

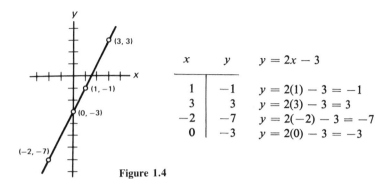

x	y	$y = 2x - 3$
1	-1	$y = 2(1) - 3 = -1$
3	3	$y = 2(3) - 3 = 3$
-2	-7	$y = 2(-2) - 3 = -7$
0	-3	$y = 2(0) - 3 = -3$

Figure 1.4

The graph of an equation that is not linear is usually a curve of some kind and hence requires several points to sketch a smooth curve.

Example 7

Graph $y = x^2 - 4$. (See Fig. 1.5.)

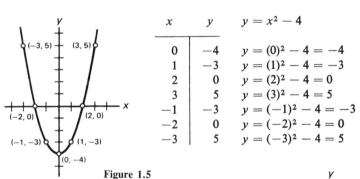

x	y	$y = x^2 - 4$
0	-4	$y = (0)^2 - 4 = -4$
1	-3	$y = (1)^2 - 4 = -3$
2	0	$y = (2)^2 - 4 = 0$
3	5	$y = (3)^2 - 4 = 5$
-1	-3	$y = (-1)^2 - 4 = -3$
-2	0	$y = (-2)^2 - 4 = 0$
-3	5	$y = (-3)^2 - 4 = 5$

Figure 1.5

Example 8

Graph $y = 2x^2 + x - 5$. (See Fig. 1.6.)

x	y	$y = 2x^2 + x - 5$
0	-5	$y = 2(0)^2 + (0) - 5 = -5$
1	-2	$y = 2(1)^2 + (1) - 5 = -2$
2	5	$y = 2(2)^2 + (2) - 5 = 5$
-1	-4	$y = 2(-1)^2 + (-1) - 5 = -4$
-2	1	$y = 2(-2)^2 + (-2) - 5 = 1$
-3	10	$y = 2(-3)^2 + (-3) - 5 = 10$

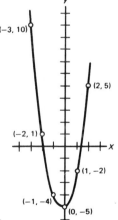

Figure 1.6

For a more complicated function more ordered pairs are usually required to obtain a smooth curve. It may also be necessary to scale the graph in order to plot enough ordered pairs to obtain a smooth curve. To scale means to enlarge or reduce the units on the axes according to a specified ratio. This ratio is chosen on the basis of fitting the necessary values on a given graph.

Exercises

For each relation determine whether or not it is a function. Write its domain and range.

1. $A = \{(2, 4), (3, 7), (9, 2)\}$ **2.** $B = \{(5, 2), (3, 3), (1, 2)\}$

3. $C = \{(2, 5), (7, 3), (2, 1), (1, 3)\}$ **4.** $D = \{(0, 2), (5, -1), (2, 7), (5, 1)\}$

5. $E = \{(3, 2), (5, 2), (2, 2), (-2, 2)\}$

6. $F = \{(3, 4), (3, -4), (-3, -4), (-3, 4)\}$

7. $y = 2x + 5$ **8.** $y = -3x$ **9.** $y = x^2 + 1$

10. $y = 2x^2 - 3$ **11.** $x = y^2 - 2$ **12.** $x = 3y^2 + 4$

13. $y = \sqrt{x + 3}$ **14.** $y = \sqrt{3 - 6x}$

15. Plot the point corresponding to each ordered pair in the number plane.

A (2, 5)	B (-3, -6)	C (-4, 0)
D (2, -1)	E (3, 4)	F (0, 2)
G (0, -3)	H (5, 0)	I (-4, -1)
J (-2, 2)	K (3, -5)	L (-3, 4)

Graph each equation.

16. $y = 3x - 4$ **17.** $y = 2x + 1$ **18.** $2y = -4x - 3$

19. $-2x - 3y = 6$ **20.** $y = x^2 + x - 6$ **21.** $y = x^2 - 9$

22. $y = x^2 + 3$ **23.** $y = x^2 - 5x + 4$ **24.** $y = -x^2 + 2x + 4$

25. $y = -2x^2 + 4x$ **26.** $y = x^2 - 4x$ **27.** $y = x^2 + 2x$

28. $y = x^3 - 8x - 3$ **29.** $y = x^3 - x^2 - 10x + 8$ **30.** $y = \sqrt{3x - 12}$

31. $y = \sqrt{x + 4}$ **32.** $y = \sqrt{3 - x}$ **33.** $y = \sqrt{12 - 6x}$

1.2 THE STRAIGHT LINE

The graph of a linear equation (of degree one) with two unknowns in the form $ax + by = c$ is always a straight line. Two ordered pairs are sufficient to graph a linear function, since two points determine a straight line. However, finding a third point provides good insurance against a careless error.

Example 1

    Graph $y = -3x + 5$. (See Fig. 1.7.)

x	y	$y = -3x + 5$
0	5	$y = -3(0) + 5 = 5$
2	-1	$y = -3(2) + 5 = -1$
-1	8	$y = -3(-1) + 5 = 8$

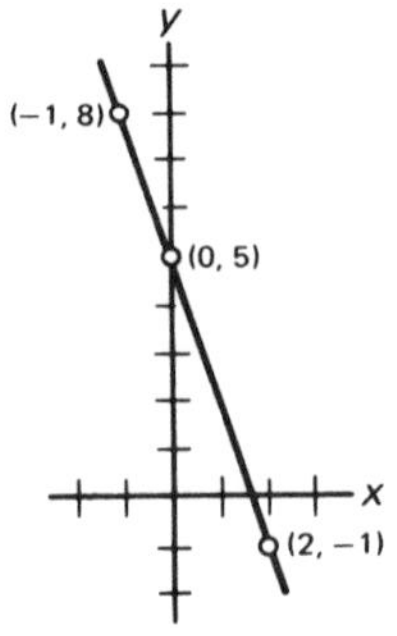

Figure 1.7

The **slope** of a nonvertical line is the ratio of the difference of the y-coordinates of any two points on the line to the difference of their x-coordinates when the differences are taken in the same order. (See Fig. 1.8.)

Slope

If $P_1(x_1, y_1)$ and $P_2(x_2, y_2)$ represent any two points on a straight line, then its slope m is

$$m = \frac{y_2 - y_1}{x_2 - x_1}$$

Example 2

Find the slope of the line passing through $(-2, 1)$ and $(3, 5)$.

If we let $x_1 = -2$, $y_1 = 1$, $x_2 = 3$, and $y_2 = 5$ as in Fig. 1.9, then

$$m = \frac{y_2 - y_1}{x_2 - x_1} = \frac{5 - 1}{3 - (-2)} = \frac{4}{5}$$

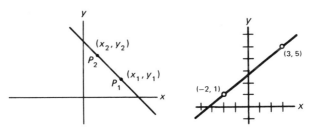

Figure 1.8   Figure 1.9

Note if we reverse the order of taking the differences of the coordinates, the result is the same.

$$\frac{y_1 - y_2}{x_1 - x_2} = \frac{1 - 5}{-2 - 3} = \frac{-4}{-5} = \frac{4}{5} = m$$

Example 3

Find the slope of the line passing through $(-2, 4)$ and $(4, -6)$.

If we let $x_1 = -2$, $y_1 = 4$, $x_2 = 4$, and $y_2 = -6$ as in Fig. 1.10, then

$$m = \frac{y_2 - y_1}{x_2 - x_1} = \frac{-6 - 4}{4 - (-2)} = \frac{-10}{6} = -\frac{5}{3}$$

Note that in Example 2, the line slopes upward from left to right, while in Example 3 the line slopes downward. In general, we have the following:

1. If a line has positive slope, then the line slopes upward from left to right.
2. If a line has negative slope, then the line slopes downward from left to right.
3. If the line has zero slope, then the line is horizontal.

4. If the line is vertical, then the line has no slope as $x_1 = x_2$, or $x_1 - x_2 = 0$. In this case, the ratio $\dfrac{y_2 - y_1}{x_2 - x_1}$ is undefined because division by zero is undefined.

We can use these facts to assist us in graphing a line if we know the slope of the line and one point P on the line. The line can be sketched by drawing a line through the given point P and a point Q which is plotted by moving one unit to the right of P, then moving vertically m units. That is, a point moving along a line will move vertically an amount equal to m, the slope, for every unit move to the right as in Fig. 1.11.

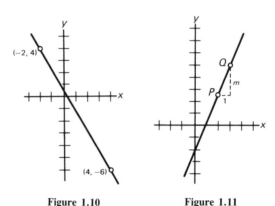

Figure 1.10   Figure 1.11

Example 4

Graph a line whose slope is -2 and which passes through the point $(1, 3)$.

Since the slope is -2, points on the line drop 2 units for every unit move to the right. The line passes through $(1, 3)$ and $(2, 1)$ as in Fig. 1.12.

Knowing the slope and one point on the line will also determine the equation of the straight line.

Let m be the slope of a given nonvertical straight line and let (x_1, y_1) be the coordinates of a point on this line. If (x, y) is any other point on the line (as in Fig. 1.13), then we have

$$\frac{y - y_1}{x - x_1} = m$$

By multiplying each side of the equation by $(x - x_1)$ we obtain

---

Point-Slope Form of a Straight Line

If m is the slope and (x_1, y_1) is any point on a nonvertical straight line, its equation is

$$y - y_1 = m(x - x_1)$$

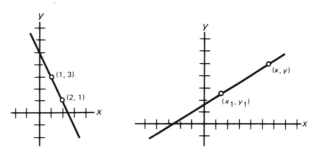

Figure 1.12   Figure 1.13

Example 5

Find the equation of the line with slope 3 that passes through the point $(-1, 2)$.

Here $m = 3$, $x_1 = -1$, and $y_1 = 2$. Using the point-slope form, we have the equation

$$y - y_1 = m(x - x_1)$$
$$y - 2 = 3[x - (-1)]$$

Simplifying, we have

$$y - 2 = 3x + 3$$
$$3x - y + 5 = 0$$

The point-slope form can also be used to find the equation of a straight line that passes through two points.

Example 6

Find the equation of the line passing through the points $(2, -3)$ and $(-2, 5)$.

First, find the slope.

$$m = \frac{y_2 - y_1}{x_2 - x_1} = \frac{5 - (-3)}{-2 - 2} = \frac{5 + 3}{-4} = \frac{8}{-4} = -2$$

Substituting $m = -2$ and the point $(2, -3)$ in the point-slope form, we obtain the equation

$$y - y_1 = m(x - x_1)$$
$$y - (-3) = -2(x - 2)$$
$$y + 3 = -2x + 4$$
$$2x + y - 1 = 0$$

Note: We could have used the other point $(-2, 5)$ in the point-slope form to obtain the equation

$$y - 5 = -2[x - (-2)]$$

which simplifies to

$$2x + y - 1 = 0$$

A nonvertical line will intersect the y-axis at some point in the form $(0, b)$ as in Fig. 1.14.

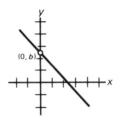

Figure 1.14

This ordinate b is called the **y-intercept** of the line. If the slope of the line is m, then

$$y - y_1 = m(x - x_1)$$
$$y - b = m(x - 0)$$
$$y - b = mx$$

or

Slope-Intercept Form of a Straight Line

If m is the slope and $(0, b)$ is the y-intercept of a nonvertical straight line, its equation is

$$y = mx + b$$

Example 7

Find the equation of the line whose slope is $\frac{1}{2}$ and which crosses the y-axis at $b = -3$.

Using the slope-intercept form we have

$$y = mx + b$$
$$y = \frac{1}{2}x + (-3)$$
$$y = \frac{1}{2}x - 3$$

A line which is parallel to the x-axis has slope $m = 0$, so its equation becomes

$$y = mx + b$$
$$y = (0)x + b$$

or

Equation of a Horizontal Line

If a horizontal line passes through the point $(0, b)$, its equation is

$$y = b$$

Example 8

Find the equation of the line parallel to and 3 units above the x-axis.

The equation is $y = 3$.

Analytic Geometry   Chap. 1

By writing the equation of a nonvertical straight line in the slope-intercept form, it is possible to determine quickly the line's slope and a point on the line (the point where it crosses the y-axis). The line may then be graphed as in Example 4.

Example 9

Find the slope and the y-intercept of $3y - x + 6 = 0$. Graph the line.
Write the equation in slope-intercept form; that is, solve for y.

$$3y - x + 6 = 0$$
$$3y = x - 6$$
$$y = \frac{x}{3} - 2$$
$$y = \left(\frac{1}{3}\right)x + (-2)$$

So $m = \frac{1}{3}$ and $b = -2$. (See Fig. 1.15.)

Example 10

Describe the line whose equation is $y = -5$.
This is a line parallel to and 5 units below the x-axis as in Fig. 1.16.

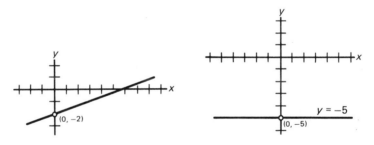

| Figure 1.15 | Figure 1.16 |

If a line is vertical, then we cannot use any of these equations since the line has no slope. However, note that in this case as in Fig. 1.17 the line crosses the x-axis at some point in the form $(a, 0)$. All points on the line have the same abscissa as the point $(a, 0)$. This characterizes the line, giving us the equation

---

Equation of a Vertical Line

If a vertical line passes through the point $(a, 0)$, its equation is

$$x = a$$

Example 11

Describe the line whose equation is $x = 2$.
This is a line perpendicular to the x-axis that crosses the x-axis at the point $(2, 0)$ as in Fig. 1.18.

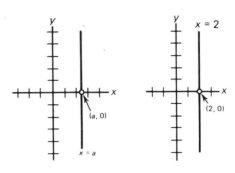

Figure 1.17   Figure 1.18

Example 12

Write the equation of the line perpendicular to the x-axis that crosses the x-axis at the point $(-3, 0)$.

The equation is $x = -3$.

Note: All the equations presented in this section can be put in the form

$$Ax + By + C = 0$$

This is known as the **general form** of the equation of the line and agrees with our definition of a linear equation whose graph is a straight line.

Parallel Lines

Two lines are parallel if either one of the following conditions holds.

1. They are both perpendicular to the x-axis. (See Fig. 1.19a.)
2. They both have the same slope (see Fig. 1.19b); that is, if the equations of the lines are

$$L_1: y = m_1x + b_1 \quad \text{and} \quad L_2: y = m_2x + b_2$$

then

$$m_1 = m_2$$

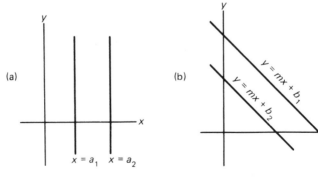

Figure 1.19

Perpendicular Lines

Two lines are perpendicular if either one of the following conditions holds.

1. One line is vertical with equation $x = a$ and the other is horizontal with equation $y = b$.
2. Neither is vertical and the slope of one line is the negative reciprocal of the other; that is, if the equations of the lines are

$$L_1: y = m_1x + b_1 \quad \text{and} \quad L_2: y = m_2x + b_2$$

then

$$m_1 = -\frac{1}{m_2}$$

To show this last relationship, consider the triangle in Fig. 1.20, where L_1 is perpendicular to L_2.

Let

$(c, 0)$ represent the point P

$(d, 0)$ represent the point R

(e, f) represent the point Q

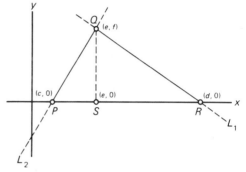

Figure 1.20

Draw QS perpendicular to the x-axis. Then S must be represented by $(e, 0)$.

Triangles PSQ and QSR are similar. (Note that angle PQS equals angle QRS.) From geometry we know that

$$\frac{PS}{QS} = \frac{QS}{SR}$$

In this case,

$PS = e - c$ (the distance from c to e on the x-axis)

$QS = f$ (the distance from 0 to f on the y-axis)

$SR = d - e$ (the distance from e to d on the x-axis)

Substituting these values in the equation above we have

$$\frac{e - c}{f} = \frac{f}{d - e}$$

Multiplying each side of the equation by $(d - e)f$ gives

$$f^2 = (d - e)(e - c)$$

We can compute the slopes m_1 and m_2 as follows.

$$m_1 = \frac{f - 0}{e - d} = \frac{f}{e - d}$$

$$m_2 = \frac{f - 0}{e - c} = \frac{f}{e - c}$$

$$(m_1)(m_2) = \frac{f}{e - d} \cdot \frac{f}{e - c} = \frac{f^2}{(e - d)(e - c)}$$

Substituting from the equation $f^2 = (d - e)(e - c)$ we have

$$(m_1)(m_2) = \frac{(d - e)(e - c)}{(e - d)(e - c)} = \frac{d - e}{e - d} = -\left(\frac{e - d}{e - d}\right) = -1$$

Dividing each side of the equation by m_2 we have

$$m_1 = \frac{-1}{m_2}$$

Example 13

Determine whether the lines given by the equations $3y + 6x - 5 = 0$ and $2y - x + 7 = 0$ are perpendicular.

Change each equation into slope-intercept form; that is, solve for y.

$$y = -2x + \frac{5}{3} \qquad \text{(Slope is } -2\text{.)}$$

$$y = \frac{1}{2}x - \frac{7}{2} \qquad \text{(Slope is } \tfrac{1}{2}\text{.)}$$

Since

$$-2 = \frac{-1}{\frac{1}{2}} \qquad (-2 \text{ is the negative reciprocal of } \tfrac{1}{2}\text{.)}$$

the lines are perpendicular.

Example 14

Determine the equation of the line passing through $(-3, 2)$ and perpendicular to $2y - 3x + 5 = 0$.

We can find the slope of the desired line by finding the negative reciprocal of the slope of the given line. First find the slope of the line $2y - 3x + 5 = 0$. Writing this equation in slope-intercept form, we have

$$y = \frac{3}{2}x - \frac{5}{2}.$$

The slope of this line is $m = \tfrac{3}{2}$. The slope of the line perpendicular to this line is then equal to $-\tfrac{2}{3}$, the negative reciprocal of $\tfrac{3}{2}$. Now using the point-slope form we have

$$y - y_1 = m(x - x_1)$$

$$y - 2 = -\frac{2}{3}[x - (-3)]$$

$$y - 2 = -\frac{2}{3}(x + 3)$$

$$2x + 3y = 0$$

Exercises

Find the slope of each line passing through the given points.

1. $(4, 2), (3, 1)$
2. $(-3, 2), (-1, -2)$
3. $(4, -5), (2, 3)$
4. $(-6, -4), (5, -3)$
5. $(-3, 2), (6, 2)$
6. $(4, -7), (4, 3)$
7. $(5, 7), (-3, 2)$
8. $(-3, 6), (-1, 3)$

Graph each line passing through the given point with the given slope.

9. $(2, -1)$, $m = 2$
10. $(0, 1)$, $m = -3$
11. $(-3, -2)$, $m = \frac{1}{2}$
12. $(4, 4)$, $m = -\frac{1}{3}$
13. $(4, 0)$, $m = -2$
14. $(-3, 1)$, $m = 4$
15. $(0, -3)$, $m = -\frac{3}{4}$
16. $(5, -2)$, $m = \frac{3}{2}$

Find the equation of the line with the given properties.

17. Passes through $(-2, 8)$ with slope of -3.
18. Passes through $(3, -5)$ with slope of 2.
19. Passes through $(-3, -4)$ with slope of $\frac{1}{2}$.
20. Passes through $(6, -7)$ with slope of $-\frac{3}{4}$.
21. Passes through $(-2, 7)$ and $(1, 4)$.
22. Passes through $(1, 6)$ and $(4, -3)$.
23. Passes through $(6, -8)$ and $(-4, -3)$.
24. Passes through $(-2, 2)$ and $(7, -1)$.
25. Crosses the y-axis at -2 with slope of -5.
26. Crosses the y-axis at 8 with slope of $\frac{1}{3}$.
27. Has y-intercept of 7 and slope of 2.
28. Has y-intercept of -4 and slope of $-\frac{3}{4}$.
29. Parallel to and 5 units above the x-axis.
30. Parallel to and 2 units below the x-axis.
31. Perpendicuar to the x-axis and crosses the x-axis at $(-2, 0)$.
32. Perpendicular to the x-axis and crosses the x-axis at $(5, 0)$.

Find the slope and y-intercept of each straight line.

33. $x + 4y = 12$
34. $-2x + 3y + 9 = 0$
35. $4x - 2y + 14 = 0$
36. $3x - 6y = 0$
37. $y = 6$
38. $x = -4$

Graph each equation.

39. $y = 3x - 2$
40. $y = -2x + 5$
41. $5x - 2y + 4 = 0$
42. $4x + 3y + 6 = 0$
43. $x = 7$
44. $x = -2$
45. $y = -3$
46. $y = 2$

Determine whether the given pairs of equations represent lines which are parallel, perpendicular, or neither.

47. $x + 3y - 7 = 0$; $-3x + y + 2 = 0$
48. $x + 2y - 11 = 0$; $x + 2y + 4 = 0$
49. $-x + 4y + 7 = 0$; $x + 4y - 5 = 0$
50. $2x + 7y + 4 = 0$; $7x - 2y - 5 = 0$
51. $y - 5x + 13 = 0$; $y - 5x + 9 = 0$
52. $-3x + 9y + 22 = 0$; $x + 3y - 17 = 0$

Determine the equation of the line that satisfies each set of conditions.

53. Passes through $(-1, 5)$ and is parallel to $-2x + y + 13 = 0$.

54. Passes through $(2, -2)$ and is perpendicular to $3x - 2y - 14 = 0$.

55. Passes through $(-7, 4)$ and is perpendicular to $5y = x$.

56. Passes through $(2, -10)$ and is parallel to $2x + 3y - 7 = 0$.

57. Passes through the origin and is parallel to $3x - 4y = 12$.

58. Passes through the origin and is perpendicular to $4x + 5y = 17$.

59. Has an x-intercept of 6 and is perpendicular to $4x + 6y = 9$.

60. Has a y-intercept of -2 and is parallel to $6x - 4y = 11$.

61. Has a y-intercept of 8 and is parallel to $y = 2$.

62. Has an x-intercept of -4 and is perpendicular to $y = 6$.

63. Has an x-intercept of 7 and is parallel to $x = -4$.

64. Has a y-intercept of -9 and is perpendicular to $x = 5$.

65. The vertices of a quadrilateral are $A(-2, 3)$, $B(2, 2)$, $C(9, 6)$, and $D(5, 7)$.
(a) Is the quadrilateral a parallelogram? Why or why not? (b) Is the quadrilateral a rectangle? Why or why not?

66. The vertices of a quadrilateral are $A(-4, 1)$, $B(0, -2)$, $C(6, 6)$, and $D(2, 9)$.
(a) Is the quadrilateral a parallelogram? Why or why not? (b) Is the quadrilateral a rectangle? Why or why not?

1.3 THE DISTANCE AND MIDPOINT FORMULAS

We now wish to compute the distance between two points on a straight line. Suppose that P has the coordinates (x_1, y_1) and Q has the coordinates (x_2, y_2). Then a triangle similar to Fig. 1.21 can be constructed. Note that R must have the coordinates (x_2, y_1). (Point R has the same x-coordinate as Q and the same y-coordinate as P.)

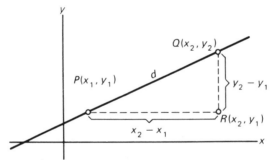

Figure 1.21

Using the Pythagorean theorem we have
$$PQ^2 = PR^2 + QR^2$$

Observe that

$PR = x_2 - x_1$ (the horizontal distance between x_1 and x_2 on the x-axis)

$QR = y_2 - y_1$ (the vertical distance between y_1 and y_2 on the y-axis)

Substituting these values for PR and QR in the equation above gives
$$PQ^2 = (x_2 - x_1)^2 + (y_2 - y_1)^2$$

Distance Formula

The distance between two points $P(x_1, y_1)$ and $Q(x_2, y_2)$ is given by the formula
$$d = PQ = \sqrt{(x_2 - x_1)^2 + (y_2 - y_1)^2}$$

Example 1

Find the distance, d, between $(3, 4)$ and $(-2, 7)$.
$$d = \sqrt{(x_2 - x_1)^2 + (y_2 - y_1)^2}$$
$$d = \sqrt{(-2 - 3)^2 + (7 - 4)^2}$$
$$= \sqrt{(-5)^2 + (3)^2} = \sqrt{25 + 9} = \sqrt{34}$$

Midpoint Formula

The coordinates of point $Q(x_m, y_m)$ which is midway between two points $P(x_1, y_1)$ and $R(x_2, y_2)$ are given by
$$x_m = \frac{x_1 + x_2}{2} \quad \text{and} \quad y_m = \frac{y_1 + y_2}{2}$$

This can be seen by looking at the points P, Q, and R plotted in Fig. 1.22. Triangles PSQ and QTR are congruent. This means that
$$PS = QT$$
Since
$$PS = x_m - x_1 \quad \text{and} \quad QT = x_2 - x_m$$
then
$$x_m - x_1 = x_2 - x_m$$
$$2x_m = x_1 + x_2$$
$$x_m = \frac{x_1 + x_2}{2}$$

The formula for y_m is determined in the same manner.

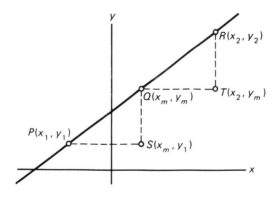

Figure 1.22

Example 2

Find the point midway between $(2, -3)$ and $(-4, 6)$.

$$x_m = \frac{x_1 + x_2}{2} = \frac{2 + (-4)}{2} = \frac{-2}{2} = -1$$

$$y_m = \frac{y_1 + y_2}{2} = \frac{-3 + 6}{2} = \frac{3}{2}$$

The midpoint is $(-1, \frac{3}{2})$.

Exercises

Find the distance between each pair of points.

1. $(4, -7)$; $(-5, 5)$ 2. $(4, 3)$; $(-2, -1)$ 3. $(3, -2)$; $(10, -2)$
4. $(6, -2)$; $(6, 4)$ 5. $(5, -2)$; $(1, 2)$ 6. $(2, -3)$; $(-1, 1)$
7. $(3, -5)$; $(3, 2)$ 8. $(2, -4)$; $(6, -4)$

Find the coordinates of the point midway between each pair of points.

9. $(2, 3)$; $(5, 7)$ 10. $(0, 5)$; $(2, -4)$ 11. $(3, -2)$; $(0, 0)$
12. $(2, -3)$; $(4, -3)$ 13. $(11, 4)$; $(-11, -9)$ 14. $(4, 10)$; $(-6, -8)$

The vertices of each $\triangle ABC$ are given below. For each triangle, find (a) the perimeter, (b) whether it is a right triangle, (c) whether it is isosceles, and (d) its area if it is a right triangle.

15. $A(2, 8)$; $B(10, 2)$; $C(10, 8)$ 16. $A(0, 0)$; $B(3, 3)$; $C(3, -3)$

17. Given $\triangle ABC$ with vertices $A(7, -1)$, $B(9, 1)$, and $C(-3, 5)$, find the distance from A to the midpoint of side BC.

18. Find the distance from B to the midpoint of side AC in Exercise 17.

19. Write the equation of the line through the midpoint of AB, where $A(4, 2)$ and $B(8, -6)$, and parallel to the line $3x - 6y = 10$.

20. Write the equation of the line through the midpoint of AB, where $A(-3, -4)$ and $B(7, -8)$, and perpendicular to the line $2x + 5y = 12$.

1.4 THE CIRCLE

Equations in two variables of second degree in the form

$$Ax^2 + Bxy + Cy^2 + Dx + Ey + F = 0$$

are called **conics**. We now begin a systematic study of conics with the circle.

The **circle** consists of the set of points located the same distance from a given point, called the **center**. The distance at which all points are located from the center is called the **radius**. A circle may thus be graphed in the plane given its center and radius.

Example 1

Graph the circle whose center is $(1, -2)$ and whose radius $r = 3$.

Plot all points in the plane located 3 units away from the point $(1, -2)$ as in Fig. 1.23. (You may wish to use a compass.)

From the definition of a circle we can determine the equation of a circle. Let (h, k) be the coordinates of the center and let r represent the radius. If any point (x, y) is located on the circle, it must be a distance r from the center (h, k) as in Fig. 1.24.

Using the distance formula,

$$\sqrt{(x_2 - x_1)^2 + (y_2 - y_1)^2} = d$$

we find that $\sqrt{(x - h)^2 + (y - k)^2} = r$. Squaring each side, we have

---

Standard Form of a Circle

$$(x - h)^2 + (y - k)^2 = r^2$$

where r is the radius and (h, k) is the center.

Any point (x, y) satisfying this equation must lie on the circle.

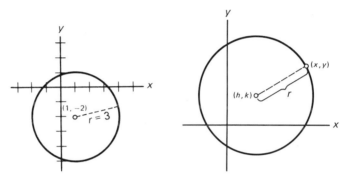

Figure 1.23   Figure 1.24

Example 2

Determine the equation of the circle with radius 3 and center $(1, -2)$. (See Example 1.)

Using the standard form of the equation of a circle, we have

$$(x - h)^2 + (y - k)^2 = r^2$$
$$(x - 1)^2 + [y - (-2)]^2 = (3)^2$$
$$(x - 1)^2 + (y + 2)^2 = 9$$

Example 3

Determine the equation of the circle with center at $(3, -2)$ and passing through $(-2, 1)$.

To write the equation, we need to know the radius r of the circle. While r has not been stated, we do know that every point on the circle is a distance r from the center, $(3, -2)$. In particular, the point $(-2, 1)$ is a distance r from $(3, -2)$. Using the distance formula, we have

$$d = r = \sqrt{(x - h)^2 + (y - k)^2}$$
$$r = \sqrt{(-2 - 3)^2 + [1 - (-2)]^2}$$
$$= \sqrt{(-5)^2 + 3^2} = \sqrt{25 + 9} = \sqrt{34}$$

Now write the equation of the circle.

$$(x - h)^2 + (y - k)^2 = r^2$$
$$(x - 3)^2 + [y - (-2)]^2 = (\sqrt{34})^2$$
$$(x - 3)^2 + (y + 2)^2 = 34$$

If we now perform the multiplications indicated in the equation
$$(x - h)^2 + (y - k)^2 = r^2$$
we have
$$x^2 - 2xh + h^2 + y^2 - 2yk + k^2 = r^2$$
Rearranging terms, we have
$$x^2 + y^2 - 2hx - 2ky + h^2 + k^2 - r^2 = 0$$
If we let $D = -2h$, $E = -2k$, and $F = h^2 + k^2 - r^2$, we obtain the equation

---

General Form of a Circle
$$x^2 + y^2 + Dx + Ey + F = 0$$

Any equation in this form represents a circle.

Example 4

Write the equation $(x - 3)^2 + (y + 2)^2 = 34$ obtained in Example 3 in general form.
$$(x - 3)^2 + (y + 2)^2 = 34$$
$$x^2 - 6x + 9 + y^2 + 4y + 4 = 34$$
$$x^2 + y^2 - 6x + 4y - 21 = 0$$

Example 5

Find the center and radius of the circle given by the equation
$$x^2 + y^2 - 4x + 2y - 11 = 0$$

Looking back at how we arrived at the general equation of a circle, we see that if we rearrange the terms of the equation
$$(x^2 - 4x \quad) + (y^2 + 2y \quad) = 11$$
then $(x^2 - 4x \quad)$ represents the first two terms of
$$(x - h)^2 = x^2 - 2hx + h^2$$
and $(y^2 + 2y \quad)$ represents the first two terms of
$$(y - k)^2 = y^2 - 2ky + k^2$$
This means that
$$-4 = -2h \quad \text{or} \quad h = 2$$
and
$$2 = -2k \quad \text{or} \quad k = -1$$
To complete the squares $(x - h)^2$ and $(y - k)^2$, we must add $h^2 = 2^2 = 4$ and $k^2 = (-1)^2 = 1$ to each side of the equation.
$$(x^2 - 4x \quad) + (y^2 + 2y \quad) = 11$$
$$(x^2 - 4x + 4) + (y^2 + 2y + 1) = 11 + 4 + 1$$

$$(x - 2)^2 + (y + 1)^2 = 16$$
$$(x - 2)^2 + [y - (-1)]^2 = 16 = 4^2$$

From this we see that we have the standard form of the equation of a circle with radius 4 and center at the point $(2, -1)$.

This process is called **completing the square** of the *x*- and *y*-terms. In general, if the coefficients of x^2 and y^2 are both equal to 1, then these values can be found as follows: Add h^2 and k^2 to each side of the equation, where

$$h^2 = (\tfrac{1}{2} \text{ the coefficient of } x)^2 = (\tfrac{1}{2}D)^2$$
$$k^2 = (\tfrac{1}{2} \text{ the coefficient of } y)^2 = (\tfrac{1}{2}E)^2$$

Example 6

Find the center and radius of the circle given by the equation

$$x^2 + y^2 + 6x - 4y - 12 = 0$$

Sketch the graph of the circle.

$$h^2 = (\tfrac{1}{2} \cdot 6)^2 = 3^2 = 9$$
$$k^2 = [\tfrac{1}{2}(-4)]^2 = (-2)^2 = 4$$

Rewrite the equation and add 9 and 4 to each side.

$$(x^2 + 6x \quad) + (y^2 - 4y \quad) = 12$$
$$(x^2 + 6x + 9) + (y^2 - 4y + 4) = 12 + 9 + 4$$
$$(x + 3)^2 + (y - 2)^2 = 25 = 5^2$$

The center is at $(-3, 2)$ and the radius is 5. Plot all points which are at a distance of 5 from the point $(-3, 2)$ as in Fig. 1.25.

If the center of a circle is at the origin, then $h = 0$ and $k = 0$, and its standard equation becomes

$$x^2 + y^2 = r^2$$

where *r* is the radius and the center is at the origin.

Example 7

Write the equation of the circle with radius 3 and center at the origin. Also, graph the circle. (See Fig. 1.26.)

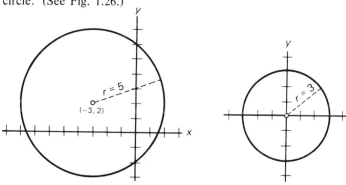

Figure 1.25   Figure 1.26

$$x^2 + y^2 = r^2$$
$$x^2 + y^2 = (3)^2$$
$$x^2 + y^2 = 9$$

Exercises

Graph the circle with the given center and radius.

1. Center at $(2, -1)$, $r = 3$.　　　　　　**2.** Center at $(3, 3)$, $r = 2$.

3. Center at $(0, 2)$, $r = 4$.　　　　　　**4.** Center at $(-4, -5)$, $r = 3$.

Determine the equation of the circle (in standard form) with the given properties.

5. Center at $(1, -1)$, radius 4.　　　　**6.** Center at $(-2, 3)$, radius $\sqrt{5}$.

7. Center at $(-2, -4)$, passing through $(1, -9)$.

8. Center at $(5, 2)$, passing through $(-2, -6)$.

9. Center at $(0, 0)$, radius 6.

10. Center at $(0, 0)$, passing through $(3, -4)$.

Find the center and radius of the given circle.

11. $x^2 + y^2 = 16$　　　　　　　　**12.** $x^2 + y^2 - 4x - 5 = 0$

13. $x^2 + y^2 + 6x - 8y - 39 = 0$　　**14.** $x^2 + y^2 - 6x + 14y + 42 = 0$

15. $x^2 + y^2 - 8x + 12y - 8 = 0$　　**16.** $x^2 + y^2 + 10x + 2y - 14 = 0$

17. $x^2 + y^2 - 12x - 2y - 12 = 0$　**18.** $x^2 + y^2 + 4x - 9y + 4 = 0$

19. $x^2 + y^2 + 7x + 3y - 9 = 0$　　**20.** $x^2 + y^2 - 5x - 8y = 0$

21. Find the equation of the circle or circles whose center is on the y-axis and contains the points $(1, 4)$ and $(-3, 2)$. Give its center and radius.

22. Find the equation of the circle with center in the first quadrant on the line $y = 2x$, which is tangent to the x-axis and radius 6 units. Give its center.

23. Find the equation of the circle containing the points $(3, 1)$, $(0, 0)$, and $(8, 4)$. Give its center and radius.

24. Find the equation of the circle containing the points $(1, -4)$, $(-3, 4)$, and $(4, 5)$. Give its center and radius.

1.5 THE PARABOLA

While the parabola may not be as familiar a geometric curve as the circle, examples of the parabola are found in many technical applications. A **parabola** consists of all points which are the same distance from a given fixed point and a given fixed line. The fixed point is called the **focus**. The fixed line is called the **directrix**. This relationship is shown in Fig. 1.27 for the points P, Q, and V, which lie on a parabola with focus F and directrix D.

　　Note:

$$RP = PF$$
$$SV = VF$$
$$TQ = QF$$

The point V midway between the directrix and the focus is called the **vertex**. The vertex and the focus lie on a line perpendicular to the directrix, which is called the **axis of symmetry**.

There are two standard forms for the equation of a parabola. The form depends upon the position of the parabola in the plane. We shall first look at the parabola with focus on the x-axis at $(p, 0)$ and directrix the line $x = -p$ as in Fig. 1.28. We shall let $P(x, y)$ represent any point on this parabola. The vertex V is then at the origin and the axis of symmetry is the x-axis.

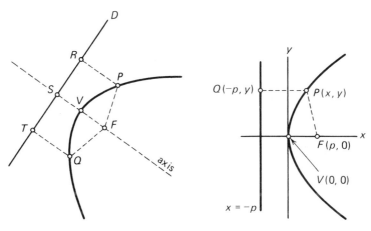

Figure 1.27   Figure 1.28

By the way we have described the parabola, the distance between P and F must equal the distance between P and Q. Using the distance formula, we have

$$PF = PQ$$
$$\sqrt{(x - p)^2 + (y - 0)^2} = \sqrt{[x - (-p)]^2 + (y - y)^2}$$

Squaring each side,

$$(x - p)^2 + y^2 = (x + p)^2$$
$$x^2 - 2xp + p^2 + y^2 = x^2 + 2xp + p^2$$

or

Standard Form of Parabola
$$y^2 = 4px$$
with focus at $(p, 0)$ and with the line $x = -p$ as the directrix.

Note that in Fig. 1.28, $p > 0$.

Example 1

Find the equation of the parabola with focus at $(3, 0)$ and directrix the line $x = -3$.

In this case $p = 3$, so we have
$$y^2 = 4(3)x$$
$$y^2 = 12x$$

Example 2

Find the focus and equation of the directrix of the parabola with equation $y^2 = 24x$.
$$y^2 = 24x$$
$$y^2 = 4(6)x$$
$$y^2 = 4px$$

Since p must be 6, the focus is $(6, 0)$ and the directrix is the line $x = -6$.

Example 3

Find the equation of the parabola with focus at $(-2, 0)$ and with directrix $x = 2$. Sketch the graph. (See Fig. 1.29.)
Here $p = -2$. The equation becomes
$$y^2 = 4(-2)x$$
$$y^2 = -8x$$

Note that when $p < 0$, the parabola opens to the left and that the coefficient of x in the equation is negative.

1. If $p > 0$, the coefficient of x in the equation $y^2 = 4px$ is *positive* and the parabola opens to the *right*. (See Fig. 1.30a.)
2. If $p < 0$, the coefficient of x in the equation $y^2 = 4px$ is *negative* and the parabola opens to the *left*. (See Fig. 1.30b.)

We obtain the other standard form when the focus lies on the y-axis and the directrix is parallel to the x-axis. Let $(0, p)$ be the focus F and $y = -p$ be the directrix. The vertex is still at the origin, but the axis of symmetry is now the y-axis.

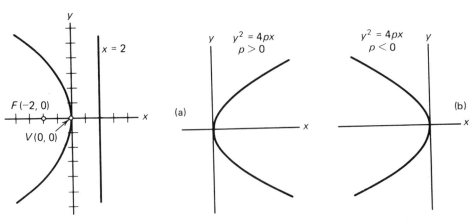

Figure 1.29   Figure 1.30

Standard Form of Parabola
$$x^2 = 4py$$
with focus at $(0, p)$ and with the line $y = -p$ as the directrix. (See Fig. 1.31.)

Example 4

Find the equation of the parabola with focus at $(0, 3)$ and with directrix $y = -3$. Sketch the graph.

Since the focus lies on the y-axis and the directrix is parallel to the x-axis, we use the equation $x^2 = 4py$ with $p = 3$. (See Fig. 1.32.)

$$x^2 = 4(3)y$$
$$x^2 = 12y$$

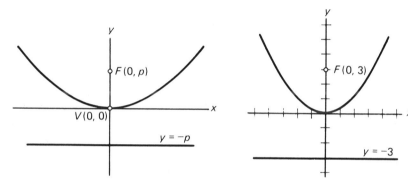

Figure 1.31   Figure 1.32

Example 5

Find the equation of the parabola with focus at $(0, -1)$ and with directrix $y = 1$. Sketch the graph.

Again the focus lies on the y-axis with directrix parallel to the x-axis, so we use the equation $x^2 = 4py$ with $p = -1$. (See Fig. 1.33.)

$$x^2 = 4(-1)y$$
$$x^2 = -4y$$

Observe the following.

1. If $p > 0$, the coefficient of y in the equation $x^2 = 4py$ is *positive* and the parabola opens *upward*. (See Fig. 1.34a.)
2. If $p < 0$, the coefficient of y in the equation $x^2 = 4py$ is *negative* and the parabola opens *downward*. (See Fig. 1.34b.)

We are now able to describe the graph of a parabola with an equation in standard form by inspection. We can also find the focus and directrix.

Sec. 1.5   The Parabola   **25**

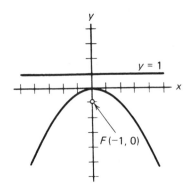

Figure 1.33

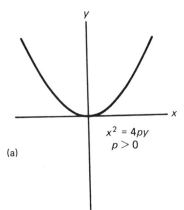

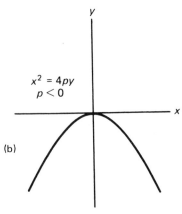

(a)

(b)

$x^2 = 4py$
$p > 0$

$x^2 = 4py$
$p < 0$

Figure 1.34

Example 6

Describe the graph of the equation $y^2 = 20x$.

This is an equation of a parabola in the form $y^2 = 4px$. Since $p = 5$, this parabola has its focus at $(5, 0)$ and its directrix is the line $x = -5$. The parabola opens to the right (since $p > 0$).

Example 7

Describe the graph of the equation $x^2 = -2y$.

This is an equation of a parabola in the form $x^2 = 4py$. In this case $p = -\frac{1}{2}$ as $[4(-\frac{1}{2}) = -2]$. The parabola has its focus at $(0, -\frac{1}{2})$ and its directrix is the line $y = \frac{1}{2}$. The parabola opens downward (since $p < 0$).

Of course, not all parabolas are given in standard position.

Example 8

Find the equation of the parabola with focus at $(1, 3)$ and with the line $y = -1$ as directrix.

We must use our definition of the parabola (see Fig. 1.35).

$$PF = PQ$$

$$\sqrt{(x - 1)^2 + (y - 3)^2} = \sqrt{(x - x)^2 + [y - (-1)]^2}$$

This simplifies to

$$x^2 - 2x - 8y + 9 = 0$$

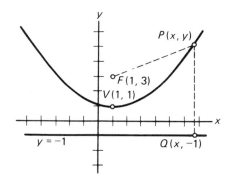

Figure 1.35

In fact, any equation of the form

$$Ax^2 + Dx + Ey + F = 0$$

or

$$Cy^2 + Dx + Ey + F = 0$$

always represents a parabola.

Exercises

Find the focus and the directrix of each parabola. Sketch each graph.

1. $x^2 = 4y$ 2. $x^2 = -8y$ 3. $y^2 = -16x$

4. $x^2 = -6y$ 5. $y^2 = x$ 6. $y^2 = -4x$

7. $x^2 = 16y$ 8. $y^2 = -12x$ 9. $y^2 = 8x$

10. $x^2 = -y$

Find the equation of the parabola with given focus and directrix.

11. $(2, 0)$, $x = -2$ 12. $(0, -3)$, $y = 3$ 13. $(-8, 0)$, $x = 8$

14. $(5, 0)$, $x = -5$ 15. $(0, 6)$, $y = -6$ 16. $(0, -1)$, $y = 1$

17. Find the equation of the parabola with focus at $(-4, 0)$ and vertex at $(0, 0)$.

18. Find the equation of the parabola with vertex at $(0, 0)$ and directrix the line $y = -2$.

19. Find the equation of the parabola with focus $(-1, 3)$ and directrix the line $x = 3$.

20. Find the equation of the parabola with focus $(2, -5)$ and directrix the line $y = -1$.

21. The surface of a roadway over a bridge follows a parabolic curve with vertex at the middle of the bridge. The span of the bridge is 400 m. The roadway is 16 m higher in the middle than at the end supports. How far above the end supports is a point 50 m from the middle? 150 m from the middle?

22. The shape of a wire hanging between two poles closely approximates a parabola. Find the equation of a wire, which is suspended between two poles 40 m apart and whose lowest point is 10 m below the level of the insulators. (Choose the lowest point as the origin of your coordinate system.)

23. A suspension bridge is supported by two cables which hang between two supports. The curve of these cables is approximately parabolic. Find the equation of this curve

if the focus lies 8 m above the lowest point of the cable. (Set up the xy-coordinate system so that the vertex is at the origin.)

24. Sketch the graph of $H = Ri^2$, where H is the rate of development of heat in a resistor measured in watts, i is the current measured in amperes, and $R = 100$ ohms is the resistance.

1.6 THE ELLIPSE

An **ellipse** consists of the set of points in a plane, the *sum* of whose distances from two fixed points is a positive constant. These two fixed points are called **foci**.

Assume for now as in Fig. 1.36 that the foci lie on the x-axis at $(-c, 0)$ and $(c, 0)$. Then any point $P(x, y)$ lies on the ellipse if its distance d_1 from P to the point $(-c, 0)$ plus its distance d_2 from P to the point $(c, 0)$ is equal to a given constant k. Let the constant be written as $k = 2a$; then

$$d_1 + d_2 = 2a$$

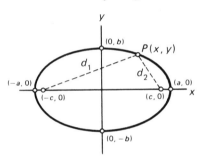

Figure 1.36

Again using the formula for computing the distance between two points, we have

$$\sqrt{[x - (-c)]^2 + (y - 0)^2} + \sqrt{(x - c)^2 + (y - 0)^2} = 2a$$

Rewrite the above equation as follows.

$$\sqrt{(x + c)^2 + y^2} = 2a - \sqrt{(x - c)^2 + y^2}$$

$$(x + c)^2 + y^2 = 4a^2 - 4a\sqrt{(x - c)^2 + y^2} + (x - c)^2 + y^2 \quad \text{(Square each side.)}$$

$$x^2 + 2cx + c^2 + y^2 = 4a^2 - 4a\sqrt{(x - c)^2 + y^2} + x^2 - 2cx + c^2 + y^2$$

$$4cx - 4a^2 = -4a\sqrt{(x - c)^2 + y^2}$$

$$a^2 - cx = a\sqrt{(x - c)^2 + y^2} \quad \text{(Divide each side by } -4.)$$

$$(a^2 - cx)^2 = a^2[(x - c)^2 + y^2] \quad \text{(Square each side.)}$$

$$a^4 - 2a^2cx + c^2x^2 = a^2(x^2 - 2cx + c^2 + y^2)$$

$$a^4 - 2a^2cx + c^2x^2 = a^2x^2 - 2a^2cx + a^2c^2 + a^2y^2$$

$$a^4 - a^2c^2 = a^2x^2 - c^2x^2 + a^2y^2$$

$$a^2(a^2 - c^2) = (a^2 - c^2)x^2 + a^2y^2 \quad \text{(Factor.)}$$

$$1 = \frac{x^2}{a^2} + \frac{y^2}{a^2 - c^2} \quad \text{[Divide each side by } a^2(a^2 - c^2).]$$

If we now let $y = 0$ in this equation, we find that $x^2 = a^2$. The points $(-a, 0)$ and $(a, 0)$ which lie on the graph are called **vertices** of the ellipse. Observe that $a > c$.

If we let $b^2 = a^2 - c^2$, the equation above then becomes

$$\frac{x^2}{a^2} + \frac{y^2}{b^2} = 1$$

The line segment connecting the vertices $(a, 0)$ and $(-a, 0)$ is called the **major axis**. The point midway between the vertices is called the **center** of the ellipse. In this case the major axis lies on the x-axis and the center is at the origin. If we let $x = 0$ in the above equation, we find $y^2 = b^2$. The line connecting $(0, b)$ and $(0, -b)$ is perpendicular to the major axis and passes through the center (see Fig. 1.36). This line is called the **minor axis** of the ellipse. In this case the minor axis lies on the y-axis. Note that $2a$ is the length of the major axis and $2b$ is the length of the minor axis.

Standard Form of Ellipse

$$\frac{x^2}{a^2} + \frac{y^2}{b^2} = 1$$

with center at the origin and with the major axis lying on the x-axis. *Note: $a > b$.*

One easy way to approximate the curve of an ellipse is to fix a string at two points (foci) on a piece of paper as in Fig. 1.37. Then using a pencil to keep the string taut, trace out the curve as illustrated. Note that $d_1 + d_2$ is always constant—namely, the length of the string. Detach the string and compare the length of the string with the length of the major axis; note that the lengths are the same, $2a$.

The relationship $b^2 = a^2 - c^2$ or $a^2 = b^2 + c^2$ can also be seen by looking at this string demonstration as in Fig. 1.38. Put a pencil inside the taut string and on an end of the minor axis; this bisects the length of string and sets up a right triangle with a as its hypotenuse and b and c as the legs.

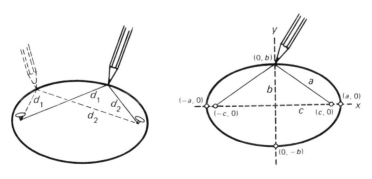

Figure 1.37   Figure 1.38

Example 1

Find the vertices, the foci, and the lengths of the major and minor axes of the ellipse

$$\frac{x^2}{25} + \frac{y^2}{9} = 1$$

Sketch the graph.

Since $a^2 = 25$ the vertices are at $(5, 0)$ and $(-5, 0)$. The length of the major axis is $2a = 2(5) = 10$. Since $b^2 = 9$, the length of the minor axis is $2b = 2(3) = 6$. We need to know the value of c to determine the foci. Since $b^2 = a^2 - c^2$, we can write

$$c^2 = a^2 - b^2$$
$$c^2 = 25 - 9 = 16$$
$$c = 4$$

The foci are thus $(4, 0)$ and $(-4, 0)$. The graph of the ellipse is shown in Fig. 1.39.

You will want to remember the equation relating a, b, and c for the ellipse.

$$\boxed{c^2 = a^2 - b^2}$$

When the major axis lies on the y-axis with center at the origin as in Fig. 1.40, the **standard form** of the equation of the ellipse becomes

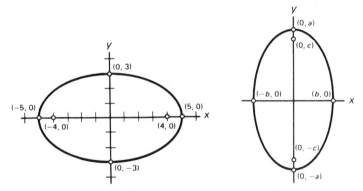

Figure 1.39   Figure 1.40

Standard Form of Ellipse

$$\frac{y^2}{a^2} + \frac{x^2}{b^2} = 1$$

with center at the origin and with the major axis lying on the y-axis.
     Note: $a > b$.

This result may be shown similarly as the derivation of the first standard form. Notice that the larger denominator now lies below y^2 instead of below x^2 as in the first case. The vertices are now $(0, a)$ and $(0, -a)$.

Example 2

Given the ellipse $25x^2 + 9y^2 = 225$, find the foci, vertices, and lengths of the major and minor axes. Sketch the graph.

First divide each side of the equation by 225 to put the equation in standard form.

$$\frac{x^2}{9} + \frac{y^2}{25} = 1$$

Since the larger denominator belongs to the y^2 term, this ellipse has its major axis on the y-axis and a^2 must then be 25. So $a = 5$ and $b = 3$. The vertices are $(0, 5)$ and $(0, -5)$. The length of the major axis is $2a = 10$ and the length of the minor axis is $2b = 6$.

$$c^2 = a^2 - b^2$$
$$= 25 - 9 = 16$$
$$c = 4$$

Thus the foci are $(0, 4)$ and $(0, -4)$. (See Fig. 1.41.)

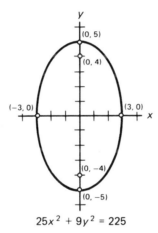

$25x^2 + 9y^2 = 225$ **Figure 1.41**

In general, a is always greater than b for an ellipse. The following are also true.

1. If the larger denominator belongs to the x^2-term, then this denominator is a^2; the major axis lies on the x-axis; and the vertices are $(a, 0)$ and $(-a, 0)$.
2. If the larger denominator belongs to the y^2-term, then this denominator is a^2; the major axis lies on the y-axis; and the vertices are $(0, a)$ and $(0, -a)$.

Example 3

Find the equation of the ellipse with vertices at $(6, 0)$ and $(-6, 0)$ and foci at $(4, 0)$ and $(-4, 0)$.

Since $a = 6$ and $c = 4$, we have $a^2 = 36$ and $c^2 = 16$. Thus

$$b^2 = a^2 - c^2 = 36 - 16 = 20$$

Since the major axis lies on the x-axis, the equation in standard form is

$$\frac{x^2}{a^2} + \frac{y^2}{b^2} = 1$$

$$\frac{x^2}{36} + \frac{y^2}{20} = 1$$

Ellipses with centers not located at the origin will be presented in Section 1.8. If we were to determine the equation of the ellipse where the sum of the distances of all points from the foci (3, 2) and (−5, 2) is always 10, we would have

$$9x^2 + 25y^2 + 18x - 100y - 116 = 0$$

In general, an equation of the form

$$Ax^2 + Cy^2 + Dx + Ey + F = 0$$

represents an ellipse with axes parallel to the coordinate axes, where A and C are both positive (or both negative) and, unlike the circle, $A \neq C$.

Exercises

Find the vertices, foci, and lengths of the major and minor axes of each ellipse. Sketch each graph.

1. $\dfrac{x^2}{25} + \dfrac{y^2}{16} = 1$ 2. $\dfrac{x^2}{36} + \dfrac{y^2}{64} = 1$ 3. $9x^2 + 16y^2 = 144$

4. $25x^2 + 16y^2 = 400$ 5. $36x^2 + y^2 = 36$ 6. $4x^2 + 3y^2 = 12$

7. $16x^2 + 9y^2 = 144$ 8. $x^2 + 4y^2 = 16$

Find the equation of each ellipse satisfying the given conditions.

9. Vertices at (4, 0) and (−4, 0); foci at (2, 0) and (−2, 0).

10. Vertices at (0, 7) and (0, −7); foci at (0, 5) and (0, −5).

11. Vertices at (0, 9) and (0, −9); foci at (0, 6) and (0, −6).

12. Vertices at (12, 0) and (−12, 0); foci at (10, 0) and (−10, 0).

13. Vertices at (6, 0) and (−6, 0); length of minor axis is 10.

14. Vertices at (0, 10) and (0, −10); length of minor axis is 18.

15. Foci at (0, 5) and (0, −5); length of major axis is 16.

16. Foci at (3, 0) and (−3, 0); length of major axis is 8.

17. A weather satellite with an orbit about the earth reaches a minimum altitude of 1000 mi and a maximum altitude of 1600 mi. The path of its orbit is approximately that of an ellipse. Find the equation of this curve. Assume the radius of the earth is 4000 mi and the x-axis is the major axis.

18. An arch is in the shape of the upper half of an ellipse with a horizontal major axis supporting a foot bridge 40 m long over a stream in a park. The center of the arch is 8 m above the bridge supports. Find an equation of the ellipse. (Choose the point midway between the bridge supports as the origin.)

1.7 THE HYPERBOLA

A **hyperbola** consists of the set of points, the *difference* of whose distances from two fixed points is a positive constant. The two fixed points are called the **foci**.

Assume now as in Fig. 1.42 that the foci lie on the x-axis at $(-c, 0)$ and $(c, 0)$. Then a point $P(x, y)$ lies on the hyperbola if the difference between its distances to the foci is equal to a given constant k. That is, $d_1 - d_2 = k$ or $d_2 - d_1 = k$. Again, this constant k equals $2a$; that is,

$$d_2 - d_1 = 2a$$

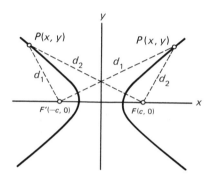

Figure 1.42

To obtain the equation of the hyperbola, use the distance formula.

$$d_2 - d_1 = 2a$$

$$\sqrt{(x - c)^2 + (y - 0)^2} - \sqrt{[x - (-c)]^2 + (y - 0)^2} = 2a$$

Rewrite the equation above as follows:

$$\sqrt{(x - c)^2 + y^2} = 2a + \sqrt{(x + c)^2 + y^2}$$

$$(x - c)^2 + y^2 = 4a^2 + 4a\sqrt{(x + c)^2 + y^2} + (x + c)^2 + y^2 \quad \text{(Square each side.)}$$

$$x^2 - 2cx + c^2 + y^2 = 4a^2 + 4a\sqrt{(x + c)^2 + y^2} + x^2 + 2cx + c^2 + y^2$$

$$-4a^2 - 4cx = 4a\sqrt{(x + c)^2 + y^2}$$

$$-a^2 - cx = a\sqrt{(x + c)^2 + y^2} \quad \text{(Divide each side by 4.)}$$

$$a^4 + 2a^2cx + c^2x^2 = a^2[(x + c)^2 + y^2] \quad \text{(Square each side.)}$$

$$a^4 + 2a^2cx + c^2x^2 = a^2(x^2 + 2cx + c^2 + y^2)$$

$$a^4 + 2a^2cx + c^2x^2 = a^2x^2 + 2a^2cx + a^2c^2 + a^2y^2$$

$$a^4 - a^2c^2 = a^2x^2 - c^2x^2 + a^2y^2$$

$$a^2(a^2 - c^2) = (a^2 - c^2)x^2 + a^2y^2 \quad \text{(Factor.)}$$

$$1 = \frac{x^2}{a^2} + \frac{y^2}{a^2 - c^2} \quad \text{[Divide each side by } a^2(a^2 - c^2)\text{.]}$$

$$1 = \frac{x^2}{a^2} - \frac{y^2}{c^2 - a^2}$$

In triangle $F'PF$

$$\overline{PF'} < \overline{PF} + \overline{FF'}$$ (The sum of any two sides of a triangle is greater than the third side.)

$$\overline{PF'} - \overline{PF} < \overline{FF'}$$

$$2a < 2c$$ ($\overline{PF'} - \overline{PF} = 2a$ by the definition of a hyperbola and $\overline{FF'} = 2c$.)

$$a < c$$

$$a^2 < c^2$$ (Since $a > 0$ and $c > 0$.)

$$0 < c^2 - a^2$$

Since $c^2 - a^2$ is positive, we may replace it by the positive number, b^2, as follows:

$$1 = \frac{x^2}{a^2} - \frac{y^2}{b^2}$$

where $b^2 = c^2 - a^2$.

The equation of the hyperbola with foci on the x-axis at $(c, 0)$ and $(-c, 0)$ is

$$\boxed{\frac{x^2}{a^2} - \frac{y^2}{b^2} = 1}$$

The points $(a, 0)$ and $(-a, 0)$ are called the **vertices**. The line segment connecting the vertices is called the **transverse axis**. The vertices and transverse axis in this case lie on the x-axis. The length of the transverse axis is $2a$. The line segment connecting the points $(0, b)$ and $(0, -b)$ is called the **conjugate axis** and in this case lies on the y-axis. The length of the conjugate axis is $2b$. The **center** lies at the intersection of the conjugate and transverse axes.

Standard Form of Hyperbola

$$\frac{x^2}{a^2} - \frac{y^2}{b^2} = 1$$

with center at the origin and with the transverse axis lying on the x-axis.

If we draw the central rectangle as in Fig. 1.43 and draw lines passing through opposite vertices of the rectangle, we obtain lines called the **asymptotes** of the hyperbola. In this case the equations of these lines are

$$\boxed{\begin{aligned} y &= \frac{b}{a}x \\ y &= -\frac{b}{a}x \end{aligned}}$$

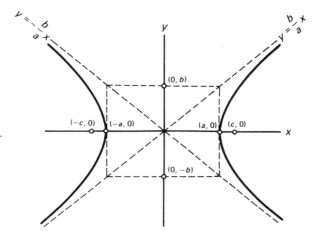

Figure 1.43

Asymptotes serve as guidelines to the branches of the hyperbola. That is, as the distance from the center of the hyperbola increases, the points on the branches get closer and closer to the asymptotes but never cross nor touch.

To sketch the graph of the hyperbola:

1. Locate the vertices $(a, 0)$ and $(-a, 0)$.
2. Locate the points $(0, b)$ and $(0, -b)$.
3. Sketch the central rectangle as in Fig. 1.43. [The coordinates of the vertices are (a, b), $(a, -b)$, $(-a, b)$, and $(-a, -b)$.]
4. Sketch the two asymptotes: the lines passing through the pairs of opposite vertices of the rectangle.
5. Sketch the branches of the hyperbola.

Example 1

Find the vertices, foci, and lengths of the transverse and conjugate axes of the hyperbola

$$\frac{x^2}{9} - \frac{y^2}{16} = 1$$

Sketch the graph. Find the equations of the asymptotes.

Since 9 is the denominator of the x^2-term, $a^2 = 9$ and $a = 3$. The vertices are therefore $(3, 0)$ and $(-3, 0)$ and the length of the transverse axis is $2a = 2(3) = 6$. Since 16 is the denominator of the y^2-term, $b^2 = 16$ and $b = 4$. So the length of the conjugate axis is $2b = 2(4) = 8$.

To find the foci we need to know c^2. Since $b^2 = c^2 - a^2$, we have

$$c^2 = a^2 + b^2$$
$$c^2 = (3)^2 + (4)^2 = 25$$
$$c = 5$$

The foci are $(5, 0)$ and $(-5, 0)$. The asymptotes are $y = \frac{4}{3}x$ and $y = -\frac{4}{3}x$. (See Fig. 1.44.)

You will want to remember the equation relating a, b, and c for the hyperbola.

$$c^2 = a^2 + b^2$$

Example 2

Write the equation of the hyperbola with foci at $(5, 0)$ and $(-5, 0)$ and whose transverse axis is 8 units in length.

Here we have $c = 5$. Since $2a = 8$, $a = 4$,

$$c^2 = a^2 + b^2$$
$$25 = 16 + b^2$$
$$b^2 = 9$$

The equation is then

$$\frac{x^2}{16} - \frac{y^2}{9} = 1$$

Standard Form of Hyperbola

$$\frac{y^2}{a^2} - \frac{x^2}{b^2} = 1$$

with center at the origin and with the transverse axis lying on the y-axis.

We obtain a graph as shown in Fig. 1.45.

Note that the difference between this equation and the first equation is that a^2 is now the denominator of the y^2-term, which is the positive term. This means that the vertices (and transverse axis) now lie on the y-axis.

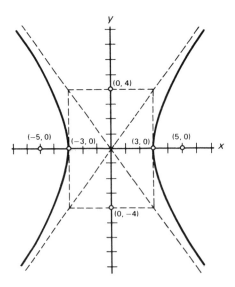

Figure 1.44

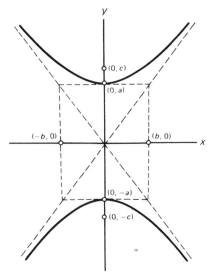

Figure 1.45

The equations of the asymptotes are

$$y = \frac{a}{b}x$$

$$y = -\frac{a}{b}x$$

In general, the positive term indicates on which axis the vertices, foci, and transverse axis lie.

1. If the x^2 term is positive, then the denominator of x^2 is a^2 and the denominator of y^2 is b^2. The transverse axis lies along the x-axis and the vertices are $(a, 0)$ and $(-a, 0)$.
2. If the y^2 term is positive, then the denominator of y^2 is a^2 and the denominator of x^2 is b^2. The transverse axis lies along the y-axis and the vertices are $(0, a)$ and $(0, -a)$.

Example 3

Sketch the graph of the hyperbola

$$\frac{y^2}{36} - \frac{x^2}{49} = 1$$

Since the y^2-term is positive, the vertices lie on the y-axis and $a^2 = 36$. Then $b^2 = 49$, $a = 6$, and $b = 7$. The graph is sketched in Fig. 1.46.

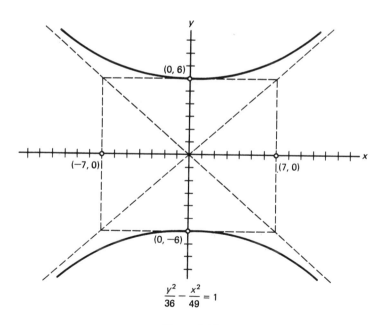

$$\frac{y^2}{36} - \frac{x^2}{49} = 1$$

Figure 1.46

Example 4

Write the equation of the hyperbola with foci at $(0, 8)$ and $(0, -8)$ and vertices at $(0, 6)$ and $(0, -6)$.

In this case $a = 6$ and $c = 8$, so $b^2 = c^2 - a^2 = 64 - 36 = 28$. Since the vertices and foci lie on the y-axis, the y^2-term is positive with denominator a^2. The equation is

$$\frac{y^2}{36} - \frac{x^2}{28} = 1$$

As with the ellipse, not all hyperbolas are located with their centers at the origin. We have seen the standard forms of the equation of the hyperbola with center at the origin and whose transverse and conjugate axes lie on the x-axis and y-axis. In general, however, the equation of a hyperbola is of the form

$$Ax^2 + Bxy + Cy^2 + Dx + Ey + F = 0$$

where either (1) $B = 0$ and A is positive and C is negative, or A is negative and C is positive, or (2) $A = 0$, $C = 0$, and $B \neq 0$.

A simple example of this last case is the equation $xy = k$. The foci and vertices lie on the line $y = x$ if $k > 0$ or on the line $y = -x$ if $k < 0$.

Example 5

Sketch the graph of the hyperbola $xy = -6$.

Since there are no easy clues for sketching this equation (unlike hyperbolas in standard position), we must set up a table of values for x and y. We then plot the corresponding points in the plane as in Fig. 1.47.

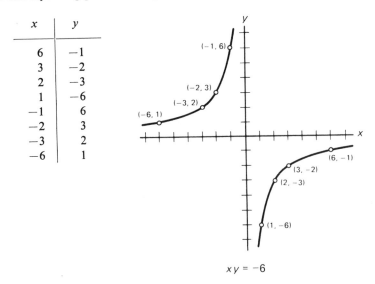

x	y
6	−1
3	−2
2	−3
1	−6
−1	6
−2	3
−3	2
−6	1

$xy = -6$

Figure 1.47

Exercises

Find the vertices, foci, and lengths of the transverse and conjugate axes of each hyperbola. Find the equations of the asymptotes and sketch each graph.

1. $\dfrac{x^2}{25} - \dfrac{y^2}{144} = 1$ 2. $\dfrac{x^2}{144} - \dfrac{y^2}{25} = 1$ 3. $\dfrac{y^2}{9} - \dfrac{x^2}{16} = 1$

4. $\dfrac{y^2}{16} - \dfrac{x^2}{9} = 1$ 5. $5x^2 - 2y^2 = 10$ 6. $3y^2 - 2x^2 = 6$

7. $4y^2 - x^2 = 4$ 8. $4x^2 - y^2 = 4$

Find the equation of the hyperbola satisfying each of the given conditions.

9. Vertices at $(4, 0)$ and $(-4, 0)$; foci at $(6, 0)$ and $(-6, 0)$.

10. Vertices at $(0, 5)$ and $(0, -5)$; foci at $(0, 7)$ and $(0, -7)$.

11. Vertices at $(0, 6)$ and $(0, -6)$; foci at $(0, 8)$ and $(0, -8)$.

12. Vertices at $(2, 0)$ and $(-2, 0)$; foci at $(5, 0)$ and $(-5, 0)$.

13. Vertices at $(3, 0)$ and $(-3, 0)$; length of conjugate axis is 10.

14. Vertices at $(0, 6)$ and $(0, -6)$; length of conjugate axis is 8.

15. Foci at $(6, 0)$ and $(-6, 0)$; length of transverse axis is 10.

16. Foci at $(0, 8)$ and $(0, -8)$; length of transverse axis is 12.

17. Sketch the graph of the hyperbola given by $xy = 8$.

18. Sketch the graph of the hyperbola given by $xy = -4$.

1.8 TRANSLATION OF AXES

We have seen the difficulty in determining the equations of the parabola, ellipse, and hyperbola when these are not in standard position in the plane. It is still possible to find the equation of these curves fairly easily if the axes of these curves lie on lines parallel to the coordinate axes. This is accomplished by the method of translating the axes. We shall demonstrate this method with four examples.

Example 1

Find the equation of the ellipse with foci at $(-2, 3)$ and $(6, 3)$ and vertices at $(-3, 3)$ and $(7, 3)$.

The center of the ellipse is at $(2, 3)$, which is midway between the foci or the vertices. The distance between the foci $(-2, 3)$ and $(6, 3)$ is 8. So $c = 4$. The distance between $(-3, 3)$ and $(7, 3)$ is 10. So $a = 5$.

$$c^2 = a^2 - b^2$$
$$16 = 25 - b^2$$
$$b^2 = 9$$
$$b = 3$$

We now sketch the graph in Fig. 1.48.

The same ellipse is now plotted in another coordinate system with center at the origin. We denote the coordinate axes of this new system by x' and y'. We know that in this $x'y'$-coordinate system the equation for this ellipse is

$$\frac{(x')^2}{a^2} + \frac{(y')^2}{b^2} = 1$$

And, since $a = 5$ and $b = 3$, we have

$$\frac{(x')^2}{25} + \frac{(y')^2}{9} = 1$$

Each point on the ellipse can now be seen as having coordinates (x, y) in the xy-plane and coordinates (x', y') in the $x'y'$-plane. If we compare coordinates in the two coordinate systems, we see, for example, that the right-hand vertex has coordinates $(7, 3)$ in the xy-plane, but the same point has coordinates $(5, 0)$ in the $x'y'$-plane. Likewise, the point at the upper end of the minor axis has coordinates $(2, 6)$ in the xy-plane, but the same point has coordinates $(0, 3)$ in the $x'y'$-plane. (See Fig. 1.49.)

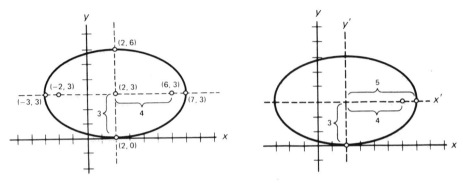

Figure 1.48   Figure 1.49

In general, the x- and x'-coordinates are related as follows.

$$x = x' + 2$$

That is, the original x-coordinates are 2 larger than the new x'-coordinates. Note that this is the distance that the new origin was moved along the x-axis: the x-coordinate of the center of the ellipse. (See Fig. 1.49.)

Similarly, the y- and y'-coordinates are related as follows.

$$y = y' + 3$$

Note that 3 is the distance that the new origin was moved along the y-axis: the y-coordinate of the center of the ellipse. (See Fig. 1.49.) We now can rearrange terms, and have

$$x' = x - 2$$
$$y' = y - 3$$

Now replace x' by $x - 2$ and y' by $y - 3$ in the equation

$$\frac{(x')^2}{25} + \frac{(y')^2}{9} = 1$$

$$\frac{(x - 2)^2}{25} + \frac{(y - 3)^2}{9} = 1$$

This is the equation of the ellipse with center at $(2, 3)$ in the xy-plane.

To write an equation for a parabola, ellipse, or hyperbola whose axes are parallel to the x-axis and y-axis:

1. For a parabola, identify (h, k) as the vertex; for an ellipse or hyperbola identify (h, k) as the center.

2. Translate *xy*-coordinates to a new *x'y'*-coordinate system by using the translation equations

$$x' = x - h$$
$$y' = y - k$$

where (h, k) has been identified as in step 1.

3. Write the equation of the conic, which is now in standard position in the *x'y'*-coordinate system.

4. Translate the equation derived in step 3 back into the original coordinate system by making the following substitutions for *x* and *y* into the derived equation.

$$x' = x - h$$
$$y' = y - k$$

The resulting equation is an equation for the conic in the *xy*-coordinate system.

Example 2

Find the equation of the parabola with focus $(-1, 2)$ and directrix $x = -7$.

Step 1: The vertex of this parabola is halfway along the line $y = 2$ between the focus $(-1, 2)$ and the directrix $x = -7$. (See Fig. 1.50.) Thus the vertex has coordinates $(-4, 2)$. This becomes the origin of the new coordinate system. So $h = -4$ and $k = 2$.

Step 2:
$$x' = x - h = x - (-4) = x + 4$$
$$y' = y - k = y - 2$$

The new *x'y'*-coordinates of the focus become

$$x' = x - h = -1 - (-4) = 3$$
$$y' = y - k = 2 - 2 = 0$$

and the equation of the directrix $x = -7$ becomes

$$x' = x - h = -7 - (-4)$$
$$x' = -3$$

Step 3: Since the parabola is now in standard position in the new coordinate system with focus $(3, 0)$, we have $p = 3$. The equation in this system becomes

$$(y')^2 = 4px'$$
$$(y')^2 = 4(3)x'$$
$$(y')^2 = 12x'$$

Step 4: We replace x' with $x + 4$ and y' with $y - 2$.
$$(y')^2 = 12x'$$
$$(y - 2)^2 = 12(x + 4)$$
$$y^2 - 4y - 12x - 44 = 0$$

We sometimes know the equation of a curve and desire to identify the curve and sketch its graph, as in the following example.

Example 3

Describe and sketch the graph of the equation

$$\frac{(y-4)^2}{9} - \frac{(x+2)^2}{16} = 1$$

If we let

$$x' = x - h = x + 2 = x - (-2)$$
$$y' = y - k = y - 4$$

we have

$$\frac{(y')^2}{9} - \frac{(x')^2}{16} = 1$$

We see that this is the equation of a hyperbola with center at $(-2, 4)$. Since $a^2 = 9$ and $b^2 = 16$, we have

$$c^2 = a^2 + b^2 = 9 + 16 = 25$$

so

$$a = 3, \quad b = 4 \quad \text{and} \quad c = 5$$

In terms of the $x'y'$-coordinates, the foci are $(0, 5)$ and $(0, -5)$; the vertices are at $(0, 3)$ and $(0, -3)$; the length of the transverse axis is 6; and the length of the conjugate axis is 8.

To translate the $x'y'$-coordinates to xy-coordinates, we use the equations

$$x = x' + h \qquad y = y' + k$$

In this case

$$x = x' + (-2) \qquad y = y' + 4$$

So in the xy-plane the foci are at $(-2, 9)$ and $(-2, -1)$; the vertices are at $(-2, 7)$ and $(-2, 1)$; the length of the transverse axis is 6; and the length of the conjugate axis is 8. (See Fig. 1.51.)

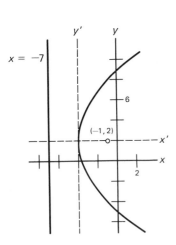

Figure 1.50

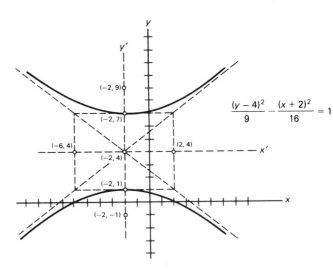

Figure 1.51

Example 4

Name the equation $16x^2 + 9y^2 + 64x + 54y + 1 = 0$. Locate the vertex if it is a parabola or the center if it is an ellipse or a hyperbola.

First complete the square for x and y (see Section 1.4).

$$(16x^2 + 64x \qquad) + (9y^2 + 54y \qquad) = -1$$

$$16(x^2 + 4x \qquad) + 9(y^2 + 6y \qquad) = -1 \qquad \text{(Factor out the coefficients of } x^2 \text{ and } y^2 \text{ before completing the square.)}$$

$$16(x^2 + 4x + 4) + 9(y^2 + 6y + 9) = -1 + 16(4) + 9(9)$$

$$16(x + 2)^2 + 9(y + 3)^2 = 144$$

$$\frac{(x + 2)^2}{9} + \frac{(y + 3)^2}{16} = 1 \qquad \text{(Divide each side by 144.)}$$

This is an equation of an ellipse. Noting that

$$x' = x - h = x + 2 = x - (-2)$$
$$y' = y - k = y + 3 = y - (-3)$$

we see that the center is at $(-2, -3)$.

General Forms of Conics with Axes Parallel to the Coordinate Axes

1. $(y - k)^2 = 4p(x - h)$
   is a parabola with vertex at (h, k) and axis parallel to the x-axis.
2. $(x - h)^2 = 4p(y - k)$
   is a parabola with vertex at (h, k) and axis parallel to the y-axis.
3. $\dfrac{(x - h)^2}{a^2} + \dfrac{(y - k)^2}{b^2} = 1, \qquad a > b$
   is an ellipse with center at (h, k) and major axis parallel to the x-axis.
4. $\dfrac{(y - k)^2}{a^2} + \dfrac{(x - h)^2}{b^2} = 1, \qquad a > b$
   is an ellipse with center at (h, k) and major axis parallel to the y-axis.
5. $\dfrac{(x - h)^2}{a^2} - \dfrac{(y - k)^2}{b^2} = 1$
   is a hyperbola with center at (h, k) and transverse axis parallel to the x-axis.
6. $\dfrac{(y - k)^2}{a^2} - \dfrac{(x - h)^2}{b^2} = 1$
   is a hyperbola with center at (h, k) and transverse axis parallel to the y-axis.

Exercises

Find the equation of each curve determined by the given information.

1. Ellipse with center at $(1, -1)$, vertices at $(5, -1)$ and $(-3, -1)$, and foci at $(3, -1)$ and $(-1, -1)$.

2. Parabola with vertex at $(-1, 3)$, focus at $(-1, 4)$, and directrix $y = 2$.

3. Hyperbola with center at $(1, 1)$, vertices at $(1, 7)$ and $(1, -5)$, and foci at $(1, 9)$ and $(1, -7)$.

4. Ellipse with center at $(-2, -3)$, vertices at $(4, -3)$ and $(-8, -3)$, and length of the minor axis is 10.

5. Parabola with vertex at $(3, -1)$, focus at $(5, -1)$, and directrix $x = 1$.

6. Hyperbola with center at $(-2, -2)$, vertices at $(1, -2)$ and $(-5, -2)$, and length of the conjugate axis is 10.

Name and graph each equation.

7. $(x - 2)^2 = 4(y + 3)$

8. $\dfrac{(x + 1)^2}{36} + \dfrac{(y - 2)^2}{64} = 1$

9. $\dfrac{y^2}{9} - \dfrac{(x + 2)^2}{16} = 1$

10. $y^2 = 8(x + 1)$

11. $9(x - 2)^2 + 16y^2 = 144$

12. $\dfrac{(x + 1)^2}{9} - \dfrac{(y + 3)^2}{16} = 1$

13. $\dfrac{(x - 3)^2}{36} + \dfrac{(y - 1)^2}{16} = 1$

14. $\dfrac{(x - 3)^2}{36} - \dfrac{(y - 1)^2}{16} = 1$

15. $(y + 3)^2 = 8(x - 1)$

16. $(x - 5)^2 = 12(y + 2)$

17. $\dfrac{(y + 1)^2}{9} - \dfrac{(x + 1)^2}{9} = 1$

18. $\dfrac{(y + 4)^2}{4} + \dfrac{(x - 2)^2}{9} = 1$

Name and sketch the graph of each equation. Locate the vertex if it is a parabola or the center if it is an ellipse or a hyperbola.

19. $x^2 - 4x + 2y + 6 = 0$

20. $9x^2 + 4y^2 - 18x + 24y + 9 = 0$

21. $x^2 + 4y^2 + 4x - 8y - 8 = 0$

22. $-2x^2 + 3y^2 + 8x - 14 = 0$

23. $4x^2 - y^2 - 8x + 2y + 3 = 0$

24. $y^2 + 6y - x + 12 = 0$

25. $25y^2 - 4x^2 - 24x - 150y + 89 = 0$

26. $25x^2 + 9y^2 - 100x - 54y - 44 = 0$

27. $x^2 + 16x - 12y + 40 = 0$

28. $9x^2 - 4y^2 + 54x + 40y - 55 = 0$

29. $4x^2 + y^2 + 48x + 4y + 84 = 0$

30. $y^2 - 10x - 6y + 39 = 0$

1.9 THE GENERAL SECOND-DEGREE EQUATION

The circle, parabola, ellipse, and hyperbola are all special cases of the second-degree equation

$$Ax^2 + Bxy + Cy^2 + Dx + Ey + F = 0$$

When $B = 0$ and at least one of the coefficients A or C is not zero, the following summarizes the conditions for each curve.

1. If $A = C$, we have a *circle*.
   In special cases, the graph of the equation may be a point or there may be no graph. (The equation may have only one or no solution.)
2. If $A = 0$ and $C \neq 0$ or $C = 0$ and $A \neq 0$, then we have a *parabola*.
3. If $A \neq C$, and A and C are either both positive or both negative, then we have an *ellipse*.
   In special cases, the graph of the equation may be a point or there may be no graph. (The equation may have only one or no solution.)

4. If A is positive and C is negative or A is negative and C is positive, then we have a *hyperbola*.

In some special cases the graph may be a pair of intersecting lines.

If $D \neq 0$ or $E \neq 0$ or both are not zero, the curve does not have its center (or vertex in the case of the parabola) at the origin (see Section 1.8). If $B \neq 0$, then the axis of the curve does not lie along the x-axis or y-axis. The hyperbola $xy = k$ is the only such example we have studied (see Section 1.7).

Example

Identify the curve

$$x^2 + 3y^2 - 2x + 4y - 7 = 0$$

Since $A \neq C$, A and C are both positive, and $B = 0$, the curve is an ellipse. (The center is not the origin since $D \neq 0$ and $E \neq 0$.)

The curves represented by the second-degree equation

$$Ax^2 + Bxy + Cy^2 + Dx + Ey + F = 0$$

are called **conic sections** since they can be obtained by cutting the cones by a plane, as is indicated in Fig. 1.52.

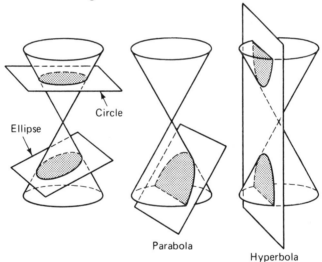

Circle

Ellipse

Parabola

Hyperbola

Figure 1.52

Exercises

Determine whether each equation represents a circle, a parabola, an ellipse, or a hyperbola.

1. $x^2 + 3y^2 + 4x - 5y - 40 = 0$

2. $x^2 + y^2 + 4x - 6y - 12 = 0$

3. $4y^2 - 8y + 3x - 2 = 0$

4. $9x^2 + 4y^2 + 36x - 8y + 4 = 0$

5. $4x^2 - 5y^2 - 16x + 10y + 20 = 0$

6. $x^2 + y^2 + 3x - 2y - 14 = 0$

7. $3x^2 + 3y^2 + x - y - 6 = 0$

8. $x^2 + 4x - 3y - 52 = 0$

9. $x^2 + y^2 + 2x - 3y - 21 = 0$

10. $x^2 - y^2 - 6x + 3y - 100 = 0$

11. $9x^2 + 4y^2 - 18x + 8y + 4 = 0$

12. $3x^2 - 2y^2 + 6x - 8y - 17 = 0$

13. $3x^2 - 3y^2 - 2x - 4y - 13 = 0$ **14.** $4x^2 + 4y^2 - 16x - 4y - 5 = 0$

15. $y^2 - 6x - 6y + 3 = 0$ **16.** $4x^2 - 4x - 4y - 5 = 0$

CHAPTER SUMMARY

1. *Basic terms*
   (a) *Analytic geometry:* Study of relationships between algebra and geometry.
   (b) *Relation:* Set of ordered pairs, usually in the form (x, y).
   (c) *Independent variable:* First element of an ordered pair, usually x.
   (d) *Dependent variable:* Second element of an ordered pair, usually y.
   (e) *Domain:* Set of all first elements in a relation or set of all x's.
   (f) *Range:* Set of all second elements in a relation or set of all y's.
   (g) *Positive integers:* 1, 2, 3, ...
   (h) *Negative integers:* $-1, -2, -3, ...$
   (i) *Integers:* . . . , $-3, -2, -1, 0, 1, 2, 3, ...$
   (j) *Rational numbers:* Numbers that can be represented as the ratio of two integers a/b, $b \neq 0$.
   (k) *Irrational numbers:* Numbers that cannot be represented as the ratio of two integers.
   (l) *Real numbers:* Set of numbers consisting of the rational numbers and the irrational numbers.
   (m) *Inequalities:* Statements involving less than or greater than and may be used to describe various intervals on the number line as follows:

Type of interval	Symbols	Meaning	Number line graph
Open	$x > a$	x is greater than a	
	$x < b$	x is less than b	
	$a < x < b$	x is between a and b	
Half-open	$x \geq a$	x is greater than or equal to a	
	$x \leq b$	x is less than or equal to b	
	$a < x \leq b$	x is between a and b, including b but excluding a	
	$a \leq x < b$	x is between a and b, including a but excluding b	
Closed	$a \leq x \leq b$	x is between a and b, including both a and b	

2. *Slope of a straight line:* If $P_1(x_1, y_1)$ and $P_2(x_2, y_2)$ represent any two points on a straight line, its slope m is

$$m = \frac{y_2 - y_1}{x_2 - x_1}$$

(a) If a line has positive slope, the line slopes upward from left to right.
(b) If a line has negative slope, the line slopes downward from left to right.
(c) If a line has zero slope, the line is horizontal.
(d) If a line has no slope, the line is vertical.

3. *Point-slope form of a straight line:* If m is the slope and $P_1(x_1, y_1)$ is any point on a nonvertical straight line, its equation is

$$y - y_1 = m(x - x_1)$$

4. *Slope-intercept form of a straight line:* If m is the slope and $(0, b)$ is the y-intercept of a nonvertical straight line, its equation is

$$y = mx + b$$

5. *Equation of a horizontal line:* If a horizontal line passes through the point $(0, b)$, its equation is

$$y = b$$

6. *Equation of a vertical line:* If a vertical line passes through the point $(a, 0)$, its equation is

$$x = a$$

7. *General form of the equation of a straight line*

$$Ax + By + C = 0$$

8. *Parallel lines:* Two lines are parallel if either one of the following conditions holds:
(a) They are both perpendicular to the x-axis.
(b) They both have the same slope; that is, if the equations of the lines are

$$L_1: y = m_1x + b_1 \quad \text{and} \quad L_2: y = m_2x + b_2$$

then

$$m_1 = m_2$$

9. *Perpendicular lines:* Two lines are perpendicular if either one of the following conditions holds:
(a) One line is vertical with equation $x = a$ and the other is horizontal with equation $y = b$.
(b) Neither is vertical and the slope of one line is the negative reciprocal of the other; that is, if the equations of the lines are

$$L_1: y = m_1x + b_1 \quad \text{and} \quad L_2: y = m_2x + b_2$$

then

$$m_1 = -\frac{1}{m_2}$$

10. *Distance formula:* The distance between two points $P(x_1, y_1)$ and $Q(x_2, y_2)$ is given by the formula

$$d = PQ = \sqrt{(x_2 - x_1)^2 + (y_2 - y_1)^2}$$

11. *Midpoint formula:* The coordinates of the point $Q(x_m, y_m)$ which is midway between two points $P(x_1, y_1)$ and $R(x_2, y_2)$ are given by

$$x_m = \frac{x_1 + x_2}{2} \quad \text{and} \quad y_m = \frac{y_1 + y_2}{2}$$

12. *Conics:* Equations in the form $Ax^2 + Bxy + Cy^2 + Dx + Ey + F = 0$ are called conics.

13. *Circle*
   (a) *Standard form:* $(x - h)^2 + (y - k)^2 = r^2$, where r is the radius and (h, k) is the center.
   (b) *General form:* $x^2 + y^2 + Dx + Ey + F = 0$.
   (c) *Center at the origin:* $x^2 + y^2 = r^2$, where r is the radius.

14. *Parabola with vertex at the origin*
   $y^2 = 4px$ with focus at $(p, 0)$ and $x = -p$ as the directrix.
   (a) When $p > 0$, the parabola opens to the right.
   (b) When $p < 0$, the parabola opens to the left.
   $x^2 = 4py$ with focus at $(0, p)$ and $y = -p$ as the directrix.
   (c) When $p > 0$, the parabola opens upward.
   (d) When $p < 0$, the parabola opens downward.

15. *Ellipse with center at the origin*
   (a) $\dfrac{x^2}{a^2} + \dfrac{y^2}{b^2} = 1$ with the major axis on the x-axis and $a > b$.
   (b) $\dfrac{y^2}{a^2} + \dfrac{x^2}{b^2} = 1$ with major axis on the y-axis and $a > b$.

16. *Hyperbola with center at the origin*
   (a) $\dfrac{x^2}{a^2} - \dfrac{y^2}{b^2} = 1$ with the transverse axis on the x-axis.
   (b) $\dfrac{y^2}{a^2} - \dfrac{x^2}{b^2} = 1$ with the transverse axis on the y-axis.

17. *Translation equations*

$$x' = x - h \quad \text{and} \quad y' = y - k$$

18. *General forms of conics with axes parallel to the coordinate axes*
   (a) $(y - k)^2 = 4p(x - h)$
   is a parabola with vertex at (h, k) and axis parallel to the x-axis.
   (b) $(x - h)^2 = 4p(y - k)$
   is a parabola with vertex at (h, k) and axis parallel to the y-axis.
   (c) $\dfrac{(x - h)^2}{a^2} + \dfrac{(y - k)^2}{b^2} = 1, \qquad a > b$
   is an ellipse with center at (h, k) and major axis parallel to the x-axis.

(d) $\dfrac{(y - k)^2}{a^2} + \dfrac{(x - h)^2}{b^2} = 1, \qquad a > b$

is an ellipse with center at (h, k) and major axis parallel to the y-axis.

(e) $\dfrac{(x - h)^2}{a^2} - \dfrac{(y - k)^2}{b^2} = 1$

is a hyperbola with center at (h, k) and transverse axis parallel to the x-axis.

(f) $\dfrac{(y - k)^2}{a^2} - \dfrac{(x - h)^2}{b^2} = 1$

is a hyperbola with center at (h, k) and transverse axis parallel to the y-axis.

19. *The general second-degree equation:* The circle, parabola, ellipse, and hyperbola are all special cases of the second-degree equation

$$Ax^2 + Bxy + Cy^2 + Dx + Ey + F = 0$$

When $B = 0$ and at least one of the coefficients A or C is not zero, the following summarizes the conditions for each curve:

(a) If $A = C$, we have a *circle*.

In special cases, the graph of the equation may be a point or there may be no graph. (The equation may have only one or no solution.)

(b) If $A = 0$ and $C \neq 0$ or $C = 0$ and $A \neq 0$, then we have a *parabola*.

(c) If $A \neq C$, and A and C are either both positive or both negative, then we have an *ellipse*.

In special cases, the graph of the equation may be a point or there may be no graph. (The equation may have only one or no solution.)

(d) If A is positive and C is negative or A is negative and C is positive, then we have a *hyperbola*. In some special cases the graph may be a pair of intersecting lines.

If $D \neq 0$ or $E \neq 0$ or both not zero, the curve does not have its center (or vertex in the case of the parabola) at the origin. If $B \neq 0$, then the axis of the curve does not lie along the x-axis or y-axis.

CHAPTER REVIEW

For each relation determine whether or not it is a function. Write its domain and range.

1. $A = \{(2, 3), (3, 4), (4, 5), (5, 6)\}$ 2. $B = \{(2, 6), (6, 4), (2, 1), (4, 3)\}$

3. $y = -4x + 3$ 4. $y = x^2 - 5$

5. $x = y^2 + 4$ 6. $y = \sqrt{4 - 8x}$

Graph each equation.

7. $y = 4x + 5$ 8. $y = x^2 + 4$

9. $y = x^2 + 2x - 8$ 10. $y = 2x^2 + x - 6$

Use the points $(3, -4)$ and $(-6, -2)$ in Exercises 11–13.

11. Find the slope of the line through the two points.

12. Find the distance between the two points.

13. Find the coordinates of the point midway between the two points.

Find the equation of the line that satisfies each condition in Exercises 14–17.

14. Passes through (4, 7) and (6, −4).

15. Passes through (−3, 1) with a slope of $\frac{2}{3}$.

16. Crosses the y-axis at −3 with a slope of $-\frac{1}{3}$.

17. Is parallel to and 3 units to the left of the y-axis.

18. Find the slope and y-intercept of $3x - 2y - 6 = 0$.

19. Graph $3x - 4y = 12$.

Determine whether the given pairs of equations represent lines that are parallel, perpendicular, or neither.

20. $x - 2y + 3 = 0; 8x + 4y - 9 = 0$

21. $2x - 3y + 4 = 0; -8x + 12y = 16$

22. $3x - 2y + 5 = 0; 2x - 3y + 9 = 0$

23. $x = 2; y = -3$ 24. $x = 4; x = 7$

25. Find the equation of the line parallel to the line $2x - y + 4 = 0$ that passes through the point (5, 2).

26. Find the equation of the line perpendicular to the line $3x + 5y - 6 = 0$ that passes through the point (−4, 0).

27. Write the equation of the circle with center at (5, −7) and with radius 6.

28. Find the center and radius of the circle $x^2 + y^2 - 8x + 6y - 24 = 0$.

29. Find the focus and directrix of the parabola $x^2 = 6y$ and sketch its graph.

30. Write the equation of the parabola with focus at (−4, 0) and directrix $x = 4$.

31. Write the equation of the parabola with focus at (4, 3) and directrix $x = 0$.

32. Find the vertices and foci of the ellipse $4x^2 + 49y^2 = 196$ and sketch its graph.

33. Write the equation of the ellipse with vertices (0, 4) and (0, −4) and with foci at $(0, 2\sqrt{3})$ and $(0, -2\sqrt{3})$.

34. Find the vertices and foci of the hyperbola $4x^2 - 9y^2 = 144$ and sketch its graph.

35. Write the equation of the hyperbola with vertices at (0, 5) and (0, −5) and with foci at $(0, \sqrt{41})$ and $(0, -\sqrt{41})$.

36. Write the equation of the ellipse with center at (3, −4), vertices at (3, 1) and (3, −9), and foci at (3, 0) and (3, −8).

37. Write the equation of the hyperbola with center at (−7, 4), vertices at (2, 4) and (−16, 4); the length of the conjugate axis is 6.

38. Name and sketch the graph of $16x^2 - 4y^2 - 64x - 24y + 12 = 0$.

2

THE DERIVATIVE

2.1 MOTION

While the topics of algebra, trigonometry, and geometry are of fundamental importance to the mathematician and technician, a wide variety of technical problems cannot be solved using only these tools of mathematics. Many problems can only be solved using the methods of the calculus. As early as the seventeenth century, scientists found the need for new techniques of mathematics. They were studying the motion of projectiles, the motion of the moon and planets, and the motion of light. Scientists, like Isaac Newton, began developing a new branch of mathematics in order to solve the problems involving motion. This new branch of mathematics came to be known as *the calculus*. Today, the calculus remains a powerful development of mathematics. While the calculus began with the study of motion, its usefulness today can be seen in many varied technical areas.

Motion is usually defined as a continual change in position. Linear motion is motion along a straight line. In this section, we limit our discussion to linear motion. You are familiar with finding the average speed of an object in motion. For example, if on a trip you drive 150 mi in 3 hours (h), then by dividing 150 mi by 3 h you find that you drove 50 mi/h on the average. This does not tell you how fast you were driving exactly 1 h and 32 minutes (min) after you began the trip. You may have been stopped at a traffic light or you may have been traveling 55 mi/h!

In attempting to solve this problem mathematically, we assume we can describe the distance traveled by an object as a function of time. That is, at each point in time t we can associate a number s representing the distance traveled by the object. For example, $s = 4t + 1$ is a function which describes the motion of an object as it moves along a straight line in terms of time t. If t is measured in seconds (s) and s in metres (m), then after 2 s, the object is at $s = 4(2) + 1 = 9$ m along the line of motion. Three seconds later, $t = 2 + 3$, the object has moved to $s = 4(2 + 3) + 1 = 4(5) + 1 = 21$ m along the line of motion. (See Fig. 2.1.)

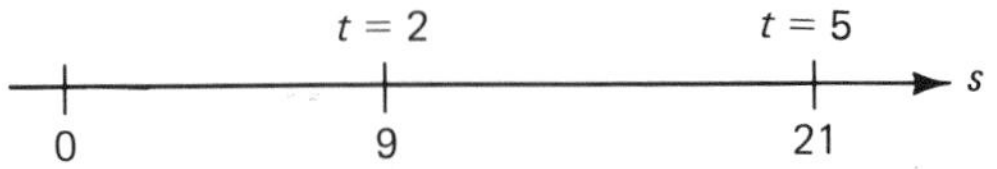

Figure 2.1

The *average speed* $\bar{v}$ of an object in motion is the ratio of distance traveled by an object to the time taken to travel that distance. In our example above, the distance traveled by the object is 21 m − 9 m = 12 m. It traveled this distance in 3 s. The average speed over this time period is then

$$\bar{v} = \frac{12 \text{ m}}{3 \text{ s}} = 4 \text{ m/s}$$

We need to introduce the mathematical symbol, the Greek letter Δ, to indicate a change between two values of a variable. In this section Δt (read "delta t") represents the change in time t and Δs (read "delta s") represents the change in distance s. In the example above, $\Delta t = 3$ s. This is the change in time needed for the object to go from 9 m to 21 m along the line of motion. The change in distance for this time interval $\Delta t = 3$ s is $\Delta s = 21$ m − 9 m = 12 m. Using this notation we can write

$$\bar{v} = \frac{\Delta s}{\Delta t}$$

When both speed and direction are needed to describe the motion, the term *velocity* is used. The **velocity** of an object is its time rate of change of its displacement. Since both magnitude (speed) and direction are required to completely describe velocity, velocity is a vector quantity. The direction of velocity along a straight line is usually given as positive or negative.

Note: Since speed is the magnitude of velocity, it is common to use v for either speed or the magnitude of velocity. In addition, the terms *speed* and *velocity* are often used interchangeably because the direction of the motion is often understood. For example, velocity that results from motion in a straight line has the same direction as the direction of the motion, such as in a freely falling body.

Recall that a function is a set of ordered pairs no two of which have the same first element. We now find it helpful to introduce a special notation, called *functional notation,* to represent a functional relationship. For example, the function $y = x^2 + 3$ is written $f(x) = x^2 + 3$ using functional notation. The symbol $f(x)$, read "f of x," is used to represent the number y which corresponds to a number x under the given functional relationship. That is, $f(x) = y$ or, as in this case, $f(x) = x^2 + 3$. The table below gives $f(x)$ for various values of x.

x	$f(x) = x^2 + 3$
-3	$f(-3) = (-3)^2 + 3 = 12$
0	$f(0) = (0)^2 + 3 = 3$
1	$f(1) = (1)^2 + 3 = 4$
2	$f(2) = (2)^2 + 3 = 7$
h	$f(h) = (h)^2 + 3 = h^2 + 3$
$3t$	$f(3t) = (3t)^2 + 3 = 9t^2 + 3$
$1 + \Delta x$	$f(1 + \Delta x) = (1 + \Delta x)^2 + 3 = 4 + 2\Delta x + (\Delta x)^2$

The use of the symbol $f(x)$ is helpful since we can use $f(x)$ to represent the number corresponding to x under the functional relationship without having to actually determine the number as done in the table above. For example $f(3)$ represents the number corresponding to $x = 3$ under any given functional relationship. For this reason, $f(x)$ is often called the *value of the function at x*.

Example 1

Write in functional notation the function which relates to each number x its cube less 2.

The relationship is $y = x^3 - 2$. Using the symbol f to represent this function, we write

$$f(x) = x^3 - 2$$

Example 2

Find the value of the function $f(x) = x^3 - 2$ for (a) $x = -2$ and (b) $x = 2 + \Delta x$.

(a) $\qquad\qquad f(-2) = (-2)^3 - 2 = -8 - 2 = -10$

(b) $\qquad f(2 + \Delta x) = (2 + \Delta x)^3 - 2$

$$= (2)^3 + 3(2)^2(\Delta x) + 3(2)(\Delta x)^2 + (\Delta x)^3 - 2$$
$$= 8 + 12(\Delta x) + 6(\Delta x)^2 + (\Delta x)^3 - 2$$
$$= 6 + 12(\Delta x) + 6(\Delta x)^2 + (\Delta x)^3$$

(See the binomial expansion formula in Appendix A.)

Example 3

Evaluate the function $g(x) = \sqrt{2x + 3}$ at $x = 3$.

$$g(3) = \sqrt{2(3) + 3}$$
$$= \sqrt{9}$$
$$= 3$$

Example 4

Evaluate the function $f(x) = x^2 + 3x - 5$ at $x = h + 2$.

$$f(h + 2) = (h + 2)^2 + 3(h + 2) - 5$$
$$= h^2 + 4h + 4 + 3h + 6 - 5$$
$$= h^2 + 7h + 5$$

On page 52 an object moved along a straight line according to the function $s = 4t + 1$. We could write this in functional notation: $s = f(t) = 4t + 1$.

Recall that Δs is the change in distance s and Δt is the change in time after $t = 2$ s. Then using functional notation,

$$\Delta s = f(2 + \Delta t) - f(2)$$

represents, as we saw earlier, the change in distance traveled for a given change in time Δt. When we look at $\Delta t = 3$ s, we have

$$\begin{aligned}
\Delta s &= f(2 + \Delta t) - f(2) \\
&= f(2 + 3) - f(2) \qquad (\Delta t = 3) \\
&= f(5) - f(2) \\
&= [4(5) + 1] - [4(2) + 1] \\
&= 21 - 9 \\
&= 12 \text{ m}
\end{aligned}$$

So the average speed over this time period is

$$\bar{v} = \frac{\Delta s}{\Delta t} = \frac{f(2 + \Delta t) - f(2)}{\Delta t}$$

$$= \frac{12 \text{ m}}{3 \text{ s}} = 4 \text{ m/s}$$

as we determined earlier.

In general, the distance traveled by an object from time t to time $t + \Delta t$ is given in functional notation by

$$\Delta s = f(t + \Delta t) - f(t)$$

The average speed of this object over the change in time Δt is then

> **Average Speed**
>
> $$\bar{v} = \frac{\Delta s}{\Delta t} = \frac{f(t + \Delta t) - f(t)}{\Delta t}$$

Example 5

Given that $s = f(t) = t^2 - 1$ describes the motion of an object moving along a straight line, where s is measured in feet (ft); (a) find Δs and $\bar{v}$; and (b) find $\bar{v}$ from 4 s of travel to 7 s of travel.

(a)

$$\begin{aligned}
\Delta s &= f(t + \Delta t) - f(t) \\
&= [(t + \Delta t)^2 - 1] - (t^2 - 1) \\
&= [t^2 + 2t(\Delta t) + (\Delta t)^2 - 1] - (t^2 - 1) \\
&= 2t(\Delta t) + (\Delta t)^2
\end{aligned}$$

$$\begin{aligned}
\bar{v} = \frac{\Delta s}{\Delta t} &= \frac{2t(\Delta t) + (\Delta t)^2}{\Delta t} \\
&= \frac{(2t + \Delta t)\, \Delta t}{\Delta t} \\
&= 2t + \Delta t
\end{aligned}$$

(b) $$\Delta t = 7 - 4 = 3 \text{ s}$$

From (a) we have

$$\bar{v} = 2t + \Delta t$$
$$= 2(4) + (3)$$
$$= 11 \text{ ft/s}$$

You should verify that this is the same result that would be obtained by calculating:

$$\bar{v} = \frac{f(4 + 3) - f(4)}{3} = \frac{\text{distance traveled}}{\text{time traveled}}$$

From Example 5 we see that to calculate $\bar{v} = (\Delta s/\Delta t)$ we need to be given the time t at which we begin measuring $\bar{v}$ as well as the change in time Δt. Note that both t and Δt can be negative. If $\Delta t = -1$, then $f(t + (-1))$ represents the position of the object 1 s before it reaches the position $f(t)$.

The use of functional notation, as well as the concept of function itself, will receive strong emphasis in the remaining material. The development of the calculus depends heavily on this concept.

To find instantaneous speeds, consider the motion of an object moving along a straight line described by $s = f(t) = 3t^2 + 1$ with s measured in feet. We will now find the "instantaneous" speed after exactly 2 s of travel.

The average speed from 2 s to $(2 + \Delta t)$ s is given by

$$\bar{v} = \frac{\text{change in distance}}{\text{change in time}} = \frac{f(2 + \Delta t) - f(2)}{\Delta t}$$
$$= \frac{[3(2 + \Delta t)^2 + 1] - [3(2)^2 + 1]}{\Delta t}$$
$$= \frac{12(\Delta t) + 3(\Delta t)^2}{\Delta t}$$
$$= \frac{[12 + 3(\Delta t)]\,\Delta t}{\Delta t} = 12 + 3(\Delta t)$$

So, for example, with a change in time $\Delta t = 4$ s the average speed is $12 + 3(4) = 24$ ft/s. We now tabulate $\bar{v}$ for different values of Δt:

Δt (s)	$\bar{v}$ (ft/s)
4	24
2	18
1	15
0.5	13.5
0.1	12.3
0.001	12.003
-0.001	11.997
-0.5	10.5
-2	6

From this table we can see that the closer Δt is to 0, the closer $\bar{v}$ is to 12 ft/s. As we average over shorter time spans we would expect that the average speed will better approximate the instantaneous speed of the object at 2 s. That is, $\bar{v} = 12.3$ ft/s after 0.1 s of travel (beyond the 2-s mark) is a better approximation than $\bar{v} = 24$ ft/s after 4 s of travel (beyond the 2-s mark). Looking at this table, we see that the instantaneous speed at time $t = 2$ s must be 12 ft/s.

Instantaneous Speed

To find the *instantaneous speed* of an object in motion at a given time t:

1. Find

$$\bar{v} = \frac{f(t + \Delta t) - f(t)}{\Delta t} = \frac{\Delta s}{\Delta t}$$

where $s = f(t)$ describes the motion of the object as a function of time.

2. Observe what number, if any, $\bar{v}$ approaches as values of Δt approach 0. If there is such a number, it is called the instantaneous speed v.

Example 6

Find the instantaneous speed of an object moving according to $s = f(t) = 5t^2 - 4$ at $t = 3$ s.

$$\text{Step 1: } \bar{v} = \frac{f(3 + \Delta t) - f(3)}{\Delta t}$$

$$= \frac{[5(3 + \Delta t)^2 - 4] - [5(3)^2 - 4]}{\Delta t}$$

$$= \frac{30(\Delta t) + 5(\Delta t)^2}{\Delta t}$$

$$= \frac{\Delta t[30 + 5(\Delta t)]}{\Delta t}$$

$$= 30 + 5(\Delta t)$$

Step 2: We see that as Δt approaches (gets close to) 0, $\bar{v}$ approaches 30. We conclude that

$$v = 30 \text{ ft/s}$$

Note: It is tempting to simply substitute $\Delta t = 0$ in the formula for $\bar{v}$. This would be an attempt to compute an average velocity over a time change of 0 s. This gives us a zero time interval over which to average! We would be attempting to divide by zero, which is undefined.

$$\frac{f(3 + 0) - f(3)}{0} = \frac{0}{0} !!!$$

As in Example 6, we must find a way to simplify the expression for $\bar{v}$ so that Δt does not remain in the denominator. Only then can we begin to see what number $\bar{v}$ approaches as Δt approaches 0.

Example 7

Find v at $t = 2$ when $s = f(t) = \dfrac{1}{t}$.

$$\text{Step 1: } \bar{v} = \frac{f(2 + \Delta t) - f(2)}{\Delta t}$$

$$= \frac{\dfrac{1}{2 + \Delta t} - \dfrac{1}{2}}{\Delta t}$$

$$= \frac{\dfrac{-\Delta t}{2(2 + \Delta t)}}{\Delta t}$$

$$= \frac{-\Delta t}{2(2 + \Delta t)\Delta t}$$

$$= \frac{-1}{2(2 + \Delta t)}$$

Step 2: As Δt approaches 0, $\bar{v}$ approaches $-\frac{1}{4}$. So $v = -\frac{1}{4}$.

Exercises

Evaluate each function at the given value.

1. $f(x) = 2x^2 + 7$, $x = 1$

2. $g(x) = x^3 - x + 3$, $x = -1$

3. $h(x) = 3x^3 - 2x + 4$, $x = -2$

4. $k(x) = 2x^3 + x - 5$, $x = 2$

5. $f(x) = \dfrac{x^2 - 3}{x + 5}$, $x = 2$

6. $g(x) = \dfrac{(x^3 - 2x + 3)\sqrt{x - 2}}{x - 4}$, $x = 3$

7. $f(z) = \sqrt{z^2 + 3}$, $z = -5$

8. $f(t) = \sqrt{t^2 - 1}$, $t = 4$

9. $f(x) = 3x - 2$, $x = h + 3$

10. $g(x) = x^2 + x - 3$, $x = w + 4$

11. $f(t) = 3t^2 + 2t - 5$, $t = 2 + \Delta t$

12. $f(t) = t^2 - 3t + 4$, $t = -3 + \Delta t$

Find (a) Δs and (b) $\bar{v}$ for each function expressing distance s in terms of time t. (Express results in terms of t and Δt.)

13. $s = 3t - 4$

14. $s = 2t + 2$

15. $s = 2t^2 + 5$

16. $s = 3t^2 - 7$

17. $s = t^2 - 2t + 8$

18. $s = 5t^2 + 3t - 9$

Find $\bar{v}$ for each function $s = f(t)$ at the values of t and Δt (s is measured in metres and t in seconds).

19. $s = 5t^2 + 6$, $t = 3$, $\Delta t = 4$

20. $s = 2t^2 - 5$, $t = 2$, $\Delta t = 3$

21. $s = 3t^2 - t + 4$, $t = 2$, $\Delta t = 2$

22. $s = 6t^2 + 2t - 7$, $t = 5$, $\Delta t = 1$

23. A charged particle moves from 0.2 m to 0.5 m from a fixed reference point in 0.5 microseconds (μs). Find its average speed during that interval (1 μs $= 10^{-6}$ s).

24. An electron moved a distance of 0.2 m in 1 μs. Find its average speed in m/s during that interval.

25. The average current in a capacitor over a time interval Δt is given by $i_{av} = C(\Delta V/\Delta t)$ amperes where C is the capacitance of the capacitor in farads and V is the voltage across the capacitor in volts (V). Find the average current (in μA) in a 10-microfarad (μF) capacitor from 2 s to 5 s where the voltage across the capacitor is given by $V = t^2 + 3t + 160$.

26. Find the average current (in μA) in a 15-μF capacitor from 3 s to 7 s where the voltage across the capacitor is given by $V = 2t^2 - 4t + 200$ (see Exercise 25).

Find the instantaneous velocity v of an object moving along a straight line for each given expression for s (measured in metres) at the given value of t (measured in seconds).

27. $s = f(t) = 3t^2 - 6t + 1, t = 2$ 28. $s = f(t) = -4t^2 + 8, t = 3$

29. $s = f(t) = 5t^2 - 7, t = 1$ 30. $s = f(t) = 8t^2 + 3t - 11, t = 4$

31. $s = f(t) = \dfrac{1}{2t}, t = 3$ 32. $s = f(t) = \dfrac{1}{t + 1}, t = 2$

33. $s = f(t) = \dfrac{1}{t - 2}, t = 4$ 34. $s = f(t) = \dfrac{1}{3t}, t = 4$

35. A free-falling object (neglecting friction) is falling from rest according to the equation $s = 16t^2$ when time t is measured in s and distance s is measured in ft. Find the speed of an object after 2 s of falling.

36. A circuit-breaker contact moves approximately $s = 200,000t^3$, where s is in centimetres (cm) and t is in seconds. Find the speed v of the contact when $t = 0.1$ s.

2.2 THE LIMIT

The process we developed in problems involving motion is very useful in other applications. The technique used is often called "the limit process."

Given any function, we may find whether the functional values approach some number when the value of the variable approaches a specified number.

Example 1

Let $f(x) = x^2 - 3x + 2$. What number, if any, does $f(x)$ approach as x approaches -1?

Since x^2 approaches $(-1)^2 = 1$ as x approaches -1 and $-3x$ approaches $(-3)(-1) = 3$ as x approaches -1, we conclude that $f(x) = x^2 - 3x + 2$ approaches $1 + 3 + 2 = 6$ as x approaches -1.

Mathematicians use symbols to describe this limit process more compactly. The symbol "$\rightarrow$" means "approaches" so that x approaches -1 would be written $x \rightarrow -1$.

Limit

If $f(x) \rightarrow L$ as $x \rightarrow a$, then L is called "the limit of the function as $x \rightarrow a$." This process is written as

$$\lim_{x \to a} f(x) = L$$

and read "the limit of f of x as x approaches a equals L."

The expression in Example 1 would be written $\lim\limits_{x \to -1} (x^2 - 3x + 2) = 6$.

Example 2

Given $f(x) = \dfrac{\tan x}{x}$. What number, if any, does $f(x)$ approach as x approaches 0?

That is, find $\lim\limits_{x \to 0} \dfrac{\tan x}{x}$.

Let's use a calculator and make a table of values for x close to zero. (Be certain that your calculator is in the radian mode.)

x	$\dfrac{\tan x}{x}$	x	$\dfrac{\tan x}{x}$
0.5	1.092605	-0.5	1.092605
0.25	1.021368	-0.25	1.021368
0.15	1.007568	-0.15	1.007568
0.10	1.003347	-0.10	1.003347
0.05	1.000834	-0.05	1.000834
0.01	1.0000332	-0.01	1.0000332
0.001	1.0000005	-0.001	1.0000005

As you can see, as x gets closer and closer to 0, $\dfrac{\tan x}{x}$ approaches 1. Thus

$$\lim_{x \to 0} \frac{\tan x}{x} = 1$$

Note: This example does not constitute a proof. It is designed to give you an intuitive idea about limits.

Example 3

Find $\lim\limits_{x \to 3} \left(\dfrac{x^2 - 9}{x - 3} \right)$

As $x \to 3$, the denominator approaches 0. We cannot divide by zero. However,

$$\frac{x^2 - 9}{x - 3} = \frac{(x + 3)(x - 3)}{x - 3} = x + 3$$

In the limit process we are not concerned about what happens at $x = 3$, but only what happens as $x \to 3$. As $x \to 3$, $x + 3 \to 6$. So

$$\lim_{x \to 3} \left(\frac{x^2 - 9}{x - 3} \right) = \lim_{x \to 3} (x + 3) = 6$$

Note that in Example 3 we can find the limit of $f(x) = \dfrac{x^2 - 9}{x - 3}$ as $x \to 3$ even though the function is not defined at $x = 3$. However, we will now see that limits do not always exist.

Example 4

Find $\lim\limits_{x \to 0} \sqrt{x - 5}$.

Since we cannot obtain a real-valued number when taking the square root of a negative number, the function $f(x) = \sqrt{x - 5}$ cannot be evaluated for x less than 5. It is impossible then to observe the values of $\sqrt{x - 5}$ as x takes on values close to 0 (since the quantity $x - 5$ would be negative).

We conclude that $\lim\limits_{x \to 0} \sqrt{x - 5}$ does not exist.

Sometimes a function approaches a limiting number L as $x \to \infty$; that is, the function approaches L as x is allowed to get large without bound.

Example 5

Find $\lim\limits_{x \to \infty} (1/x)$.

As the denominator $x \to \infty$, the fraction $1/x$ approaches 0. So

$$\lim_{x \to \infty} \frac{1}{x} = 0$$

Example 6

Find $\lim\limits_{x \to \infty} \dfrac{2x^2 + x}{7x^2 - 3}$.

As $x \to \infty$, both numerator and denominator approach ∞ separately. However, if we divide numerator and denominator by the highest power of x in the denominator, x^2, we have

$$\frac{2 + \dfrac{1}{x}}{7 - \dfrac{3}{x^2}}$$

$$\lim_{x \to \infty} \frac{2x^2 + x}{7x^2 - 3} = \lim_{x \to \infty} \frac{2 + \dfrac{1}{x}}{7 - \dfrac{3}{x^2}}$$

$$= \frac{2 + 0}{7 - 0} = \frac{2}{7}$$

Note: as $x \longrightarrow \infty$ $\quad \dfrac{1}{x} \longrightarrow 0$ and $\dfrac{3}{x^2} \longrightarrow 0$

Finding instantaneous velocity is an application of the limit process. As the time interval, Δt, decreases in the average velocity formula

$$\bar{v} = \frac{\Delta s}{\Delta t} = \frac{f(t + \Delta t) - f(t)}{\Delta t}$$

the average velocity approaches the instantaneous velocity, v—the velocity at a given instant. That is,

---

Instantaneous Velocity

$$v = \lim_{\Delta t \to 0} \frac{f(t + \Delta t) - f(t)}{\Delta t}$$

Example 7

Find the instantaneous velocity v at $t = 3$ when $s = f(t) = t^2 - 7$.

$$v = \lim_{\Delta t \to 0} \frac{f(3 + \Delta t) - f(3)}{\Delta t}$$

$$= \lim_{\Delta t \to 0} \frac{[9 + 6t(\Delta t) + (\Delta t)^2 - 7] - [9 - 7]}{\Delta t}$$

$$= \lim_{\Delta t \to 0} \frac{6(\Delta t) + (\Delta t)^2}{\Delta t}$$

$$= \lim_{\Delta t \to 0} \frac{\Delta t(6 + \Delta t)}{\Delta t}$$

$$= \lim_{\Delta t \to 0} (6 + \Delta t)$$

$$= 6$$

It can be shown that the limit process obeys the following formulas:

A.
$$\boxed{\lim_{x \to a} [f(x) \pm g(x)] = \lim_{x \to a} f(x) \pm \lim_{x \to a} g(x)}$$

Example 8

$$\lim_{x \to 3} (x^3 + x^2) = \lim_{x \to 3} x^3 + \lim_{x \to 3} x^2$$

$$36 = 27 + 9$$

B.
$$\boxed{\lim_{x \to a} [k \cdot f(x)] = k \cdot \lim_{x \to a} f(x), \text{ where } k \text{ is a constant.}}$$

Example 9

$$\lim_{x \to -2} 12x^2 = 12 \lim_{x \to -2} x^2$$

$$48 = 12(4)$$

C.
$$\boxed{\lim_{x \to a} k = k, \text{ where } k \text{ is a constant.}}$$

Example 10

$$\lim_{x \to 2} 8 = 8$$

Note: No matter what x approaches, $f(x) = 8$; so $f(x)$ not only approaches 8, but actually is 8.

D.
$$\boxed{\lim_{x \to a} [f(x) \cdot g(x)] = \lim_{x \to a} f(x) \cdot \lim_{x \to a} g(x)}$$

Example 11

$$\lim_{x \to 3} [x^2(x - 1)] = \lim_{x \to 3} x^2 \cdot \lim_{x \to 3} (x - 1)$$

$$18 = 9 \cdot 2$$

Using rule D, one could show that

$$\lim_{x \to a} x^n = a^n \qquad \text{(where } n \text{ is a positive integer)}$$

E.

$$\lim_{x \to a} \frac{f(x)}{g(x)} = \frac{\lim\limits_{x \to a} f(x)}{\lim\limits_{x \to a} g(x)}, \quad \text{where} \quad \lim_{x \to a} g(x) \neq 0$$

Example 12

$$\lim_{x \to 1} \frac{x^2 - 4}{x + 2} = \frac{\lim\limits_{x \to 1} (x^2 - 4)}{\lim\limits_{x \to 1} (x + 2)}$$

$$-1 = \frac{-3}{3}$$

The idea of continuity is very closely related to the limit idea. A function is *continuous* if its graph is an unbroken curve.

The function $f(x) = x^2$ is continuous, as you can see from its graph in Fig. 2.2.

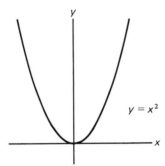

$y = x^2$

Figure 2.2

The function $f(x) = \begin{cases} x & \text{if } x \geqslant 0 \\ 1 & \text{if } x < 0 \end{cases}$

is not continuous because its graph in Fig. 2.3 is broken at $x = 0$.

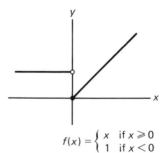

$f(x) = \begin{cases} x & \text{if } x \geqslant 0 \\ 1 & \text{if } x < 0 \end{cases}$ **Figure 2.3**

More formally,

Continuity
A function is continuous at $x = a$ if and only if

(a) $f(a)$ is defined,

(b) $\lim\limits_{x \to a} f(x)$ exists, and

(c) $\lim\limits_{x \to a} f(x) = f(a)$.

Figure 2.4 shows the graphs of three functions that are not continuous at $x = a$.

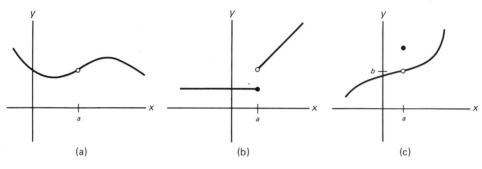

(a)   (b)   (c)

Figure 2.4

Figure 2.4a is not continuous because $f(a)$ is not defined.

Figure 2.4b is not continuous because $\lim\limits_{x \to a} f(x)$ does not exist.

Figure 2.4c is not continuous because $\lim\limits_{x \to a} f(x) = b \neq f(a)$.

If a function is continuous at all points in a given interval, it is continuous in that interval.

Exercises

Find each limit using a calculator. (For trigonometric functions be certain that your calculator is in the radian mode.)

1. $\lim\limits_{x \to 2} \dfrac{x^2 - 4}{x - 2}$

2. $\lim\limits_{x \to -\frac{1}{2}} \dfrac{4x^2 - 1}{2x + 1}$

3. $\lim\limits_{x \to \infty} \dfrac{3x + 2}{x}$

4. $\lim\limits_{x \to \infty} \dfrac{3x^2 + 4x}{2x^2 - 1}$

5. $\lim\limits_{x \to 0} \dfrac{\sin x}{x}$

6. $\lim\limits_{x \to 0} \dfrac{\cos x - 1}{x}$

Find each limit.

7. $\lim\limits_{x \to 2} (x^2 - 5x)$

8. $\lim\limits_{x \to -1} (3x^2 + 7x + 1)$

9. $\lim\limits_{x \to -1} (2x^3 + 5x^2 - 2)$

10. $\lim\limits_{x \to 2} (x^3 - 3x^2 + x + 4)$

11. $\lim\limits_{x \to 1} \dfrac{x^2 - 1}{x - 1}$

12. $\lim\limits_{x \to -3} \dfrac{x^2 - 9}{x + 3}$

13. $\lim\limits_{x \to -3/2} \dfrac{4x^2 - 9}{2x + 3}$

14. $\lim\limits_{x \to 4/3} \dfrac{9x^2 - 16}{3x - 4}$

15. $\lim\limits_{x \to -1} \sqrt{2x + 3}$

16. $\lim\limits_{x \to 4} \sqrt{3x - 3}$

17. $\lim\limits_{x \to 6} \sqrt{4 - x}$

18. $\lim\limits_{x \to -1} \sqrt{2x + 1}$

19. $\lim\limits_{x \to \infty} \dfrac{1}{2x}$

20. $\lim\limits_{x \to \infty} \dfrac{1}{x^2}$

21. $\lim\limits_{x \to \infty} \dfrac{3x^2 - 5x + 2}{4x^2 + 8x - 11}$

22. $\lim\limits_{x \to \infty} \dfrac{7x^3 + 2x - 13}{4x^3 + x^2}$

Find the instantaneous velocity v for each expression of s and value of t.

23. $s = f(t) = 4t^2 - 3t,\ t = 2$

24. $s = f(t) = 3t^2 - 5t + 2,\ t = 3$

25. $s = f(t) = t^2 + 3t - 10,\ t = 4$

26. $s = f(t) = 5t^2 - 7t + 8,\ t = 2$

Find each limit using formulas A–E.

27. $\lim\limits_{x \to 2} (x^2 + x)$

28. $\lim\limits_{x \to 3} (x^3 + x^2)$

29. $\lim\limits_{x \to 1} (4x^2 + 100x - 2)$

30. $\lim\limits_{x \to -1} (3x^2 + 5x - 8)$

31. $\lim\limits_{x \to 1} (x + 3)(x - 4)$

32. $\lim\limits_{x \to 4} (2x + 1)(x - 3)$

33. $\lim\limits_{x \to -2} (x^2 + 3x + 1)(x^4 - 2x^2 + 3)$

34. $\lim\limits_{x \to 2} (x^2 + 5x - 10)(x^3 + 6x^2 - x)$

35. $\lim\limits_{x \to 2} \dfrac{x^2 + 3x + 2}{x^2 + 1}$

36. $\lim\limits_{x \to 3} \dfrac{x^2 - 4x + 5}{x^2 + 2x}$

37. $\lim\limits_{x \to -7} \dfrac{x^2 - 49}{x + 7}$

38. $\lim\limits_{x \to 2} \dfrac{x^2 - 4}{x - 2}$

39. $\lim\limits_{x \to 5/2} \dfrac{4x^2 - 25}{2x - 5}$

40. $\lim\limits_{x \to -4/3} \dfrac{9x^2 - 16}{3x + 4}$

41. $\lim\limits_{x \to 3} \dfrac{(x^2 + 3x + 1)(x + 5)}{x - 2}$

42. $\lim\limits_{x \to -2} \dfrac{(x^2 + x - 5)(x - 3)}{x + 3}$

43. $\lim\limits_{h \to 0} \dfrac{(x + h)^2 - x^2}{h}$

44. $\lim\limits_{x \to a} \dfrac{x^3 - a^3}{x - a}$

45. $\lim\limits_{h \to 0} \dfrac{\dfrac{1}{x + h} - \dfrac{1}{x}}{h}$

46. $\lim\limits_{x \to a} \dfrac{\dfrac{1}{x^2} - \dfrac{1}{a^2}}{x - a}$

47. $\lim\limits_{x \to a} \dfrac{\sqrt{x} - \sqrt{a}}{x - a}$ (*Hint:* Rationalize the numerator.)

48. $\lim\limits_{h \to 0} \dfrac{\sqrt{x + h} - \sqrt{x}}{h}$

In Exercises 49–56, find $\lim\limits_{x \to a} f(x)$ if it exists.

49.

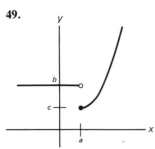

Figure 2.5

50.

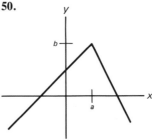

Figure 2.6

51.

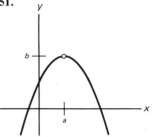

Figure 2.7

The Derivative   Chap. 2

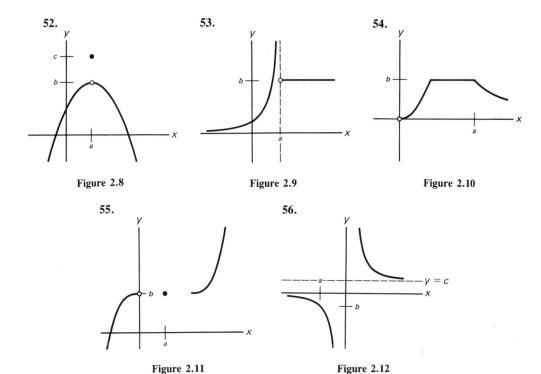

52.

53.

54.

Figure 2.8

Figure 2.9

Figure 2.10

55.

56.

Figure 2.11

Figure 2.12

57–64. In Exercises 49–56, is the function continuous at $x = a$?

65. In Exercise 49, find $\lim\limits_{x \to +\infty} f(x)$ if it exists.

66. In Exercise 53, find $\lim\limits_{x \to +\infty} f(x)$ if it exists.

67. In Exercise 54, find $\lim\limits_{x \to +\infty} f(x)$ if it exists.

68. In Exercise 55, find $\lim\limits_{x \to +\infty} f(x)$ if it exists.

69. In Exercise 49, find $\lim\limits_{x \to -\infty} f(x)$ if it exists.

70. In Exercise 53, find $\lim\limits_{x \to -\infty} f(x)$ if it exists.

71. In Exercise 54, find $\lim\limits_{x \to -\infty} f(x)$ if it exists.

72. In Exercise 55, find $\lim\limits_{x \to -\infty} f(x)$ if it exists.

2.3 THE SLOPE OF A TANGENT LINE TO A CURVE

The limit process so far has only been applied to motion problems. We will now look at its geometric application.

We saw in Section 1.2 how to find the slope of a line. But how can we describe the slope of a tangent line to a nonlinear curve at a given point? As in Fig. 2.13 assume the curve is the graph of a given function $y = f(x)$. We

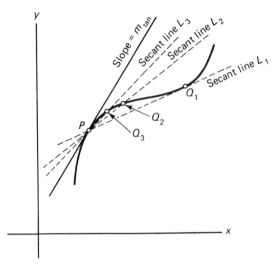

Figure 2.13

wish to find the slope of the tangent line, m_{tan}, at the point P with coordinates $(x, f(x))$.

As in Fig. 2.13 we can determine the slope of a line passing through P and any other point Q on the curve (the secant line). Observe the slopes of these secant lines as we choose points Q closer and closer to the point P. As Q approaches P, the values of the slopes of these secant lines will be coming closer and closer to the slope of the tangent line, m_{tan}. We can express this process in terms of the coordinates of P and Q as in Fig. 2.14.

In this figure $\Delta y = f(x + \Delta x) - f(x)$.

As we choose values of Δx closer to 0, point Q moves closer to P along the curve. Thus the slope of the secant line approaches m_{tan}, the slope of the tangent line. The slope of the secant line through P and Q is given by

$$\frac{f(x + \Delta x) - f(x)}{(x + \Delta x) - x} = \frac{f(x + \Delta x) - f(x)}{\Delta x} = \frac{\Delta y}{\Delta x}$$

so

Slope of Tangent Line

$$m_{\text{tan}} = \lim_{\Delta x \to 0} \frac{\Delta y}{\Delta x} = \lim_{\Delta x \to 0} \frac{f(x + \Delta x) - f(x)}{\Delta x}$$

Example 1

Find the slope of the tangent line to the curve $y = x^2 + 3$ at $(1, 4)$.

$$m_{\text{tan}} = \lim_{\Delta x \to 0} \frac{\Delta y}{\Delta x} = \lim_{\Delta x \to 0} \frac{f(x + \Delta x) - f(x)}{\Delta x}$$

$$= \lim_{\Delta x \to 0} \frac{[(1 + \Delta x)^2 + 3] - [(1)^2 + 3]}{\Delta x}$$

The Derivative   Chap. 2

$$= \lim_{\Delta x \to 0} \frac{2(\Delta x) + (\Delta x)^2}{\Delta x}$$

$$= \lim_{\Delta x \to 0} \frac{\Delta x(2 + \Delta x)}{\Delta x}$$

$$= \lim_{\Delta x \to 0} (2 + \Delta x)$$

$$= 2$$

The curve and the tangent line appear in Fig. 2.15.

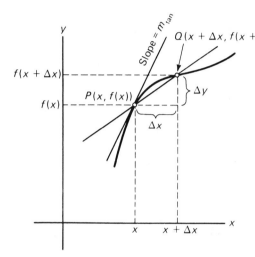

Figure 2.14　　　　　　　　　　**Figure 2.15**

The process used to solve the geometric problem is the same as used for the motion problem. This process, the limit, is the foundation of the calculus.

Example 2

Find the equation of the tangent line to the curve $y = 2x^2 - 5$ at $(2, 3)$.

Step 1: Find m_{tan}:

$$m_{\text{tan}} = \lim_{\Delta x \to 0} \frac{\Delta y}{\Delta x} = \lim_{\Delta x \to 0} \frac{f(x + \Delta x) - f(x)}{\Delta x}$$

$$= \lim_{\Delta x \to 0} \frac{[2(2 + \Delta x)^2 - 5] - [2(2)^2 - 5]}{\Delta x}$$

$$= \lim_{\Delta x \to 0} \frac{\Delta x[8 + 8(\Delta x)]}{\Delta x}$$

$$= \lim_{\Delta x \to 0} [8 + 8(\Delta x)]$$

$$= 8$$

Step 2: Find the equation of the line:

Using the point-slope formula, we have

$$y - y_1 = m(x - x_1)$$
$$y - 3 = 8(x - 2)$$
$$y = 8x - 13$$

Exercises

Find the slope of the tangent line to each curve at the given point.

1. $y = x^2$; (3, 9)
2. $y = 3x^2$; (1, 3)
3. $y = 3x^2 - 4$; (-1, -1)
4. $y = 4x^2 + 3$; (2, 19)
5. $y = 2x^2 + x - 3$; (2, 7)
6. $y = 3x^2 + 8x - 10$; (-2, -16)
7. $y = -4x^2 + 3x - 2$; (1, -3)
8. $y = -5x^2 - 3x - 1$; (2, -27)
9. $y = x^3$; (2, 8)
10. $y = x^3 + 1$; (-1, 0)

Find the equation of the tangent line to each curve at the given point.

11. $y = x^2$; (-2, 4)
12. $y = -5x^2$; (2, -20)
13. $y = 2x^2 - 3$; (-2, 5)
14. $y = -4x^2 + 2$; (-1, -2)
15. $y = 5x^2 - 3x + 2$; (-1, 10)
16. $y = 4x^2 - 7x + 5$; (3, 20)
17. $y = -3x^2 + 5x + 2$; $x = 3$
18. $y = -2x^2 + 4x - 7$; $x = -2$
19. $y = x^3 + x - 1$; $x = 1$
20. $y = x^3 - 3x + 4$; $x = 2$

Find a point or points on the curve where the slope of the tangent line is the given value.

21. $y = x^2$; $m = -\frac{1}{3}$
22. $y = 3x^2 - 4$; $m = 6$
23. $y = x^3 + x$; $m = 4$
24. $y = x^3 - 3x^2 - 5x$; $m = 4$

2.4 THE DERIVATIVE

Instantaneous velocity and the slope of the tangent line to a curve are only two examples of using the concept of limit. A more general use of the limit is found in the derivative.

When we investigated instantaneous velocity and the slope of the tangent line, we applied the limit concept to related functions as follows:

$$v = \lim_{\Delta t \to 0} \frac{\Delta s}{\Delta t} = \lim_{\Delta t \to 0} \frac{f(t + \Delta t) - f(t)}{\Delta t}$$

and

$$m_{\text{tan}} = \lim_{\Delta x \to 0} \frac{\Delta y}{\Delta x} = \lim_{\Delta x \to 0} \frac{f(x + \Delta x) - f(x)}{\Delta x}$$

For any function $y = f(x)$,

$$\frac{\Delta y}{\Delta x} = \frac{f(x + \Delta x) - f(x)}{\Delta x}$$

is the *average rate of change* of the function f at x, and

$$\lim_{\Delta x \to 0} \frac{\Delta y}{\Delta x} = \lim_{\Delta x \to 0} \frac{f(x + \Delta x) - f(x)}{\Delta x}$$

is the *derivative* of the function f at x. (The derivative could also be thought of as the instantaneous rate of change of f at x.) The process of finding this limit, the derivative, is called *differentiation*.

Example 1

Find the derivative of $f(x) = x^2$ at $x = 3$.

$$\lim_{\Delta x \to 0} \frac{\Delta y}{\Delta x} = \lim_{\Delta x \to 0} \frac{f(x + \Delta x) - f(x)}{\Delta x}$$

$$= \lim_{\Delta x \to 0} \frac{(3 + \Delta x)^2 - (3)^2}{\Delta x}$$

$$= \lim_{\Delta x \to 0} \frac{9 + 6(\Delta x) + (\Delta x)^2 - 9}{\Delta x}$$

$$= \lim_{\Delta x \to 0} \frac{6(\Delta x) + (\Delta x)^2}{\Delta x}$$

$$= \lim_{\Delta x \to 0} \frac{\Delta x(6 + \Delta x)}{\Delta x}$$

$$= \lim_{\Delta x \to 0} (6 + \Delta x)$$

$$= 6$$

In Example 1 we determined the derivative at $x = 3$. However, we can still determine the derivative without specifying a value for x.

Example 2

Find the derivative of $f(x) = x^2$.

$$\lim_{\Delta x \to 0} \frac{\Delta y}{\Delta x} = \lim_{\Delta x \to 0} \frac{f(x + \Delta x) - f(x)}{\Delta x}$$

$$= \lim_{\Delta x \to 0} \frac{(x + \Delta x)^2 - x^2}{\Delta x}$$

$$= \lim_{\Delta x \to 0} \frac{x^2 + 2x(\Delta x) + (\Delta x)^2 - x^2}{\Delta x}$$

$$= \lim_{\Delta x \to 0} \frac{\Delta x(2x + \Delta x)}{\Delta x}$$

$$= \lim_{\Delta x \to 0} (2x + \Delta x)$$

$$= 2x$$

Note that when determining the above limit, x is considered to be held constant. Only Δx is approaching 0. Note also that if we let $x = 3$ in Example 2, the derivative has the value $2x = 2(3) = 6$ as in Example 1.

The following notations for the derivative are often used in place of the symbol $\lim_{\Delta x \to 0} \frac{\Delta y}{\Delta x}$:

$$\frac{dy}{dx}, \quad y', \quad \frac{d}{dx}[f(x)], \quad f'(x), \quad \text{and} \quad D_x y$$

So for Example 1 we could write

$$\frac{dy}{dx} = 6 \quad \text{at} \quad x = 3$$

$$y' = 6 \quad \text{at} \quad x = 3$$

$$\frac{d}{dx}[f(3)] = 6$$

$$f'(3) = 6$$

or

$$D_x y = 6 \quad \text{at} \quad x = 3$$

For Example 2 we could write

$$\frac{dy}{dx} = 2x$$

$$y' = 2x$$

$$\frac{d}{dx}[f(x)] = 2x$$

$$f'(x) = 2x$$

or

$$D_x y = 2x$$

Note that the expressions above for the derivative determined in Example 2 show that the derivative becomes a new function of x when the value for x is not specified. That is, $y' = 2x$ is a function which relates to each x the value $2x$.

Example 3

Find the derivative of $y = x^3$.

$$\frac{dy}{dx} = \lim_{\Delta x \to 0} \frac{f(x + \Delta x) - f(x)}{\Delta x}$$

$$= \lim_{\Delta x \to 0} \frac{(x + \Delta x)^3 - x^3}{\Delta x}$$

$$= \lim_{\Delta x \to 0} \frac{x^3 + 3x^2(\Delta x) + 3x(\Delta x)^2 + (\Delta x)^3 - x^3}{\Delta x}$$

$$= \lim_{\Delta x \to 0} \frac{3x^2(\Delta x) + 3x(\Delta x)^2 + (\Delta x)^3}{\Delta x}$$

$$= \lim_{\Delta x \to 0} \frac{\Delta x[3x^2 + 3x(\Delta x) + (\Delta x)^2]}{\Delta x}$$

$$= \lim_{\Delta x \to 0} [3x^2 + 3x(\Delta x) + (\Delta x)^2]$$

$$= 3x^2$$

Example 4

Find the derivative of $y = \dfrac{1}{x + 2}$.

$$\frac{dy}{dx} = \lim_{\Delta x \to 0} \frac{f(x + \Delta x) - f(x)}{\Delta x}$$

$$= \lim_{\Delta x \to 0} \frac{\dfrac{1}{x + \Delta x + 2} - \dfrac{1}{x + 2}}{\Delta x}$$

$$= \lim_{\Delta x \to 0} \frac{\dfrac{(x + 2) - (x + \Delta x + 2)}{(x + \Delta x + 2)(x + 2)}}{\Delta x}$$

$$= \lim_{\Delta x \to 0} \frac{\dfrac{-\Delta x}{(x + \Delta x + 2)(x + 2)}}{\Delta x}$$

$$= \lim_{\Delta x \to 0} \frac{-1}{(x + \Delta x + 2)(x + 2)}$$

$$= \frac{-1}{(x + 2)^2}$$

Example 5

Find the derivative of $y = 3x^2 + 2x - 4$.

$$\frac{dy}{dx} = \lim_{\Delta x \to 0} \frac{f(x + \Delta x) - f(x)}{\Delta x}$$

$$= \lim_{\Delta x \to 0} \frac{[3(x + \Delta x)^2 + 2(x + \Delta x) - 4] - (3x^2 + 2x - 4)}{\Delta x}$$

$$= \lim_{\Delta x \to 0} \frac{3x^2 + 6x(\Delta x) + 3(\Delta x)^2 + 2x + 2(\Delta x) - 4 - 3x^2 - 2x + 4}{\Delta x}$$

$$= \lim_{\Delta x \to 0} \frac{6x(\Delta x) + 3(\Delta x)^2 + 2(\Delta x)}{\Delta x}$$

$$= \lim_{\Delta x \to 0} \frac{\Delta x[6x + 3(\Delta x) + 2]}{\Delta x}$$

$$= \lim_{\Delta x \to 0} [6x + 3(\Delta x) + 2]$$

$$= 6x + 2$$

Example 6

Find the derivative of $f(x) = \sqrt{x}$.

$$\frac{dy}{dx} = \lim_{\Delta x \to 0} \frac{f(x + \Delta x) - f(x)}{\Delta x}$$

$$= \lim_{\Delta x \to 0} \frac{\sqrt{x + \Delta x} - \sqrt{x}}{\Delta x} \qquad (\textit{Note: } \text{To evaluate this limit, you must rationalize the numerator.})$$

$$= \lim_{\Delta x \to 0} \frac{\sqrt{x + \Delta x} - \sqrt{x}}{\Delta x} \cdot \frac{\sqrt{x + \Delta x} + \sqrt{x}}{\sqrt{x + \Delta x} + \sqrt{x}}$$

$$= \lim_{\Delta x \to 0} \frac{x + \Delta x - x}{\Delta x(\sqrt{x + \Delta x} + \sqrt{x})}$$

$$= \lim_{\Delta x \to 0} \frac{\Delta x}{\Delta x(\sqrt{x + \Delta x} + \sqrt{x})}$$

$$= \lim_{\Delta x \to 0} \frac{1}{\sqrt{x + \Delta x} + \sqrt{x}}$$

$$= \frac{1}{\sqrt{x} + \sqrt{x}} = \frac{1}{2\sqrt{x}}$$

Computer Example: Derivative of a Function

The derivative of a function $f(x)$ at point $x = a$ is defined by

$$f'(a) = \lim_{d \to 0} \frac{f(a + d) - f(a)}{d}$$

The following computer example compares the quotient

$$Q = \frac{f(a + d) - f(a)}{d}$$

to the derivative $f'(a)$ as d varies from 0.1 to 0.01 in steps of 0.005. The function used is $f(x) = x^3 - 6x^2 + 9x + 4$ (defined on line 30) and its derivative is $f'(x) = 3x^2 - 12x + 9$ (defined on line 40).

This program can be used for other functions by making the appropriate changes in lines 30 and 40. The range of values for d can be adjusted by making the appropriate changes in line 80. (See Fig. 2.16.)

Exercises

Find the derivative of each function.

1. $y = 3x + 4$
2. $y = 2 - 6x$
3. $y = 1 - 2x$
4. $y = 4x - 5$
5. $y = 3x^2$
6. $y = -4x^2$
7. $y = x^2 - 2x$
8. $y = x^2 + 1$
9. $y = 3x^2 - 4x + 1$
10. $y = 6x^2 - 8x + 2$
11. $y = 1 - 6x^2$
12. $y = 3 - 4x - 7x^2$

13. $y = x^3 + 4x$
14. $y = 1 - 2x^3$
15. $y = \dfrac{1}{x}$

16. $y = \dfrac{1}{x - 1}$
17. $y = \dfrac{2}{x - 3}$
18. $y = \dfrac{-1}{2x + 1}$

19. $y = \dfrac{1}{x^2}$
20. $y = \dfrac{1}{x^2 - 1}$
21. $y = \dfrac{1}{4 - x^2}$

22. $y = \dfrac{4}{x^2 + 1}$
23. $y = \sqrt{x + 1}$
24. $y = \sqrt{x - 2}$

25. $y = \sqrt{1 - 2x}$
26. $y = \sqrt{x^2 + 1}$
27. $y = \dfrac{1}{\sqrt{x - 1}}$

The Derivative   Chap. 2

```
10 CLS
20 F1$ = " #.###   ##.####   ##.#####   ###.###"
30 DEF FN F(A) = A * A * A - 6 * A * A + 9 * A + 4
40 DEF FN F1(A) = 3 * A * A - 12 * A + 9
50 PRINT "ENTER THE VALUE FOR A"
60 INPUT A
70 PRINT " DELTA   QUOTIENT    ERROR    % ERROR"
80 FOR D= .1 TO .005 STEP -.005
90 Q = (FN F(A + D) - FN F(A))/D
100 E = Q - FN F1(A)
110 IF FN F1(A)=0 THEN 150
120 P = E / FN F1(A) * 100
130 PRINT USING F1$; D; Q; E; P
140 GOTO 160
150 PRINT USING F1$; D; Q; E
160 NEXT D
170 END
```

Sample output for A=2.

DELTA	QUOTIENT	ERROR	% ERROR
0.100	-2.9900	0.01000	-0.333
0.095	-2.9910	0.00900	-0.300
0.090	-2.9919	0.00809	-0.270
0.085	-2.9928	0.00724	-0.241
0.080	-2.9936	0.00639	-0.213
0.075	-2.9944	0.00561	-0.187
0.070	-2.9951	0.00486	-0.162
0.065	-2.9958	0.00420	-0.140
0.060	-2.9964	0.00359	-0.120
0.055	-2.9970	0.00300	-0.100
0.050	-2.9976	0.00245	-0.082
0.045	-2.9980	0.00203	-0.068
0.040	-2.9984	0.00160	-0.053
0.035	-2.9988	0.00116	-0.039
0.030	-2.9992	0.00082	-0.027
0.025	-2.9994	0.00058	-0.019
0.020	-2.9996	0.00041	-0.014
0.015	-2.9999	0.00012	-0.004
0.010	-3.0001	-0.00007	0.002

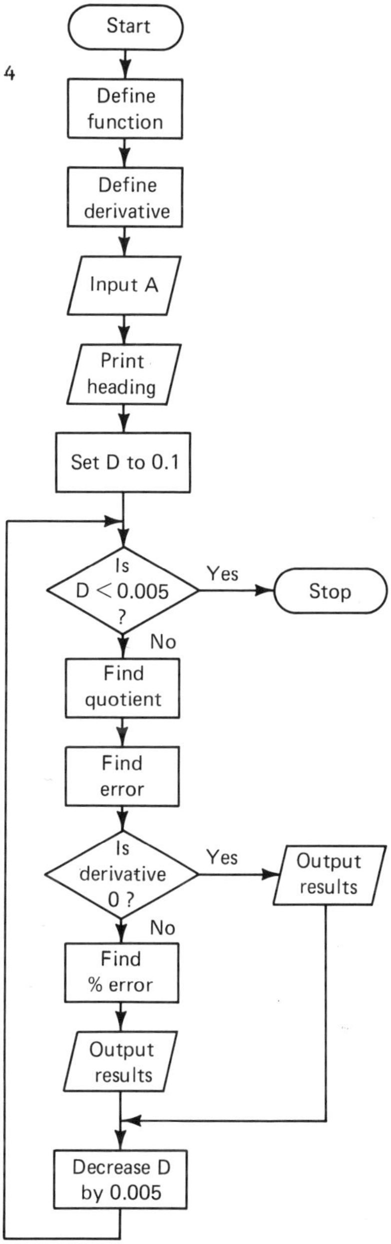

Figure 2.16

28. $y = \dfrac{1}{\sqrt{4 - x}}$

Find the slope of the tangent line to each curve at the given point.

29. $y = 3x^2 - 4$; $(1, -1)$ 30. $y = \dfrac{1}{x - 4}$; $(1, -\frac{1}{3})$ 31. $y = \dfrac{1}{1 - 3x^2}$; $(1, -\frac{1}{2})$

32. $y = \sqrt{x + 5}$; $(4, 3)$

Sec. 2.4   The Derivative

Find the equation of the tangent line to each curve at the given point.

33. $y = x^2 - 3x$; $(2, -2)$ **34.** $y = \dfrac{6}{x}$; $(-2, -3)$ **35.** $y = \sqrt{x - 7}$; $(11, 2)$

36. $y = \dfrac{2}{x^2 + 1}$; $(-1, 1)$

Find a point or points on the curve where the slope of the tangent line is the given value.

37. $y = \dfrac{1}{x - 3}$; $m = -1$ **38.** $y = \dfrac{1}{x^2}$; $m = \frac{1}{4}$ **39.** $y = \sqrt{x + 4}$; $m = \frac{1}{2}$

40. $y = \dfrac{2}{\sqrt{x}}$; $m = -\frac{1}{8}$

2.5 DIFFERENTIATION OF POLYNOMIALS

Recall from algebra that a *polynomial* is defined as follows:

$$a_n x^n + a_{n-1} x^{n-1} + a_{n-2} x^{n-2} + \cdots + a_2 x^2 + a_1 x + a_0$$

where n is a positive integer and the coefficients $a_n, a_{n-1}, a_{n-2}, \ldots, a_2, a_1, a_0$ are real numbers.

As we saw in Section 2.4, the derivative definition can be used to find the derivative of a polynomial. However, the process is tedious. We now need to develop some formulas that will shorten the process to find such a derivative.

First, let's find the derivative of a constant function.

Example 1

Find $\dfrac{dy}{dx}$ when $y = C$, where C is a constant.

$$\frac{dy}{dx} = \lim_{\Delta x \to 0} \frac{f(x + \Delta x) - f(x)}{\Delta x}$$

$$= \lim_{\Delta x \to 0} \frac{C - C}{\Delta x}$$

$$= \lim_{\Delta x \to 0} \frac{0}{\Delta x}$$

$$= 0$$

So, if $f(x)$ is a constant function, i.e., $f(x) = C$ for all x, then $f'(x) = 0$ for all x.

Example 2

Find the derivative of $f(x) = 10$.
Since this is a constant function, $f'(x) = 0$ for all x.

Suppose that $y = x^n$, where n is some positive integer. By using the binomial theorem (see Appendix A) we have

$$(x + \Delta x)^n = x^n + nx^{n-1}(\Delta x) + \frac{n(n - 1)}{2!} x^{n-2}(\Delta x)^2 \qquad \text{(cont. on next page)}$$

$$+ \frac{n(n-1)(n-2)}{3!} x^{n-3}(\Delta x)^3 + \cdots$$

$$+ nx(\Delta x)^{n-1} + (\Delta x)^n$$

where the right-hand side consists of $(n+1)$ terms, each of which has a factor of Δx except for the first term. So if

$$\Delta y = (x + \Delta x)^n - x^n$$

$$= x^n + nx^{n-1}(\Delta x) + \frac{n(n-1)}{2!} x^{n-2}(\Delta x)^2$$

$$+ \frac{n(n-1)(n-2)}{3!} x^{n-3}(\Delta x)^3 + \cdots + nx(\Delta x)^{n-1} + (\Delta x)^n - x^n$$

then

$$\frac{\Delta y}{\Delta x} = nx^{n-1} + \frac{n(n-1)}{2!} x^{n-2}(\Delta x) + \frac{n(n-1)(n-2)}{3!} x^{n-3}(\Delta x)^2 + \cdots$$

$$+ nx(\Delta x)^{n-2} + (\Delta x)^{n-1}$$

We then have

$$\frac{dy}{dx} = \lim_{\Delta x \to 0} \frac{\Delta y}{\Delta x} = nx^{n-1}$$

Note that every term but the first term has a factor of Δx, which will make each of these other terms approach 0 as Δx approaches 0.

We now add the restriction that $x \neq 0$ in the formula above to avoid later complications. That is,

$$\boxed{\frac{d}{dx} x^n = nx^{n-1}, \qquad x \neq 0}$$

Example 3

Find the derivative of $y = x^{32}$.

$$\text{Since } n = 32, \frac{dy}{dx} = 32x^{32-1}$$

$$= 32x^{31}$$

We now make the following observation.

$$\boxed{\frac{d}{dx}[c\,f(x)] = c\,\frac{d}{dx}[f(x)]}$$

This formula is shown below:

$$\frac{d}{dx}[c\,f(x)] = \lim_{\Delta x \to 0} \frac{c[f(x + \Delta x)] - c[f(x)]}{\Delta x}$$

$$= \lim_{\Delta x \to 0} \frac{c[f(x + \Delta x) - f(x)]}{\Delta x}$$

$$= c \lim_{\Delta x \to 0} \frac{f(x + \Delta x) - f(x)}{\Delta x}$$

$$= c \frac{d}{dx}[f(x)]$$

Example 4

Find the derivative of $y = -5x^{12}$.

$$\frac{dy}{dx} = \frac{d}{dx}(-5x^{12})$$

$$= -5 \frac{d}{dx}(x^{12})$$

$$= -5(12x^{12-1})$$

$$= -60x^{11}$$

We have seen how to differentiate functions of the form $y = cx^n$, where n is a positive integer, by use of a simple formula. We will now develop formulas to differentiate other functions.

If $h(x) = f(x) + g(x)$, the sum of two functions, then

$$\frac{d}{dx}h(x) = \frac{d}{dx}[f(x) + g(x)]$$

$$= \lim_{\Delta x \to 0} \frac{[f(x + \Delta x) + g(x + \Delta x)] - [f(x) + g(x)]}{\Delta x}$$

$$= \lim_{\Delta x \to 0} \frac{[f(x + \Delta x) - f(x)] + [g(x + \Delta x) - g(x)]}{\Delta x}$$

$$= \lim_{\Delta x \to 0} \left\{ \frac{[f(x + \Delta x) - f(x)]}{\Delta x} + \frac{[g(x + \Delta x) - g(x)]}{\Delta x} \right\}$$

$$= \lim_{\Delta x \to 0} \frac{[f(x + \Delta x) - f(x)]}{\Delta x} + \lim_{\Delta x \to 0} \frac{[g(x + \Delta x) - g(x)]}{\Delta x}$$

$$= \frac{d}{dx}[f(x)] + \frac{d}{dx}[g(x)]$$

That is, the derivative of a sum of functions is the sum of the derivatives of the functions:

$$\boxed{\frac{d}{dx}[f(x) + g(x)] = \frac{d}{dx}[f(x)] + \frac{d}{dx}[g(x)]}$$

Similarly,

$$\frac{d}{dx}[f(x) - g(x)] = \frac{d}{dx}[f(x)] - \frac{d}{dx}[g(x)]$$

That is, the derivative of a difference of functions is the difference of the derivatives of the functions.

We now have the formulas we need to differentiate polynomials.

Example 5

Differentiate $y = 7x^5 - 2x^3 + x^2 - 8$.

$$\frac{dy}{dx} = \frac{d}{dx}(7x^5 - 2x^3 + x^2 - 8)$$

$$= \frac{d}{dx}(7x^5) - \frac{d}{dx}(2x^3) + \frac{d}{dx}(x^2) - \frac{d}{dx}(8)$$

$$= 7\frac{d}{dx}(x^5) - 2\frac{d}{dx}(x^3) + \frac{d}{dx}(x^2) - \frac{d}{dx}(8)$$

$$= 7 \cdot 5x^{5-1} - 2 \cdot 3x^{3-1} + 2x^{2-1} - 0$$

$$= 35x^4 - 6x^2 + 2x$$

Example 6

Find the equation of the tangent line to the curve of the function $y = 5x^3 - 2x^2 + 3$ at the point $(1, 6)$.

The slope of the tangent line, m_{tan}, is the derivative of $y = 5x^3 - 2x^2 + 3$ at $x = 1$.

$$\frac{dy}{dx} = \frac{d}{dx}(5x^3 - 2x^2 + 3)$$

$$= 5 \cdot 3x^{3-1} - 2 \cdot 2x^{2-1} + 0$$

$$= 15x^2 - 4x$$

Thus the derivative $\frac{dy}{dx}$ at $x = 1$ is $15(1)^2 - 4(1) = 11$. So, $m_{tan} = 11$. The equation of the tangent line can be written in point-slope form as

$$y - y_1 = m(x - x_1)$$

$$y - 6 = 11(x - 1)$$

$$y = 11x - 5$$

In this example we needed to evaluate the derivative of a function after applying the differentiation formulas. To show that the derivative is to be evaluated at $x = a$, we usually write

$$\frac{dy}{dx}\bigg|_{x=a} \qquad \text{or} \qquad f'(a)$$

So in Example 6 we would write

$$m_{\text{tan}} = \frac{dy}{dx}\bigg|_{x=1} = 15(1)^2 - 4(1) = 11 \quad \text{or} \quad m_{\text{tan}} = f'(1) = 15(1)^2 - 4(1) = 11$$

Example 7

Find the instantaneous velocity v of an object falling freely from a building 100 ft tall after 2 s. The position of this free-falling object is given by $s = -16t^2 + 100$ (measured in feet.)

v is the derivative of $s = -16t^2 + 100$ at $t = 2$. So

$$v = \frac{ds}{dt}\bigg|_{t=2} = \frac{d}{dt}(-16t^2 + 100)\bigg|_{t=2}$$

$$= -32t + 0|_{t=2}$$

$$= -32(2)$$

$$= -64 \text{ ft/s}$$

Note: The negative sign indicates that the velocity of the object is in the negative direction (downward). A positive velocity would indicate motion in the positive direction (upward).

While we showed that $\frac{d}{dx}(x^n) = nx^{n-1}$ based on n being a positive integer, in fact n can be any real number.

Example 8

Find the derivative of $y = \sqrt{x}$.

Since $y = \sqrt{x} = x^{1/2}$, then

$$\frac{dy}{dx} = \tfrac{1}{2}x^{(1/2)-1}$$

$$= \tfrac{1}{2}x^{-1/2}$$

$$= \frac{1}{2\sqrt{x}}$$

Example 9

Find the derivative of $y = 1/x^3$.

$$y = \frac{1}{x^3} = x^{-3}$$

so

$$\frac{dy}{dx} = -3x^{-3-1}$$

$$= -3x^{-4}$$

$$= -\frac{3}{x^4}$$

Example 10

Find the derivative of $y = 3/\sqrt{x}$.

$$y = \frac{3}{\sqrt{x}} = 3x^{-1/2}$$

so

$$\frac{dy}{dx} = \frac{d}{dx}(3x^{-1/2})$$

$$= 3\frac{d}{dx}(x^{-1/2})$$

$$= 3(-\tfrac{1}{2}x^{-(1/2)-1})$$

$$= -\tfrac{3}{2}x^{-3/2}$$

Exercises

Find the derivative of each polynomial function.

1. $y = 32$

2. $y = -100$

3. $y = x^5$

4. $y = x^{16}$

5. $y = 4x + 1$

6. $y = 6x - 2$

7. $y = 1 - 3x$

8. $y = x$

9. $y = 5x^2$

10. $y = -3x^2$

11. $y = x^2 - 3x$

12. $y = x^2 - 4x + 2$

13. $y = 4x^2 - 3x - 2$

14. $y = 9x^2 + 8x - 7$

15. $y = 1 - 8x^2$

16. $y = 4 - 6x - 5x^2$

17. $y = 3x^3 + 2x^2 - 6x$

18. $y = 13x^4 - \dfrac{7x^2}{2} + 2$

19. $y = 4x^5 - 2x^3 + x + 3$

20. $y = 5x^6 - 3x^4 + 2x^3 - 8x^2 + 4x - 7$

21. $y = \tfrac{5}{2}x^8 - \tfrac{6}{5}x^5 + \tfrac{15}{2}x^4 - x^3 + \sqrt{2}$

22. $y = \tfrac{8}{3}x^6 - \tfrac{7}{3}x^5 + 4x^4 - 3x^3 - 7x + \pi$

23. $y = \sqrt{7}x^4 - \sqrt{5}x^3 - \sqrt{3}x + \sqrt{7}$

24. $y = \sqrt{10}x^{10} - \sqrt{7}x^7 + \sqrt{5}x^5 - \sqrt{3}$

Find $f'(a)$ for each function.

25. $y = 3x^2 + 2x - 1$; $a = -1$

26. $y = 4x^2 + 6x$; $a = 2$

27. $y = 2x^3 - 6x^2 + 2x + 9$; $a = -3$

28. $y = 6x^3 - 5x + 3$; $a = -2$

29. $y = 4x^5 + 3x^2 - 2$; $a = 1$

30. $y = 5x^4 - 2x^2 + 18$; $a = 2$

31. $y = 5x^4 + 8x^3 + 2x - 1$; $a = 0$

32. $y = 9x^6 + 6x^5 - 7x^4 + 2$; $a = 0$

33. $y = 1 - 8x^2 - 5x^3 + 5x^6$; $a = 3$

34. $y = 3 - 4x^2 - 6x^4 + 5x^9$; $a = -2$

35. Find the equation of the tangent line to the curve $y = x^3 + 4x^2 - x + 2$ at $(-2, 12)$.

36. Find the equation of the tangent line to the curve $y = x^{100} + 5$ at $(-1, 6)$.

37. Given that $p = Ri^2$, find the rate of change dp/di of the power p in a 30-ohm (Ω) resistor when $i = 2$ amperes (A).

38. An object is falling according to $h = 5000 - 3.28t^2$ where h is in metres and t is in seconds. Find the speed (instantaneous velocity) of the object after 5 s.

39. Given that $V = ir$, find the rate of change dV/dr of the voltage drop for a 0.4-A current when $r = 4\ \Omega$.

40. In a certain circuit the voltage V varies according to $V = t^4$ where t is time in seconds. Find dV/dt when $t = 3$ s.

Sec. 2.5   Differentiation of Polynomials

Find the derivative of each function.

41. $y = x^{3/2}$

42. $y = \sqrt[3]{x}$

43. $y = \dfrac{1}{x^4}$

44. $y = \dfrac{4}{x^3}$

45. $y = 6x^{20}$

46. $y = 35x^{10}$

47. $y = \dfrac{14}{x^8}$

48. $y = \dfrac{52}{x^{13}}$

49. $y = \dfrac{5}{\sqrt[3]{x}}$

50. $y = \dfrac{2}{x\sqrt{x}}$

51. Given that $V = ir$, find the (instantaneous) rate of change dV/dr of the voltage drop V with respect to r (measured in ohms) for a 0.5-A current.

52. Given that $p = Ri^2$, find the rate of change dp/di of the power p in a 20-Ω resistor where p is in watts (W) and i is in amperes.

2.6 DERIVATIVES OF PRODUCTS AND QUOTIENTS

We need to differentiate products and quotients of functions. For example, the techniques developed so far would not apply to $y = \dfrac{32x^2 - 9}{5x^5 - 47x^2}$. We would be forced to find the derivative $\dfrac{dy}{dx}$ by using the limit process method of Section 2.4. Fortunately, we can develop formulas to handle this example.

Derivative of a Product

If $y = f(x)$ can be written as the product of two functions, $u = g(x)$ and $v = h(x)$, then we can write

$$y = g(x) \cdot h(x)$$

or

$$y = u \cdot v$$

In Section 2.9 we show that

$$\frac{dy}{dx} = u\frac{dv}{dx} + v\frac{du}{dx}$$

That is, *the derivative of a product of two functions is the product of the first function and the derivative of the second function plus the product of the second function and the derivative of the first function.*

Example 1

Find $\dfrac{d}{dx}[(4x + 3)(7x - 1)]$.

Let $u = 4x + 3$, $v = 7x - 1$, and $y = u \cdot v$.

$$\frac{dy}{dx} = u\frac{dv}{dx} + v\frac{du}{dx}$$

The Derivative   Chap. 2

$$= (4x + 3)\frac{d}{dx}(7x - 1) + (7x - 1)\frac{d}{dx}(4x + 3)$$

$$= (4x + 3)(7) + (7x - 1)(4)$$

$$= 28x + 21 + 28x - 4$$

$$= 56x + 17$$

Example 2

Find $\frac{d}{dx}[(3x^2 - 4)(5x^3 - 7)]$.

Let $u = 3x^2 - 4$, $v = 5x^3 - 7$, and $y = u \cdot v$.

$$\frac{dy}{dx} = u\frac{dv}{dx} + v\frac{du}{dx}$$

$$= (3x^2 - 4)\frac{d}{dx}(5x^3 - 7) + (5x^3 - 7)\frac{d}{dx}(3x^2 - 4)$$

$$= (3x^2 - 4)(15x^2) + (5x^3 - 7)(6x)$$

$$= 45x^4 - 60x^2 + 30x^4 - 42x$$

$$= 75x^4 - 60x^2 - 42x$$

Derivative of a Quotient

If $y = f(x)$ can be written as the quotient of two functions, $u = g(x)$ and $v = h(x)$, then we can write

$$y = \frac{g(x)}{h(x)}$$

or

$$y = \frac{u}{v}$$

In Section 2.9 we show that

$$\frac{d}{dx}\left(\frac{u}{v}\right) = \frac{v\dfrac{du}{dx} - u\dfrac{dv}{dx}}{v^2}$$

That is, *the derivative of a quotient of two functions is the denominator times the derivative of the numerator minus the numerator times the derivative of the denominator all divided by the square of the denominator.*

Example 3

Find $\frac{d}{dx}\left(\frac{5x + 3}{4x^2 - 7}\right)$.

Here $u = 5x + 3$ and $v = 4x^2 - 7$. Then

$$\frac{d}{dx}\left(\frac{u}{v}\right) = \frac{v\dfrac{du}{dx} - u\dfrac{dv}{dx}}{v^2}$$

$$= \frac{(4x^2 - 7)\dfrac{d}{dx}(5x + 3) - (5x + 3)\dfrac{d}{dx}(4x^2 - 7)}{(4x^2 - 7)^2}$$

$$= \frac{(4x^2 - 7)(5) - (5x + 3)(8x)}{(4x^2 - 7)^2}$$

$$= \frac{20x^2 - 35 - 40x^2 - 24x}{(4x^2 - 7)^2}$$

$$= \frac{-20x^2 - 24x - 35}{(4x^2 - 7)^2}$$

Example 4

Find $\dfrac{dy}{dx}$ where $y = \dfrac{32x^2 - 9}{5x^5 - 7x^2}$.

$$\frac{dy}{dx} = \frac{d}{dx}\left(\frac{32x^2 - 9}{5x^5 - 7x^2}\right)$$

$$= \frac{(5x^5 - 7x^2)(64x) - (32x^2 - 9)(25x^4 - 14x)}{(5x^5 - 7x^2)^2}$$

$$= \frac{320x^6 - 448x^3 - 800x^6 + 225x^4 + 448x^3 - 126x}{(5x^5 - 7x^2)^2}$$

$$= \frac{-480x^6 + 225x^4 - 126x}{(5x^5 - 7x^2)^2}$$

Exercises

Find the derivative of each function (a) by finding the derivative of the product as given using the derivative of the product and (b) by first multiplying the given expression then find the derivative of the resulting polynomial.

1. $y = x^2(2x + 1)$ **2.** $y = x^3(3x^2 + 2x - 1)$ **3.** $y = 2x(4x^2 + 3x - 5)$

4. $y = 4x^2(5 - 6x - 3x^2)$ **5.** $y = (2x + 3)(5x - 4)$ **6.** $y = (6x - 2)(4x - 3)$

7. $y = (4x + 7)(x^2 - 1)$ **8.** $y = (3x^3 - 1)(2x - 1)$

Find the derivative of each product.

9. $y = (x^2 + 3x + 4)(x^3 - 4x)$ **10.** $y = (x^3 + 2x - 7)(x^2 + 4x - 2)$

11. $y = (x^4 - 3x^2 - x)(2x^3 - 4x)$ **12.** $y = (x^2 - 3x + 8)(1 - 3x^4)$

Find the derivative of each quotient.

13. $y = \dfrac{x}{2x + 5}$ **14.** $y = \dfrac{3x}{x - 4}$ **15.** $y = \dfrac{1}{x^2 + x}$

16. $y = \dfrac{3}{x^2 - 1}$ **17.** $y = \dfrac{3x - 1}{2x + 4}$ **18.** $y = \dfrac{5x - 1}{3x + 2}$

19. $y = \dfrac{x^2}{2x + 1}$ **20.** $y = \dfrac{4x^2}{3x^2 - 1}$ **21.** $y = \dfrac{x - 1}{x^2 + x + 1}$

22. $y = \dfrac{3x - 7}{x^2 - 2}$ **23.** $y = \dfrac{4x^2 + 9}{3x^3 - 4x^2}$ **24.** $y = \dfrac{4x^3 - 1}{6x^2 + 3}$

25. Find $f'(2)$ when $f(x) = (x^2 - 4x + 3)(x^3 - 5x)$.

26. Find $f'(2)$ when $f(x) = (3x^2 - 1)(x^4 - 6x)$.

82   The Derivative   Chap. 2

27. Find $f'(-1)$ when $f(x) = \dfrac{3x - 4}{x + 2}$.   **28.** Find $f'(3)$ when $f(x) = \dfrac{x - 2}{x^2 + 5}$.

29. Find the slope of the tangent line to the curve $y = \dfrac{x - 3}{2 - 5x}$ at $(2, \frac{1}{8})$.

30. Find the slope of the tangent line to the curve $y = \dfrac{2x^2 + 1}{x - 2}$ at $x = 1$.

31. Find the equation of the tangent line to the curve $y = \dfrac{x + 3}{x - 2}$ at $x = 3$.

32. Find the equation of the tangent line to the curve $y = \dfrac{1}{x + 3}$ at $x = -1$.

33. Given that $V = ir$, the voltage across a resistor, find $\dfrac{dV}{dt}$ at $t = 3$ s if the current $i = 6 + 0.02t^3$ A and the resistance $r = 20 - 0.05t$ Ω.

34. Given that $i = V/r$, find di/dt at $t = 10$ s if $V = 40 + 0.1t^2$ V and $r = 60 - 0.01t$ Ω.

2.7 THE DERIVATIVE OF A POWER

If $y = (3x + 2)^{50}$, then $\dfrac{dy}{dx}$ could be found by applying the product formula repeatedly. But this would be quite a long process. Also, there is a need to be able to differentiate such a function as $y = \sqrt{3x + 2}$. We will now use a formula for such powers.

We show in Section 2.9 that if $y = u^n$ where $u = f(x)$, then

---

Power Rule

$$\frac{dy}{dx} = \frac{d}{dx}(u^n) = nu^{n-1}\frac{du}{dx}$$

The power rule should not be confused with the formula considered in Section 2.5:

$$\frac{d}{dx}(x^n) = nx^{n-1}$$

Note that the formula for $\dfrac{d}{dx}(u^n)$ involves an additional factor $\dfrac{du}{dx}$. This is because u itself is a function of x. Be sure to keep this distinction in mind. For example, if $y = u^3$ and $u = x$, then

$$y = x^3$$
$$\frac{dy}{dx} = 3x^{3-1} = 3x^2$$

But if $u = x^2 + 1$, then

$$y = (x^2 + 1)^3$$

$$\frac{dy}{dx} = 3u^{3-1}\frac{du}{dx}$$

$$= 3(x^2 + 1)^2 \frac{d}{dx}(x^2 + 1)$$

$$= 3(x^2 + 1)^2(2x)$$

$$= 6x(x^2 + 1)^2$$

Thus, only when $u = x$ are the formulas the same; for then the additional factor $\frac{du}{dx} = 1$.

Example 1

Find $\frac{dy}{dx}$ where $y = (3x + 2)^{50}$.

$$\frac{dy}{dx} = \frac{d}{dx}(3x + 2)^{50}$$

$$= 50(3x + 2)^{50-1}\frac{d}{dx}(3x + 2)$$

$$= 50(3x + 2)^{49}(3)$$

$$= 150(3x + 2)^{49}$$

Compare this result with the following example.

Example 2

Find $\frac{dy}{dx}$ where $y = x^{50}$.

$$\frac{dy}{dx} = 50x^{50-1}$$

$$= 50x^{49}$$

The formula $\frac{d}{dx}(u^n) = nu^{n-1}\frac{du}{dx}$ is also valid for n any real number. We will also assume that $u \neq 0$ to avoid later complications.

The power rule is a special case of the *chain rule*. The chain rule is used to find derivatives of a function, which is a function of another function.

A **composite function** is a function of a function. For example,

$$y = (3x - 2)^4$$

may be written as

$$y = u^4 \qquad \text{where} \qquad u = 3x - 2$$

Here y is a function of u and u is a function of x. So y is a function of a function of x.

More generally, let

$$y = f(g(x)) \qquad \text{or}$$

$$y = f(u) \qquad \text{where} \qquad u = g(x)$$

be any composite function. Then

$$\frac{dy}{dx} = \lim_{\Delta x \to 0} \frac{\Delta y}{\Delta x}$$

$$= \lim_{\Delta x \to 0} \left(\frac{\Delta y}{\Delta u} \cdot \frac{\Delta u}{\Delta x} \right) \qquad (\Delta u \neq 0)$$

$$= \lim_{\Delta x \to 0} \frac{\Delta y}{\Delta u} \cdot \lim_{\Delta x \to 0} \frac{\Delta u}{\Delta x}$$

$$= \lim_{\Delta u \to 0} \frac{\Delta y}{\Delta u} \cdot \lim_{\Delta x \to 0} \frac{\Delta u}{\Delta x} \qquad (\textit{Note: } \Delta u \to 0 \text{ as } \Delta x \to 0.)$$

$$= \frac{dy}{du} \cdot \frac{du}{dx}$$

Thus

Chain Rule

For any function $y = f(g(x))$, where $u = g(x)$,

$$\frac{dy}{dx} = \frac{dy}{du} \frac{du}{dx}$$

The chain rule is used extensively beginning with the transcendental functions in Chapter 6.

Using the chain rule to find the derivative of $y = (3x - 2)^4$, let

$$y = u^4 \qquad \text{and} \qquad u = 3x - 2$$

$$\frac{dy}{du} = 4u^3 \qquad \frac{du}{dx} = 3$$

Then

$$\frac{dy}{dx} = \frac{dy}{du} \frac{du}{dx}$$

$$= (4u^3)(3)$$

$$= 12(3x - 2)^3$$

Example 3

Find $\frac{dy}{dx}$ where $y = \sqrt{3x^2 + 2}$.

$$\frac{dy}{dx} = \frac{d}{dx}(\sqrt{3x^2 + 2})$$

$$= \frac{d}{dx}[(3x^2 + 2)^{1/2}]$$

$$= \frac{1}{2}(3x^2 + 2)^{(1/2)-1} \frac{d}{dx}(3x^2 + 2) \qquad \left(n = \frac{1}{2} \right)$$

$$= \tfrac{1}{2}(3x^2 + 2)^{-1/2}(6x)$$

$$= \frac{3x}{\sqrt{3x^2 + 2}}$$

Example 4

Find $\dfrac{dy}{dx}$ where $y = (8x + 1)^{1/4}(6x^2 + 2)$.

Using the derivative of a product form,

$$\frac{dy}{dx} = (8x + 1)^{1/4}\frac{d}{dx}(6x^2 + 2) + (6x^2 + 2)\frac{d}{dx}(8x + 1)^{1/4}$$

Note: the factor $(8x + 1)^{1/4}$ is in the form $u^{1/4}$ where $u = 8x + 1$. Using the chain rule on this factor, we have

$$\frac{1}{4}u^{-3/4}\frac{du}{dx}$$

$$\frac{1}{4}(8x + 1)^{-3/4}(8)$$

$$\frac{dy}{dx} = (8x + 1)^{1/4}(12x) + (6x^2 + 2)\left(\frac{1}{4}\right)(8x + 1)^{-3/4}(8)$$

$$= 12x(8x + 1)^{1/4} + (12x^2 + 4)(8x + 1)^{-3/4}$$

$$= (8x + 1)^{-3/4}[12x(8x + 1) + (12x^2 + 4)] \qquad \text{(Factor out the power with negative exponent.)}$$

$$= \frac{96x^2 + 12x + 12x^2 + 4}{(8x + 1)^{3/4}}$$

$$= \frac{108x^2 + 12x + 4}{(8x + 1)^{3/4}}$$

Example 5

Find $\dfrac{dy}{dx}$ where $y = \dfrac{(2x - 3)^3}{(3x + 1)^4}$.

Using the derivative of a quotient form,

$$\frac{dy}{dx} = \frac{(3x + 1)^4\frac{d}{dx}(2x - 3)^3 - (2x - 3)^3\frac{d}{dx}(3x + 1)^4}{[(3x + 1)^4]^2}$$

$$= \frac{(3x + 1)^4(3)(2x - 3)^2(2) - (2x - 3)^3(4)(3x + 1)^3(3)}{[(3x + 1)^4]^2}$$

$$= \frac{6(3x + 1)^3(2x - 3)^2[(3x + 1) - 2(2x - 3)]}{(3x + 1)^8} \qquad \text{(Factor.)}$$

$$= \frac{6(2x - 3)^2(-x + 7)}{(3x + 1)^5}$$

Example 6

Find $\dfrac{dy}{dx}$ when $y = \dfrac{(3x + 1)^{2/3}}{(2x - 3)^{1/2}}$.

$$\frac{dy}{dx} = \frac{(2x - 3)^{1/2}(\frac{2}{3})(3x + 1)^{-1/3}(3) - (3x + 1)^{2/3}(\frac{1}{2})(2x - 3)^{-1/2}(2)}{[(2x - 3)^{1/2}]^2}$$

$$= \frac{2(2x - 3)^{1/2}(3x + 1)^{-1/3} - (3x + 1)^{2/3}(2x - 3)^{-1/2}}{2x - 3}$$

Next, factor the negative exponent factors in the numerator as follows:

$$= \frac{(3x + 1)^{-1/3}(2x - 3)^{-1/2}[2(2x - 3) - (3x + 1)]}{(2x - 3)^1}$$

$$= \frac{4x - 6 - 3x - 1}{(3x + 1)^{1/3}(2x - 3)^{3/2}}$$

$$= \frac{x - 7}{(3x + 1)^{1/3}(2x - 3)^{3/2}}$$

Example 7

Find $\dfrac{dy}{dx}$ when $y = (2x + 1)^{1/2}(4x - 5)^{3/4}$.

$$\frac{dy}{dx} = (2x + 1)^{1/2}(\tfrac{3}{4})(4x - 5)^{-1/4}(4) + (4x - 5)^{3/4}(\tfrac{1}{2})(2x + 1)^{-1/2}(2)$$

$$= 3(2x + 1)^{1/2}(4x - 5)^{-1/4} + (4x - 5)^{3/4}(2x + 1)^{-1/2}$$

Next, factor the negative exponent factors as follows:

$$= (4x - 5)^{-1/4}(2x + 1)^{-1/2}[3(2x + 1) + (4x - 5)]$$

$$= (4x - 5)^{-1/4}(2x + 1)^{-1/2}[6x + 3 + 4x - 5]$$

$$= \frac{10x - 2}{(4x - 5)^{1/4}(2x + 1)^{1/2}}$$

Exercises

Find the derivative of each function.

1. $y = (4x + 3)^{40}$
2. $y = 5(x^2 - 7)^{16}$
3. $y = (3x^2 - 7x + 4)^5$
4. $y = (x^4 - 3x^2 + 1)^6$
5. $y = \dfrac{1}{(x^3 + 3)^4}$
6. $y = \dfrac{1}{(3x^4 - 2x + 3)^3}$
7. $y = \sqrt{5x^2 - 7x + 2}$
8. $y = \sqrt{4x^3 + 2x^2 - x}$
9. $y = (8x^3 + 3x)^{2/3}$
10. $y = (6x^4 - 5x + 2)^{3/8}$
11. $y = (2x + 3)^{-3/4}$
12. $y = (4x^2 - 1)^{-1/2}$
13. $y = (x^2 + 1)\sqrt{9x^2 - 2x}$
14. $y = (x + 1)^{3/2}x^4$
15. $y = (3x + 4)^{3/4}(4x^2 + 8)$
16. $y = (3x - 1)^{2/3}\sqrt{4x + 5}$
17. $y = \dfrac{(x^3 + 2)^4}{4x^2 - 3x}$
18. $y = \dfrac{8x^2 + 2}{\sqrt{1 - x}}$
19. $y = \dfrac{(3x + 2)^5}{(2x - 1)^3}$
20. $y = \dfrac{(4x^2 - 1)^3}{(3x + 4)^5}$
21. $y = \dfrac{(3x - 1)^{2/3}}{\sqrt{4x + 3}}$
22. $y = \dfrac{\sqrt[4]{8x^2 + 4x}}{(3x^2 + 1)^{5/3}}$

23. Find the velocity of an object after 3 s of travel where the distance s (in metres) traveled by the object is given by $s = \dfrac{t + 1}{\sqrt{t^2 - 1}}$

2.8 IMPLICIT DIFFERENTIATION

Often a function $y = f(x)$ is defined *implicitly*. That is, y is not directly expressed in terms of x. Such a function is called an *implicit function* of x. For example, the relation $y^2 = x$ actually defines two implicit functions of x:

$$y = f(x) = \sqrt{x} \quad \text{and} \quad y = g(x) = -\sqrt{x}$$

since both $[f(x)]^2 = x$ and $[g(x)]^2 = x$. In this example, we were able to determine the two functions of x defined by the relation $y^2 = x$. Often it is not possible or, at least, not easy to determine directly the functions defined implicitly by a given relation, say, $2y^3 + 3x^2y - 8x^3 + 2x - y = 0$. That is, could you solve this equation for y in terms of x? It is still possible, however, to obtain an expression for the derivative $\dfrac{dy}{dx}$.

Example 1

Find $\dfrac{dy}{dx}$ when $y^2 = x$.

$$\frac{d}{dx}(y^2) = \frac{d}{dx}(x)$$

$$2y^{2-1}\frac{d}{dx}(y) = 1 \qquad \text{(Use the power or chain rule.)}$$

$$2y\frac{dy}{dx} = 1 \qquad \left(\text{Solve for } \frac{dy}{dx}.\right)$$

$$\frac{dy}{dx} = \frac{1}{2y}$$

This process is called *implicit differentiation*.

Example 2

Find $\dfrac{dy}{dx}$ when $2y^3 + 3x^2y - 8x^3 + 2x - y = 0$ defines y as an implicit function of x.

$$\frac{d}{dx}(2y^3 + 3x^2y - 8x^3 + 2x - y) = \frac{d}{dx}(0) \qquad \begin{array}{l}\text{(Use Product Rule}\\ \text{on second term.)}\end{array}$$

$$6y^2\frac{dy}{dx} + 3x^2\frac{dy}{dx} + 6xy - 24x^2 + 2 - \frac{dy}{dx} = 0 \qquad \left(\text{Solve for } \frac{dy}{dx}.\right)$$

$$6y^2\frac{dy}{dx} + 3x^2\frac{dy}{dx} - \frac{dy}{dx} = -6xy + 24x^2 - 2$$

$$(6y^2 + 3x^2 - 1)\frac{dy}{dx} = -6xy + 24x^2 - 2 \qquad \text{(Factor.)}$$

$$\frac{dy}{dx} = \frac{-6xy + 24x^2 - 2}{6y^2 + 3x^2 - 1}$$

Note: When differentiating the term $3x^2y$, we must consider this the product of two functions of x: $u = 3x^2$ and $v = y$. Then apply the product formula for differentiating.

Example 3

Find $\dfrac{dy}{dx}$ when $(y + 3)^3 = (5x^2 - 4)^2$.

$$\frac{d}{dx}[(y + 3)^3] = \frac{d}{dx}[(5x^2 - 4)^2]$$

$$3(y + 3)^{3-1} \frac{d}{dx}(y + 3) = 2(5x^2 - 4)^{2-1} \frac{d}{dx}(5x^2 - 4)$$

$$3(y + 3)^2 \left(\frac{dy}{dx} + 0 \right) = 2(5x^2 - 4)(10x)$$

$$3(y + 3)^2 \frac{dy}{dx} = 20x(5x^2 - 4) \qquad \left(\text{Solve for } \frac{dy}{dx}. \right)$$

$$\frac{dy}{dx} = \frac{20x(5x^2 - 4)}{3(y + 3)^2}$$

Example 4

Find $\dfrac{dy}{dx}$ when $(2y^3 + 1)^4 = 8x^2 + 4y^2$.

Note the left side of the equation is in the form u^4, where $u = 2y^3 + 1$. Using the chain rule on this term, we have

$$4u^3 \frac{du}{dx}$$

$$4(2y^3 + 1)^3 \left(6y^2 \frac{dy}{dx} \right)$$

So,

$$\frac{d}{dx}(2y^3 + 1)^4 = \frac{d}{dx}(8x^2 + 4y^2)$$

$$4(2y^3 + 1)^3 \left(6y^2 \frac{dy}{dx} \right) = 16x + 4 \left(2y \frac{dy}{dx} \right)$$

$$24y^2(2y^3 + 1)^3 \frac{dy}{dx} = 16x + 8y \frac{dy}{dx}$$

$$24y^2(2y^3 + 1)^3 \frac{dy}{dx} - 8y \frac{dy}{dx} = 16x$$

$$[24y^2(2y^3 + 1)^3 - 8y] \frac{dy}{dx} = 16x$$

$$\frac{dy}{dx} = \frac{16x}{24y^2(2y^3 + 1)^3 - 8y} = \frac{2x}{3y^2(2y^3 + 1)^3 - y}$$

Exercises

Find the derivative of each expression using implicit differentiation.

1. $4x + 3y = 7$ 2. $10y - 8x = 4$
3. $x^2 - y^2 = 9$ 4. $3x^2 + 4y^2 = 12$
5. $x^2 + y^2 + 4y = 0$ 6. $x^2 - y^2 + 8x = 0$
7. $3x^2 - y^3 - 3xy = 0$ 8. $y^3 = 3x^2y - 4$
9. $y^4 - y^2x + x^2 = 0$ 10. $y^2 - 2xy + x^3 = 0$
11. $y^4 - 2y^2x^2 + 3x^2 = 0$ 12. $3x^2y^2 + 4y^5 + 8x^2y^3 + xy = 5$
13. $(y^2 + 2)^2 = (x^3 - 4x)^3$ 14. $(y + 4)^4 = (2x^2 - 3x + 2)^5$
15. $(x + y)^3 = (x - y + 4)^2$ 16. $(x^2 + y^2)^2 = 3x^2 + 4y^2$
17. $\dfrac{x + y}{x - y} = y^2$ 18. $(x + y)^{2/3} = x^2$

Find the slope of the tangent line to each curve at the given point.

19. $4x^2 + 5y^2 = 36$; $(2, -2)$ **20.** $x^2 - y^2 = 16$; $(5, 3)$

21. $x^2 + y^2 - 6x - 2y = 0$; $(2, 4)$ **22.** $y^2 + 3x - 8y + 3 = 0$; $(4, 3)$

Find the equation of the tangent line to each curve at the given point.

23. $y = 3x^2 + 4x + 9$; $(0, 9)$ **24.** $y = \dfrac{(x + 2)^2}{\sqrt{3x^2 + 1}}$; $(0, 4)$

25. $y^2 + 3xy = 4$; $(1, -4)$ **26.** $xy + y^2 = 6$; $(1, 2)$

2.9 PROOFS OF DERIVATIVE FORMULAS

Derivative of a Product

NOT
RESPONSIBLE
FOR 2.9

In Section 2.6 we claimed that if $y = f(x) = g(x) \cdot h(x)$ then

$$\frac{dy}{dx} = u\frac{dv}{dx} + v\frac{du}{dx}$$

where $u = g(x)$ and $v = h(x)$. We will now demonstrate this result.
Using the notation of increments we have that

$$\Delta u = g(x + \Delta x) - g(x)$$
$$\Delta v = h(x + \Delta x) - h(x)$$

and

$$\Delta y = g(x + \Delta x)h(x + \Delta x) - g(x)h(x)$$

$$\frac{dy}{dx} = \lim_{\Delta x \to 0} \frac{\Delta y}{\Delta x} = \lim_{\Delta x \to 0} \frac{g(x + \Delta x)h(x + \Delta x) - g(x)h(x)}{\Delta x}$$

$$= \lim_{\Delta x \to 0} \frac{g(x + \Delta x)h(x + \Delta x) - g(x + \Delta x)h(x) + g(x + \Delta x)h(x) - g(x)h(x)}{\Delta x}$$

[*Note:* $-g(x + \Delta x)h(x) + g(x + \Delta x)h(x)$, which is zero, is added to the numerator.]

$$= \lim_{\Delta x \to 0} \frac{g(x + \Delta x)[h(x + \Delta x) - h(x)] + h(x)[g(x + \Delta x) - g(x)]}{\Delta x}$$

$$= \lim_{\Delta x \to 0} \left[g(x + \Delta x)\frac{h(x + \Delta x) - h(x)}{\Delta x} + h(x)\frac{g(x + \Delta x) - g(x)}{\Delta x} \right]$$

$$= \lim_{\Delta x \to 0} \left[g(x + \Delta x)\frac{h(x + \Delta x) - h(x)}{\Delta x} \right]$$

$$+ \lim_{\Delta x \to 0} \left[h(x)\frac{g(x + \Delta x) - g(x)}{\Delta x} \right] \quad \text{(By limit formula A, page 61.)}$$

$$= \left[\lim_{\Delta x \to 0} g(x + \Delta x) \right]\left[\lim_{\Delta x \to 0} \frac{h(x + \Delta x) - h(x)}{\Delta x} \right] \quad \text{(cont. on next page)}$$

The Derivative   Chap. 2

$$+ \left[\lim_{\Delta x \to 0} h(x) \right] \left[\lim_{\Delta x \to 0} \frac{g(x + \Delta x) - g(x)}{\Delta x} \right] \quad \text{(By limit formula D, page 61.)}$$

$$= g(x) \frac{dv}{dx} + h(x) \frac{du}{dx}$$

[*Note:* $h(x)$ is not affected by Δx and so is constant as we apply the limit as $\Delta x \to 0$.]

$$= u \frac{dv}{dx} + v \frac{du}{dx}$$

or

$$\frac{d}{dx}[g(x)h(x)] = g(x) \frac{d}{dx}[h(x)] + h(x) \frac{d}{dx}[g(x)]$$

Derivative of a Quotient

The formula for $\dfrac{d}{dx}\left(\dfrac{u}{v}\right)$ from Section 2.6 can be found as follows:

Let $y = f(x) = \dfrac{g(x)}{h(x)}$ where $u = g(x)$ and $v = h(x)$. Using the same increment notation as before, we have

$$\Delta u = g(x + \Delta x) - g(x)$$
$$\Delta v = h(x + \Delta x) - h(x)$$

and

$$\Delta y = \frac{g(x + \Delta x)}{h(x + \Delta x)} - \frac{g(x)}{h(x)}$$

Then

$$\frac{dy}{dx} = \lim_{\Delta x \to 0} \frac{\Delta y}{\Delta x} = \lim_{\Delta x \to 0} \frac{\dfrac{g(x + \Delta x)}{h(x + \Delta x)} - \dfrac{g(x)}{h(x)}}{\Delta x}$$

$$= \lim_{\Delta x \to 0} \frac{h(x)g(x + \Delta x) - h(x + \Delta x)g(x)}{\Delta x h(x)h(x + \Delta x)}$$

$$= \lim_{\Delta x \to 0} \frac{\dfrac{h(x)g(x + \Delta x) - h(x + \Delta x)g(x)}{\Delta x}}{h(x)h(x + \Delta x)}$$

$$= \lim_{\Delta x \to 0} \frac{\dfrac{h(x)g(x + \Delta x) - h(x)g(x) + h(x)g(x) - h(x + \Delta x)g(x)}{\Delta x}}{h(x)h(x + \Delta x)}$$

[*Note:* $-h(x)g(x) + h(x)g(x)$, which is zero, is added to the numerator.]

$$= \lim_{\Delta x \to 0} \frac{h(x)\dfrac{g(x + \Delta x) - g(x)}{\Delta x} - g(x)\dfrac{h(x + \Delta x) - h(x)}{\Delta x}}{h(x)h(x + \Delta x)}$$

$$= \frac{\left[\lim_{\Delta x \to 0} h(x)\right]\left[\lim_{\Delta x \to 0} \dfrac{g(x + \Delta x) - g(x)}{\Delta x}\right] - \left[\lim_{\Delta x \to 0} g(x)\right]\left[\lim_{\Delta x \to 0} \dfrac{h(x + \Delta x) - h(x)}{\Delta x}\right]}{\left[\lim_{\Delta x \to 0} h(x)\right]\left[\lim_{\Delta x \to 0} h(x + \Delta x)\right]}$$

(By limit formulas A, D, and E, page 61.)

$$= \frac{h(x)\dfrac{du}{dx} - g(x)\dfrac{dv}{dx}}{h(x)h(x)}$$

$$= \frac{v\dfrac{du}{dx} - u\dfrac{dv}{dx}}{v^2}$$

Derivative of Power Rule

In Section 2.7 we introduced the formula $\dfrac{d}{dx}u^n = nu^{n-1}\dfrac{du}{dx}$.

We now look at a demonstration of this result. If $u = f(x)$ and $y = u^n$ where n is a positive integer, then

$$\Delta y = (u + \Delta u)^n - u^n$$

$$= u^n + nu^{n-1}(\Delta u) + \frac{n(n-1)}{2}u^{n-2}(\Delta u)^2 + \cdots + (\Delta u)^n - u^n$$

$$= nu^{n-1}(\Delta u) + \frac{n(n-1)}{2}u^{n-2}(\Delta u)^2 + \cdots + (\Delta u)^n$$

$$= nu^{n-1}(\Delta u)$$
$$\quad + \left[\frac{n(n-1)}{2}u^{n-2} + (\text{other terms involving } u \text{ and } \Delta u) + (\Delta u)^{n-2}\right](\Delta u)^2$$

So

$$\frac{\Delta y}{\Delta x} = nu^{n-1}\frac{\Delta u}{\Delta x}$$
$$\quad + \left[\frac{n(n-1)}{2}u^{n-2} + (\text{other terms involving } u \text{ and } \Delta u) + (\Delta u)^{n-2}\right]\left(\frac{\Delta u}{\Delta x}\right)(\Delta u)$$

Then, as $\Delta x \to 0$, all terms except $nu^{n-1}\dfrac{\Delta u}{\Delta x}$ approach 0 since

$$\lim_{\Delta x \to 0}\left(\frac{\Delta u}{\Delta x} \cdot \Delta u\right) = \lim_{\Delta x \to 0}\frac{\Delta u}{\Delta x} \cdot \lim_{\Delta x \to 0} \Delta u$$

$$= \frac{du}{dx} \cdot \lim_{\Delta x \to 0}[f(x + \Delta x) - f(x)]$$

$$= \frac{du}{dx} \cdot 0$$

$$= 0$$

and $\left(\frac{\Delta u}{\Delta x} \cdot \Delta u \right)$ is a factor of all the remaining terms. So

$$\frac{dy}{dx} = \lim_{\Delta x \to 0} \frac{\Delta y}{\Delta u}$$

$$= \lim_{\Delta x \to 0} nu^{n-1} \frac{\Delta u}{\Delta x}$$

$$= n \lim_{\Delta x \to 0} u^{n-1} \cdot \lim_{\Delta x \to 0} \frac{\Delta u}{\Delta x}$$

$$= nu^{n-1} \frac{du}{dx}$$

2.10 HIGHER DERIVATIVES

The derivative of a function is also a function. For example, given

$$f(x) = 5x^4$$

then

$$f'(x) = 20x^3$$

Since $f'(x)$ is a function, its derivative (if it exists) is also a function. This result, called the second derivative of $f(x)$, is written $f''(x)$. Thus

$$f''(x) = 60x^2$$

Continuing, we have

$$f'''(x) = 120x$$
$$f^{(4)}(x) = 120$$
$$f^{(5)}(x) = 0$$
$$f^{(6)}(x) = 0$$

and so forth.

Some of the various notations used for derivatives are given below.

First derivative	Second derivative	Third derivative	Fourth derivative
$f'(x)$	$f''(x)$	$f'''(x)$	$f^{(4)}(x)$
y'	y''	y'''	$y^{(4)}$
$\dfrac{dy}{dx}$	$\dfrac{d^2y}{dx^2}$	$\dfrac{d^3y}{dx^3}$	$\dfrac{d^4y}{dx^4}$
$\dfrac{d}{dx}f(x)$	$\dfrac{d^2}{dx^2}f(x)$	$\dfrac{d^3}{dx^3}f(x)$	$\dfrac{d^4}{dx^4}f(x)$
Dy	D^2y	D^3y	D^4y
$Df(x)$	$D^2f(x)$	$D^3f(x)$	$D^4f(x)$

Example 1

Find the first four derivatives for $y = x^6 - 8x^4 + 3x^3 - 2$.

$$\frac{dy}{dx} = 6x^5 - 32x^3 + 9x^2$$

$$\frac{d^2y}{dx^2} = 30x^4 - 96x^2 + 18x$$

$$\frac{d^3y}{dx^3} = 120x^3 - 192x + 18$$

$$\frac{d^4y}{dx^4} = 360x^2 - 192$$

Example 2

Find the first three derivatives of $y = x^2 + x^{1/2}$.

$$y' = 2x + \frac{1}{2}x^{-1/2} = \frac{4x^{3/2} + 1}{2\sqrt{x}}$$

$$y'' = 2 - \frac{1}{4}x^{-3/2} = \frac{8x^{3/2} - 1}{4x^{3/2}}$$

$$y''' = \frac{3}{8}x^{-5/2} = \frac{3}{8x^{5/2}}$$

The second derivative of an implicit function may be found as shown in the following example.

Example 3

Given $y^2 - xy = 5$, find $\dfrac{d^2y}{dx^2}$.

First, find y' as shown in Section 2.8.

$$2yy' - xy' - y = 0$$

Then solve for y'.

$$(2y - x)y' = y$$

$$y' = \frac{y}{2y - x}$$

Next, find y'' using the quotient formula.

$$y'' = \frac{(2y - x)y' - y(2y' - 1)}{(2y - x)^2}$$

Now substitute for y' in y'' and simplify.

$$y'' = \frac{(2y - x)\left(\dfrac{y}{2y - x}\right) - y\left[2\left(\dfrac{y}{2y - x}\right) - 1\right]}{(2y - x)^2}$$

$$= \frac{(2y - x)\left(\dfrac{y}{2y - x}\right) - y\left[2\left(\dfrac{y}{2y - x}\right) - 1\right]}{(2y - x)^2} \cdot \left(\frac{2y - x}{2y - x}\right)$$

$$= \frac{(2y - x)y - y[2y - (2y - x)]}{(2y - x)^3}$$

$$= \frac{2y^2 - xy - y[x]}{(2y - x)^3}$$

$$= \frac{2y^2 - 2xy}{(2y - x)^3}$$

$$= \frac{2(y^2 - xy)}{(2y - x)^3}$$

$$= \frac{2(5)}{(2y - x)^3} \qquad (Note: y^2 - xy = 5.)$$

$$= \frac{10}{(2y - x)^3}$$

Acceleration is the instantaneous rate of change of velocity with respect to time. So, acceleration is the first derivative of velocity with respect to time and the second derivative of displacement (distance) with respect to time. That is,

Acceleration
$a = \dfrac{dv}{dt} = \dfrac{d^2s}{dt^2}$

Example 4

The displacement of an object (in metres) is described by $s = 5t^4 - 6t^3 + 8t + 3$. Find the equation that describes the acceleration of the object. Find the acceleration at $t = 2$ s.

First, find the velocity equation:

$$v = \frac{ds}{dt} = 20t^3 - 18t^2 + 8$$

Then find the acceleration equation:

$$a = \frac{dv}{dt} = 60t^2 - 36t$$

At $t = 2$ s,

$$a = 60(2)^2 - 36(2) = 168 \text{ m/s}^2$$

Exercises

Find the first four derivatives of each function.

1. $y = x^5 + 3x^2$ **2.** $y = 3x^6 - 8x^3 + 2$ **3.** $y = 5x^5 + 2x^3 - 8x$

4. $y = 3x^2 + 4x - 7$

Find the indicated derivative.

5. $y = \dfrac{1}{x}; \dfrac{d^3y}{dx^3}$ **6.** $y = \dfrac{3}{x^2}; y'''$ **7.** $y = (3x - 5)^3; y'''$

8. $y = (4x^2 - 2)^4; \dfrac{d^2y}{dx^2}$ **9.** $y = \sqrt{3x + 2}; \dfrac{d^4y}{dx^4}$ **10.** $y = (2x + 1)^{3/2}; \dfrac{d^4y}{dx^4}$

11. $y = \dfrac{1}{x^2 + 1}$; y'' **12.** $y = \dfrac{4}{(x^2 + 5)^2}$; y'' **13.** $y = \dfrac{x + 1}{x - 1}$; $\dfrac{d^2y}{dx^2}$

14. $y = \dfrac{3x + 1}{2x - 4}$; $\dfrac{d^2y}{dx^2}$

Find y' and y'' implicitly and express the result in terms of x and y.

15. $x^2 + y^2 = 1$ **16.** $(x - 3)^2 + y^2 = 9$ **17.** $x^2 - xy + y^2 = 1$

18. $y^2 + xy = 1$ **19.** $\sqrt{x} + \sqrt{y} = 1$ **20.** $x + \dfrac{x}{y} = 8$

21. $\dfrac{1}{x} - \dfrac{1}{y} = 1$ **22.** $x = \dfrac{y + 1}{y - 1}$ **23.** $x = (1 + y)^2$

24. $\sqrt{1 + y^2} = x$

Each equation describes the displacement of an object. Find the equation that describes the acceleration of the object.

25. $s = 0.5t^4 - 6t^3 - 4t^2 - 1$ **26.** $s = \dfrac{4}{2t + 1}$

27. $s = \sqrt{6t - 4}$ **28.** $s = \dfrac{8t + 3}{5 - 4t}$

29. The curve $x^2 - xy + y^2 = 7$ has two tangents at $x = 1$. Find their equations.

30. Show that the tangent to the parabola $y^2 = 4cx$ at (x_0, y_0) is $y_0y = 2c(x + x_0)$.

31. Show that the tangent to the ellipse $\dfrac{x^2}{a^2} + \dfrac{y^2}{b^2} = 1$ at (x_0, y_0) is $\dfrac{x_0x}{a^2} + \dfrac{y_0y}{b^2} = 1$.

32. Show that the tangent to the hyperbola $\dfrac{x^2}{a^2} - \dfrac{y^2}{b^2} = 1$ at (x_0, y_0) is $\dfrac{x_0x}{a^2} - \dfrac{y_0y}{b^2} = 1$.

CHAPTER SUMMARY

1. *Average velocity*

$$\bar{v} = \frac{\text{change in distance}}{\text{change in time}} = \frac{\Delta s}{\Delta t} = \frac{f(t + \Delta t) - f(t)}{\Delta t}$$

2. *Limit:* If $f(x) \to L$ as $x \to a$, then L is called the "limit of the function as $x \to a$." This is written

$$\lim_{x \to a} f(x) = L$$

and read "the limit of f of x as x approaches a equals L."

3. *Instantaneous velocity*

$$v = \lim_{\Delta t \to 0} \frac{f(t + \Delta t) - f(t)}{\Delta t}$$

4. *Limit formulas*
   (a) $\lim_{x \to a} [f(x) + g(x)] = \lim_{x \to a} f(x) + \lim_{x \to a} g(x)$
   (b) $\lim_{x \to a} [k \cdot f(x)] = k \cdot \lim_{x \to a} f(x)$, where k is a constant
   (c) $\lim_{x \to a} k = k$, where k is a constant

(d) $\lim_{x \to a} [f(x) \cdot g(x)] = \lim_{x \to a} f(x) \cdot \lim_{x \to a} g(x)$

(e) $\lim_{x \to a} \dfrac{f(x)}{g(x)} = \dfrac{\lim_{x \to a} f(x)}{\lim_{x \to a} g(x)}$, where $\lim_{x \to a} g(x) \neq 0$

5. *Continuity*
   (a) A function is continuous at $x = a$ if
      (i) $f(a)$ is defined,
      (ii) $\lim_{x \to a} f(x)$ exists, and
      (iii) $\lim_{x \to a} f(x) = f(a)$.
   (b) If a function is continuous at all points in a given interval, it is continuous in that interval.

6. *Slope of a tangent line*
$$m_{\text{tan}} = \lim_{\Delta x \to 0} \frac{\Delta y}{\Delta x} = \lim_{\Delta x \to 0} \frac{f(x + \Delta x) - f(x)}{\Delta x}$$

7. *Derivative:* The derivative of the function f of x, $f(x)$, is
$$\frac{dy}{dx} = \lim_{\Delta x \to 0} \frac{\Delta y}{\Delta x} = \lim_{\Delta x \to 0} \frac{f(x + \Delta x) - f(x)}{\Delta x}$$

8. *Basic differentiation formulas*

(a) $\dfrac{d}{dx}(c) = 0$

(b) $\dfrac{d}{dx}(x) = 1$

(c) $\dfrac{d}{dx}(x^n) = nx^{n-1}, \qquad x \neq 0$

(d) $\dfrac{d}{dx}(cu) = c\,\dfrac{du}{dx}$

(e) $\dfrac{d}{dx}(u + v) = \dfrac{du}{dx} + \dfrac{dv}{dx}$

(f) $\dfrac{d}{dx}(u - v) = \dfrac{du}{dx} - \dfrac{dv}{dx}$

(g) $\dfrac{d}{dx}(uv) = u\,\dfrac{dv}{dx} + v\,\dfrac{du}{dx}$

(h) $\dfrac{d}{dx}\left(\dfrac{u}{v}\right) = \dfrac{v\,\dfrac{du}{dx} - u\,\dfrac{dv}{dx}}{v^2}, \qquad v \neq 0$

(i) $\dfrac{d}{dx}(u^n) = nu^{n-1}\,\dfrac{du}{dx}, \qquad u \neq 0$

9. *Acceleration*
$$a = \frac{dv}{dt} = \frac{d^2s}{dt^2}$$

CHAPTER REVIEW

Find (a) Δs and (b) $\bar{v}$ for each function expressing distance s in terms of t.

1. $s = 3t^2 + 4$ **2.** $s = 5t^2 - 6$ **3.** $s = t^2 - 3t + 5$
4. $s = 3t^2 - 6t + 8$

Find $\bar{v}$ for each function $s = f(t)$ at the values of t and Δt (s is measured in metres and t in seconds).

5. $s = 3t^2 - 7$, $t = 2$, $\Delta t = 5$ **6.** $s = 5t^2 - 3$, $t = 1$, $\Delta t = 2$
7. $s = 2t^2 - 4t + 7$, $t = 2$, $\Delta t = 3$ **8.** $s = 4t^2 - 7t + 2$, $t = 3$, $\Delta t = 4$

9–12. Find the instantaneous velocity v for the functions given in Exercises 5–8 at the values of t.

Find each limit.

13. $\lim\limits_{x \to 3} (2x^2 - 5x + 1)$ **14.** $\lim\limits_{x \to -2} (x^2 + 4x - 7)$ **15.** $\lim\limits_{x \to -2} \dfrac{x^2 - 4}{x + 2}$

16. $\lim\limits_{x \to 5} \dfrac{25 - x^2}{5 - x}$ **17.** $\lim\limits_{x \to 2} \sqrt{3 - 4x}$ **18.** $\lim\limits_{x \to -3} \sqrt{6 + x^2}$

19. $\lim\limits_{x \to 2} \dfrac{5x^2 + 2}{3x^2 - 2x + 1}$ **20.** $\lim\limits_{x \to -3} \dfrac{2x^2 - 4x + 7}{x^3 - x}$

21. $\lim\limits_{x \to -3} (x^2 - 4x + 3)(2x^2 + 5x + 4)$ **22.** $\lim\limits_{x \to 2} (x^3 + x - 2)(x^3 + x^2 + x)$

23. $\lim\limits_{h \to 0} \dfrac{\dfrac{1}{2 + h} - \dfrac{1}{2}}{h}$ **24.** $\lim\limits_{h \to 0} \dfrac{\sqrt{1 + h} - 1}{h}$ **25.** $\lim\limits_{x \to \infty} \dfrac{5x^2 - 2x + 3}{2x^2 - 4}$

26. $\lim\limits_{x \to \infty} \dfrac{7x^3 - 4x + 2}{10x^3 - x^2 + 5}$

In Exercises 27–32, use Fig. 2.17.

27. Find $\lim\limits_{x \to a} f(x)$ if it exists.

28. Find $\lim\limits_{x \to d} f(x)$ if it exists.

29. Find $\lim\limits_{x \to +\infty} f(x)$ if it exists.

30. Find $\lim\limits_{x \to -\infty} f(x)$ if it exists.

31. Is the function continuous at $x = a$?
32. Is the function continuous at $x = d$?

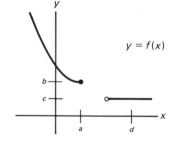

Figure 2.17

Find the slope and equation of the tangent line to each curve at the given point.

33. $y = 3x^2 - 4x + 5$; $(-1, 12)$ **34.** $y = x^2 - 5x - 12$; $(2, -18)$
35. $y = 2x^2 + 2x + 7$; $(3, 31)$ **36.** $y = 4x^2 - 8x + 3$; $(-2, 35)$

37. An object is moving along a straight-line path according to the equation $s = 3/t^2$, where s is measured in centimetres. Find the velocity after 5 s.

38. An object is falling freely according to the equation $s = 150 - 16t^2$, where s is measured in feet. Find the velocity of the falling object after 2 s.

Find each derivative dy/dx:

39. $y = 5x^4 - 3x^3 + 2x^2 + 5x - 9$ **40.** $y = x^{100} + 80x^5 + 16$
41. $y = (x^3 + 4)(x^3 - x + 1)$ **42.** $y = (3x^2 - 5)(x^5 + x^2 - 4x)$

The Derivative Chap. 2

43. $y = \dfrac{x^2 + 1}{3x - 4}$

44. $y = \dfrac{2x - x^2}{3x^4 + 2}$

45. $y = (3x^2 - 8)^5$

46. $y = (x^4 + 2x^3 + 7)^{3/4}$

47. $y = \dfrac{1}{(3x + 5)^4}$

48. $y = \dfrac{\sqrt{7x^2 - 5}}{(x + 3)^2}$

49. $y = \dfrac{x\sqrt{2 - 3x}}{x + 5}$

50. $x^2 - 4xy^3 + y^2 = 0$

51. $y^4 - y^2 = 2xy$

52. $(y^2 + 1)^3 = 4x^2 + 3$

53. $(y + 2)^4 = (2x^3 - 3)^3$

Find $f'(a)$ for each.

54. $f(x) = (x^2 + 2)^3(x + 1);\ a = -2$

55. $f(x) = \dfrac{x^2 - 8}{x + 3};\ a = -4$

56. $f(x) = \dfrac{x^2 + 3x - 2}{x - 2};\ a = 3$

57. $f(x) = \dfrac{\sqrt{3x^2 - 1}}{x + 5};\ a = 1$

Find the equation of the tangent line to each curve at the given point.

58. $y = 3x^2 + x - 2$ at $(-2, 8)$

59. $y = x^3 - x + 8$ at $(-3, -16)$

60. $y = \dfrac{x^2 - 2}{\sqrt{x - 3}}$ at $(4, 14)$

61. $y = \dfrac{\sqrt{x^2 + 7}}{x - 2}$ at $(3, 4)$

Find the instantaneous velocity of an object moving along a straight line according to each function $s = f(t)$ at the given time t (s is measured in metres and t in seconds).

62. $s = 3t^2 - 8t + 4;\ t = 2$

63. $s = t^3 - 9t^2 + 3;\ t = -2$

64. $s = \dfrac{t^2 - 5}{\sqrt{t + 1}};\ t = 3$

65. $s = \dfrac{\sqrt{t^2 - 3}}{t + 5};\ t = 4$

66. Find the equation of the tangent line to the curve $y = \dfrac{x^2 - 6}{x + 3}$ at $(-4, -10)$.

67. The instantaneous current $i = dq/dt$ at any point in an electrical circuit where q is the charge (in coulombs, C) and t is the time (in seconds). Find i (in amperes) where $q = 1000t^3 + 50t$ when $t = 0.01$ s.

68. The specific heat c of a gas as a function of the temperature T is given by

$$c = 8.40 + 0.5T + 0.000006T^2$$

Find the rate of change of the specific heat with respect to temperature, dc/dT.

69. The resonant frequency of a series a-c circuit is given by

$$f = \dfrac{1}{2\pi\sqrt{LC}}$$

where L and C are the inductance and capacitance in the circuit, respectively. Find the rate of change f with respect to C, assuming L remains constant.

70. Find the first four derivatives of $y = 4x^6 - 8x^4 + 9x^3 - 6x + 9$.

Find the indicated derivative.

71. $y = \sqrt{2x - 3};\ y''$

72. $y = \dfrac{3}{2x^2 + 1};\ \dfrac{d^2y}{dx^2}$

Find y' and y'' implicitly and express the result in terms of x and y.

73. $y^2 + 2xy = 4$

74. $\dfrac{1}{\sqrt{x}} - \dfrac{1}{\sqrt{y}} = 1$

75. The equation $s = (2t + 3)^{1/4}$ describes the displacement of an object. Find the equation that describes its acceleration.

3

APPLICATIONS
OF THE DERIVATIVE

3.1 CURVE SKETCHING

In this course, as well as in your other technical courses, you will find it most helpful to sketch quickly the graph of an equation without plotting several points. In this section you will be given a series of graphing aids, each designed to give you valuable information about the graph, that will enable you to sketch quickly the graph of an equation.

Intercepts

The intercepts are those points where the graph crosses either the x-axis or the y-axis.

1. To find the x-intercepts, substitute $y = 0$ into the given equation and solve for x. Each resulting value of x is an x-intercept.
2. To find the y-intercepts, substitute $x = 0$ into the given equation and solve for y. Each resulting value of y is a y-intercept.

Example 1

Find the intercepts for the graph of $y = (x - 1)(x + 3)^2$.

Step 1: To find the x-intercepts, substitute $y = 0$ and solve for x:

$$y = (x - 1)(x + 3)^2$$
$$0 = (x - 1)(x + 3)^2$$
$$x - 1 = 0 \quad \text{or} \quad (x + 3)^2 = 0$$
$$x = 1 \quad \text{and} \quad x = -3 \quad \text{are } x\text{-intercepts}$$

100

Step 2: To find the *y*-intercepts, substitute $x = 0$ and solve for *y*:

$$y = (x - 1)(x + 3)^2$$
$$y = (0 - 1)(0 + 3)^2$$
$$y = -9 \text{ is the } y\text{-intercept}$$

Location of the Curve above and below the x-Axis

Knowing in which regions the curve is above and below the *x*-axis can be most important information. If the expression can be factored, this information can be quickly found as follows:

1. Solve the given equation for *y*.
2. Factor the result from step 1 as much as possible. [If this is a rational (fractional) expression, factor both the numerator and the denominator as much as possible.]
3. Set each factor from step 2 equal to zero. These solutions are called *points of division* because they divide the number line (*x*-axis) into regions where the graph is above or below the *x*-axis. Plot each solution on the *x*-axis.
4. Find the sign of *y* in each region by choosing a test point. Substitute this test point value into each factor of the given equation and determine if each factor is positive or negative in the region. Determine the sign of *y* by using the rules for multiplication and/or division of signed numbers. If the sign of *y* is *positive,* the curve is *above* the *x*-axis in this region. If the sign of *y* is *negative,* the curve is *below* the *x*-axis in the region.

Example 2

Determine where the graph of $y = (x - 1)(x + 3)^2$ is above and below the *x*-axis.

From Example 1, the points of division are 1 and -3. Plot them on the *x*-axis.

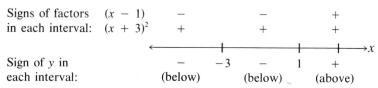

Signs of factors $(x - 1)$ $-$ $-$ $+$
in each interval: $(x + 3)^2$ $+$ $+$ $+$

Sign of *y* in $-$ -3 $-$ 1 $+$
each interval: (below)   (below)   (above)

As you can see, we have three regions.

For $x > 1$, the product is *positive,* which means the curve is *above* the *x*-axis.

For $-3 < x < 1$, the product is *negative,* which means the curve is *below* the *x*-axis.

For $x < -3$, the product is *negative,* which means the curve is *below* the *x*-axis.

From the information gained from these first two graphing aids, we can now make a quick sketch of the graph of the equation $y = (x - 1)(x + 3)^2$ as in Fig. 3.1. We will be able to refine this procedure even more as we progress through the chapter.

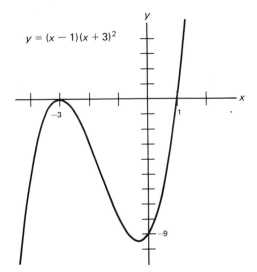

$$y = (x - 1)(x + 3)^2$$

Figure 3.1

Symmetry

Another graphing aid or characteristic that is helpful in sketching a curve is symmetry. There are two kinds of symmetry: symmetry about a line and symmetry about a point.

A curve is symmetric about a line if one-half of the curve is the mirror image of the other half on opposite sides of the given line. That is, for every point P on the curve on one side of the given line, there is a point P' on the curve on the opposite side of the line so that PP' is perpendicular to the given line and bisected by it. For example, the graph of $y = 1/x^2$ is symmetric about the y-axis. (See Fig. 3.2.)

A curve is symmetric about a point, say O, if for every other point $P \neq$ O on the curve, there is a point P' on the curve so that PP' is bisected by point O. For example, the graph of $y = 1/x$ is symmetric about the origin. (See Fig. 3.3.)

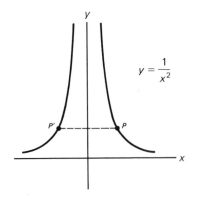

Figure 3.2

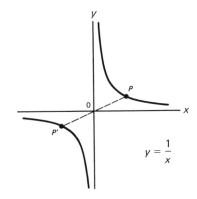

Figure 3.3

Of the many possible lines and points of symmetry, we shall present the three most common: symmetry about the y-axis, symmetry about the x-axis, and symmetry about the origin.

1. A curve is symmetric about the y-axis if for every point $P(x, y)$ on the curve, there is a point $P'(-x, y)$ on the curve. This symmetry can be tested as follows:

Symmetry about the y-Axis

The graph of an equation is symmetric about the y-axis if replacement of x by $-x$ in the original equation results in an equivalent equation.

Note that if we replace x by $-x$ in the earlier equation $y = 1/x^2$, we get

$$y = \frac{1}{(-x)^2}$$

which is equivalent to the original equation.

2. A curve is symmetric about the x-axis if for every point $P(x, y)$ on the curve, there is a point $P'(x, -y)$ on the curve. This symmetry can be tested as follows:

Symmetry about the x-Axis

The graph of an equation is symmetric about the x-axis if replacement of y by $-y$ in the original equation results in an equivalent equation.

Note that if we replace y by $-y$ in the equation $x = y^2$, we get

$$x = (-y)^2$$

which is equivalent to the original equation. (See Fig. 3.4.)

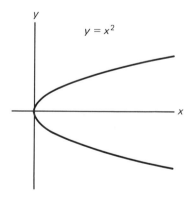

Figure 3.4

3. A curve is symmetric about the origin if for every point $P(x, y)$ on the curve, there is a point $P'(-x, -y)$ on the curve. This symmetry can be tested as follows:

Note that if we replace x by $-x$ and y by $-y$ in the earlier equation

$$y = \frac{1}{x}$$

we get

$$-y = \frac{1}{-x}$$

which is equivalent to the original equation.

If any two of these three symmetry tests are satisfied, then the graph contains all three symmetries. Do you see why?

Asymptotes

An *asymptote* to a curve is a line that the curve approaches (gets closer to) as the distance from the origin increases without bound.

A *rational function* is an equation in the form

$$y = \frac{f(x)}{g(x)}$$

where $f(x)$ and $g(x)$ are polynomials. Such rational functions may have asymptotes. We will present three possible asymptotes. In all cases, first solve the given equation for y in terms of x.

1. *Vertical asymptotes:* Set the denominator equal to zero and solve for x. These solutions give the points where the vertical asymptotes cross the x-axis. For example, if $x = a$ is a solution, one vertical asymptote crosses the x-axis at a and $x = a$ is the equation of the vertical asymptote.

2. *Horizontal asymptotes:* If $\lim\limits_{x \to +\infty} f(x) = b$ or $\lim\limits_{x \to -\infty} f(x) = b$, then $y = b$ is a horizontal asymptote. We have two cases:

    (a) If the degree of the numerator is less than the degree of the denominator, then $y = 0$ is a horizontal asymptote.
    Given the equation $y = 1/x^2$, do you see that as $x \to \infty$, $y \to 0$?

    (b) If the rational function is in the form

$$f(x) = \frac{a_n x^n + \cdots + a_0}{b_n x^n + \cdots + b_0} \qquad b_n \neq 0$$

then $y = \dfrac{a_n}{b_n}$ is a horizontal asymptote. That is, if the numerator and the denominator are of the same degree, the horizontal asymptote is the ratio of the coefficients of the terms of highest degree.

Given the equation

$$y = \frac{3x^2 - 10x - 8}{x^2 + 3x - 10}$$

divide numerator and denominator by x^2, which is the largest power of x:

$$\lim_{x \to \infty} \frac{3 - \dfrac{10}{x} - \dfrac{8}{x^2}}{1 + \dfrac{3}{x} - \dfrac{10}{x^2}} = 3$$

So $y = 3$ is a horizontal asymptote.

3. *Slant asymptotes:* If the degree of the numerator is one greater than the degree of the denominator, there may be a slant asymptote. To find it, divide the numerator by the denominator using long or synthetic division and drop the remainder. And $y =$ the quotient is the equation of the slant asymptote.

Given the equation

$$y = \frac{x^2 - 1}{x - 2}$$

divide the numerator by the denominator:

$$
\begin{array}{r}
x + 2 + \dfrac{3}{x-2} \\[2pt]
x - 2 \overline{\smash{\big)}\, x^2 \qquad\quad - 1} \\
\underline{x^2 - 2x} \\
2x - 1 \\
\underline{2x - 4} \\
3
\end{array}
$$

The slant asymptote is $y = x + 2$.

Do you see that as $x \to \infty$, the remainder, $\dfrac{3}{x - 2}$, approaches zero?

Restricted Domains

Some equations have restricted domains by the nature of their definitions. For example, the domain of

$$y = \sqrt{3 - x}$$

is $x \le 3$ since the square root of a negative number is not real. The domain of

$$y = \log_{10}(x + 2)$$

is $x > -2$ since the logarithm of only a positive number is defined.

These graphing aids will now be illustrated with the following examples. You will find it most helpful to follow a consistent pattern of applying each aid.

Example 3

Graph $y = x^2(x^2 - 9)$.

Factor:

$$y = x^2(x + 3)(x - 3)$$

Intercepts: If $y = 0$, then $x = 0, 3, -3$.
If $x = 0$, then $y = 0$.

Plot points of division on the x-axis:

Signs of factors in each interval:	x^2	+	+	+	+
	$(x + 3)$	$-$	+	+	+
	$(x - 3)$	$-$	$-$	$-$	+

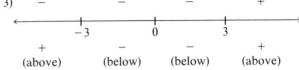

Sign of y in each interval:

+	$-$	$-$	+
(above)	(below)	(below)	(above)

Symmetry: y-axis   (Replacement of x by $-x$ results in an equivalent equation.)
Asymptotes: None
Restricted domain: None
(See Fig. 3.5.)

Example 4

Graph $y = \dfrac{3}{(x + 1)^2}$.

Intercepts: Note that $y \neq 0$ because $\dfrac{3}{(x + 1)^2} \neq 0$; therefore, there is no x-intercept.

If $x = 0$, then $y = 3$.

Points of division on the x-axis:

Signs of factor in each interval: $(x + 1)^2$ +   +

$$\leftarrow\!\!\!\!\rule{5cm}{0.4pt}\underset{-1}{|}\rule{6cm}{0.4pt}\!\!\!\!\rightarrow x$$

Note: y is always positive; therefore, this curve always stays above the x-axis.

Symmetry: None

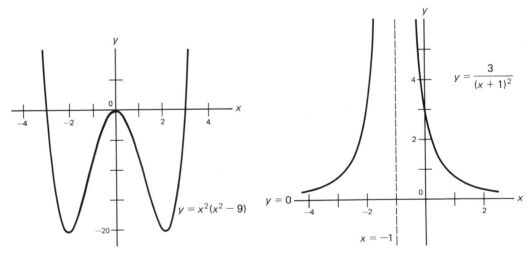

Figure 3.5   Figure 3.6

Asymptotes:

Vertical: $x = -1$ [*Note:* $(x + 1)^2 = 0$ when $x = -1$.]

Horizontal: $y = 0$ Degree of numerator is less than degree of denominator.

Slant: None

Restricted domain: None except $x \neq -1$ when the denominator is zero.

(See Fig. 3.6.)

Example 5

Graph $y = \dfrac{3x^2 - 10x - 8}{x^2 + 3x - 10}$.

Factor:

$$y = \frac{(3x + 2)(x - 4)}{(x + 5)(x - 2)}$$

Intercepts: If $y = 0$, then $x = -\frac{2}{3}$, 4.

If $x = 0$, then $y = \frac{4}{5}$.

Plot points of division on the x-axis: (Set each factor equal to zero.)

Signs of factors in each interval:					
$(3x + 2)$	−	−	+	+	+
$(x - 4)$	−	−	−	−	+
$(x + 5)$	−	+	+	+	+
$(x - 2)$	−	−	−	+	+

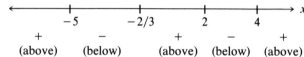

Sign of y in each interval:	+	−	+	−	+
	(above)	(below)	(above)	(below)	(above)

Symmetry: None

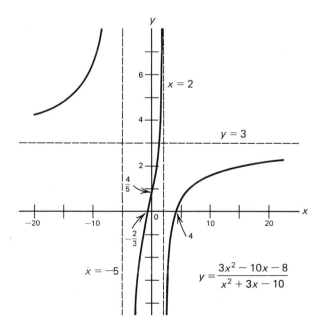

$$y = \frac{3x^2 - 10x - 8}{x^2 + 3x - 10}$$

Figure 3.7

Asymptotes:
Vertical: $x = -5$, $x = 2$.   Set the denominator equal to zero.
Horizontal: $y = 3$.   Degree of numerator equals degree of denomi-
nator; the ratio of coefficients of highest de-
gree is 3/1 or 3.
Slant: None
Restricted domain: None except $x \neq -5, 2$.
(See Fig. 3.7.)

Example 6

Graph $y = \dfrac{x^2 - 1}{x - 2}$.

Factor:

$$y = \frac{(x + 1)(x - 1)}{x - 2}$$

Intercepts: If $y = 0$, then $x = -1, 1$.
If $x = 0$, then $y = \frac{1}{2}$.
Plot points of division on the x-axis:

Signs of factors	$(x + 1)$	$-$	$+$	$+$	$+$
in each interval:	$(x - 1)$	$-$	$-$	$+$	$+$
	$(x - 2)$	$-$	$-$	$-$	$+$

$$\xleftarrow{\qquad\qquad} \underset{-1}{+} \quad \underset{1}{+} \quad \underset{2}{+} \xrightarrow{\qquad} x$$

Sign of y in	$-$	$+$	$-$	$+$
each interval:	(below)	(above)	(below)	(above)

Symmetry: None
Asymptotes:
Vertical: $x = 2$
Horizontal: None
Slant: $y = x + 2$ Degree of numerator is one greater than degree of de-
nominator. Divide the numerator by the denominator as
shown on page 105.
Restricted domain: None except $x \neq 2$.
(See Fig. 3.8.)

Example 7

Graph $y = \sqrt{\dfrac{x + 4}{1 - x}}$.

Intercepts: If $y = 0$, then $x = -4$.
If $x = 0$, then $y = 2$.
Plot points of division on the x-axis:

Signs of factors	$(x + 4)$	$-$	$+$	$+$
in each interval:	$(1 - x)$	$+$	$+$	$-$

$$\xleftarrow{\qquad\qquad} \underset{-4}{+} \quad \underset{1}{+} \xrightarrow{\qquad} x$$

Sign under radical	$-$	$+$	$-$
in each interval:			

Note: This function has a restricted domain of $-4 \leqslant x < 1$ and the graph is never below the *x*-axis.

Symmetry: None
Asymptotes:
Vertical: $x = 1$
(See Fig. 3.9.)

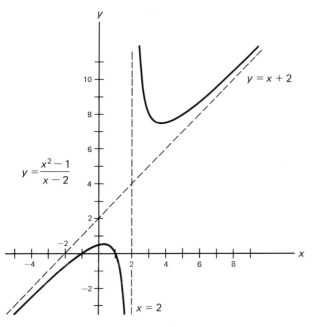

$y = \dfrac{x^2 - 1}{x - 2}$

$y = x + 2$

$x = 2$

Figure 3.8

$y = \sqrt{\dfrac{x + 4}{1 - x}}$

$x = 1$

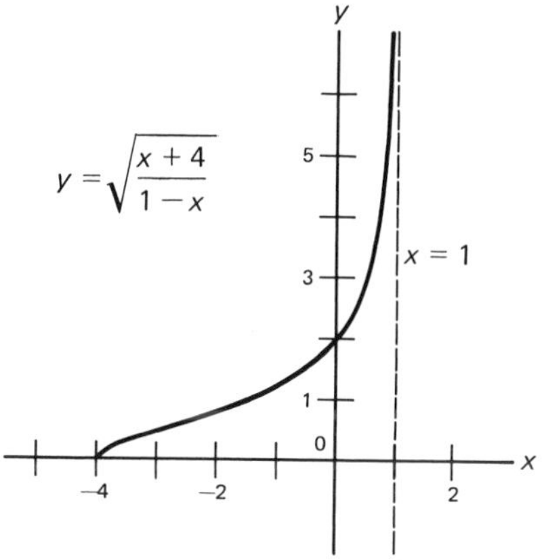

Figure 3.9

Example 8

Graph $y^2 = \dfrac{x^2}{x^2 - 4}$.

Solve for *y* and factor:

$$y = \pm \sqrt{\dfrac{x^2}{(x + 2)(x - 2)}}$$

Intercepts: If $y = 0$, then $x = 0$.
If $x = 0$, then $y = 0$.
Plot points of division on the *x*-axis:

Signs of factors in each interval:				
x^2	+	+	+	+
$(x + 2)$	−	+	+	+
$(x - 2)$	−	−	−	+

$\xleftarrow{\hspace{2cm}}\underset{-2}{+}\hspace{1.5cm}\underset{0}{+}\hspace{1.5cm}\underset{2}{+}\hspace{1cm}\xrightarrow{} x$

Sign under radical
in each interval: +   −   −   +

Note: This equation has a restricted domain of
$x < -2, x > 2, x = 0$.

Symmetry: y-axis, x-axis, and origin.

Asymptotes:

    Vertical: $x = 2, x = -2$.

    Horizontal: $y = 1, y = -1$ Divide numerator and denominator of the given

equation by x^2: $y^2 = \dfrac{1}{1 - \dfrac{4}{x^2}}$. Do you see that

$y^2 \to 1$ as $x \to \infty$?

(See Fig. 3.10.)

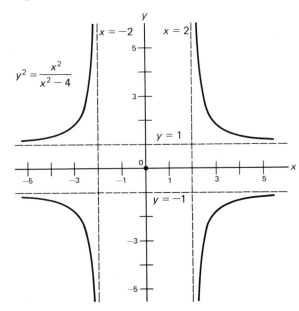

Figure 3.10

Exercises

Graph each equation.

1. $y = 2x(x + 1)(x - 4)$

2. $y = (x + 2)(x - 3)(x + 4)$

3. $y = (x - 1)(x - 3)(x + 5)$

4. $y = (3 - x)(x^2 - 4)$

5. $y = x^3 + 2x^2 - 15x$

6. $y = x^3 - 10x^2 + 24x$

7. $y = x^2(x + 1)(3 - 2x)$

8. $y = x(x + 4)(x - 1)^2$

9. $y = x^4 - 13x^2 + 36$

10. $y = (x + 1)^3(x - 7)^2(x + 2)$

11. $y = x^2(x - 2)^2(x + 4)^2$

12. $y = (x - 3)(2x + 1)(x - 2)^2$

13. $y = \dfrac{3}{2x + 1}$

14. $y = \dfrac{2}{(x - 4)^2}$

15. $y = \dfrac{2x}{(x + 1)(x - 3)}$

16. $y = \dfrac{2x - 1}{(x - 2)(x - 4)}$

17. $y = \dfrac{3}{x^2 + 4}$

18. $y = \dfrac{4x}{x^2 - 9}$

19. $y = \dfrac{4x}{x - 2}$

20. $y = \dfrac{3x}{1 - 2x}$

21. $y = \dfrac{3x^2}{x^2 - 4}$

22. $y = \dfrac{x - 5}{x + 2}$

23. $y = \dfrac{x^2 - 6}{x + 2}$

24. $y = \dfrac{2x^2 - 11x + 12}{x - 1}$

    Applications of the Derivative   Chap. 3

25. $y = \dfrac{2x^2 - x - 3}{x - 4}$

26. $y = \dfrac{x^2 - x}{x + 3}$

27. $y = \sqrt{x + 4}$

28. $y = \sqrt{1 - 2x}$

29. $y = \sqrt{\dfrac{x}{x - 3}}$

30. $y = -\sqrt{\dfrac{9}{x + 4}}$

31. $y^2 = x + 9$

32. $y^2 = (x + 1)(x - 3)$

33. $y^2 = \dfrac{x}{x + 4}$

34. $y^2 = \dfrac{x}{(x + 1)(x - 3)}$

35. $y^2 = \dfrac{x^2}{x^2 + 4}$

36. $y^2 = \dfrac{x^2}{(1 - x)(x + 2)}$

3.2 USING DERIVATIVES IN CURVE SKETCHING

The concept of a derivative is very useful in curve sketching. We can obtain information about the curve $y = f(x)$ that would be unavailable without the use of differentiation.

By looking at the derivative of a function $y = f(x)$ we can determine where a function is increasing and where it is decreasing.

Increasing and Decreasing Functions

A function $y = f(x)$ is *increasing* in an interval if $f(x_2) > f(x_1)$ for any two points $x_2 > x_1$ in the interval.
A function $y = f(x)$ is *decreasing* in an interval if $f(x_2) < f(x_1)$ for any two points $x_2 > x_1$ in the interval.

In Fig. 3.11 the function is increasing for $x < a$ and for $x > b$. The function is decreasing for $a < x < b$.

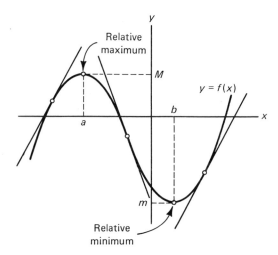

Figure 3.11

Observe that $f'(x)$, the slope of the tangent line at x, is positive when $y = f(x)$ is increasing and negative when the function is decreasing. That is,

$$\boxed{\begin{array}{l} \quad f(x) \text{ is increasing when } f'(x) > 0 \\ \text{and} \\ \quad f(x) \text{ is decreasing when } f'(x) < 0 \end{array}}$$

Observe also, in Fig. 3.11, that at $x = a$ the function changes from increasing to decreasing. At such a point the value of the function, $f(a) = M$, is called a *relative maximum*. At $x = b$, the function changes from decreasing to increasing. The value of the function, $f(b) = m$, is called a *relative minimum*.

In Fig. 3.11 note that at both $x = a$ and $x = b$ the derivative of $y = f(x)$ is zero. That is, $f'(a) = 0$ and $f'(b) = 0$. Note that if the function $y = f(x)$ is to change from increasing to decreasing or from decreasing to increasing, then the derivative $f'(x)$ must be zero at that point.

A relative maximum or relative minimum can also occur at a point x where the curve has no slope (the derivative does not exist). In Fig. 3.12, $y = f(x)$ has a relative maximum at $x = a$ but there is no derivative.

Critical Points

Critical points are those points $(x, f(x))$ on $y = f(x)$ where $f'(x) = 0$ or where $f'(x)$ is undefined.

The following test can be used to determine if a given critical point is a relative maximum, a relative minimum, or neither.

First Derivative Test

Given $(c, f(c))$ is a critical point on $y = f(x)$.

(a) If $f'(x)$ changes from positive to negative at $x = c$, then $f(x)$ is changing from increasing to decreasing and the point $(c, f(c))$ is a relative maximum.
(b) If $f'(x)$ changes from negative to positive at $x = c$, then $f(x)$ is changing from decreasing to increasing and the point $(c, f(c))$ is a relative minimum.
(c) If $f'(x)$ does not change sign at $x = c$, then the point $(c, f(c))$ is neither a relative maximum nor a relative minimum.

Example 1

Find any relative maximum or minimum values of the function $y = x^3 - 3x$ and sketch the curve.

$$\frac{dy}{dx} = 3x^2 - 3 = 3(x^2 - 1)$$

so

$$\frac{dy}{dx} = 0 \quad \text{when} \quad 3(x^2 - 1) = 0 \quad \text{or}$$

         Applications of the Derivative   Chap. 3

$$3(x + 1)(x - 1) = 0 \quad \text{or}$$
$$x = -1 \quad \text{or} \quad x = 1$$

So $x = -1$ and $x = 1$ are critical points. Check to see if the function has relative maximums or minimums at these points. Observe that

$$\text{if } x < -1, \qquad \frac{dy}{dx} = 3(x + 1)(x - 1) > 0$$

$$\text{if } -1 < x < 1, \frac{dy}{dx} = 3(x + 1)(x - 1) < 0$$

$$\text{if } x > 1, \qquad \frac{dy}{dx} = 3(x + 1)(x - 1) > 0$$

Note: The diagram below can help determine the sign of the derivative. In this example, the expression for the derivative involves two factors. Determine the sign of each factor for each interval determined by the critical points. Applying the rule of signs for multiplication, determine the sign of the derivative:

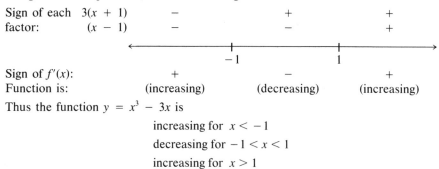

Sign of each $3(x + 1)$ — + +
factor: $(x - 1)$ — — +

-1 1

Sign of $f'(x)$: + — +
Function is: (increasing) (decreasing) (increasing)

Thus the function $y = x^3 - 3x$ is

increasing for $x < -1$

decreasing for $-1 < x < 1$

increasing for $x > 1$

We conclude that $f(-1) = (-1)^3 - 3(-1) = 2$ or $(-1, 2)$ is a relative maximum and $f(1) = (1)^3 - 3(1) = -2$ or $(1, -2)$ is a relative minimum. (See Fig. 3.13.) *Note:* When the *original equation factors easily,* we may use the discussion in Section 3.1 to determine where the curve is above and below the x-axis:

$$y = x^3 - 3x = x(x^2 - 3) = x(x + \sqrt{3})(x - \sqrt{3})$$

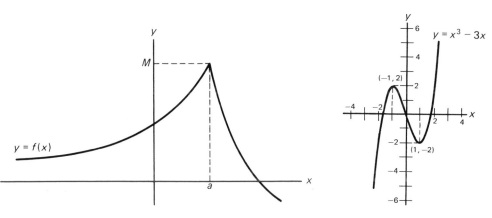

Figure 3.12   Figure 3.13

Signs of factors in each interval:	x	$-$		$-$	$+$	$+$
	$x + \sqrt{3}$	$-$		$+$	$+$	$+$
	$x - \sqrt{3}$	$-$		$-$	$-$	$+$

$$\xleftarrow{\hspace{3cm}} \quad \underset{-\sqrt{3}}{+} \quad \underset{0}{\quad} \quad \underset{\sqrt{3}}{+} \quad \xrightarrow{\hspace{1cm}} x$$

Sign of y in each interval:	$-$	$+$	$-$	$+$
	(below)	(above)	(below)	(above)

Example 2

Find any relative maximum or minimum values of $y = x^3$ and sketch the curve.

$$\frac{dy}{dx} = 3x^2$$

so

$$\frac{dy}{dx} = 0 \quad \text{when} \quad 3x^2 = 0 \quad \text{or}$$

$$x = 0$$

We find that $x = 0$ is the only critical point. But since $\dfrac{dy}{dx} = 3x^2$ is never negative, the function $y = x^3$ is always increasing. Thus, there is no relative maximum or minimum. We observe, however, in the sketch for $y = x^3$ in Fig. 3.14 that while the curve is always increasing there is a difference in the shape of the curve for $x < 0$ and for $x > 0$.

The number line diagram confirms this discussion.

Sign of $f'(x)$: $3x^2$ $\qquad +$ $\qquad\qquad +$

$$\xleftarrow{\hspace{4cm}} \underset{0}{\quad} \xrightarrow{\hspace{3cm}} x$$

Function is: (increasing) (increasing)

Example 3

Find any relative maximum or minimum values of the function $f(x) = x^4 + \frac{8}{3}x^3$ and sketch the curve.

$$f'(x) = 4x^3 + 8x^2 = 4x^2(x + 2) = 0$$

when

$$x = 0 \quad \text{or} \quad x = -2$$

Using the number line diagram, we have:

Sign of each factor:	$4x^2$	$+$	$+$	$+$
	$(x + 2)$	$-$	$+$	$+$

$$\xleftarrow{\hspace{3cm}} \underset{-2}{\quad} \quad \underset{0}{\quad} \xrightarrow{\hspace{1cm}} x$$

Sign of $f'(x)$:	$-$	$+$	$+$
Function is:	(decreasing)	(increasing)	(increasing)

We conclude that $f(-2) = (-2)^4 + \frac{8}{3}(-2)^3 = 16 - \frac{64}{3} = -\frac{16}{3}$ or $(-2, -\frac{16}{3})$ is a relative minimum, and that $x = 0$ or $(0, 0)$ is neither a relative maximum nor a relative minimum. (See Fig. 3.15.)

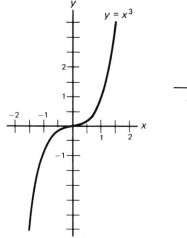

Figure 3.14

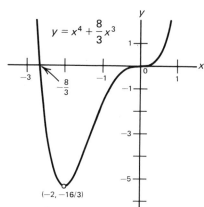

Figure 3.15

Exercises

Find any relative maximum or minimum values of each function and sketch the curve.

1. $y = x^2 + 6x - 16$
2. $y = 3x^2 + 12x + 9$
3. $y = 4 - 4x - 3x^2$
4. $y = 24 + 4x - 4x^2$
5. $y = x^3 + 3x^2 + 4$
6. $y = x^3 - 3x - 6$
7. $y = \frac{1}{3}x^3 - 9x - 4$
8. $y = \frac{5}{3}x^3 + 10x^2 + 8$
9. $y = 3x^5 - 5x^3$
10. $y = x^4 - 4x^2 + 3$
11. $y = (x - 2)^5$
12. $y = (x + 1)^4$
13. $y = (x^2 - 1)^4$
14. $y = (x^2 - 4)^3$
15. $y = \dfrac{x^2}{x - 4}$
16. $y = \dfrac{x^2}{x + 9}$
17. $y = \dfrac{x}{x + 1}$
18. $y = \dfrac{x}{x - 3}$
19. $y = x + 1/x$
20. $y = 4x + 16/x$
21. $y = \sqrt{x}$
22. $y = \sqrt[3]{x}$
23. $y = x^{2/3}$
24. $y = x^{-1/3}$

3.3 MORE ON CURVE SKETCHING

As for any function, the derivative tells us where the function increases and where it decreases. The derivative of $\dfrac{dy}{dx}, \dfrac{d^2y}{dx^2}$, which is the second derivative of $y = f(x)$, will now tell us where the derivative $\dfrac{dy}{dx} = f'(x)$ increases and decreases.

As with any derivative, those values of x where $f''(x) = 0$ are possible points where the function $f'(x)$ can possibly change from increasing to decreasing or from decreasing to increasing.

For example, in $y = x^3$, $\dfrac{d^2y}{dx^2} = f''(x) = 6x$.

1. For $x < 0$, $\dfrac{d^2y}{dx^2} = 6x < 0$. (This corresponds to our observation in Fig. 3.16 that the derivative is decreasing for $x < 0$.)

2. For $x > 0$, $\dfrac{d^2y}{dx^2} = 6x > 0$. (This tells us that the derivative $\dfrac{dy}{dx}$ is increasing as shown in Fig. 3.17.) The curve $y = f(x)$ is *concave upward* (opens up) where the derivative $\dfrac{dy}{dx} = f'(x)$ is increasing, that is, where $\dfrac{d^2y}{dx^2} = f''(x) > 0$. The curve is *concave downward* (opens down) where the derivative $\dfrac{dy}{dx} = f'(x)$ is decreasing, that is, where $\dfrac{d^2y}{dx^2} = f''(x) < 0$. Any point $(x, f(x))$ where the curve changes from concave upward to concave downward or from concave downward to concave upward is called a *point of inflection*.

So $y = x^3$ is concave downward for $x < 0$ and concave upward for $x > 0$ and $(0, 0)$ is a point of inflection.

The curve in Fig. 3.18 is concave upward (opens up) for $a < x < b$. This curve is concave downward (opens down) for $b < x < c$.

Example 1

Find the concavity of the function $y = x^3 - 3x$.

$$\frac{dy}{dx} = 3x^2 - 3, \quad \text{so} \quad \frac{d^2y}{dx^2} = 6x$$

As in $y = x^3$, the function is concave downward for $x < 0$, concave upward for $x > 0$, and $(0, 0)$ is a point of inflection. The sketch of the curve appears in Example 1 in Section 3.2.

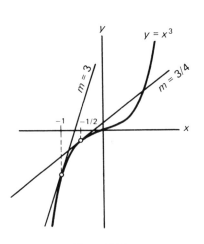

Figure 3.16

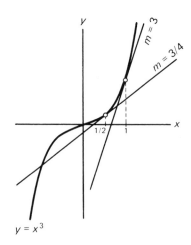

Figure 3.17

Applications of the Derivative   Chap. 3

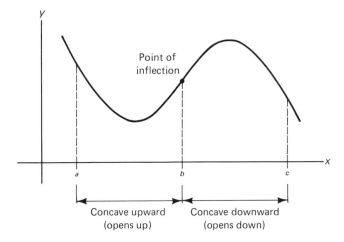

Concave upward   Concave downward
(opens up)   (opens down)   **Figure 3.18**

The following method is used to aid us in sketching a curve $y = f(x)$:

Curve Sketching

1. Find any x- and y-intercepts. The curve crosses the x-axis at an x-intercept. That is, an x-intercept is a point where $y = 0$. The curve crosses the y-axis at a y-intercept. That is, a y-intercept is a point where $x = 0$. If the original equation factors easily, you may use the discussion in Section 3.1 to determine where the curve is above and below the x-axis.

2. Find all critical points for $y = f(x)$. That is, find all values of x where $f'(x) = 0$ or where $f'(x)$ is undefined.

3. Knowing the critical points, determine where the function is increasing and where it is decreasing.

$$y = f(x) \text{ is increasing where } f'(x) > 0.$$
$$y = f(x) \text{ is decreasing where } f'(x) < 0.$$

4. Find the relative maximum and minimum values. That is, find all values of x where $f'(x) = 0$. Check the critical points to see which critical points are possible relative maximums and relative minimums using the information obtained in step 3.

5. Find the x-coordinate of each possible point of inflection. That is, find all values of x where $f''(x) = 0$.

6. Knowing the possible points of inflection, determine the concavity of $y = f(x)$.

$$y = f(x) \text{ is concave upward where } f''(x) > 0.$$
$$y = f(x) \text{ is concave downward where } f''(x) < 0.$$

7. Find which points from step 5 are points of inflection using the information obtained in step 6. Note that it is possible for the second derivative to be zero at $x = a, f''(a) = 0$, and not have a point of inflection at $x = a$.

That is, the curve does not necessarily change concavity where $f''(x) = 0$. This must be checked in step 6.

8. Sketch the curve.

Example 2

Find the maximum and minimum values of the function $y = 3x^2 - 2x^3$. Determine the concavity and sketch the curve.

Step 1: Setting $y = 0$, we have

$$3x^2 - 2x^3 = 0$$

$$x^2(3 - 2x) = 0$$

so that $x = 0$ and $x = \frac{3}{2}$ are x-intercepts.
Setting $x = 0$, we have

$$y = 3(0)^2 - 6(0)^3 = 0$$

so that $y = 0$ is the only y-intercept.
To find where the curve is above and below the x-axis:

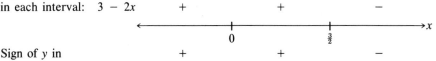

| Signs of factors in each interval: | x^2 | $+$ | $+$ | $+$ |
| | $3 - 2x$ | $+$ | $+$ | $-$ |

Sign of y in each interval:

| $+$ | $+$ | $-$ |
| (above) | (above) | (below) |

Step 2: Setting $\dfrac{dy}{dx} = 0$, we have

$$\frac{dy}{dx} = 6x - 6x^2 = 0$$

$$6x(1 - x) = 0$$

so that $x = 0$ and $x = 1$ are the critical points.

Step 3: $\dfrac{dy}{dx} = 6x(1 - x) > 0$ for $0 < x < 1$.

$\dfrac{dy}{dx} = 6x(1 - x) < 0$ for $x < 0$ or $x > 1$.

To find where the curve is increasing and decreasing:

| Signs of factors in each interval: | $6x$ | $-$ | $+$ | $+$ |
| | $1 - x$ | $+$ | $+$ | $-$ |

Sign of $f'(x)$:

| $-$ | $+$ | $-$ |
| (decreasing) | (increasing) | (decreasing) |

So $y = f(x)$ is increasing for $0 < x < 1$ and decreasing for $x < 0$ and for $x > 1$.

Step 4: Since $y = f(x)$ changes from decreasing to increasing at $x = 0$, $f(0) = 0$ is a relative minimum. Since $y = f(x)$ changes from increasing to decreasing at $x = 1$, $f(1) = 1$ is a relative maximum.

Step 5: Setting $\dfrac{d^2y}{dx^2} = 0$, we have

$$\frac{d^2y}{dx^2} = 6 - 12x = 0$$

so that $x = \frac{1}{2}$ will give us a possible point of inflection.

Step 6: $\dfrac{d^2y}{dx^2} = 6 - 12x$

To find where the curve is concave upward and concave downward:

Sign of $f''(x)$
in each interval: $6 - 12x$ $+$ $-$

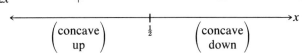

$$\begin{pmatrix} \text{concave} \\ \text{up} \end{pmatrix} \qquad \tfrac{1}{2} \qquad \begin{pmatrix} \text{concave} \\ \text{down} \end{pmatrix}$$

So $y = f(x)$ is concave upward for $x < \frac{1}{2}$ and concave downward for $x > \frac{1}{2}$.

Step 7: Since $y = f(x)$ changes concavity at $x = \frac{1}{2}$, and $f(\frac{1}{2}) = 3(\frac{1}{2})^2 - 2(\frac{1}{2})^3 = \frac{1}{2}$, $(\frac{1}{2}, \frac{1}{2})$ is a point of inflection.

Step 8: The curve is sketched in Fig. 3.19.

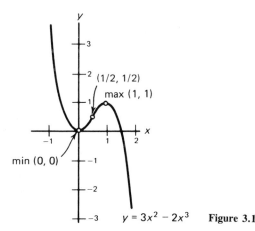

$y = 3x^2 - 2x^3$ **Figure 3.19**

From Fig. 3.19 note that the curve is concave downward at a relative maximum and concave upward at a relative minimum. We can use this information to locate relative minimums and relative maximums in place of the first derivative test. This procedure for locating relative maximums and minimums is known as the *second derivative test.* That is,

---

Second Derivative Test

1. If $f'(a) = 0$ and $f''(a) < 0$,
   then $M = f(a)$ is a relative maximum.
2. If $f'(a) = 0$ and $f''(a) > 0$,
   then $m = f(a)$ is a relative minimum.

Note: If $f''(a) = 0$, the test fails.

In Example 2, we saw that $\dfrac{d^2y}{dx^2} = f''(x) = 6 - 12x$. At $x = 0$, $f''(0) = 6 - 12(0) = 6 > 0$, so $0 = f(0)$ is a relative minimum (as already determined by the first derivative test). At $x = 1$, $f''(1) = 6 - 12(1) = -6$, so that $1 = f(1)$ is a relative maximum.

Example 3

Find any relative maximum and minimum values for $y = x^4 - 2x^2$.

$$\frac{dy}{dx} = 4x^3 - 4x = 4x(x + 1)(x - 1)$$

Setting $\dfrac{dy}{dx} = 0$, we have $4x(x + 1)(x - 1) = 0$ so that $x = 0$, $x = -1$ and $x = 1$ are critical points.

$$\frac{d^2y}{dx^2} = f''(x) = 12x^2 - 4$$

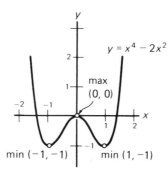

min $(-1, -1)$ min $(1, -1)$

Figure 3.20

Since $f''(0) = -4 < 0$, $f(0) = 0$ is a relative maximum.
Since $f''(-1) = 8 > 0$, $f(-1) = -1$ is a relative minimum.
Since $f''(1) = 8 > 0$, $f(1) = -1$ is a relative minimum.
(See Fig. 3.20.)

Figure 3.21 compares the relationship between the curve $y = f(x)$, its first derivative $y' = f'(x)$ and its second derivative $y'' = f''(x)$.

From A to B, $y = f(x)$ is decreasing and $y' = f'(x) < 0$.
From B to D, y is increasing and $y' > 0$.
From D to E, y is decreasing and $y' < 0$.
At B, $y' = 0$ and $y'' > 0$ (relative minimum).
At D, $y' = 0$ and $y'' < 0$ (relative maximum).
From A to C, y' is increasing, $y'' > 0$, and y is concave upward.
From C to E, y' is decreasing, $y'' < 0$ and y is concave downward.
At C and E, $y'' = 0$ (points of inflection).

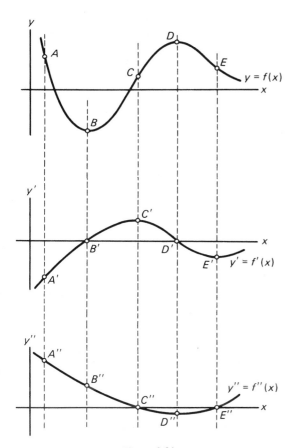

Figure 3.21

Figure 3.22

The graphs of functions that describe most physical relationships are both continuous and "smooth"; that is, they contain neither breaks (missing points or displaced points) nor make sudden or abrupt changes. Physical changes are most often smooth and gradual. Functions whose graphs are both continuous and smooth are differentiable at each and every point. Polynomial functions are both continuous and smooth, and the derivative exists at every value of x. The function in Fig. 3.22 is not smooth at $x = a$ and the derivative does not exist at $x = a$. Why not?

Exercises

For each function determine (a) intervals for which $f(x)$ is increasing and decreasing, (b) relative maximums and relative minimums, (c) intervals for which $f(x)$ is concave upward and concave downward, (d) points of inflection, and (e) sketch the curve.

1. $y = x^2 - 4x$
2. $y = x^2 - x - 2$
3. $y = x^4$
4. $y = 5 - x^2$
5. $y = 3x - x^3$
6. $y = 3x^4 - 4x^3$
7. $y = -x^2 + 4x - 1$
8. $y = x^4 - 18x^2 + 40$
9. $y = x^4 - 8x^2 + 5$

10. $y = x^7$

11. $y = \dfrac{1}{x^2}$

12. $y = \dfrac{1}{x^3}$

13. $y = \dfrac{x}{x + 2}$

14. $y = \dfrac{x}{x - 1}$

15. $y = \dfrac{-1}{(x + 1)^3}$

16. $y = \dfrac{1}{(x - 2)^2}$

17. $y = \dfrac{x}{x^2 + 1}$

18. $y = \dfrac{x^2 + 1}{x}$

19. $y = \dfrac{x - 2}{x + 3}$

20. $y = \dfrac{x + 3}{x - 2}$

21. $y = \dfrac{4}{x^2 + 4}$

22. $y = \dfrac{4}{x^2 - 4}$

23. $y = \dfrac{x - 1}{x^2}$

24. $y = \dfrac{x + 4}{x^2}$

3.4 MAXIMUM AND MINIMUM PROBLEMS

Many applications involve finding the maximum or minimum values of a variable quantity under given conditions. If we can express this quantity as a function of some other variable, then we can use the techniques of the last two sections to locate the possible relative maximum or relative minimum values.

In addition to the techniques developed in those sections, we must

1. Translate the information of a stated word problem into a mathematical function.
2. Determine that maximum (or minimum) value which is the largest (or smallest) value that the quantity can attain.

Example 1

Enclose a rectangular area with 120 m of fencing. Find the largest possible area that can be enclosed.

The area of a rectangle is given by $A = xy$, where x is the length of one side and y is the length of the other side (see Fig. 3.23.) The perimeter P of a rectangle is given by

$$P = 2x + 2y$$

which in this case must be 120 m.

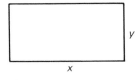

Figure 3.23

Possible choices for x and y are $x = 10$ and $y = 50$, where $A = xy = (10)(50) = 500$ m^2 or $x = 20$ and $y = 40$ where $A = xy = (20)(40) = 800$ m^2. But what is the maximum area A? To find the maximum area we will use the techniques of the previous sections. We must express A as a function of one variable only. To do this, solve for x in $2x + 2y = P$ where $P = 120$:

$$2x + 2y = 120$$
$$2x = 120 - 2y$$
$$x = 60 - y$$

and substitute for x in the formula for area,

$$A = xy$$
$$A = (60 - y)y$$
$$= 60y - y^2$$

The area is now expressed as a function of one variable y. Using the techniques of the last sections, locate the maximum value:

$$\frac{dA}{dy} = \frac{d}{dy}(60y - y^2)$$
$$= 60 - 2y$$

Setting $\dfrac{dA}{dy} = 0$, we have

$$0 = 60 - 2y$$
$$2y = 60$$
$$y = 30$$

Since $\dfrac{d^2A}{dy^2} = -2$, the curve $A = 60y - y^2$ is always concave downward.

So $y = 30$ is a relative maximum and the maximum area is obtained when $y = 30$ m. To find the value of the area, find the corresponding value of x when $y = 30$ m:

$$x = 60 - y$$
$$= 60 - (30)$$
$$= 30 \text{ m}$$

The maximum value of the area occurs when the rectangle is a square and its value is:

$$A = xy$$
$$= (30 \text{ m})(30 \text{ m})$$
$$= 900 \text{ m}^2$$

When an expression that is to be maximized or minimized is given in more than one other variable, a formula, the Pythagorean theorem, or using similar triangles may be used to eliminate a variable. In Example 1 it was necessary to combine the formula for perimeter with the formula for area. Only then could the area be expressed in terms of one variable.

It is usually not necessary to test as to whether one has a maximum or minimum value. The information contained in the problem will usually make it clear which situation will result.

Example 2

Each side of a square piece of tin is 12 cm. A small square is to be cut from each corner and the tin folded so as to form a box without a top. Determine the dimensions of the box which will have the largest volume.

If a square with dimension s cm is cut out from each corner, the resulting box will have a square base with each side $12 - 2s$ cm and a height of s cm. (See Fig. 3.24.) The volume of the box will be

$$V = (12 - 2s)^2 s$$

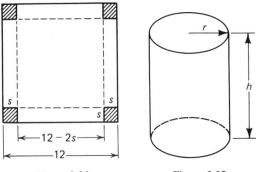

Figure 3.24   **Figure 3.25**

From the conditions of the problem, the maximum value will occur when $0 \leqslant s \leqslant$ 6 (since $2s \leqslant 12$). Differentiating,

$$\frac{dV}{ds} = (12 - 2s)^2(1) + s(2)(12 - 2s)(-2)$$

$$= (12 - 2s)[(12 - 2s) - 4s]$$
$$= (12 - 2s)(12 - 6s)$$
$$= 12(6 - s)(2 - s)$$

Alternatively, since $V = (12 - 2s)^2 s = 144s - 48s^2 + 4s^3$,

$$\frac{dV}{ds} = 144 - 96s + 12s^2$$

$$= 12(12 - 8s + s^2)$$
$$= 12(6 - s)(2 - s)$$

Setting $\dfrac{dV}{ds} = 12(6 - s)(2 - s) = 0$, we have $s = 6$ or $s = 2$ as solutions. The volume $V = 0$ cm^3 when $s = 6$ cm. Since $\dfrac{d^2V}{ds^2} = -96 + 24s$,

$$\left. \frac{d^2V}{ds^2} \right|_{s=2} = -96 + 24(2) = -48$$

While consideration of the information in the problem would indicate that there must be a maximum volume, the second derivative test guarantees a relative maximum at $s = 2$. So indeed, the maximum volume must occur when $s = 2$ cm. That is,

$$V = [12 - 2(2)]^2(2)$$
$$= 128 \text{ cm}^3$$

Note that the second derivative test cannot be used if the value of the second derivative is zero at a critical point. Should this occur, the first derivative test must be used.

Example 3

The height of a bullet fired vertically upward is given by the equation $s = 1200t -$ $16t^2$ (initial velocity is 1200 ft/s). Find the maximum height that the bullet will rise (neglecting air resistance).

124   Applications of the Derivative   Chap. 3

Differentiate and set $\dfrac{ds}{dt} = 1200 - 32t = 0.$

$$t = 37.5$$

Since $\dfrac{d^2s}{dt^2} = -32$, there is a relative maximum at $t = 37.5$ s. The maximum height is then

$$s = 1200(37.5) - 16(37.5)^2$$
$$= 45,000 - 22,500$$
$$= 22,500 \text{ ft}$$

Example 4

Make a cylindrical can with 24π cm^2 of metal. Find the dimensions of the can that gives the maximum volume. (See Fig. 3.25.)

The volume of a cylinder is given by

$$V = \pi r^2 h$$

The lateral surface area of a cylinder is given by $A = 2\pi rh$. The area of the two circular bases or ends is $2\pi r^2$. The total surface area is then

$$A = 2\pi rh + 2\pi r^2$$

We are to maximize the volume while the total surface area is given. We must express the volume as a function of only one variable. So substitute $A = 24\pi$ into

$$A = 2\pi rh + 2\pi r^2$$
$$24\pi = 2\pi rh + 2\pi r^2$$

and solve for h. First, divide both sides of this equation by 2π:

$$12 = rh + r^2$$
$$12 - r^2 = rh$$
$$\frac{12 - r^2}{r} = h$$

Then substitute for h into the volume formula:

$$V = \pi r^2 h$$
$$V = \pi r^2 \left(\frac{12 - r^2}{r} \right)$$
$$V = \pi r(12 - r^2)$$
$$V = \pi(12r - r^3)$$

Then differentiate:

$$\frac{dV}{dr} = \pi(12 - 3r^2)$$

and set $\dfrac{dV}{dr} = 0.$

$$\frac{dV}{dr} = \pi(12 - 3r^2) = 0$$

$$12 - 3r^2 = 0$$
$$12 = 3r^2$$
$$4 = r^2$$
$$\pm 2 = r$$

Since $r > 0$, $r = 2$. Then substitute into

$$h = \frac{12 - r^2}{r}$$

$$h = \frac{12 - 2^2}{2} = \frac{12 - 4}{2} = \frac{8}{2} = 4$$

Thus $r = 2$ cm and $h = 4$ cm gives the maximum volume.
Do you see that these values give a maximum and not a minimum volume?

Solving Maximum and Minimum Problems

1. Make a sketch of the conditions given in the problem whenever possible.
2. Label appropriate quantities and assign symbols for variable quantities.
3. Determine which variable quantity is to be maximized or minimized.
4. Write an equation that expresses the quantity identified in step 3 as a function of one other variable. You may need to combine different equations to obtain the desired function of one variable. These equations will be derived either from given conditions or from your understanding of the relationship between quantities involved in the problem.
5. Differentiate the function obtained in step 4.
6. Set the derivative from step 5 equal to zero. Solve for the values of the variable that make the derivative zero or undefined.
7. Determine which values of the variable from step 6 provide the desired maximum or minimum.

Exercises

1. The sum of two positive numbers is 56. Find the two numbers if their product is to be at a maximum.

2. Find a positive number such that the sum of this number and its reciprocal is a minimum.

3. An open box is to be made from a square piece of aluminum, 3 cm on a side, by cutting equal squares from each corner and then folding up the sides. Determine the dimensions of the box which will have the largest volume.

4. An open rectangular box is to be made from a rectangular piece of metal 16 in. by 10 in. by cutting equal squares from each corner and then folding up the sides. Determine the maximum volume of the box.

5. A man wishes to fence in a rectangular plot lying next to a river. No fencing is required along the river bank. If he has 800 m of fence, and he wishes the area to be at a maximum, find the dimensions of the desired enclosed plot.

6. A long rectangular sheet of metal, 12 in. wide, is to be made into a trough by turning up two sides forming right angles. Find the dimensions of the trough to give it maximum capacity.

7. Find the maximum possible area of a rectangle whose perimeter is 36 cm.

8. A rancher wants to fence in two rectangular corrals, equal in area, using an inner fence as shown in Fig. 3.26. If 300 ft of fencing is used, find the maximum area of the combined corrals.

9. A farmer wants to fence in 80,000 m² of land and then divide it into three plots of equal area as shown in Fig. 3.27. Find the minimum amount of fence needed.

10. Four neighbors want to fence in an 8100-ft² garden plot into equal plots as shown in Fig. 3.28. Find the minimum amount of fence needed.

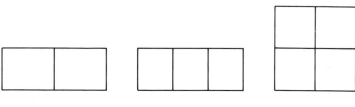

Figure 3.26   Figure 3.27   Figure 3.28

11. Find the maximum area of the rectangle with two vertices on the parabola $y = 36 - x^2$ and the remaining two vertices on the x-axis.

12. Find the dimensions of the largest rectangle that can be inscribed in a semicircle of radius 4.

13. Find the area of the largest rectangle that can be inscribed in the ellipse $x^2 + 9y^2 = 16$.

14. The charge transmitted through a circuit varies according to $q = \dfrac{4t}{t^2 + 1}$ coulombs (C). Find the maximum charge at time t in milliseconds (ms).

15. The power P in watts (W) in a circuit with resistance R in ohms (Ω) varies according to $P = \dfrac{36R}{(R + 4)^2}$. Find the resistance that gives the maximum power.

16. A rectangular yard 24,300 ft² in area is bounded on three sides by a fence that costs \$6/ft and in the front by a fence that costs \$10/ft. Find the most economical dimensions of the yard.

17. Show that the rectangle of largest area that can be inscribed in a given circle is a square.

18. Find the minimum slope of the curve $y = 4x^2 + 8x^3$.

19. Find the maximum slope of the curve $y = 3x^2 - 2x^3$.

20. The work done by a solenoid in moving an armature varies according to $w = 2t^3 - 3t^4$ joules (J). Find the greatest power in watts (W) developed in t seconds $\left(\text{power } p = \dfrac{dw}{dt}\right)$.

21. The charge transmitted through a circuit varies according to $q = t^4 - 4t^3$ coulombs

(C). Determine the time t in seconds when the current i in amperes (A)$i = \dfrac{dq}{dt}$ reaches a minimum.

22. Find the greatest current in a capacitor with capacitance C equal to $\frac{1}{3} \times 10^{-6}$ farad (F) if the applied voltage is given by $V = 250t^2 - 200t^3$ volts (V) $\left(i = C\dfrac{dV}{dt}\right)$.

23. A rectangular box, open at the top, with a square base is to have a volume of 4000 cm³. Find the dimensions if the box is to contain the least amount of material.

24. If the box in Exercise 23 has a closed top, find its dimensions.

25. The total cost C of making x units of a certain commodity is given by $C = 0.005x^3 + 0.45x^2 + 12.75x$. All units made are sold at \$36.75 per unit. The profit P is then given by $P = 36.75x - C$. Determine the number of units that should be made in order to obtain maximum profit.

26. Find the volume of the largest right circular cylinder that can be inscribed in a sphere of radius 3 in.

27. A cylindrical can with one end is to be made with 24π cm² of metal. Find the dimensions of the can that will give it the maximum volume.

28. The current I in a cell is given by $I = \dfrac{E}{R + r}$, where E is the electromotive force, R is the external resistance, and r is the internal resistance. In a given cell, E and r are fixed and determined by the cell. The power developed is $P = RI^2$. Show that P is a maximum when $R = r$.

29. The strength of a rectangular beam is proportional to the product of its width and the square of its depth; that is, $S = kwd^2$, where k is a constant. Find the dimensions of the strongest wooden beam that can be cut from a circular log of radius r in Fig. 3.29.

30. A person is in a boat 4 km off a straight coast. The person needs to reach a point 10 km down and along the coast in the least possible time. If the person can row at 6 km/h and run at 8 km/h, how far, x, down the coast should he land the boat? (See Fig. 3.30.)

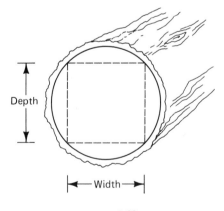

Depth

Width

Figure 3.29

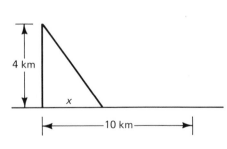

4 km

x

10 km

Figure 3.30

Applications of the Derivative   Chap. 3

31. Two ships are anchored off a straight coast as shown in Fig. 3.31. A boat from ship A is to land a passenger and then proceed to ship B in the shortest distance. Find x.

32. Show that any cylindrical can with two ends made with a given amount of material has a maximum volume when the height equals the diameter.

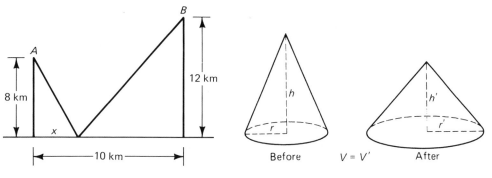

| Figure 3.31 | Figure 3.32 |

3.5 RELATED RATES

When two or more variables are related, the rate of change of one variable will affect the rate of change of the remaining variables. For example, if the volume of a cone must remain constant, then enlarging the radius will affect the height of the cone in that the height will become smaller as in Fig. 3.32.

Problems which involve observing the effects that the rate of change of one variable has on other related variables are called *related rate problems*.

Example 1

A hot air balloon (spherical) is being inflated at the rate of 3 m³/s. Find the rate at which the radius is increasing when the radius is 10 m.

First determine the relationship between the volume of the balloon and the radius. The volume of a sphere is given by $V = \frac{4}{3}\pi r^3$. Differentiate each side of the equation with respect to time.

$$\frac{dV}{dt} = \frac{d}{dt}\left(\frac{4}{3}\pi r^3\right)$$

$$= \frac{4}{3}\pi \frac{d}{dt}(r^3)$$

$$= \frac{4}{3}\pi\left(3r^2\frac{dr}{dt}\right) \quad \text{(note that } r \text{ is a function of time)}$$

$$= 4\pi r^2 \frac{dr}{dt}$$

$\dfrac{dV}{dt}$ is given as 3 m³/s and we wish to find $\dfrac{dr}{dt}$ when $r = 10$ m. Substituting,

$$3 \text{ m}^3/\text{s} = 4\pi(10 \text{ m})^2\frac{dr}{dt}$$

$$3 \text{ m}^3/\text{s} = (400\pi \text{ m}^2)\frac{dr}{dt}$$

$$\frac{3 \text{ m}^3/\text{s}}{400\pi \text{ m}^2} = \frac{dr}{dt}$$

$$\frac{dr}{dt} = \frac{3}{400\pi} \text{ m/s} \quad \text{or } 0.00239 \text{ m/s or } 2.39 \text{ mm/s}$$

Solving Related Rate Problems

1. Determine the relation for the variables whose rates are to be related. This relation may or may not be directly stated in the problem.
2. Differentiate each side of the resulting equation with respect to time (or other desired variable).
3. *After* differentiating in step 2, substitute all given rates and given values of variables into the equation.
4. Solve for the unknown rate of change quantity.

Example 2

Two airplanes pass each other in flight at 9:00 A.M. A Piper Cub is traveling east at 90 mi/h. A Beechcraft Bonanza is traveling south at 180 mi/h. How fast are they separating at 11:00 A.M.?

Step 1: Relate the distance that each plane has traveled with the distance traveled since separation at 9:00 A.M. From Fig. 3.33 note that the distances are related as the sides of a right triangle. The Pythagorean theorem provides the relationship:

$$x^2 + y^2 = z^2$$

where x is the distance traveled by A, y is the distance traveled by B, and z is the distance of separation.

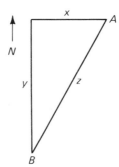

Figure 3.33

Applications of the Derivative   Chap. 3

Step 2:
$$\frac{d}{dt}(x^2 + y^2) = \frac{d}{dt}(z^2)$$

$$2x\frac{dx}{dt} + 2y\frac{dy}{dt} = 2z\frac{dz}{dt}$$

$$x\frac{dx}{dt} + y\frac{dy}{dt} = z\frac{dz}{dt}$$

Step 3: We are given that $\dfrac{dx}{dt} = 90$ mi/h and $\dfrac{dy}{dt} = 180$ mi/h.

Now, after 2 h of travel,
$$x = (90 \text{ mi/h})(2 \text{ h}) = 180 \text{ mi } (d = rt)$$
$$y = (180 \text{ mi/h})(2 \text{ h}) = 360 \text{ mi}$$

So, again from the Pythagorean theorem,
$$z = \sqrt{x^2 + y^2} = \sqrt{(180 \text{ mi})^2 + (360 \text{ mi})^2} = 402 \text{ mi}$$

Step 4:
$$\frac{dz}{dt} = \frac{x\dfrac{dx}{dt} + y\dfrac{dy}{dt}}{z} \quad \text{[from step 2]}$$

$$= \frac{(180 \text{ mi})(90 \text{ mi/h}) + (360 \text{ mi})(180 \text{ mi/h})}{402 \text{ mi}}$$

$$= 201 \text{ mi/h}$$

Example 3

The electric resistance of a resistor is given by $R = 6.000 + 0.002T^2$ (in ohms, Ω) where T is the temperature (in °C). If the temperature is increasing at the rate of 0.2°C/s, find how fast the resistance is changing when $T = 120$°C.

Step 1: The relationship is given:
$$R = 6.000 + 0.002T^2$$

Step 2:
$$\frac{dR}{dt} = 2(0.002T)\frac{dT}{dt}$$

$$= 0.004T\frac{dT}{dt}$$

Step 3:
$$\frac{dR}{dt} = 0.004(120)(0.2)$$

Step 4:
$$\frac{dR}{dt} = 0.096 \ \Omega/\text{s}$$

Exercises

1. The electric resistance of a resistor is given by $R = 3.000 + 0.001T^2$ (in ohms, Ω) where T is the temperature (in °C). If the temperature is increasing at the rate of 0.3°C/s, find how fast the resistance is changing when $T = 100$°C.
2. A man begins walking due north at 3 mi/h. At the same time a woman begins walking east at 4 mi/h. Find the rate of change of the distance between them after 1 h of walking.

3. A square plate is being heated so that the length of a side increases at a rate of 0.08 cm/min. How fast is the area increasing when the length of a side is 12 cm long?

4. A circular plate is being heated. Find the rate of increase in the area of the plate as it expands if the radius is increasing 0.2 cm/h, when the radius is 5 cm.

5. The side of a cube due to heating is increasing at the rate of 0.1 cm/min. Find how fast the volume of the container is changing when a side measures 12 cm.

6. A stone is dropped into a lake and forms ripples of concentric circles of increasing radii. Find the rate at which the area of one of these circles is increasing when its radius is 3 ft and when its radius is increasing at the rate of 0.5 ft/s.

7. The sides of an equilateral triangular plate, being cooled, are decreasing at a rate of 0.04 cm/min. At what rate is the area changing when the sides are 8 cm in length? The area of an equilateral triangle is given by $A = \dfrac{\sqrt{3}}{4}s^2$.

8. The area of an equilateral triangular plate, being heated, is increasing at a rate of 150 mm^2/min. At what rate is the length of a side changing when the sides are 250 mm long?

9. Mineral waste is falling into a conical pile at the rate of 5 m^3/min. The height is equal to $\frac{3}{4}$ the radius of the base. Find how fast the radius is changing when the pile is 10 m high.

10. The relation between the voltage V which produces a current i in a wire of radius r is given by $V = 0.02\ i/r^2$. Find the rate at which the voltage is increasing when the current is increasing at the rate of 0.04 A/s in a wire with radius 0.02 in.

11. The power in a circuit is given by $P = I^2R$. Find how fast the power changes in W/s when the current is 8 A and decreasing at a rate of 0.4 A/s and the resistance is 75 Ω and increasing at a rate of 5 Ω/s.

12. Boyle's law states that the pressure of a gas varies inversely as its volume with constant temperature. We begin with 800 cm^3 at a pressure of 250 kilopascals (kPa). Then the volume is decreased at a rate of 20 cm^3/min. How is the pressure changing when the volume is 500 cm^3?

13. A cylinder with an inside radius of 75 mm is sealed at one end and a piston is at the other end. At what rate is the piston moving if fluid is being pumped into the cylinder at the rate of 90 cm^3/s?

14. A man on a dock throws a rope to a woman in a boat, who fastens the rope to the boat. The man then pulls the boat in. If the man's hands are 8 ft above the water and if he hauls the rope in at 2 ft/s, find how fast the boat is approaching the base of the dock when there is still 10 ft of rope remaining to be pulled in.

15. A 5-m ladder is leaning against a wall. Its upper end is sliding down the wall at rate of 1 m/s. Find how fast the bottom end of the ladder is moving at the point 3 m from the wall.

16. The work W done by an electromagnet varies according to $W = 36t^3 - t^2$. Find the power $\dfrac{dW}{dt}$ at $t = \frac{1}{4}$ s.

17. When a gas is compressed adiabatically (with no loss or gain of heat), its behavior is described by the equation $PV^{1.4} = k$, where P is the pressure, V is the volume, and k is a constant. At a given instant, the pressure is 60 lb/in^2 and the volume is 56 in^3 and decreasing at a rate of 8 in^3/min. At what rate is the pressure changing?

Applications of the Derivative   Chap. 3

18. In Exercise 17, find the rate at which the volume is changing when the volume is 5600 cm^3 and the pressure is 400 kilopascals (kPa) and the pressure is increasing at a rate of 50 kPa/min.

19. The equivalent resistance R of a parallel circuit with two resistances R_1 and R_2 is given by

$$R = \frac{R_1 R_2}{R_1 + R_2}$$

If R_1 is a constant 120 Ω and R_2 is decreasing at a rate of 15 Ω/s, find the rate at which R is changing when R_2 is 180 Ω.

20. In Exercise 19, assume that R_1 is 150 Ω and increasing at 15 Ω/s while R_2 is 300 Ω and decreasing at 25 Ω/s. Find the rate at which R is changing.

21. Given that $P = Ri^2$, find the rate of change of $\dfrac{dP}{di}$ in a 30-Ω resistor when the current $i = 4$ A.

22. The power P, in watts, in a circuit varies according to $P = Ri^2$, where R is the resistance in ohms and i is the current in amperes. If $R = 100 \ \Omega$ and i varies according to $i = t^2 + 3t$, find the rate of change of the power with respect to time when $t = 2$ s.

23. The voltage V which produces a current i in a wire of radius r is given by $V = 0.02i/r^2$. Find the rate at which the voltage is increasing when the current is increasing at the rate of 0.04 A/s in a wire of radius 0.01 in.

24. A light inside a garage is 10 ft above the floor and 6 ft from the door opening. If the overhead garage door opener lets the door down at a rate of 1 ft/s, at what rate is the door's shadow approaching the garage when the bottom of the door is 2 ft above the floor? Assume that the bottom of the garage door stays in the same vertical plane as the door opening.

25. A tank is in the shape of an inverted cone with a height of 8 m and radius 2 m. Water is being pumped in at a rate of 2π m^3/min. How fast is the depth of the water changing when the depth is (a) 2 m and (b) 6 m? (*Hint:* Use similar triangles to express the volume in terms of the depth only.)

26. The weight in pounds of an object varies according to

$$W = W_e\left(1 + \frac{r}{3960}\right)^{-2}$$

where W_e is the object's weight on the earth's surface and r is the distance above the earth's surface in miles. Find the rate at which a person's weight is decreasing in a space shuttle 500 miles above the earth when its altitude is increasing at 16 mi/s. The person's weight on earth is 175 lb.

3.6 DIFFERENTIALS

Up to now, we have defined and treated dy/dx as the limit:

$$\frac{dy}{dx} = \lim_{\Delta x \to 0} \frac{\Delta y}{\Delta x}$$

Now we need to discuss the separate meanings of dy and dx.

First, solve the expression $\dfrac{dy}{dx} = f'(x)$ for dy.

Differential

$$dy = f'(x)\, dx$$

The expression dy is called the differential of y and dx is called the differential of x.

Figure 3.34 shows the geometric relationships between dy, dx, Δy, and Δx. Note that L_1 is the line tangent to the curve at the point (x, y) and L_2 is a secant line through the point (x, y) and a second point "close" to the point (x, y). Note that:

1. $f'(x)$ is the slope of the tangent line to $y = f(x)$.

2. $\dfrac{dy}{dx}$ is the instantaneous rate of change of the function y with respect to x.

3. $\dfrac{\Delta y}{\Delta x}$ is the slope of the secant line and the rate of change of the function between the points (x, y) and $(x + \Delta x, y + \Delta y)$.

4. dx and Δx are each the amount of change in x.

5. Δy is the amount of change in y as measured along the secant line.

6. dy is the amount of change in y as measured along the tangent line.

7. If Δx is "small," dy is approximately equal to Δy; that is, $|dy - \Delta y|$ is approximately zero.

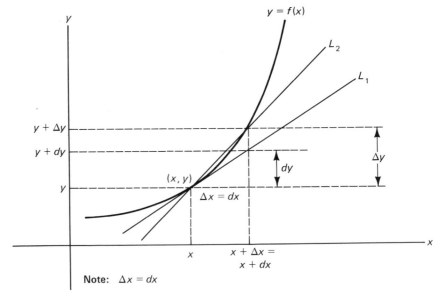

Note: $\Delta x = dx$

Figure 3.34

Applications of the Derivative   Chap. 3

To find the differential, dy, of a function, simply find the derivative, dy/dx, and then multiply both sides of the equation by dx.

Example 1

Given $y = 8x^3 - 4x + 2$, find dy.

$$y = 8x^3 - 4x + 2$$

$$\frac{dy}{dx} = 24x^2 - 4$$

$$dy = (24x^2 - 4)dx$$

Example 2

Given $y = \dfrac{8x + 4}{2x - 1}$, find dy.

$$y = \frac{8x + 4}{2x - 1}$$

$$\frac{dy}{dx} = \frac{(2x - 1)(8) - (8x + 4)(2)}{(2x - 1)^2}$$

$$\frac{dy}{dx} = \frac{-16}{(2x - 1)^2}$$

$$dy = \frac{-16}{(2x - 1)^2} dx \quad \text{or} \quad \frac{-16\, dx}{(2x - 1)^2}$$

Approximations can be made using differentials.

Example 3

The radius of a sphere is measured to be 12.00 cm with a possible error of 0.05 cm. Find the maximum error in the volume.

The volume of a sphere is found by

$$V = \tfrac{4}{3}\pi r^3$$

Find dV:

$$dV = \tfrac{4}{3}\pi(3r^2)\, dr$$

$$dV = 4\pi r^2\, dr$$

Here $r = 12.00$ cm and $dr = 0.05$ cm:

$$dV = 4\pi(12.00 \text{ cm})^2(0.05 \text{ cm})$$

$$= 90.48 \text{ cm}^3$$

which is the differential approximation of the maximum error.

Note: $\Delta V = \tfrac{4}{3}\pi(12.05 \text{ cm})^3 - \tfrac{4}{3}\pi(12.00 \text{ cm})^3 = 90.86 \text{ cm}^3$, which is the calculated maximum error by substitution into the volume formula. That is, when Δr or dr is small, $|dV - \Delta V|$ is small.

An expression that better describes some approximations is relative error or percentage error.

$$\text{relative error of } x = \frac{\text{error in } x}{x} = \frac{dx}{x}$$

$$\text{percentage error of } x = \frac{\text{error in } x}{x} \cdot 100\% = \frac{dx}{x} \cdot 100\%$$

Example 4

Find the percentage error in Example 3.

$$\frac{dV}{V} \cdot 100\% = \frac{4\pi r^2 \, dr}{\frac{4}{3}\pi r^3} \cdot 100\%$$

$$= \frac{3 \, dr}{r} \cdot 100\%$$

$$= \frac{3(0.05 \text{ cm})}{12.00 \text{ cm}} \cdot 100\%$$

$$= 1.25\%$$

Example 5

The current in a circuit changes according to $i = (3t + 2)^{1/3}$. Find the approximate change in current using differentials as t changes from 2.00 s to 2.05 s.

First, find the differential di:

$$i = (3t + 2)^{1/3}$$

$$di = \frac{1}{3}(3t + 2)^{-2/3}(3) \, dt$$

$$= (3t + 2)^{-2/3} \, dt$$

$$= [3(2.00) + 2]^{-2/3}(0.05) \qquad (Note: dt = 0.05.)$$

$$= \left(\frac{1}{4}\right)(0.05) = 0.0125 \text{ A or } 12.5 \text{ mA}$$

Exercises

Find the differential for each expression.

1. $y = 5x^2 - 8x^3$

2. $y = 4t^5 - 6\sqrt{t} + 7$

3. $y = \dfrac{x + 3}{2x - 1}$

4. $s = \dfrac{2t + 1}{t + 6}$

5. $y = (2t^2 + 1)^4$

6. $i = 4t^2(t^2 + 1)^3$

7. $s = (t^4 - t^{-2})^{-2}$

8. $m = \left(3t - \dfrac{1}{3t}\right)^{-3}$

Find dy for each expression.

9. $x^2 + 4y^2 = 6$

10. $y^2 + 4xy = x$

11. $(x + y)^3 = \sqrt{x} + \sqrt{y}$

12. $(x^2 + y^2)^{1/2} = x + y$

Using a differential expression, find the change in each expression for the given change in the given independent variable.

13. $y = 8x^4$ from $x = 3.00$ to 3.05

14. $y = 4t^{2/3}$ from $t = 8.0$ to 8.1

15. $v = r^3 - 3r^2$ from $r = 4.00$ to 4.05

16. $y = (x - 2)^3$ from $x = 5.0$ to 5.1

17. $V = \frac{4}{3}\pi r^3$ from $r = 15.00$ to 15.10

18. $A = 4\pi r^2$ from $r = 21.0$ to 21.5

19. The side of a square is measured to be 12.00 cm with a possible error of 0.05 cm. (a) Find the maximum error in the area using differentials. (b) Find the maximum error by substituting into the formula for the area of a square. (c) Find the percentage error.

20. The side of a cube is measured to be 12.00 cm with a possible error of 0.05 cm. (a) Find the maximum error in the volume using differentials. (b) Find the maximum error by substituting into the formula for the volume of a cube. (c) Find the percentage error.

21. Suppose that you wanted to build a spherical water tower with an inner diameter of 26.00 m and sides of thickness 4.00 cm. (a) Find the approximate volume of steel needed using differentials. (b) If the density of steel is 7800 kg/m³, find the approximate amount of steel used.

22. The current in a resistor varies according to $i = 0.08t^6 - 0.04t^2$. Find the approximate change in current using differentials as t changes from 2.00 s to 2.10 s.

23. The horsepower of an internal combustion engine is given by the formula $P = nd^2$, where n is the number of cylinders and d is the diameter of each bore. Find the approximate increase in horsepower using differentials for an engine with eight cylinders when the bore of each cylinder is increased from 3.750 in. to 3.755 in.

24. A freely falling body drops according to $s = \frac{1}{2} gt^2$, where s is the distance in metres, $g = 9.80$ m/s², and t is the time in seconds. Approximate the distance, ds, that an object falls from $t = 10.00$ s to $t = 10.03$ s.

25. The voltage v in volts varies according to $v = 10P^{2/3}$, where P is the power in watts. Find the change dv when the power changes from 125 W to 128 W.

26. The impedance Z in an ac circuit varies according to $Z = \sqrt{R^2 + X^2}$, where R is the resistance and X is the reactance. If $R = 300 \ \Omega$ and $X = 225 \ \Omega$, find dZ when R changes to 310 Ω.

CHAPTER SUMMARY

1. *Algebraic curve sketching aids*
   (a) *Intercepts*
      (i) To find the *x*-intercepts, substitute $y = 0$ into the given equation and solve for x. Each resulting value of x is an *x*-intercept.
      (ii) To find the *y*-intercepts, substitute $x = 0$ into the given equation and solve for y. Each resulting value of y is a *y*-intercept.
   (b) *Location of the curve above and below the x-axis*
      (i) Solve the given equation for y.
      (ii) Factor the result from step (i) as much as possible.
      (iii) Set each factor from step (ii) equal to zero.
      (iv) Find the sign of y in each region by choosing a test point. Substitute this test point value into each factor of the given equation and determine if each factor is positive or negative in the region. If the sign of y is *positive,* the curve is *above* the *x*-axis in this region. If the sign of y is *negative,* the curve is *below* the *x*-axis in this region.
   (c) *Symmetry*
      (i) *Symmetry about the y-axis:* The graph of an equation is symmetric about the *y*-axis if replacement of x by $-x$ in the original equation results in an equivalent equation.
      (ii) *Symmetry about the x-axis:* The graph of an equation is symmetric

about the x-axis if replacement of y by $-y$ in the original equation results in an equivalent equation.

(iii) *Symmetry about the origin:* The graph of an equation is symmetric about the origin if replacement of both x by $-x$ and y by $-y$ in the original equation results in an equivalent equation.

(d) *Asymptotes*

(i) *Vertical:* Set the denominator equal to zero and solve for x. Each solution, a, gives the equation of a vertical asymptote, $x = a$.

(ii) *Horizontal:* $y = 0$ if the degree of the numerator is less than the degree of the denominator. $y = a_n/b_n$ if the degree of the numerator equals the degree of the denominator of the rational function in the form

$$f(x) = \frac{a_n x^n + \cdots + a_0}{b_n x^n + \cdots + b_0}$$

(iii) *Slant:* If the degree of the numerator is one greater than the degree of the denominator, the slant asymptote is found by dividing the numerator by the denominator. (Drop the remainder.) Then $y = $ the quotient is the equation of the slant asymptote.

(e) *Restricted domains:* Check for any restrictions in the domain of certain functions, such as factors that make the denominator zero in a rational function, values of the variable that make the radicand negative in an even root function, values that make the logarithm function undefined, and so forth.

2. *Increasing and decreasing functions*

(a) A function $y = f(x)$ is *increasing* in an interval if $f(x_2) > f(x_1)$ for any two points $x_2 > x_1$ in the interval.

(b) A function $y = f(x)$ is *decreasing* in an interval if $f(x_2) < f(x_1)$ for any two points $x_2 > x_1$ in the interval.

(c) $f(x)$ is increasing when $f'(x) > 0$.

(d) $f(x)$ is decreasing when $f'(x) < 0$.

3. *Critical points:* Those points $(x, f(x))$ on $y = f(x)$ where $f'(x) = 0$ or where $f'(x)$ is undefined.

4. *First derivative test:* Given $(c, f(c))$ is a critical point on $y = f(x)$.

(a) If $f'(x)$ changes from positive to negative at $x = c$, then $f(x)$ is changing from increasing to decreasing and the point $(c, f(c))$ is a relative maximum.

(b) If $f'(x)$ changes from negative to positive at $x = c$, then $f(x)$ is changing from decreasing to increasing and the point $(c, f(c))$ is a relative minimum.

(c) If $f'(x)$ does not change sign at $x = c$, then the point $(c, f(c))$ is neither a relative maximum nor a relative minimum.

5. *Calculus curve sketching aids:* To sketch $y = f(x)$:

(a) Find all critical points for $y = f(x)$. That is, find all values of x where $f'(x) = 0$ or where $f'(x)$ is undefined.

(b) Knowing the critical points, determine where the function is increasing and where it is decreasing.

Applications of the Derivative   Chap. 3

$$y = f(x) \text{ is increasing where } f'(x) > 0.$$
$$y = f(x) \text{ is decreasing where } f'(x) < 0.$$

(c) Determine the relative maximum and minimum values. Check the critical points to see which critical points are possible relative maximums and relative minimums using the information obtained in step (b).

(d) Determine the x-coordinate of each possible point of inflection. That is, find all values of x where $f''(x) = 0$ or is undefined.

(e) Knowing the possible points of inflection, determine the concavity of $y = f(x)$.

$$y = f(x) \text{ is concave upward where } f''(x) > 0.$$
$$y = f(x) \text{ is concave downward where } f''(x) < 0.$$

(f) Determine which points from step (d) are points of inflection using the information obtained in step (e). Note that it is possible for the second derivative to be zero at $x = a$, $f''(a) = 0$, and not have a point of inflection at $x = a$. That is, the curve does not necessarily change concavity where $f''(x) = 0$. This must be checked in step (e).

(g) Sketch the curve.

6. *Second derivative test*

(a) If $f'(a) = 0$ and $f''(a) < 0$, then $M = f(a)$ is a relative maximum.

(b) If $f'(a) = 0$ and $f''(a) > 0$, then $m = f(a)$ is a relative minimum.

7. *Maximum and minimum problems*

(a) Make a sketch of the conditions given in the problem whenever possible.

(b) Label appropriate quantities and assign symbols for variable quantities.

(c) Determine which variable quantity is to be maximized or minimized.

(d) Write an equation that expresses the quantity identified in step (c) as a function of one other variable. You may need to combine different equations to obtain the desired function in one variable. These equations will be derived either from given conditions or from your understanding of the relationship between quantities involved in the problem.

(e) Differentiate the function obtained from step (d).

(f) Set the derivative from step (e) equal to zero. Solve for the values of the variable that make the derivative zero or undefined.

(g) Determine which values of the variable from step (f) provide the desired maximum or minimum.

8. *Related rate problems*

(a) Determine the relation for the variables whose rates are to be related. This relation may or may not be directly stated in the problem.

(b) Differentiate each side of the resulting equation with respect to time (or other desired variable).

(c) *After* differentiating in step (b), substitute all given rates and given values of variables into the equation.

(d) Solve for the unknown rate of change quantity.

9. *Differential: $dy = f'(x)\, dx$*

10. *Errors*

(a) Relative error of $x = \dfrac{\text{error in } x}{x} = \dfrac{dx}{x}$

(b) Percentage error of $x = \dfrac{\text{error in } x}{x} \cdot 100\% = \dfrac{dx}{x} \cdot 100\%$

CHAPTER REVIEW

Graph each equation.

1. $y = x^3 - 16x$

2. $y = \sqrt{-2 - 4x}$

3. $y = (x + 2)(25 - x^2)$

4. $y = (x^2 + 4x)(x - 1)^2$

5. $x = y^2 + 4$

6. $y = \dfrac{x - 2}{(x + 4)(x - 1)}$

7. $y = \dfrac{2x^2}{x^2 - 1}$

8. $y = \dfrac{x^2 + x - 12}{x + 3}$

9. $y = \dfrac{2x}{x^2 + 9}$

10. $y^2 = \dfrac{x}{(1 - x)(x + 4)}$

For each function determine (a) intervals for which it is increasing and decreasing, (b) relative maximums and relative minimums, (c) intervals for which it is concave upward and concave downward, (d) points of inflection, and (e) sketch the curve.

11. $y = 6x - x^3$

12. $y = x^2 - 3x - 4$

13. $y = x^3 - 7$

14. $y = 2x^3 - 9x^2 - 24x - 2$

15. $y = \dfrac{1}{(x + 1)^2}$

16. $y = \dfrac{x^2 - 1}{x^2 + 4}$

17. $y = \dfrac{10}{x^2 + 1}$

18. $y = \dfrac{x + 1}{x^2}$

19. The height of a missile fired vertically upward is given by the equation $y = 240t - 16t^2$. Find the maximum height (in feet) that the missile will reach.

20. Find the dimensions of a right triangle if the area is to be a maximum when the hypotenuse is 20.

21. The electrical power (in watts, W) produced by a certain source is given by $p = 3r - r^3$ where r is the resistance (in ohms, Ω) in the circuit. Find the value of r which provides the maximum power.

22. Find the area of the largest rectangle that can be inscribed in the right triangle in Fig. 3.35.

23. A long rectangular sheet of metal, 12 cm wide, is bent lengthwise to form a V-shaped trough. Find the depth of the trough that gives the maximum cross-sectional area and hence the greatest volume of flow.

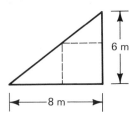

6 m

8 m

Figure 3.35

Applications of the Derivative   Chap. 3

24. Find the point on the curve $y = \sqrt{x}$ which is nearest the point $(1, 0)$.

25. A hot air balloon (spherical) is being inflated at the rate of 2 ft^3/s. Find the rate at which the radius is increasing when the radius is 12 ft.

26. The voltage V which produces a current i in a wire of radius r is given by $V = 0.06$ i/r^2. Find the rate at which the voltage is increasing when the current is increasing at the rate of 0.03 A/s in a wire with radius 0.01 in.

27. The current i through a circuit with resistance R and a battery whose voltage is E and whose internal resistance is r is given by $i = \dfrac{E}{R + r}$. A certain circuit has a variable resistor changing at the rate of 0.4 Ω/min with a 1.5-V battery whose resistance is 0.3 Ω. Find how fast the current is changing when the resistance of the variable resistor is 8 Ω.

28. A circular plate is being cooled. Find the rate of decrease in the area of the plate as it contracts if the radius is decreasing 0.05 cm/min when the radius is 8 cm.

29. Oil is leaking from an offshore well into a circular slick. If the area of the oil slick is increasing at a rate of 4 km^2/day, at what rate is the radius increasing when the radius is 2.5 km?

30. A ground TV camera is 12 km from where a rocket is to be launched vertically. At what rate is the distance from the camera to the rocket increasing when the rocket is at an altitude of 10 km and rising at a rate of 3 km/s?

Find the differential for each expression.

31. $y = 4x^5 - 6x^3 + 2x$ 32. $y = (3x - 5)^{-2/3}$ 33. $s = \dfrac{3t^2 - 4}{5t + 1}$

34. Find dy: $(x^2 + y^2)^2 = y + 2x$

Using a differential expression, find the change in each expression for the given change in the given independent variable.

35. $s = 3t^2 - 5t + 6$ from $t = 9.50$ to 9.55

36. $y = (8x + 3)^{-3/4}$ from $x = 10.00$ to 10.06

37. Find the increase in volume if the radius of a sphere is increased from 6.0 in. to 6.1 in. using differentials.

38. A particle moves along a straight line according to $s = \frac{1}{3}t^3 - 3t + 5$, where s is the distance, in metres, and t is the time, in seconds. Find the approximate distance covered by the particle between 3.00 s and 3.05 s.

39. Using differentials, approximate how much paint is needed to paint a cube 16 ft on a side if the thickness of the paint is to be 0.02 in. (One gallon contains 231 in^3.)

40. A metal sphere is 8.00 cm in diameter. How much nickel is needed to plate the sphere with a thickness of 0.4 mm?

41. The attractive force between two unlike charged particles is given by $F = k/x^2$, where k is a constant and x is the distance between the particles. If x increases from 0.030 m to 0.031 m, find the approximate decrease in F (in newtons) using differentials.

4

THE INTEGRAL

4.1 THE INDEFINITE INTEGRAL

There is a large class of extremely important problems in science, engineering, and technology that involves the notion of finding the "sum" of an "infinite number" of "infinitesimal" quantities. The integral calculus develops the mathematics needed to solve such problems.

As we have seen, many applications involve finding the derivative of a function. Following are some examples:

Function	Derivative
Position or displacement	Velocity
Velocity	Acceleration
Function	Slope of tangent line
Amount of quantity	Rate of increase or decrease
Work	Power

Often we have the derivative function and need to find the original function, which requires that we perform the inverse operation of differentiation.

Now consider finding the derivative in reverse; that is, given the derivative of a function, find the function. This is called *antidifferentiation* or *integration*. That is, we look for an unknown function whose derivative is known.

Antiderivative

If $f(x)$ is a function, then $F(x)$ is an *antiderivative* of $f(x)$ if $F'(x) = f(x)$.

Example 1

Find a function whose derivative is $10x^4$.

We are looking for a function $y = f(x)$ where $\dfrac{dy}{dx} = 10x^4$. Recall that when differentiating a polynomial in x, the exponent of x is reduced by one. This means that the original function $y = f(x)$ must have involved a power of x with exponent 5, x^5. However, when we differentiate x^5, we obtain $5x^4$. If we rewrite $\dfrac{dy}{dx} = 10x^4 = 2(5x^4)$, observe that the given derivative is a multiple of two times the derivative of x^5.

Since $\dfrac{d}{dx}(2x^5) = 2\dfrac{d}{dx}(x^5) = 2(5x^4)$, we conclude that $y = 2x^5$ is a solution.

Example 1 immediately leads us into difficulty. While $y = 2x^5$ is a solution, so also is $y = 2x^5 + 9$ since

$$\frac{d}{dx}(2x^5 + 9) = 10x^4 + 0 = 10x^4$$

In fact, any function of the form $y = 2x^5 + C$, where C is a constant, is a solution.

The process of inverse differentiation, unlike differentiation, does not lead to unique solutions. There are, in fact, an infinite number of solutions. Each solution depends on the choice of the value for C.

Example 2

Find a function whose derivative is $\dfrac{dy}{dx} = x^2 + 3x$.

The term x^2 requires that the desired function have a term involving x^3 since differentiation decreases the power of x by one. Since $\dfrac{d}{dx}(x^3) = 3x^2$, the coefficient of x^3 must be changed to $\frac{1}{3}$. Then

$$\frac{d}{dx}\left(\frac{x^3}{3}\right) = 3\frac{(x^2)}{3} = x^2$$

In a similar manner, the term $3x$ indicates that the desired function must also have a term involving x^2. Observe that the coefficient of x^2 must be $\frac{3}{2}$ so that $\dfrac{d}{dx}(\frac{3}{2}x^2) = (\frac{3}{2})(2x) = 3x$.

We conclude that the desired function is

$$y = \frac{x^3}{3} + \frac{3x^2}{2}$$

Note:

$$y = \frac{x^3}{3} + \frac{3x^2}{2} + 11$$

is also a solution for Example 2. The general solution is

$$y = \frac{x^3}{3} + \frac{3x^2}{2} + C$$

where C is constant.

The process of inverse differentiation can be easily checked. Differentiate the solution obtained and compare the result with the original given derivative. They should be equal. In Example 2,

$$\frac{d}{dx}\left(\frac{x^3}{3} + \frac{3x^2}{2}\right) = x^2 + 3x$$

Any solution $y = F(x)$ resulting from performing the integration process is called an antiderivative. In Example 1, $y = 2x^5 + 9$ was found to be an antiderivative of $10x^4$. That is, $F(x)$ is an antiderivative for a given function $f(x)$ if $\frac{d}{dx}[F(x)] = f(x)$.

Using any antiderivative, $F(x)$, for a given function $f(x)$, $F(x) + C$ is a general solution; that is, $\frac{d}{dx}[F(x) + C] = f(x)$. This general solution is called the *indefinite integral* of $f(x)$ where $f(x)$ is called the integrand. The symbolism for the indefinite integral is:

$$\int f(x)dx = F(x) + C$$

where C is called the *constant of integration*. $F(x) + C$ can be considered to be the family of all curves whose derivatives for any given x are all equal to each other. This means that for any given x, the slopes of any two of these curves are equal.

Example 3

Find the indefinite integral for $f(x) = 2x$.

Since $\frac{d}{dx}(x^2) = 2x$, we have

$$\int f(x)\,dx = \int 2x\,dx = x^2 + C$$

This is a family of parabolas $y = x^2 + C$. (See Fig. 4.1.) At $x = 1$, each curve

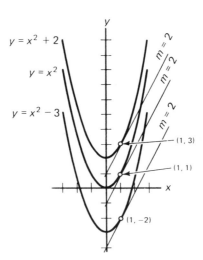

Figure 4.1

The Integral   Chap. 4

has slope

$$\frac{d}{dx}(x^2 + C)\bigg|_{x=1} = 2x\bigg|_{x=1} = 2(1) = 2$$

Just as there are basic differentiation formulas, there are basic integration formulas that can be developed. These make the integration process easier to perform.

First,

$$\int x^n \, dx = \frac{x^{n+1}}{n+1} + C, \quad n \neq -1$$

This is verified as follows:

$$\frac{d}{dx}\left(\frac{x^{n+1}}{n+1} + C\right) = \frac{(n+1)x^{(n+1)-1}}{n+1} + 0$$

$$= x^n$$

The case for $n = -1$ is discussed in Section 6.7.

Example 4

Integrate $y = x^{16}$.

$$\int x^{16} dx = \frac{x^{16+1}}{16+1} + C = \frac{x^{17}}{17} + C$$

Example 5

Find $\int dx$.

$$\int dx = \int x^0 \, dx \quad (\text{since } x^0 = 1)$$

$$= \frac{x^{0+1}}{0+1} + C = x + C$$

Example 6

Find $\int \frac{dx}{x^3}$.

$$\int \frac{dx}{x^3} = \int x^{-3} \, dx$$

$$= \frac{x^{-3+1}}{-3+1} + C$$

$$= \frac{x^{-2}}{-2} + C$$

$$= -\frac{1}{2x^2} + C$$

Example 7

Find $\int \sqrt{x} \, dx$.

Sec. 4.1   The Indefinite Integral   **145**

$$\int \sqrt{x} \, dx = \int x^{1/2} \, dx$$

$$= \frac{x^{(1/2)+1}}{\frac{1}{2} + 1} + C$$

$$= \frac{x^{3/2}}{\frac{3}{2}} + C$$

$$= \tfrac{2}{3} x^{3/2} + C$$

Next,

$$\int [f(x) + g(x)] \, dx = \int f(x) \, dx + \int g(x) \, dx$$

This follows from the fact that the derivative of the sum of two functions equals the sum of their derivatives.

Example 8

Find $\int (x^3 + x^2) \, dx$.

$$\int (x^3 + x^2) \, dx = \int x^3 \, dx + \int x^2 \, dx$$

$$= \frac{x^{3+1}}{3+1} + \frac{x^{2+1}}{2+1} + C$$

$$= \frac{x^4}{4} + \frac{x^3}{3} + C$$

Finally,

$$\int k f(x) \, dx = k \int f(x) \, dx, \qquad \text{where } k \text{ is a constant}$$

This follows from the fact that the derivative of a constant times a function is equal to that constant times the derivative of the function.

Example 9

Find $\int 12x^3 \, dx$.

$$\int 12x^3 \, dx = 12 \int x^3 \, dx$$

$$= 12 \frac{x^{3+1}}{3+1} + C$$

$$= 12 \frac{x^4}{4} + C$$

$$= 3x^4 + C$$

Example 10

Find $\int \left(3x^4 + 2x + \frac{1}{x^2} \right) dx$.

$$\int \left(3x^4 + 2x + \frac{1}{x^2} \right) dx = \int 3x^4 \, dx + \int 2x \, dx + \int \frac{1}{x^2} \, dx$$

$$= 3 \int x^4 \, dx + 2 \int x \, dx + \int x^{-2} \, dx$$

$$= 3 \frac{x^{4+1}}{4+1} + 2 \frac{x^{1+1}}{1+1} + \frac{x^{-2+1}}{-2+1} + C$$

$$= \frac{3x^5}{5} + x^2 - \frac{1}{x} + C$$

Example 11

Find $\int 5(x^3 + 4)^4(3x^2) \, dx$.

This does not seem to fit any of the formulas developed so far. While we could perform the indicated multiplication $5(x^3 + 4)^4(3x^2)$ and integrate the resulting sum of terms, there is an easier method.

Observe that $3x^2$ is the derivative of $x^3 + 4$. Also recall that when differentiating the power formula u^{n+1} we obtain $(n + 1)u^n \frac{du}{dx}$. Then $5(x^3 + 4)^4(3x^2)$ is seen to be the derivative of $(x^3 + 4)^5$. That is,

$$\int 5(x^3 + 4)^4(3x^2) \, dx = (x^3 + 4)^5 + C$$

In Example 11, if we let $u = x^3 + 4$, then $du = 3x^2 \, dx$, then

$$\int 5(x^3 + 4)^4(3x^2) \, dx = \int 5u^4 \, du$$

Success in obtaining a solution is then having the ability to see the function to be integrated as a product of two factors. One factor should be a power of u, that is, u^n. The other factor should be the differential of u, that is, du. Then use the formula

$$\boxed{\int u^n \, du = \frac{u^{n+1}}{n+1} + C}$$

Example 12

Find $\int 6x^2 \sqrt{2x^3 + 1} \, dx$.

Observe that $6x^2$ is the *derivative* of $2x^3 + 1$ (or $6x^2 \, dx$ is the *differential* of $2x^3 + 1$). Setting $u = 2x^3 + 1$ and $du = 6x^2 \, dx$, we have

$$\int 6x^2 \sqrt{2x^3 + 1} \, dx = \int \sqrt{u} \, du = \int u^{1/2} \, du$$

$$= \tfrac{2}{3} u^{3/2} + C$$

$$= \tfrac{2}{3}(2x^3 + 1)^{3/2} + C$$

Example 13

Find $\int x^3 \sqrt{3x^4 + 2} \, dx$.

If we let $u = 3x^4 + 2$, then $du = \dfrac{d}{dx}(3x^4 + 2)\ dx = 12x^3\ dx.$

Then

$$\sqrt{u}\ du = 12x^3\ \sqrt{3x^4 + 2}\ dx$$

Note that the only difference between the integrand $x^3\ \sqrt{3x^4 + 2}$ and $\sqrt{u}\ du$ is the factor 12.

So write

$$x^3\ \sqrt{3x^4 + 2}\ dx = \sqrt{3x^4 + 2}\left(\frac{12x^3\ dx}{12}\right) = \sqrt{u}\left(\frac{du}{12}\right)$$

and

$$\int x^3\ \sqrt{3x^4 + 2}\ dx = \int \frac{\sqrt{u}\ du}{12}$$

$$\boxed{\begin{aligned} u &= 3x^4 + 2 \\ du &= 12x^3\ dx \end{aligned}}$$

$$= \tfrac{1}{12}\int (u)^{1/2}\ du$$

$$= \frac{1}{12}\frac{u^{3/2}}{\frac{3}{2}} + C$$

$$= \tfrac{1}{18}(3x^4 + 2)^{3/2} + C$$

We will continue to use a box to show the appropriate substitutions for u and du whenever the formula $\int u^n\ du$ is applied. The box will appear at the right of the integral in which the substitutions have been made.

In summary,

A. $\displaystyle\int du = u + C$

B. $\displaystyle\int u^n\ du = \frac{u^{n+1}}{n+1} + C, \qquad n \neq -1$

C. $\displaystyle\int k\ du = k\int du, \qquad$ where k is a constant

D. $\displaystyle\int (du + dv) = \int du + \int dv$

where u and v may be functions of another variable such as x.

Exercises

Integrate.

1. $\displaystyle\int x^7\ dx$

2. $\displaystyle\int x^{24}\ dx$

3. $\displaystyle\int 3x^8\ dx$

4. $\displaystyle\int 200x^9\ dx$

5. $\displaystyle\int 4\ dx$

6. $\displaystyle\int 8\sqrt{x^3}\ dx$

7. $\displaystyle\int 9\sqrt[6]{x^5}\ dx$

8. $\displaystyle\int \frac{3\ dx}{x^2}$

9. $\displaystyle\int \frac{6\ dx}{x^3}$

10. $\displaystyle \int \frac{dx}{\sqrt[6]{x^5}}$ **11.** $\displaystyle \int (5x^2 - 12x + 8)\, dx$ **12.** $\displaystyle \int (7x^{10} + 8x^4 - 11x^3)\, dx$

13. $\displaystyle \int \left(3x^2 - x + \frac{5}{x^3}\right) dx$ **14.** $\displaystyle \int \left(2x^3 + 3x - \frac{5}{x^4}\right) dx$ **15.** $\displaystyle \int (2x^2 - 3)^2\, dx$

16. $\displaystyle \int (x^2 - 5)^2\, dx$ **17.** $\displaystyle \int \sqrt{6x + 2}\, dx$ **18.** $\displaystyle \int \sqrt[3]{8x - 1}\, dx$

19. $\displaystyle \int 8x(x^2 + 3)^3\, dx$ **20.** $\displaystyle \int 18x^2(x^3 + 2)^5\, dx$ **21.** $\displaystyle \int x\sqrt[3]{5x^2 - 1}\, dx$

22. $\displaystyle \int 3x\sqrt{6x^2 + 5}\, dx$ **23.** $\displaystyle \int x(x^2 - 1)^4\, dx$ **24.** $\displaystyle \int x^4(x^5 + 3)^7\, dx$

25. $\displaystyle \int \frac{2x\, dx}{\sqrt{x^2 + 1}}$ **26.** $\displaystyle \int \frac{24x\, dx}{\sqrt{8x^2 - 1}}$

27. $\displaystyle \int (3x^2 + 2)(x^3 + 2x)^3\, dx$ **28.** $\displaystyle \int (4x^3 - 6x)(x^4 - 3x^2)^4\, dx$

29. $\displaystyle \int \frac{x^2\, dx}{(x^3 - 4)^2}$ **30.** $\displaystyle \int \frac{(6x + 1)\, dx}{(3x^2 + x - 2)^3}$

31. $\displaystyle \int (10x - 1)\sqrt{5x^2 - x}\, dx$ **32.** $\displaystyle \int (3x^2 - 2x)\sqrt{x^3 - x^2 + 2}\, dx$

33. $\displaystyle \int \frac{(2x + 1)\, dx}{\sqrt{x^2 + x}}$ **34.** $\displaystyle \int \frac{(x - 4)\, dx}{\sqrt{x^2 - 8x + 3}}$ **35.** $\displaystyle \int (2x + 3)^2\, dx$

36. $\displaystyle \int (4x - 5)^2\, dx$ **37.** $\displaystyle \int (2x - 1)^4\, dx$ **38.** $\displaystyle \int (5x + 3)^6\, dx$

39. $\displaystyle \int (x^2 + 1)^3\, dx$ **40.** $\displaystyle \int (x^2 - 4)^2\, dx$ **41.** $\displaystyle \int 4x(x^2 + 1)^3\, dx$

42. $\displaystyle \int 6x(x^2 - 4)^2\, dx$ **43.** $\displaystyle \int 30x^2(5x^3 + 1)^4\, dx$ **44.** $\displaystyle \int 12x^3(6x^4 - 1)^5\, dx$

45. $\displaystyle \int \frac{6x^2\, dx}{\sqrt{x^3 + 1}}$ **46.** $\displaystyle \int \frac{24x^3\, dx}{\sqrt{2x^4 - 1}}$ **47.** $\displaystyle \int \frac{(6x^2 + 6)\, dx}{\sqrt[3]{x^3 + 3x}}$

48. $\displaystyle \int \frac{(60x^3 + 36x)\, dx}{\sqrt[4]{5x^4 + 6x^2}}$ **49.** $\displaystyle \int \frac{(x - 1)\, dx}{x^3}$ **50.** $\displaystyle \int \frac{(4 - 3x^2)\, dx}{6x^4}$

4.2 THE CONSTANT OF INTEGRATION

When additional information about the antiderivative is known, we can determine the constant of integration C.

Note: This section is actually a brief introduction to the study of differential equations, which are more extensively developed in Chapters 11 and 12.

Example 1

Find the antiderivative of $\dfrac{dy}{dx} = 3x^2$ where $y = 10$ when $x = 2$.

$$\int 3x^2 \, dx = F(x) + C$$
$$= x^3 + C$$

We need that function $y = x^3 + C$ where $y = 10$ when $x = 2$. So

$$y = x^3 + C$$
$$(10) = (2)^3 + C$$
$$2 = C$$

The antiderivative is then $y = x^3 + 2$.

Example 2

Find the equation of the curve that passes through $(3, -2)$ and whose slope is given by the equation $\dfrac{dy}{dx} = 2x - 3$.

$$y = \int (2x - 3) \, dx = x^2 - 3x + C$$
$$(-2) \qquad\qquad = (3)^2 - 3(3) + C$$
$$-2 = C$$

So the equation is $y = x^2 - 3x - 2$.

Example 3

Find the equation describing the motion of an object moving along a straight line when the acceleration is constantly 2 m/s^2, when the velocity at time $t = 3$ s is 10 m/s, and when the object has traveled 70 m from the origin at $t = 5$ s.

Since $\dfrac{dv}{dt} = a$, we have

$$v = \int a \, dt = \int 2 \, dt = 2t + C_1$$

since

$$v = 2t + C_1 \qquad \text{and } v = 10 \text{ when } t = 3$$
$$(10) = 2(3) + C_1$$
$$4 = C_1$$

so

$$v = 2t + 4$$

Also, since $\dfrac{ds}{dt} = v$,

$$s = \int v \, dt$$
$$= \int (2t + 4) \, dt$$

we have

$$s = t^2 + 4t + C_2 \qquad \text{and } s = 70 \text{ when } t = 5$$
$$(70) = (5)^2 + 4(5) + C_2$$
$$25 = C_2$$

The equation is $s = t^2 + 4t + 25$.

The Integral   Chap. 4

Example 4

A bullet is fired vertically from a gun with an initial velocity of 250 m/s. (a) Find the equation that describes its motion. (b) How long does it take to reach its maximum height? (c) How high does the bullet go?

In a problem involving a freely falling body, the acceleration is

$$a = g = -32 \text{ ft/s}^2 = -9.80 \text{ m/s}^2$$

An upward direction is positive and a downward direction is negative.

(a) Since $\dfrac{dv}{dt} = a$, we have

$$v = \int a \, dt = \int -9.80 \, dt = -9.80t + C_1$$

At $t = 0$, $v = 250$. Substituting, we have

$$250 = -9.80(0) + C_1$$
$$250 = C_1$$

So

$$v = -9.80t + 250$$

and since $\dfrac{ds}{dt} = v$,

$$s = \int v \, dt = \int (-9.80t + 250) \, dt = -4.90t^2 + 250t + C_2$$

At $t = 0$, $s = 0$. Substituting, we have

$$0 = -4.90(0)^2 + 250(0) + C_2$$
$$0 = C_2$$

So

$$s = -4.90t^2 + 250t$$

(b) At the bullet's maximum height, $v = 0$.

So

$$v = -9.80t + 250 = 0$$
$$t = \frac{250}{9.80} = 25.5 \text{ s}$$

(c) At $t = 25.5$ s,

$$s = -4.90t^2 + 250t$$
$$s = -4.90(25.5)^2 + (250)(25.5) = 3190 \text{ m}$$

In general, the position of an object moving freely (no air resistance) along a straight line with constant acceleration a, initial velocity v_0, and initial position s_0 may be written

$$s = \tfrac{1}{2} at^2 + v_0 t + s_0$$

This equation is derived in Exercise 23.

The voltage V_C across a capacitor at any time t is given by the formula

$$V_C = \frac{1}{C} \int i \, dt$$

where C is the capacitance in farads and i is the current in amperes.

Sec. 4.2   The Constant of Integration   **151**

Example 5

A 1-microfarad capacitor (1 microfarad $= 10^{-6}$ farad (F)) is measured to have a voltage of 86 volts (V) across it. At a given instant ($t = 0$) we connect this capacitor to a source that sends a current $i = 3t^2$ amperes (A) through the circuit. Find the voltage across the capacitor when $t = 0.1$ s, so

$$V_C = \frac{1}{C} \int i \, dt = \frac{1}{10^{-6}} \int 3t^2 \, dt$$

$$V_C = \frac{1}{10^{-6}} \cdot \frac{3t^3}{3} + C = t^3 \cdot 10^6 + C$$

When $t = 0$, $V_C = 86$ volts, so

$$(86) = (0)^3 \cdot 10^6 + C = C$$

We then have

$$V_C = t^3 \cdot 10^6 + 86$$
$$V_C = (0.1)^3 \cdot 10^6 + 86$$
$$= 10^{-3} \cdot 10^6 + 86$$
$$= 1086 \text{ V}$$

The current i in a circuit at any instant is given by

$$i = \frac{dq}{dt}$$

where q is the charge. That is,

$$q = \int i \, dt$$

Example 6

The current in a circuit is given by $i = t^3 + 3t^2 - 4$ amperes (A). Find the charge in coulombs (C) that passes a given point in the circuit after 2 s.

$$q = \int i \, dt = \int (t^3 + 3t^2 - 4) \, dt$$
$$q = \frac{t^4}{4} + t^3 - 4t + C$$

If we assume that $q = 0$ when $t = 0$, then

$$0 = 0 + 0 - 0 + C = C$$

and

$$q = \frac{t^4}{4} + t^3 - 4t$$

At $t = 2$ s,

$$q = \frac{(2)^4}{4} + (2)^3 - 4(2)$$
$$= 4 + 8 - 8 = 4 \text{ C}$$

Exercises

Find the equation of the curve $y = f(x)$ satisfying the given conditions.

1. $\frac{dy}{dx} = 3x$, passing through $(0, 1)$

2. $\dfrac{dy}{dx} = 5x^2$, passing through $(1, -\frac{1}{3})$

3. $\dfrac{dy}{dx} = 3x^2 + 3$, passing through $(-1, 2)$

4. $\dfrac{dy}{dx} = 4x^3 - 2x + 2$, passing through $(1, -2)$

5. $\dfrac{dy}{dx} = x(x^2 - 3)^2$, passing through $(2, \frac{7}{6})$

6. $\dfrac{dy}{dx} = \dfrac{x}{\sqrt{x^2 - 1}}$, passing through $(-3, 2\sqrt{2})$

7. Find the equation describing the motion of an object moving along a straight line when the acceleration is $a = 3t$, when the velocity at $t = 4$ s is 40 m/s, and when the object has traveled 86 m from the origin at $t = 2$ s.

8. Find the equation describing the motion of an object moving along a straight line when the acceleration is $a = 4t - 2$, when the velocity at $t = 5$ s is 25 m/s, and when the object has traveled 238 m from the origin at $t = 12$ s.

9. An object is dropped from a height of 100 ft. For a free-falling object, the acceleration is $a = -32$ ft/s^2 (the effect of gravity). Find the distance the stone has traveled after 2 s. Note that the initial velocity is 0 since the stone was dropped, not thrown down. Find also the velocity of the object when it hit the ground.

10. An object is dropped from a stationary balloon at 500 m. (a) Express the object's height above the ground as a function of time. (b) How long does it take to hit the ground? $(a = -9.80$ m/s^2.)

11. An airplane starts from rest and travels 3600 ft down a runway before lifting off in 30 s. Find its velocity at the moment of lift-off.

12. A ball rolls from a rest position down a 200-cm inclined plane in 4 s. Find its acceleration in cm/s^2.

13. A stone is hurled straight up from the ground at a velocity of 25 m/s. (a) Find the maximum height that the stone reaches. (b) How long does it take for the stone to hit the ground? (c) Find the speed at which the stone hits the ground.

14. A ball is thrown vertically upward with an initial velocity of 40 ft/s. (a) Find the maximum height of the ball. (b) How long does it take for the ball to hit the ground? (c) Find the speed at which the ball hits the ground.

15. A stone is thrown vertically upward from the roof of a 200-ft tall building with an initial velocity of 30 ft/s. (a) Find the equation describing the altitude of the stone from the ground. (b) How long does it take for the stone to hit the ground?

16. A stone is thrown straight down from an 80-m tall building with an initial velocity of 10 m/s. (a) Find the equation describing the height of the stone from the ground. (b) How long does it take for the stone to hit the ground?

17. A flywheel is turning at a rate given by the equation $\omega = 80 - 12t + 3t^2$ where ω is the angular speed in revolutions per minute (rpm). Find the number of revolutions that the flywheel makes in the first 3 s. $\omega = \dfrac{d\theta}{dt}$ (Assume that $\theta = 0$ when $t = 0$).

18. The power in a system equals the rate at which energy (work) is expanded. That is,

$p = \dfrac{dw}{dt}$. At $t = 0$, the energy in an electrical system is 4 joules (J). Find the energy in the system after 4 s if $p = 2\sqrt{t}$ W.

19. A capacitor with capacitance 10^{-4} F is measured to have a voltage of 100 V across it. At a given instant ($t = 0$) the capacitor is connected to a source that sends a current $i = (0.5)\sqrt{t} + 0.2$ A through the circuit. Find the voltage across the capacitor when $t = 0.16$ s.

20. A 0.1-F capacitor measures 150 V across it. At $t = 0$ the capacitor is connected to a source that sends current through the circuit according to $i = \dfrac{16t}{\sqrt{4t^2 + 9}}$. Find the voltage across the capacitor when $t = 2$ s.

21. The current in a circuit is given by $i = t\sqrt{t^2 + 1}$ A. Find the charge (in coulombs) that passes a given point in the circuit after 1 s. (Assume that $q = 0$ when $t = 0$).

22. The current in a circuit is given by $i = t^{3/2} + 4t$ in amperes. Find the charge in coulombs that passes a given point in the circuit after 4 s. (Assume that $q = 0$ when $t = 0$.)

23. In general, the position of an object moving freely (no air resistance) along a straight line with constant acceleration a, initial velocity v_0, and initial position s_0 may be written

$$s = \tfrac{1}{2} at^2 + v_0 t + s_0$$

For freely falling bodies ($a = g = -32$ ft/s^2 = -9.80 m/s^2), this equation becomes

$$s = -16t^2 + v_0 t + s_0 = -4.9t^2 + v_0 t + s_0$$

Derive this equation.

4.3 AREA UNDER A CURVE

Another application of integration is finding the area under a curve, which is actually the geometric interpretation of the integral. We will be able to find the area of regions which would be impossible to determine if we had to rely only on geometric methods.

The area under a curve $y = f(x)$ refers to the area of the region bounded by the curves $y = f(x)$, $x = a$, $x = b$ and the x-axis ($y = 0$) (see Fig. 4.2).

We could approximate this area by forming the rectangles in Fig. 4.3. The sum of the areas of each rectangle could then be used as an approximation for the desired area.

The height of each rectangle is the value of the function $f(x)$ for some point t along the base of the rectangle. We have chosen Δx to be the base, or the width of each rectangle.

The described area is then approximately equal to the sum

$$S_n = f(t_1)\,\Delta x + f(t_2)\,\Delta x + f(t_3)\,\Delta x + \cdots + f(t_n)\,\Delta x$$

when using n rectangles with base Δx and t_k as a point along the base of the kth rectangle.

Observe that the smaller we choose the width Δx, the better the approximation for the area under the curve. Also observe that as $\Delta x \to 0$, the number of terms n in the approximating sum S_n will increase. In fact, as $\Delta x \to 0$, $n \to \infty$ and the sums S_n appear to approach the exact area A under the curve.

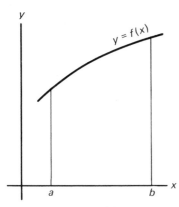

Figure 4.2   Figure 4.3

This summation process described before is symbolized as follows:

$$\lim_{n \to \infty} S_n = A$$

While this method can be used to find area as illustrated below in Example 1, the method is generally difficult to use. In practice an easier approach is used for finding area which is shown after Example 1.

Example 1

Find the area under the curve $y = 3x$ from $x = 0$ to $x = 2$.

We will form our approximating sum S_n by choosing the width of each rectangle to be

$$\Delta x = \frac{b - a}{n} = \frac{2 - 0}{n} = \frac{2}{n}$$

In Fig. 4.4 $n = 4$, so $\Delta x = \frac{2}{4} = \frac{1}{2}$ and $t_1 = a = 0$, $t_2 = \frac{1}{2}$, $t_3 = 1$, and $t_4 = \frac{3}{2}$. Then

$$S_4 = f(0)\, \Delta x + f(\tfrac{1}{2})\, \Delta x + f(1)\, \Delta x + f(\tfrac{3}{2})\, \Delta x$$

$$= (0)(\tfrac{1}{2}) + (\tfrac{3}{2})(\tfrac{1}{2}) + (3)(\tfrac{1}{2}) + (\tfrac{9}{2})(\tfrac{1}{2})$$

$$= \tfrac{18}{4} = \tfrac{9}{2}$$

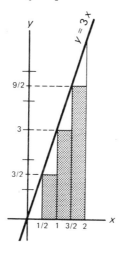

Figure 4.4

In general, since $f(x) = 3x$,

$$S_n = f(0) \Delta x + f\left(\frac{2}{n}\right) \Delta x + f\left(\frac{4}{n}\right) \Delta x$$

$$+ f\left(\frac{6}{n}\right) \Delta x + \cdots + f\left[\frac{2(n-1)}{n}\right] \Delta x$$

$$= 3(0)\left(\frac{2}{n}\right) + 3\left(\frac{2}{n}\right)\left(\frac{2}{n}\right) + 3\left(\frac{4}{n}\right)\left(\frac{2}{n}\right)$$

$$+ 3\left(\frac{6}{n}\right)\left(\frac{2}{n}\right) + \cdots + 3\left[\frac{2(n-1)}{n}\right]\left(\frac{2}{n}\right)$$

$$= \frac{12}{n^2}[1 + 2 + 3 + \cdots + (n-1)]$$

Since $1 + 2 + 3 + \cdots + (n-1)$ is the sum of an arithmetic progression whose last term is $n - 1$, we have

$$1 + 2 + 3 + \cdots + (n-1) = \frac{n-1}{2}[1 + (n-1)]$$

$$= \frac{(n-1)n}{2} \qquad \text{(see Section 10.1)}$$

Then

$$S_n = \frac{12}{n^2}\left[\frac{(n-1)n}{2}\right]$$

$$= \frac{6(n-1)}{n}$$

$$= 6\left(1 - \frac{1}{n}\right)$$

So

$$A = \lim_{n\to\infty} S_n = \lim_{n\to\infty} 6\left(1 - \frac{1}{n}\right)$$

$$= 6(1 - 0)$$

$$= 6$$

Note that this is the same area as obtained by using the formula for the area of a triangle with height $h = 6$ and base $b = 2$:

$$A = \tfrac{1}{2}bh$$

$$= \tfrac{1}{2}(2)(6)$$

$$= 6$$

To avoid the complicated summation process, let us for the moment consider only the area under the curve $y = f(x)$ from a to x in Fig. 4.5.

Note that the area increases as x increases. We can consider the area determined by x as a function of x which we will denote by $A(x)$. The difference ΔA between $A(x + \Delta x)$ and $A(x)$ is the incremental change in the area for an incremental change Δx.

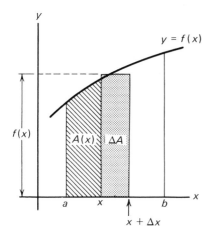

Figure 4.5   Figure 4.6

For some t between x and $x + \Delta x$ we have
$$\Delta A = f(t)\, \Delta x$$
From Fig. 4.6 we see that
$$f(x)\, \Delta x \leqslant \Delta A \leqslant f(x + \Delta x)\, \Delta x$$
Dividing by Δx,
$$f(x) \leqslant \frac{\Delta A}{\Delta x} \leqslant f(x + \Delta x)$$

We can interpret $\dfrac{\Delta A}{\Delta x}$ as the average rate of change of the area as the area increases from x to $x + \Delta x$. As $\Delta x \to 0$, $f(x + \Delta x) \to f(x)$. Since $\dfrac{\Delta A}{\Delta x}$ is squeezed between $f(x)$ and $f(x + \Delta x)$, we have $\dfrac{\Delta A}{\Delta x} \to f(x)$ as $\Delta x \to 0$. That is,
$$\lim_{\Delta x \to 0} \frac{\Delta A}{\Delta x} = f(x)$$

We can now interpret this limit. Formally, $\lim\limits_{\Delta x \to 0} \dfrac{\Delta A}{\Delta x}$ represents the instantaneous rate of change of the area $A(x)$ at x. Since $\lim\limits_{\Delta x \to 0} \dfrac{\Delta A}{\Delta x} = f(x)$, we have that
$$\frac{dA}{dx} = f(x)$$

To find $A(x)$ we integrate $\dfrac{dA}{dx} = f(x)$:

Sec. 4.3   Area Under a Curve   **157**

$$A(x) = \int f(x)\, dx = F(x) + C$$

where $F(x)$ is an antiderivative of $f(x)$. However, we need to determine the appropriate value for the constant of integration C.

Observe that $A(a) = 0$. (When $x = a$, there is no area under the curve.) We then have

$$0 = F(a) + C \quad \text{or} \quad C = -F(a)$$

The desired expression for $A(x)$ is then $A(x) = F(x) - F(a)$.

Finally, the area A under the curve, $y = f(x)$ from $x = a$ to $x = b$ is the value $A = F(b) - F(a)$. That is,

$$A = F(b) - F(a)$$
where
$$F(x) + C = \int f(x)\, dx$$

Note that the constant of integration is not involved in the formula for A. Any antiderivative $F(x)$ for $f(x)$ can be used.

Example 2

Find the area under the curve $y = x^2$ from $x = 1$ to $x = 2$.

$$\int x^2\, dx = \frac{x^3}{3} + C$$

Using the antiderivative $F(x) = x^3/3$ with $C = 0$, we have the desired area in Fig. 4.7.

$$A = F(2) - F(1) = \frac{(2)^3}{3} - \frac{(1)^3}{3} = \frac{8}{3} - \frac{1}{3} = \frac{7}{3}$$

Example 3

Find the area under the curve $y = 3x$ from $x = 0$ to $x = 2$ as in Fig. 4.8.

Note: We have already computed this area by the summation method in Example 1.

$$\int 3x\, dx = \frac{3x^2}{2} + C$$

Using the antiderivative $F(x) = \frac{3x^2}{2}$, we have

$$A = F(2) - F(0) = \frac{3(2)^2}{2} - \frac{3(0)^2}{2} = 6 - 0 = 6$$

Example 4

Find the area bounded by $y = 2 + x - x^2$ and the x-axis.

First, graph the equation by finding the x-intercepts (see Fig. 4.9):

$$y = 2 + x - x^2 = 0$$
$$(2 - x)(1 + x) = 0$$
$$x = 2, -1$$

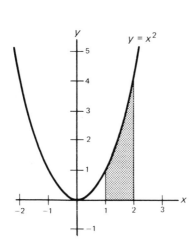

Figure 4.7

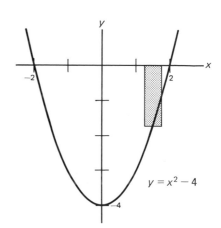

Figure 4.8

Then

$$\int (2 + x - x^2)\, dx = 2x + \frac{x^2}{2} - \frac{x^3}{3} + C$$

and

$$A = F(2) - F(-1) = \left[2(2) + \frac{2^2}{2} - \frac{2^3}{3} \right] - \left[2(-1) + \frac{(-1)^2}{2} - \frac{(-1)^3}{3} \right]$$

$$= \left[4 + 2 - \frac{8}{3} \right] - \left[-2 + \frac{1}{2} + \frac{1}{3} \right] = 4\frac{1}{2}$$

Example 5

Find the area bounded by $y = x^2 - 4$ and the x-axis.

First, graph the equation (see Fig. 4.10).

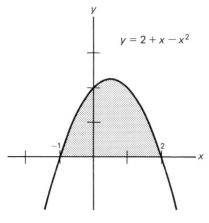

Figure 4.9

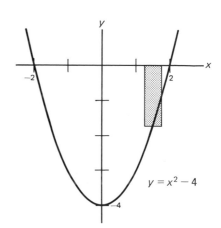

Figure 4.10

Sec. 4.3　Area Under a Curve

159

Then

$$\int (x^2 - 4)\, dx = \frac{x^3}{3} - 4x + C$$

and

$$A = F(2) - F(-2) = \left[\frac{(2)^3}{3} - 4(2)\right] - \left[\frac{(-2)^3}{3} - 4(-2)\right]$$

$$= \left[\frac{8}{3} - 8\right] - \left[-\frac{8}{3} + 8\right]$$

$$= -\frac{32}{3}$$

Note: An area below the x-axis consists of approximating rectangles whose heights, $f(x) < 0$, and whose widths, $\Delta x > 0$, yield negative products. In such cases, the area is found by finding the absolute value of this result. The area in Example 5 is thus $\left|-\frac{32}{3}\right|$ or $\frac{32}{3}$. As we shall see in Section 5.1, complications occur when finding the area between the x-axis and a curve that crosses the x-axis.

Exercises

Find the area under each curve.

1. $y = x$ from $x = 0$ to $x = 2$
2. $y = 5x$ from $x = 1$ to $x = 4$
3. $y = 2x^2$ from $x = 1$ to $x = 3$
4. $y = x^2 + 3$ from $x = 0$ to $x = 2$
5. $y = 3x^2 - 2x$ from $x = 1$ to $x = 2$
6. $y = 4 - 3x^2$ from $x = 0$ to $x = 1$
7. $y = \frac{3}{x^2}$ from $x = 1$ to $x = 2$
8. $y = \frac{1}{x^3}$ from $x = 2$ to $x = 3$
9. $y = \sqrt{3x - 2}$ from $x = 1$ to $x = 2$
10. $y = \sqrt{2x + 3}$ from $x = 0$ to $x = 3$
11. $y = 4x - x^3$ from $x = 0$ to $x = 2$
12. $y = x^3 - 16x$ from $x = -4$ to $x = 0$
13. $y = 1 - x^4$ from $x = -1$ to $x = 1$
14. $y = 16 - x^4$ from $x = -2$ to $x = 2$
15. $y = \sqrt{2x + 1}$, $x = 4$ to $x = 12$
16. $y = \sqrt{4 - 3x}$ bounded by the coordinate axes
17. $y = \sqrt[3]{x - 1}$ bounded by the coordinate axes
18. $y = \sqrt[3]{8x + 27}$ bounded by the coordinate axes
19. $y = 1/x^2$ from $x = 1$ to $x = 5$
20. $y = x^4$ from $x = 1$ to $x = 3$

21. $y = \frac{1}{(2x - 1)^2}$ from $x = -3$ to the y-axis

22. $y = \frac{1}{(2x + 3)^3}$ from $x = 3$ to the y-axis

Find the area bounded by each curve and the x-axis.

23. $y = 9 - x^2$
24. $y = 12 - 3x^2$
25. $y = 2x - x^2$
26. $y = 5x - 2x^2$
27. $y = x^2 - x^3$
28. $y = 4x^2 - x^3$
29. $y = x^2 - x^4$
30. $y = 9x^2 - x^4$

4.4 THE DEFINITE INTEGRAL

In the preceding section, we first found the area A under the curve $y = f(x)$ from $x = a$ to $x = b$ by the summation method, that is;

$$A = \lim_{n \to \infty} S_n$$

where $S_n = f(t_1) \, \Delta x + f(t_2) \, \Delta x + \cdots + f(t_n) \, \Delta x$, where Δx is the width of each rectangle used to approximate the area under the curve, and $f(t_k)$ is the height of the kth rectangle where t_k is some point along its base.

We then found that we could determine this same area A using integration, that is:

$$A = F(b) - F(a)$$

where

$$\int f(x) \, dx = F(x) + C \qquad [F'(x) = f(x)]$$

so that

$$A = \lim_{n \to \infty} S_n = F(b) - F(a)$$

Thus the summation process is a special integration process. In fact, the elongated symbol $\int$ indicates a type of summation. The summation process can be applied to situations not involvng areas. In fact, with few restrictions on the function $y = f(x)$ we can ask whether $\lim_{n \to \infty} S_n$ exists when we consider values of x between a and b. If the limit does exist (unlike our examples in Section 4.3, it need not exist), then we make the following definition: the *definite integral* of a function $y = f(x)$ from $x = a$ to $x = b$ is the number $\lim_{n \to \infty} S_n$. We symbolize this number by

$$\int_a^b f(x) \, dx$$

That is,

$$\int_a^b f(x) \, dx = \lim_{n \to \infty} S_n$$

Note: The definite integral $\int_a^b f(x) \, dx$ is a number and is not to be confused with the indefinite integral $\int f(x) \, dx$, which is a family of functions. The number a is called the *lower limit* of the integral and b is called the *upper limit*.

A fundamental result of the calculus (as found in Section 4.3) states that:

Definite Integral

$$\int_a^b f(x) \, dx = F(b) - F(a)$$

where

$$F'(x) = f(x)$$

Thus the technique for evaluating a definite integral is the same as for finding the area under a curve.

Example 1

Evaluate the integral $\int_1^3 x^3 \, dx$.

Since $\int x^3 \, dx = \dfrac{x^4}{4} + C$, then $F(x) = \dfrac{x^4}{4}$, and

$$\int_1^3 x^3 \, dx = F(3) - F(1)$$

$$= \frac{(3)^4}{4} - \frac{(1)^4}{4}$$

$$= \tfrac{81}{4} - \tfrac{1}{4}$$

$$= 20$$

We now introduce a shorthand notation for evaluating $F(b) - F(a)$:

$$F(x)|_a^b = F(b) - F(a)$$

Remember that $F(x)$ can be any antiderivative of a given function $f(x)$ since $[F(b) + C] - [F(a) + C] = F(b) - F(a)$.

Example 2

Evaluate $\int_1^2 (x^2 + 3x) \, dx$.

Since $\int (x^2 + 3x) \, dx = \dfrac{x^3}{3} + \dfrac{3x^2}{2} + C$, we have

$$\int_1^2 (x^2 + 3x) \, dx = \frac{x^3}{3} + \frac{3x^2}{2} \Big|_1^2$$

$$= \left[\frac{(2)^3}{3} + \frac{3(2)^2}{2} \right] - \left[\frac{(1)^3}{3} + \frac{3(1)^2}{2} \right]$$

$$= (\tfrac{8}{3} + \tfrac{12}{2}) - (\tfrac{1}{3} + \tfrac{3}{2})$$

$$= \tfrac{41}{6}$$

Example 3

Evaluate $\int_1^3 \dfrac{dx}{x^3}$.

$$\int_1^3 \frac{dx}{x^3} = \int_1^3 x^{-3} \, dx$$

$$= \frac{x^{-3+1}}{-3 + 1} \Big|_1^3 = -\frac{1}{2x^2} \Big|_1^3$$

$$= \left[-\frac{1}{2(3)^2} \right] - \left[-\frac{1}{2(1)^2} \right]$$

$$= -\tfrac{1}{18} + \tfrac{1}{2}$$

$$= \tfrac{8}{18} = \tfrac{4}{9}$$

Example 4

Evaluate $\int_{-2}^1 x^2 \, dx$.

$$\int_{-2}^{1} x^2 \, dx = \left. \frac{x^3}{3} \right|_{-2}^{1}$$

$$= \frac{(1)^3}{3} - \frac{(-2)^3}{3}$$

$$= \tfrac{1}{3} + \tfrac{8}{3} = 3$$

Example 5

Evaluate $\displaystyle\int_{1}^{0} x(x^2 + 1)^2 \, dx$.

Let $u = x^2 + 1$, then

$$du = \frac{d}{dx}(x^2 + 1) \, dx = 2x \, dx$$

So

$$\int x(x^2 + 1)^2 \, dx = \int u^2 \frac{du}{2}$$

$$\boxed{\begin{array}{l} u = x^2 + 1 \\ du = 2x \, dx \end{array}}$$

$$= \frac{1}{2} \cdot \frac{u^3}{3} + C$$

$$= \frac{u^3}{6} + C = \frac{(x^2 + 1)^3}{6} + C$$

Then

$$\int_{1}^{0} x(x^2 + 1)^2 \, dx = \left. \frac{(x^2 + 1)^3}{6} \right|_{1}^{0}$$

$$= \left[\frac{(0 + 1)^3}{6} \right] - \left[\frac{(1 + 1)^3}{6} \right]$$

$$= \tfrac{1}{6} - \tfrac{8}{6}$$

$$= -\tfrac{7}{6}$$

Note that in Example 5, the definite integral is negative. There are no restrictions on the value of a definite integral. The choice of upper and lower limits depends upon the application of the definite integral. One can show, however, that

$$\boxed{\int_{a}^{b} f(x) \, dx = -\int_{b}^{a} f(x) \, dx}$$

Example 6

Evaluate

$$\int_{0}^{2} \frac{2x + 3}{(x^2 + 3x - 7)^2} \, dx$$

$$\int \frac{2x + 3}{(x^2 + 3x - 7)^2} \, dx = \int \frac{du}{u^2}$$

$$\boxed{\begin{array}{l} u = x^2 + 3x - 7 \\ du = (2x + 3) \, dx \end{array}}$$

$$= \int u^{-2} \, du$$

$$= -u^{-1} + C$$

$$= -\frac{1}{x^2 + 3x - 7} + C$$

So

$$\int_0^2 \frac{2x + 3}{(x^2 + 3x - 7)^2} \, dx = -\frac{1}{x^2 + 3x - 7}\Big|_0^2$$

$$= \left(-\frac{1}{4 + 6 - 7}\right) - \left(-\frac{1}{-7}\right)$$

$$= -\tfrac{1}{3} - \tfrac{1}{7}$$

$$= -\tfrac{10}{21}$$

Exercises

Evaluate each definite integral.

1. $\displaystyle\int_0^1 5x \, dx$

2. $\displaystyle\int_0^1 x^4 \, dx$

3. $\displaystyle\int_1^2 (x^2 + 3) \, dx$

4. $\displaystyle\int_2^4 (3x^2 + x - 1) \, dx$

5. $\displaystyle\int_2^0 (x^3 + 1) \, dx$

6. $\displaystyle\int_1^0 (3x^4 - 8) \, dx$

7. $\displaystyle\int_{-1}^1 (x^2 + x + 2) \, dx$

8. $\displaystyle\int_{-2}^0 (2x^3 - 4x) \, dx$

9. $\displaystyle\int_4^9 (3x^{1/2} + x^{-1/2}) \, dx$

10. $\displaystyle\int_8^{27} (5x^{2/3} + 4x^{-1/3}) \, dx$

11. $\displaystyle\int_1^9 \frac{x + 3}{\sqrt{x}} \, dx$

12. $\displaystyle\int_1^4 \frac{x^2 + 3x - 2}{\sqrt{x}} \, dx$

13. $\displaystyle\int_1^2 (3x + 4)^4 \, dx$

14. $\displaystyle\int_0^1 (2x + 1)^5 \, dx$

15. $\displaystyle\int_0^{16} \sqrt{2x + 4} \, dx$

16. $\displaystyle\int_1^{44} \sqrt[3]{5x - 4} \, dx$

17. $\displaystyle\int_1^2 4x(x^2 - 3)^3 \, dx$

18. $\displaystyle\int_0^1 6x^2(x^3 + 1)^4 \, dx$

19. $\displaystyle\int_{-1}^0 x(1 - x^2)^{2/3} \, dx$

20. $\displaystyle\int_{-2}^0 x\sqrt{4 - x^2} \, dx$

21. $\displaystyle\int_0^1 x\sqrt{x^2 + 1} \, dx$

22. $\displaystyle\int_0^2 x^2\sqrt{x^3 + 2} \, dx$

23. $\displaystyle\int_0^4 \frac{6x}{\sqrt{x^2 + 9}} \, dx$

24. $\displaystyle\int_{-1}^1 \frac{16x \, dx}{\sqrt[3]{3x^2 + 5}}$

25. $\displaystyle\int_1^2 \frac{3x^2 + 1}{\sqrt{x^3 + x}} \, dx$

26. $\displaystyle\int_2^3 \frac{2x + 1}{\sqrt{x^2 + x - 2}} \, dx$

CHAPTER SUMMARY

1. If $f(x)$ is a function, then $F(x)$ is the antiderivative of $f(x)$ if $F'(x) = f(x)$.

2. Integration formulas:

   (a) $\displaystyle\int x^n \, dx = \frac{x^{n+1}}{n + 1} + C,\, n \neq -1$

   (b) $\displaystyle\int [f(x) + g(x)] \, dx = \int f(x) \, dx + \int g(x) \, dx$

(c) $\int k\,f(x)\,dx = k\int f(x)\,dx$, where k is a constant

(d) $\int u^n\,du = \dfrac{u^{n+1}}{n+1} + C$, where $u = f(x)$ and $n \neq -1$

3. The area under the curve $y = f(x)$ in Fig. 4.11 from $x = a$ to $x = b$ is $A = F(b) - F(a)$.

4. $\displaystyle\int_a^b f(x)\,dx = F(b) - F(a)$

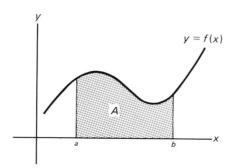

Figure 4.11

CHAPTER REVIEW

Integrate.

1. $\displaystyle\int (5x^2 - x)\,dx$

2. $\displaystyle\int (3x^7 + 2x + 4)\,dx$

3. $\displaystyle\int 6\sqrt{x^7}\,dx$

4. $\displaystyle\int 4x^{2/3}\,dx$

5. $\displaystyle\int \frac{3\,dx}{x^5}$

6. $\displaystyle\int \frac{dx}{\sqrt{x^3}}$

7. $\displaystyle\int (6x^3 + 1)(3x^4 + 2x - 1)^3\,dx$

8. $\displaystyle\int (7x + 4)(7x^2 + 8x + 2)^{3/5}\,dx$

9. $\displaystyle\int \frac{2x + 5}{\sqrt{x^2 + 5x}}\,dx$

10. $\displaystyle\int (15x^2 + 4)(5x^3 + 4x)^{-(2/3)}\,dx$

11. Find the equation of the curve $y = f(x)$ passing through $(1, -3)$ whose slope is given by $\dfrac{dy}{dx} = 3x^2$.

12. A stone is thrown vertically upward from a cliff 100 ft high with an initial velocity of 25 ft/s. Find the equation describing the altitude of the stone from the ground $(a = -32 \text{ ft/s}^2)$.

13. The rate of change of resistance with respect to temperature of an electrical resistor is given by $\dfrac{dR}{dT} = 0.009T^2 + 0.02T - 0.7$. Find the resistance when the temperature is 30°C if $R = 0.2\ \Omega$ when $T = 0°$ C.

14. The current in an electrical circuit is given by $i = \dfrac{3t^2 + 1}{\sqrt{t^3 + t + 2}}$ amp. Find the charge (in coulombs) that passes a given point in the circuit after 0.2 s. (Assume $q = 2\sqrt{2}$ at $t = 0$.)

Find the area under each curve.

15. $y = x^2 + 1$ from $x = 0$ to $x = 2$

16. $y = 8 - 6x^2$ from $x = 0$ to $x = 1$

17. $y = \dfrac{1}{x^5}$ from $x = 1$ to $x = 2$

18. $y = \dfrac{4x}{(x^2 + 1)^2}$ from $x = 0$ to $x = 1$

19. $y = \sqrt{5x + 6}$ from $x = 0$ to $x = 6$

20. $y = x\sqrt{x^2 + 4}$ from $x = 0$ to $x = 2$

Evaluate each definite integral.

21. $\displaystyle\int_0^1 (x^3 + 2x^2 + x)\, dx$

22. $\displaystyle\int_1^2 (3x^4 - 2x^3 + 7x)\, dx$

23. $\displaystyle\int_0^2 x(x^2 + 1)^2\, dx$

24. $\displaystyle\int_1^2 \dfrac{5x^2}{(x^3 + 2)^2}\, dx$

25. $\displaystyle\int_1^2 (2x + 1)\sqrt{x^2 + x}\, dx$

26. $\displaystyle\int_0^3 \dfrac{x^2}{\sqrt{x^3 + 1}}\, dx$

27. $\displaystyle\int_2^1 \left(3x^2 - x + \dfrac{1}{x^2}\right) dx$

28. $\displaystyle\int_0^{1/2} \dfrac{3x}{\sqrt{2x^2 + \frac{1}{2}}}\, dx$

The Integral   Chap. 4

APPLICATIONS
OF INTEGRATION

5.1 AREA BETWEEN CURVES

In Chapter 4 we studied the process of computing the area between a given curve and the x-axis. Let us now consider the problem of determining the area *between* two given curves. First, observe that the area between one given curve $y = f(x)$ and the x-axis between $x = a$ and $x = b$ is a definite integral:

$$A = \int_a^b f(x)\, dx = F(b) - F(a)$$

where $F'(x) = f(x)$.

Further recall that definite integration as applied to area is a summation process. That is, the definite integral is the limit of sums of approximating rectangles where the area of a typical rectangle is $f(t)\, \Delta x$ as shown in Fig. 5.1. The typical rectangle, which is shaded, is called an *element* of the area. Note the correspondence between the form of the expression for the area of the element, $f(t)\, \Delta x$, and for the differential, $f(x)\, dx$, which is to be integrated:

$$f(t)\, \Delta x \longleftrightarrow f(x)\, dx$$

Noting this correspondence will be a visual aid in setting up the appropriate integrals for computing areas between curves. The correct form for $f(x)\, dx$ can be found by determining the appropriate expression $f(t)\, \Delta x$ from viewing a sketch of the area.

Now let's determine the area bounded by the curves $y = f(x)$, $y = g(x)$, $x = a$, and $x = b$ as shown in Fig. 5.2.

167

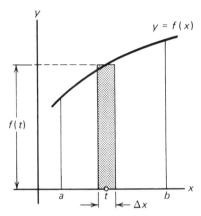

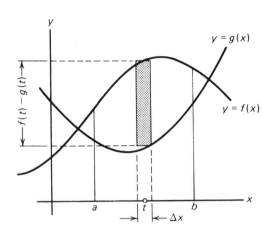

Figure 5.1   Figure 5.2

Between a and b, $g(x) \leqslant f(x)$; that is, the curve $y = g(x)$ lies below the curve $y = f(x)$. The area of the element shown used in approximating the area between the two curves is

$$[f(t) - g(t)] \, \Delta x$$

This corresponds to the differential

$$[f(x) - g(x)] \, dx.$$

We therefore use the definite integral

$$\int_a^b [f(x) - g(x)] \, dx$$

to determine the desired area.

Example 1

Find the area between the curves $y = 4 - x^2$ and $y = x + 2$.

The area is bounded by a parabola and a straight line. First, find the points where the two curves intersect by solving the two equations simultaneously.

$$4 - x^2 = x + 2$$
$$0 = x^2 + x - 2$$
$$0 = (x - 1)(x + 2)$$

or

$$x = 1, \, x = -2$$

The curves intersect at the points $(-2, 0)$ and $(1, 3)$. Observe that between -2 and 1, the line $y = x + 2$ is below the parabola $y = 4 - x^2$. The length of the element shown in Fig. 5.3 is the difference between the upper curve $y = 4 - x^2$ and the lower curve $y = x + 2$ at a given value of x.

The area A between the curves is the value of the definite integral.

$$A = \int_{-2}^{1} [(4 - x^2) - (x + 2)] \, dx$$
$$= \int_{-2}^{1} (2 - x - x^2) \, dx$$

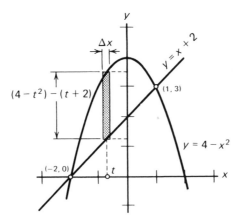

Figure 5.3

$$= 2x - \frac{x^2}{2} - \frac{x^3}{3}\Big|_{-2}^{1}$$

$$= (2 - \tfrac{1}{2} - \tfrac{1}{3}) - (-4 - 2 + \tfrac{8}{3})$$

$$= \tfrac{9}{2}$$

The limits of integration are determined by the points where the curves intersect. The lower limit should be the smallest value of x where the curves intersect, and the upper limit should be the largest value of x where the curves intersect.

Example 2

Find the area between the line $y = -x + 2$ and the parabola $x = 4 - y^2$.

The points of intersection of these two curves are $(0, 2)$ and $(3, -1)$. In this example, we run into a problem. If we use vertical elements to approximate the area as in Fig. 5.4, then between $x = 0$ and $x = 3$ the height of the element is the difference between the curve $x = 4 - y^2$ ($y = \sqrt{4 - x}$) and $y = -x + 2$. And between $x = 3$ and $x = 4$ the height of the element is the difference between $y = \sqrt{4 - x}$ and $y = -\sqrt{4 - x}$. Because of the change in boundaries at $x = 3$, we would have to find the desired area A by separately computing the areas

$$A_1 = \int_0^3 [(\sqrt{4 - x}) - (-x + 2)]\, dx$$

and

$$A_2 = \int_3^4 [(\sqrt{4 - x}) - (-\sqrt{4 - x})]\, dx$$

Then $A = A_1 + A_2$. By integrating, we find $A_1 = \tfrac{19}{6}$ and $A_2 = \tfrac{4}{3}$. So

$$A = \tfrac{19}{6} + \tfrac{4}{3} = \tfrac{9}{2}$$

Sometimes, as in this example, it is easier to set up the problem using horizontal elements as in Fig. 5.5. Then express the given curves as functions of y instead of x: $x = 4 - y^2$ and $x = -y + 2$, and integrate with respect to the y variable and use limits of integration based on the y-coordinates of the points of intersection: $y = -1$ and $y = 2$. With respect to the x-axis, the curve $x = 4 - y^2$ lies above

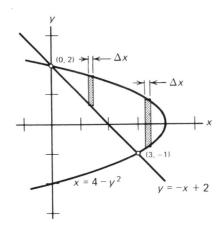

Figure 5.4

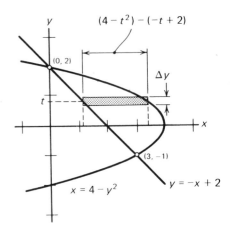

Figure 5.5

$x = -y + 2$ from $y = -1$ to $y = 2$. The length of a typical horizontal element is then $(4 - y^2) - (-y + 2)$.

$$A = \int_{-1}^{2} [(4 - y^2) - (-y + 2)] \, dy$$

$$= \int_{-1}^{2} (2 + y - y^2) \, dy$$

$$= 2y + \frac{y^2}{2} - \frac{y^3}{3} \Big|_{-1}^{2}$$

$$= (4 + 2 - \tfrac{8}{3}) - (-2 + \tfrac{1}{2} + \tfrac{1}{3})$$

$$= \tfrac{9}{2}$$

Note: Using horizontal elements and working in terms of dy is usually simpler than using vertical elements and working in terms of dx when one of the curves does not represent a function (contains a y^2-term).

In summary, to find the area between two given curves between $x = a$ and $x = b$:

1. Find the points of intersection of the two curves, if necessary.
2. Sketch the two curves:
   (a) Determine whether to use vertical elements with the curves expressed as functions of x or horizontal elements with curves expressed as functions of y.
   (b) Determine which curve lies above the other.
3. Determine the height of a typical element based in step 2b.
4. Write the definite integral:

$$\boxed{\int_a^b [f(x) - g(x)] \, dx}$$

Applications of Integration   Chap. 5

where $f(x) - g(x)$ is the length of vertical elements between $x = a$ and $x = b$ with $a < b$, or

$$\int_c^d [f(y) - g(y)]\, dy$$

where $f(y) - g(y)$ is the length of horizontal elements between $y = c$ and $y = d$ with $c < d$.

Note that the choice of using vertical or horizontal elements depends upon the difficulty of the resulting definite integral. Also, the curves do not need to lie above the x- or y-axis in order to determine the area between them.

Example 3

Find the area between the curves $x = y^3$ and $x = -y^2$.

The points of intersection are $(0, 0)$ and $(-1, -1)$. We could use either vertical or horizontal elements. We will use horizontal elements as in Fig. 5.6. Since the curve $x = y^3$ lies above $x = -y^2$ (in the positive direction along the x-axis) from $y = -1$ to $y = 0$, we have

$$A = \int_{-1}^{0} [y^3 - (-y^2)]\, dy$$

$$= \frac{y^4}{4} + \frac{y^3}{3}\Big|_{-1}^{0}$$

$$= 0 - (\tfrac{1}{4} - \tfrac{1}{3})$$

$$= \tfrac{1}{12}$$

Example 4

Find the area between the curves $y = x$ and $y = x^3$.

In this example there are three points of intersection $(-1, -1)$, $(0, 0)$ and

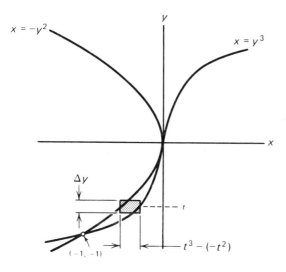

Figure 5.6

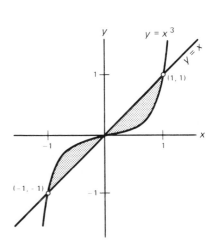

Figure 5.7

(1, 1). Between $x = -1$ and $x = 0$, $y = x^3$ lies above $y = x$. Between $x = 0$ and $x = 1$, $y = x$ lies above $y = x^3$. In this case we need to compute the two areas as in Fig. 5.7.

$$A_1 = \int_{-1}^{0} (x^3 - x)\, dx \qquad \text{and} \qquad A_2 = \int_{0}^{1} (x - x^3)\, dx$$

$$A_1 = \frac{x^4}{4} - \frac{x^2}{2} \Big|_{-1}^{0} = \frac{1}{4} \qquad \text{and} \qquad A_2 = \frac{x^2}{2} - \frac{x^4}{4} \Big|_{0}^{1} = \frac{1}{4}$$

The desired area $A = A_1 + A_2 = \frac{1}{4} + \frac{1}{4} = \frac{1}{2}$.

We could have also observed from the symmetry of the two areas that $A_1 = A_2$, so that we could also compute

$$A = 2 \int_{0}^{1} (x - x^3)\, dx$$

Example 5

Find the area between the x-axis and $y = x^2 - 4$.

The desired region lies between $x = -2$ and $x = 2$. Note that in this region the curve $y = x^2 - 4$ lies below the x-axis in Fig. 5.8. Since the x-axis is the curve $y = 0$, we have

$$A = \int_{-2}^{2} [0 - (x^2 - 4)]\, dx$$

$$= \int_{-2}^{2} (4 - x^2)\, dx$$

$$= \left(4x - \frac{x^3}{3}\right)\Big|_{-2}^{2}$$

$$= \tfrac{32}{3}$$

Example 6

Find the area between the x-axis and $y = x^3$ from $x = -1$ to $x = 1$.

In this example, there may be the temptation to simply form the integral $\int_{-1}^{1} x^3\, dx$. But

$$\int_{-1}^{1} x^3\, dx = \frac{x^4}{4}\Big|_{-1}^{1} = \frac{1}{4} - \frac{1}{4} = 0$$

We have incorrectly obtained the value zero since we failed to observe that at $x = 0$ the curve $y = x^3$ crosses the x-axis as in Fig. 5.9. To the left of $x = 0$ the curve is below the x-axis, but to the right of $x = 0$ the curve is above the x-axis. We must therefore separate the computation into two integrals as in Example 4:

$$A = \int_{-1}^{0} [0 - (x^3)]\, dx + \int_{0}^{1} [(x^3) - 0]\, dx$$

$$= -\int_{-1}^{0} x^3\, dx + \int_{0}^{1} x^3\, dx$$

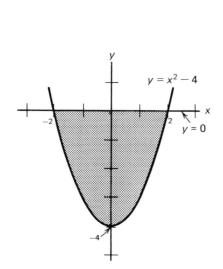

Figure 5.8

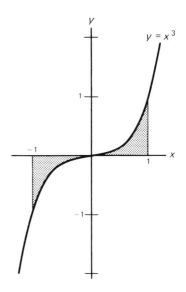

Figure 5.9

$$= -\left(\frac{x^4}{4}\right)\Bigg|_{-1}^{0} + \frac{x^4}{4}\Bigg|_{0}^{1}$$

$$= -(-\tfrac{1}{4}) + \tfrac{1}{4}$$

$$= \tfrac{1}{2}$$

Exercises

Find each area bounded by the curves.

1. $y = x^2$, $y = 0$, and $x = 1$
2. $y = x^2$, $y = 0$, $x = 1$, and $x = 2$
3. $y = 1 - x$, $x = 0$, and $y = 0$
4. $y = 2x$, $y = 0$, $x = 1$, and $x = 2$
5. $y = 2 - x^2$ and $y + x = 0$
6. $y = x^2 - 2x$ and $y = 3$
7. $y^2 = x$ and $x = 4$
8. $y^2 = 4x$, $x = 0$, $y = -1$, and $y = 4$
9. $y = x^2$ and $y = x$
10. $y = 2x^2$ and $y^2 = 4x$
11. $x = y^2 - 2y$ and $y = x$
12. $y = x^3$, $y = 2 - x^2$, $x = 0$, and $x = 1$
13. $y = x^3 - x$, $y = 0$, $x = -1$, and $x = 1$
14. $y^3 = x$, $y = 1$, and $x = -1$
15. $x = y + 1$ and $x = 3 - y^2$
16. $x = y^2$ and $x = y + 2$
17. $y = 4 - 4x^2$ and $y = 1 - x^2$
18. $y = x^2$ and $y = 8 - x^2$
19. $x = y^4$ and $x = 2 - y^2$
20. $x = y^2$ and $x = 4 + 2y - y^2$
21. $y = x^2 - 3x - 4$ and $y = 6$
22. $y = x^2$ and $y = \sqrt{x}$
23. $x^2 y = 8$, $y = x$, $x = 5$, and $y = 0$
24. $y = x$, $x + y = 6$, and $2y = x$
25. $y = x(x - 1)(x - 3)$ and the x-axis
26. $y = x(x + 3)(x - 2)$ and the x-axis

5.2 VOLUMES OF REVOLUTION: DISK METHOD

Another application of integration is finding the volume of a solid resulting from rotating an area about an axis. For example, consider the region bounded by the curves $y = f(x)$, $x = a$, $x = b$, and the x-axis. Revolving this region about the x-axis determines a solid figure as in Fig. 5.10

The area ΔA of a typical rectangle used in Chapter 4 to approximate the area under the curve $y = f(x)$ was $\Delta A = f(t) \Delta x$ where t is a point at the base of the rectangle, Δx is the width, and $f(t)$ is the height. If we now rotate this area about the x-axis as in Fig. 5.11, we obtain a cylindrical disk with volume

$$\Delta V = \pi r^2 h$$

where r, the radius, is $f(t)$ and h, the width, is Δx. So

$$\Delta V = \pi [f(t)]^2 \Delta x$$

By a method similar to approximating the area under a curve by rectangles, we can approximate the volume of revolution by using the sum of the volumes of approximating or differential disks. (See Fig. 5.12.)

For areas, the integral

$$A = \int_a^b f(x)\, dx$$

was found to give the exact area for regions which were approximated by the area of rectangles:

$$\Delta A = f(t) \Delta x$$

In a similar manner, the integral

$$V = \pi \int_a^b [f(x)]^2\, dx = \pi \int_a^b y^2\, dx$$

is found to give the exact volume for the solid of revolution about the x-axis which would be approximated by the volume of disks: $\Delta V = \pi [f(t)]^2 \Delta x$. This method of computing the volume of a solid is called the *disk method*.

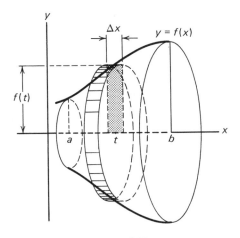

Figure 5.10

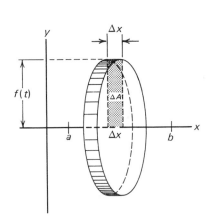

Figure 5.11

174

Applications of Integration   Chap. 5

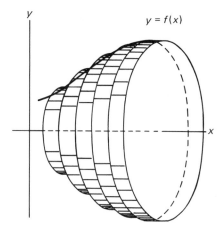

$y = f(x)$

Figure 5.12

Circular Disk Method

V = sum of circular disks

          radius   thickness
             ↓   ↓

$V = \pi \displaystyle\int_a^b [f(x)]^2 \quad dx \qquad$ (revolved about x-axis)

$V = \pi \displaystyle\int_c^d [f(y)]^2 \quad dy \qquad$ (revolved about y-axis)

Example 1

Find the volume of the solid formed by revolving the curve $y = x$ from $x = 0$ to $x = 2$ about the x-axis.

The area being rotated about the x-axis is a triangle as in Fig. 5.13. The resulting solid is a cone. The volume of a typical differential disk in Fig. 5.14 is

$$\Delta V = \pi [f(t)]^2 \,\Delta x = \pi (t)^2 \,\Delta x$$

So the integral giving the exact volume is

$$V = \pi \int_a^b y^2 \, dx = \pi \int_0^2 x^2 \, dx, \qquad (y = x)$$

$$V = \pi \left[\frac{x^3}{3} \Big|_0^2 \right]$$

$$= \pi [\tfrac{8}{3} - \tfrac{0}{3}]$$

$$= \frac{8\pi}{3}$$

Note that this is the same volume found by using the formula for the volume of a cone from geometry, $V = \tfrac{1}{3}\pi r^2 h$. In this example, $r = 2$ (radius of the base of the cone) and $h = 2$ (the altitude). So

$$V = \frac{1}{3}\pi(2)^2(2) = \frac{8\pi}{3}$$

Although the problem above could also be solved using a geometrical formula, this is not always the case. In the following example integration provides the only solution.

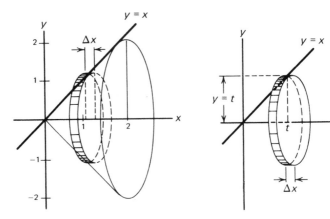

Figure 5.13   **Figure 5.14**

Example 2

Find the volume of the solid obtained by revolving the region bounded by the curves $y = x^2$, $y = 0$, and $x = 2$ about the x-axis.

The volume of a differential disk in Fig. 5.15 is

$$\Delta V = \pi[f(t)]^2 \, \Delta x = \pi[t^2]^2 \, \Delta x$$

The exact volume is then

$$V = \pi \int_a^b y^2 \, dx = \pi \int_0^2 (x^2)^2 \, dx, \qquad (y = x^2)$$

$$= \pi \int_0^2 x^4 \, dx$$

$$= \pi \frac{x^5}{5} \Big|_0^2$$

$$= \frac{32\pi}{5}$$

When a solid is formed by revolving an area about the y-axis, the integral giving the volume becomes

$$V = \pi \int_c^d [f(y)]^2 \, dy = \pi \int_c^d x^2 \, dy$$

This may be seen by observing that the radius of a differential disk is now expressed as a distance x from the y-axis and the width of the disk is an increment of y, Δy. (See Fig. 5.16.)

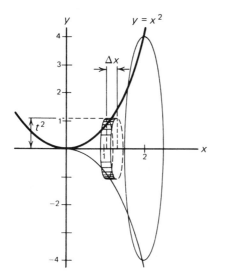

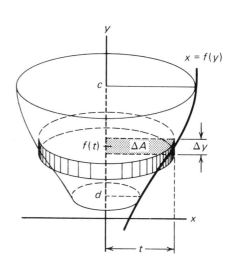

Figure 5.15

Figure 5.16

Example 3

Find the volume of the solid obtained by revolving the region bounded by the curves $y = x^2$, $x = 0$, and $y = 4$ about the y-axis.

When revolving the area about the y-axis as in Fig. 5.17, the boundary curve $(y = x^2)$ must be determined in a manner which will express x as a function of y. That is,

$$x = \sqrt{y} \quad \text{(for } 0 \leq y \leq 4)$$

Note that the boundary curve determines the radius of the differential disks.
The volume of the solid is then given by

$$V = \pi \int_c^d x^2 \, dy = \pi \int_0^4 (\sqrt{y})^2 \, dy, \qquad (x = \sqrt{y})$$

$$= \pi \int_0^4 y \, dy$$

$$= \pi \frac{y^2}{2} \Big|_0^4 = \pi \left(\frac{16}{2} - 0 \right)$$

$$= 8\pi$$

Example 4

Find the volume of the solid formed by revolving the region bounded by the curves $y = 4 - x^2$, $x = 0$, and $y = 0$ about the y-axis.

Expressing the radius of a differential disk in terms of x, we have

$$y = 4 - x^2$$

$$x^2 = 4 - y$$

$$x = \sqrt{4 - y}$$

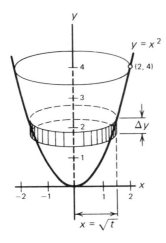

Figure 5.17

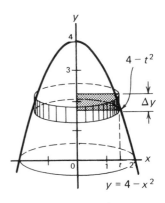

Figure 5.18

The desired volume in Fig. 5.18 is then

$$V = \pi \int_c^d x^2 \, dy = \pi \int_0^4 (\sqrt{4 - y})^2 \, dy, \qquad (x = \sqrt{4 - y})$$

$$= \pi \int_0^4 (4 - y) \, dy$$

$$= \pi \left(4y - \frac{y^2}{2} \right) \Big|_0^4$$

$$= \pi[(16 - \tfrac{16}{2}) - (0)]$$

$$= 8\pi$$

When the region between two curves is rotated as shown in Fig. 5.19 about the x-axis, the rotation results in a *washer*-type solid. If $y = f(x)$ is the outer radius and $y = g(x)$ is the inner radius of the region being revolved, the volume of the resulting washer is

$$V = \pi \int_a^b \{[f(x)]^2 - [g(x)]^2\} \, dx$$

Example 5

Find the volume of the solid formed by revolving the region bounded by $y = x^2$ and $y = x$ about the x-axis.

From Fig. 5.20, we have

$$V = \pi \int_0^1 [(x)^2 - (x^2)^2] \, dx$$

$$= \pi \int_0^1 (x^2 - x^4) \, dx$$

$$= \pi \left[\frac{x^3}{3} - \frac{x^5}{5} \right] \Big|_0^1$$

Applications of Integration   Chap. 5

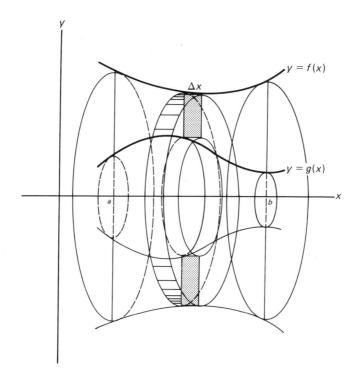

Figure 5.19

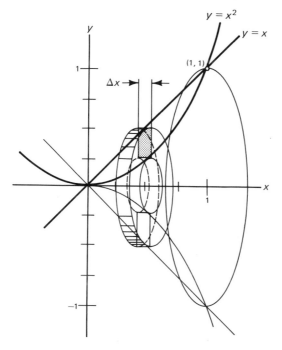

Figure 5.20

$$= \pi[(\tfrac{1}{3} - \tfrac{1}{5}) - (0)]$$

$$= \frac{2\pi}{15}$$

Example 6

Drill a hole of radius 3 in. through the center of a metal sphere of radius 5 in. Find the volume of the resulting ring.

First, rotate the shaded portion of the circle $x^2 + y^2 = 25$ about the x-axis as shown in Fig. 5.21. The outer radius is $y = \sqrt{25 - x^2}$ while its inner radius (radius of hole) is $y = 3$. The result is shown below.

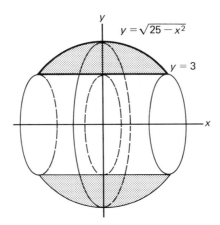

Figure 5.21

Thus

$$V = \pi \int_{-4}^{4} [(\sqrt{25 - x^2})^2 - (3)^2] \, dx$$

$$= \pi \int_{-4}^{4} (16 - x^2) \, dx$$

$$= \pi \left(16x - \frac{x^3}{3} \right) \Big|_{-4}^{4}$$

$$= \pi[(64 - \tfrac{64}{3}) - (-64 + \tfrac{64}{3})]$$

$$= \frac{256\pi}{3} \text{ in}^3$$

Example 7

Find the volume of the solid obtained by revolving the region bounded by the curves $y = x^2$, $x = 2$, and the x-axis about the line $x = 2$.

Note that the radius of the differential disk in Fig. 5.22 is $2 - x$ and its thickness is Δy. And $0 \leq y \leq 4$.

$$V = \pi \int_{0}^{4} (2 - x)^2 \, dy$$

$$= \pi \int_{0}^{4} (4 - 4x + x^2) \, dy$$

$$= \pi \int_0^4 (4 - 4\sqrt{y} + y)\, dy \qquad (x^2 = y \quad \text{and} \quad x = \sqrt{y})$$

$$= \pi \left[4y - \frac{8}{3} y^{3/2} + \frac{y^2}{2} \right] \Bigg|_0^4$$

$$= \pi[(16 - \tfrac{8}{3} \cdot 8 + 8) - (0)]$$

$$= \frac{8\pi}{3}$$

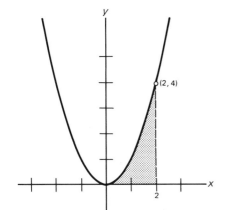
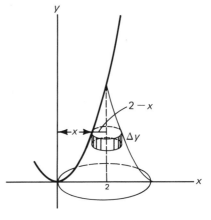

Figure 5.22

Exercises

Find the volume of each solid formed by revolving the region bounded by the given curves about the given line.

1. $y = x + 1$, $y = 0$, $x = 0$, and $x = 2$ about x-axis
2. $y = x$, $y = 0$, $x = 2$, and $x = 4$ about x-axis
3. $y = x^2 + 1$, $y = 0$, $x = 1$, and $x = 2$ about x-axis
4. $y = \sqrt{x}$, $y = 0$, $x = 1$, and $x = 4$ about x-axis
5. $y = x - 1$, $x = 0$, and $y = 1$ about y-axis
6. $y^2 = 2x$, $x = 0$, and $y = 2$ about y-axis
7. $y = 4x^2$, $x = 0$, and $y = 4$ about y-axis
8. $y = 4 - x^2$, $x = 0$, $y = 1$, and $y = 2$ about y-axis
9. $y = x$, $x = 1$, and $y = 0$ about $x = 1$
10. $y = x^2$, $x = 2$, and $y = 0$ about $x = 2$
11. $y = x$, $x = 1$, $x = 2$, and $y = 1$ about $y = 1$
12. $y = x^2$, $x = 1$, $x = 2$, and $y = 1$ about $y = 1$
13. $4y = x^2$, $y = 0$, and $x = 2$ about y-axis
14. $y^2 = x$, $y = 0$ and $x = 4$ about y-axis
15. $y^2 = x$, $y = 0$ and $x = 4$ about x-axis
16. $y = x^3$ and $y = x$ from $x = 0$ to $x = 1$ about x-axis
17. $y = x^3$ and $y = x$ from $x = 0$ to $x = 1$ about y-axis

18. $y = 2 - x^2$, $y = x$ and $x = 0$ about x-axis
19. $y = 2 - x^2$, $y = x$ and $x = 0$ about y-axis
20. $2y = x$ and $y^2 = x$ about y-axis
21. $2y = x$ and $y^2 = x$ about x-axis
22. $x = y^2$ and $x = 2 - y^2$ about y-axis
23. $y = 3 - x^2$ and $y = x^2 + 1$ about x-axis
24. $x = 2y^2 + 1$ and $x = 4 - y^2$ about y-axis
25. The area bounded in the first and second quadrants of the ellipse $9x^2 + 25y^2 = 225$ is revolved about the x-axis. Find the volume.
26. The area bounded in the first and fourth quadrants of the ellipse $9x^2 + 25y^2 = 225$ is revolved about the y-axis. Find the volume.
27. Drill a hole of radius 2 in. through the center of the solid (along the x-axis) described in Exercise 25. Find the volume of the resulting solid.
28. Drill a hole of radius 2 in. through the center of the solid (along the y-axis) described in Exercise 26. Find the volume of the resulting solid.
29. Use the disk method to verify that the volume of a sphere of radius r is $V = \frac{4}{3}\pi r^3$.
30. Use the disk method to verify that the volume of a right circular cone is $V = \frac{1}{3}\pi r^2 h$, where r is the radius of the base and h is the height.

5.3 VOLUMES OF REVOLUTION: SHELL METHOD

A second method of obtaining the volume of a solid uses cylindrical shells instead of disks. Let's use this method to find the volume of the solid described in Example 4 of Section 5.2.

The volume ΔV of a typical shell in Fig. 5.23 is

$$\Delta V = \pi r_2^2 h - \pi r_1^2 h$$

where

$$r_1 = t, \qquad \text{radius of inside wall of the shell}$$

$$r_2 = t + \Delta x, \qquad \text{radius of outside wall of the shell}$$

Now

$$\Delta V = \pi r_2^2 h - \pi r_1^2 h$$
$$= \pi h (r_2^2 - r_1^2)$$
$$= \pi h (r_2 + r_1)(r_2 - r_1)$$
$$= 2\pi h \left(\frac{r_2 + r_1}{2} \right)(r_2 - r_1) \qquad \text{(Multiply numerator and denominator by 2.)}$$

or

$$\Delta V = 2\pi y \left(\frac{2t + \Delta x}{2} \right)(\Delta x) \qquad [r_2 + r_1 = (t + \Delta x) + t = 2t + \Delta x]$$
$$= 2\pi y \left(t + \frac{\Delta x}{2} \right) \Delta x$$
$$= 2\pi f(x) x \, \Delta x, \qquad \text{where we let } x = t + \frac{\Delta x}{2}$$

This expression for ΔV is the product of the circumference of the shell of radius x, its height y, and its thickness Δx. By taking the sum of the volumes of all such approximating shells as in Fig. 5.24, we obtain another approximation for the desired volume.

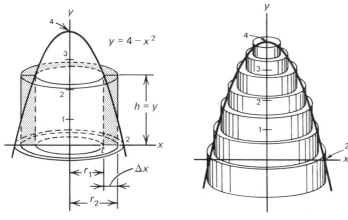

Figure 5.23   Figure 5.24

Again using the methods of Chapter 4, this approximation leads to the integral

$$V = 2\pi \int_a^b x f(x)\, dx,$$

which gives the exact volume of the solid.

Expressing y as a function of x we have $y = f(x) = 4 - x^2$ and

$$V = 2\pi \int_0^2 x(4 - x^2)\, dx$$

$$= 2\pi \int_0^2 (4x - x^3)\, dx$$

$$= 2\pi \left(2x^2 - \frac{x^4}{4} \right) \Big|_0^2 = 2\pi[(8 - 4) - (0)] = 8\pi$$

This method of computing the volume of a solid is called the *shell method*.

Cylindrical Shell Method

V = sum of concentric cylindrical shells

$$\begin{array}{cccc} & \text{radius} & \text{height} & \text{thickness} \\ & \downarrow & \downarrow & \downarrow \end{array}$$

$$V = 2\pi \int_a^b x \quad f(x) \quad dx \qquad \text{(shells parallel to } y\text{-axis)}$$

$$V = 2\pi \int_c^d y \quad f(y) \quad dy \qquad \text{(shells parallel to } x\text{-axis)}$$

Example 1

Find the volume of the solid formed by revolving the region bounded by $y = x^2$, $y = 0$, and $x = 1$ about the line $x = 1$.

In this example the shell method is more convenient than the disk method. The volume of a typical shell in Fig. 5.25 is

$$\Delta V = 2\pi(1 - x)y\,\Delta x$$

where $1 - x$ is the radius of the shell, y is the height, and Δx is its thickness. The integral for V then becomes

$$V = 2\pi \int_0^1 (1 - x)y\,dx$$

$$V = 2\pi \int_0^1 (1 - x)(x^2)\,dx \qquad (y = x^2)$$

$$= 2\pi \int_0^1 (x^2 - x^3)\,dx$$

$$= 2\pi \left(\frac{x^3}{3} - \frac{x^4}{4}\right)\Big|_0^1$$

$$= 2\pi[(\tfrac{1}{3} - \tfrac{1}{4}) - (0)]$$

$$= \frac{\pi}{6}$$

Example 2

Find the volume of the solid formed by revolving the region bounded by $y = x^2$ and $y = x$ about the x-axis.

The volume of a typical shell in Fig. 5.26 is

$$\Delta V = 2\pi(x_2 - x_1)y\,\Delta y$$

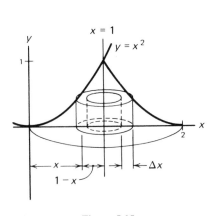

Figure 5.25

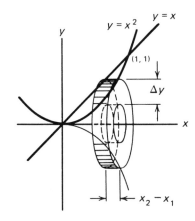

Figure 5.26

where y is the radius of the shell, $x_2 - x_1$ is the height of the shell, and Δy is its thickness. In this example the height of the shell, $x_2 - x_1$, is found by subtracting the curve $x_1 = f(y) = y$ from curve $x_2 = g(y) = \sqrt{y}$.

The integral for V then becomes

$$V = 2\pi \int_c^d [g(y) - f(y)]y \, dy$$

$$= 2\pi \int_0^1 (x_2 - x_1)y \, dy$$

$$= 2\pi \int_0^1 (\sqrt{y} - y)y \, dy$$

$$= 2\pi \int_0^1 (y^{3/2} - y^2) \, dy$$

$$= 2\pi \left(\frac{2}{5} y^{5/2} - \frac{y^3}{3} \right) \Big|_0^1$$

$$= 2\pi (\tfrac{2}{5} - \tfrac{1}{3})$$

$$= \frac{2\pi}{15}$$

Exercises

Find the volume of each solid formed by revolving the region bounded by the given curves about the given line using the shell method.

1. $y = 4x^2$, $x = 0$ and $y = 4$ about y-axis
2. $y = 4x^2$, $x = 0$ and $y = 4$ about x-axis
3. $4y = x^2$, $y = 0$ and $x = 2$ about y-axis
4. $y^2 = x$, $y = 0$ and $x = 4$ about y-axis
5. $y^2 = 2x$, $x = 0$ and $y = 2$ about x-axis
6. $y = x^3$, $y = 0$ and $x = 2$ about y-axis
7. $y = x^3$, $y = 0$ and $x = 2$ about x-axis
8. $y = \sqrt{x}$, $x = 0$ and $y = 2$ about x-axis
9. $y = 2x - x^2$ and the x-axis about y-axis
10. $x = 3y - y^2$ and the y-axis about x-axis
11. $y = x$, $x = 1$ and $y = 0$ about $x = 1$
12. $y = x^2$, $x = 2$ and $y = 0$ about $x = 2$
13. $x = y^2$, $y = 1$ and $x = 0$ about $y = 2$
14. $y = x^2$ and $y = 4$ about $y = -2$
15. $y = x^3$ and $y = x$ from $x = 0$ to $x = 1$ about x-axis
16. $y = x^3$ and $y = x$ from $x = 0$ to $x = 1$ about y-axis
17. $y = x^2 - 3x + 2$ and $y = 0$ about y-axis
18. $x = y^2 - 6y + 8$ and $x = 0$ about x-axis
19. $y = x(x - 2)^2$ and $y = 0$ about $x = 2$
20. $y = x(x - 2)^2$ and $y = 0$ about y-axis

5.4 CENTER OF MASS OF A SYSTEM OF PARTICLES

The next application of integration involves finding the center of mass, which is discussed in the next two sections. We find the center of mass of a system of particles in this section and the center of mass of a thin plate and of a solid of revolution in the next section. Finding the center of mass is of fundamental importance in the study of mechanics. The *center of mass* or *center of gravity* of an object or system of objects is the point at which the object or system balances or at which the entire mass can be considered to be concentrated.

Before finding the center of mass of a system of particles, we must first introduce the concept of a moment. The *moment* about a point P produced by some mass m is given by

$$\text{moment} = md,$$

where d is the length of the *moment arm*, which is the distance between the mass and point P (See Fig. 5.27.).

Suppose we have a 15-kg sign hanging from a support 0.8 m from a building as shown in Fig. 5.28. The length of the moment arm is 0.8 m, the distance from the mass to P. The mass is 15 kg. Therefore,

$$\text{moment} = md = (15 \text{ kg})(0.8 \text{ m}) = 12 \text{ kg m}$$

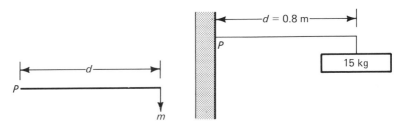

Figure 5.27   Figure 5.28

Consider the mobile that is balanced by the five weights as shown in Fig. 5.29. Let the masses m_1, m_2, m_3, m_4, and m_5 be at distances d_1, d_2, d_3, d_4, and d_5, respectively, from a point P. The moment of the system about P is

$$\text{moment} = m_1d_1 + m_2d_2 + m_3d_3 + m_4d_4 + m_5d_5$$

If we let P be the origin of the x-axis, $d_1 = -10$, $d_2 = -4$, $d_3 = 10$, $d_4 = 14$, and $d_5 = 20$. Then the moment about P is

$$\text{moment} = (200)(-10) + (150)(-4) + (140)(10) + (50)(14) + (25)(20)$$
$$= -2000 - 600 + 1400 + 700 + 500$$
$$= 0$$

If the moment is zero, the system is in equilibrium; that is, the mobile balances.

Note: Point P may be any point from which the lengths of the moment arms are measured. We usually choose point P to be the center of mass or the pivot point.

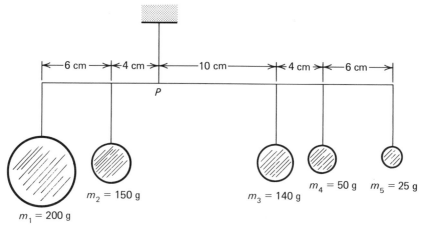

-6 cm $\quad-4$ cm $\quad-10$ cm $\quad-4$ cm $\quad-6$ cm $-$

P

$m_1 = 200$ g

$m_2 = 150$ g

$m_3 = 140$ g

$m_4 = 50$ g $\quad m_5 = 25$ g

Figure 5.29

Moment in a Linear System along the x-axis

Let $\bar{x}$ be the center of mass or balancing point of a linear system along the x-axis with n masses. That is, $\bar{x}$ is the point where all the mass seems to be concentrated. So

$$(m_1 + m_2 + m_3 + \cdots + m_n)\bar{x} = m_1x_1 + m_2x_2 + m_3x_3 + \cdots + m_nx_n$$

Then

$$\bar{x} = \frac{m_1x_1 + m_2x_2 + m_3x_3 + \cdots + m_nx_n}{m_1 + m_2 + m_3 + \cdots + m_n}$$

If we let M_o be the moment about the origin, $\bar{x}$ be the center of mass, and m be the total mass of the system, then

$$\bar{x} = \frac{M_o}{m}$$

Example 1

Find the center of mass of the linear system:

$$m_1 = 10, \ x_1 = -4; \ m_2 = 25, \ x_2 = 2; \ m_3 = 40, \ x_3 = 5; \ m_4 = 15, \ x_4 = 10$$

The moment about the origin is

$$M_o = m_1x_1 + m_2x_2 + m_3x_3 + m_4x_4$$
$$= (10)(-4) + (25)(2) + (40)(5) + (15)(10)$$
$$= -40 + 50 + 200 + 150 = 360$$

The total mass of the system is

$$m = 10 + 25 + 40 + 15 = 90$$

The center of mass is

$$\bar{x} = \frac{M_o}{m} = \frac{360}{90} = 4$$

To extend these concepts to two dimensions as in Fig. 5.30,

Moments of a Two-Dimensional System

Consider n masses m_1, m_2, . . ., m_n located in the xy-plane at points (x_1, y_1), (x_2, y_2), . . ., (x_n, y_n), respectively. Their moments with respect to the x-axis and the y-axis are defined as follows: The moment about the y-axis, M_y, is

$$M_y = m_1x_1 + m_2x_2 + \cdots + m_nx_n$$

and the moment about the x-axis, M_x, is

$$M_x = m_1y_1 + m_2y_2 + \cdots + m_ny_n$$

If we let m be the total mass of the system, the center of mass $(\bar{x}, \bar{y})$ is given by

$$\bar{x} = \frac{M_y}{m} \quad \text{and} \quad \bar{y} = \frac{M_x}{m}$$

The quantities $m\bar{x}$ and $m\bar{y}$ are regarded as the moments about the y-axis and the x-axis, respectively, of a mass m located at $(\bar{x}, \bar{y})$. That is, $(\bar{x}, \bar{y})$ is the point where the total mass seems to be concentrated that would give the same moments M_y and M_x.

Example 2

Find the center of mass of the system:

$$m_1 = 10 \text{ at } (9, 3); \quad m_2 = 6 \text{ at } (-1, 11); \quad m_3 = 8 \text{ at } (-6, 6); \quad m_4 = 12 \text{ at } (-6, 0)$$

From Fig. 5.31,

$$m = 10 + 6 + 8 + 12 = 36$$
$$M_y = (10)(9) + (6)(-1) + (8)(-6) + (12)(-6) = -36$$
$$M_x = (10)(3) + (6)(11) + (8)(6) + (12)(0) = 144$$

Then

$$\bar{x} = \frac{M_y}{m} = \frac{-36}{36} = -1$$

$$\bar{y} = \frac{M_x}{m} = \frac{144}{36} = 4$$

So the center of mass of this system is $(-1, 4)$.

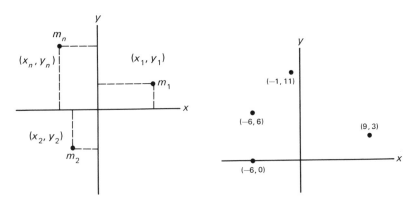

Figure 5.30   Figure 5.31

Exercises

Find the center of mass of each linear system.

1. $m_1 = 3$, $x_1 = -5$; $m_2 = 7$, $x_2 = 3$; $m_3 = 4$, $x_3 = 6$

2. $m_1 = 6$, $x_1 = -12$; $m_2 = 3$, $x_2 = -3$; $m_3 = 10$, $x_3 = 0$; $m_4 = 5$, $x_4 = 9$

3. $m_1 = 24$, $x_1 = -15$; $m_2 = 15$, $x_2 = -9$; $m_3 = 12$, $x_3 = 3$; $m_4 = 9$, $x_4 = 6$

4. $m_1 = 8$, $x_1 = -15$; $m_2 = 15$, $x_2 = -9$; $m_3 = 7$, $x_3 = -1$; $m_4 = 20$, $x_4 = 8$; $m_5 = 24$, $x_5 = 12$

5. There is a mass of 6 at $(9, 0)$ and a mass of 18 at $(-2, 0)$. Find where a mass of 3 should be placed on the x-axis so that the origin is the center of mass.

6. There is a mass of 30 at $(-4, 0)$, a mass of 9 at $(2, 0)$, and a mass of 15 at $(8, 0)$. Find where a mass of 3 should be placed on the x-axis so that the origin is the center of mass.

7. There is a mass of 24 at $(-8, 0)$ and a mass of 36 at $(12, 0)$. Find where a mass of 9 should be placed on the x-axis so that $(3, 0)$ is the center of mass.

8. There is a mass of 15 at $(-8, 0)$, a mass of 5 at $(-4, 0)$, and a mass of 12 at $(3, 0)$. Find where a mass of 8 should be placed on the x-axis so that $(-3, 0)$ is the center of mass.

9. There is a mass of 6 at $(-3, 0)$ and a mass of 9 at $(12, 0)$. Find what mass should be placed at $(-6, 0)$ so that the origin is the center of mass.

10. There is a mass of 4 at $(-5, 0)$, a mass of 16 at $(3, 0)$, and a mass of 24 at $(8, 0)$. Find what mass should be placed at $(-4, 0)$ so that the origin is the center of mass.

11. There is a mass of 25 at $(-6, 0)$, a mass of 45 at $(8, 0)$, and a mass of 40 at $(10, 0)$. Find what mass should be placed at $(-4, 0)$ so that $(3, 0)$ is the center of mass.

12. There is a mass of 18 at $(3, 0)$, a mass of 54 at $(9, 0)$, a mass of 24 at $(12, 0)$, and a mass of 36 at $(15, 0)$. Find what mass should be placed at $(4, 0)$ so that $(6, 0)$ is the center of mass.

13. A straight road connects Flatville (population 75,000), Pleasant Hill (population 50,000), and Harristown (population 25,000). Pleasant Hill is 18 mi north of Flatville while Harristown is 30 mi north of Flatville. Where is the best place to locate an airport to serve these three communities?

14. A straight road connects Leadville (population 1750), Branburg (population 2800), Princeton (population 970), and Four Oaks (population 480). The distance from Leadville to: Branburg is 4 mi, Princeton is 10 mi, and Four Oaks is 13 mi. Where is the best place to locate a hospital to serve these four communities?

Find the center of mass of each two-dimensional system.

15. $m_1 = 6$ at $(1, 4)$; $m_2 = 3$ at $(6, 2)$; $m_3 = 12$ at $(3, 3)$

16. $m_1 = 20$ at $(-5, 10)$; $m_2 = 15$ at $(-10, -15)$; $m_3 = 40$ at $(0, -5)$

17. $m_1 = 8$ at $(8, 12)$; $m_2 = 16$ at $(-12, 8)$; $m_3 = 20$ at $(-16, -4)$; $m_4 = 36$ at $(4, -20)$

18. $m_1 = 9$ at $(3, 6)$; $m_2 = 12$ at $(-6, 12)$; $m_3 = 18$ at $(0, -9)$; $m_4 = 30$ at $(15, 0)$; $m_5 = 15$ at $(-12, -9)$

19. There is a mass of 6 at $(4, 2)$ and a mass of 9 at $(-5, 8)$. Find where a mass of 10 should be placed so that the origin is the center of mass.

20. There is a mass of 18 at $(-4, 6)$, a mass of 12 at $(1, -2)$, and a mass of 9 at $(8, 0)$. Find where a mass of 6 should be placed so that $(2, -3)$ is the center of mass.

21. There is a mass of 15 at $(10, 3)$, a mass of 25 at $(-6, -1)$, and a mass of 40 at

(8, −2). Find what mass should be placed at (−5, −3) so that (−1, −2) is the center of mass.

22. There is a mass of 4 at (−5, −3), a mass of 16 at (−4, 3), and a mass of 12 at (6, −4). Find what mass should be placed at (2, 2) so that the origin is the center of mass.

23. Three towns plan to build a new hospital to serve all three communities. Town B (population 820) is 6 mi east and 3 mi south of Town A (population 1250). Town C (population 520) is 2 mi west and 8 mi south of Town A. Find the best location for the new hospital. Do not consider new roads in determining the best location.

24. Four cities plan to build a new airport to serve all four communities. City B (population 18,000) is 4 mi north and 3 mi west of City A (population 7500). City C (population 24,000) is 6 mi east and 12 mi south of City A. City D (population 10,500) is 15 mi due south of City A. Find the best location for the airport. Do not consider new roads in determining the best location.

5.5 CENTER OF MASS OF CONTINUOUS MASS DISTRIBUTIONS

As we saw in Section 5.4, the center of mass of any system of finite particles may be found arithmetically by summing the moments and the masses and dividing. Recall that the center of mass of a linear system along the x-axis with n masses was given by

$$\bar{x} = \frac{M_o}{m} = \frac{m_1 x_1 + m_2 x_2 + m_3 x_3 + \cdots + m_n x_n}{m_1 + m_2 + m_3 + \cdots + m_n}.$$

For a continuous mass distribution, the center of mass is found by integration. For example, to find the center of mass of a straight thin wire of constant density ρ, place the wire on the x-axis as shown in Fig. 5.32. Then, subdivide the wire into n equal lengths, each of length Δx and mass Δm. The mass of the ith length is

$$\text{mass} = (\text{density})(\text{length})$$
$$m = \rho \quad \Delta x$$

The total mass of the wire is the integral

$$m = \rho \int_a^b dx$$

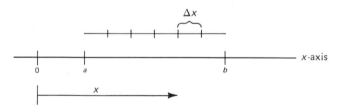

Figure 5.32

Applications of Integration   Chap. 5

Next, find the moment of the *i*th length:

$$\text{moment} = (\text{mass})(\text{length of moment arm})$$
$$= (\rho \, \Delta x)(x)$$

So, the moment of the entire wire about the origin is the integral

$$M_o = \rho \int_a^b x \, dx.$$

Then, the center of mass is

Center of Mass of a Continuous Thin Uniform Mass

$$\bar{x} = \frac{M_o}{m} = \frac{\displaystyle\int_a^b x \, dx}{\displaystyle\int_a^b dx}$$

Note: The density ρ cancels in all such cases when the density is constant or uniform. For any homogeneous mass distribution having constant density (constant mass per unit length, per unit area, or per unit volume), the *centroid* is the same as the center of mass. The centroid often refers to the geometric center.

Example 1

Find the center of mass of a straight wire 12 cm long and of uniform density.

$$\bar{x} = \frac{\displaystyle\int_a^b x \, dx}{\displaystyle\int_a^b dx} = \frac{\displaystyle\int_0^{12} x \, dx}{\displaystyle\int_0^{12} dx} = \frac{\left. \dfrac{x^2}{2} \right|_0^{12}}{\left. x \right|_0^{12}} = \frac{72}{12} = 6$$

This result should be of no surprise. For a uniform linear object, the center of mass is at its center, the point at which the object can be supported. For example, a metre stick is supported on one's finger at the 50-cm mark as in Fig. 5.33.

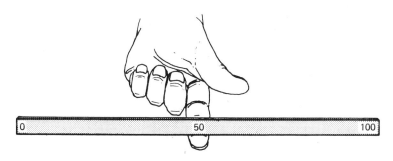

Figure 5.33

For a nonuniform object, the center of mass is usually not at its geometric center, but at the point at which it can be supported in equilibrium by a single force or the point about which it spins if allowed to spin freely in space as in Fig. 5.34.

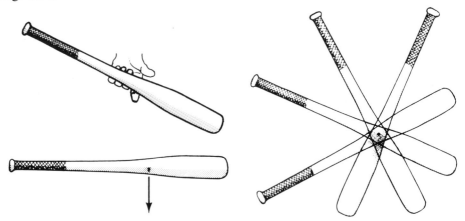

Figure 5.34

When the density of a continuous thin mass is not uniform, the center of mass is found as follows:

Center of Mass of a Continuous Thin Mass of Variable Density

$$\bar{x} = \frac{M_o}{m} = \frac{\displaystyle\int_a^b \rho(x)\, x\, dx}{\displaystyle\int_a^b \rho(x)\, dx},$$

where $\rho(x)$ is the density expressed as a function of x. Density is mass per unit length.

Example 2

Find the center of mass of a straight wire 16 cm long and whose density is given by $\rho(x) = 4\sqrt{x}$, where x is the distance from one end of the wire.

$$\bar{x} = \frac{M_o}{m} = \frac{\displaystyle\int_a^b \rho(x)\, x\, dx}{\displaystyle\int_a^b \rho(x)\, dx} = \frac{\displaystyle\int_0^{16} 4\sqrt{x}\, x\, dx}{\displaystyle\int_0^{16} 4\sqrt{x}\, dx} = \frac{\displaystyle\int_0^{16} x^{3/2}\, dx}{\displaystyle\int_0^{16} x^{1/2}\, dx} = \frac{\dfrac{2}{5} x^{5/2}\Big|_0^{16}}{\dfrac{2}{3} x^{3/2}\Big|_0^{16}}$$

$$= \frac{\dfrac{2}{5}[16^{5/2} - 0]}{\dfrac{2}{3}[16^{3/2} - 0]} = \frac{\dfrac{2}{5}(1024)}{\dfrac{2}{3}(64)} = \frac{48}{5} = 9.6 \text{ cm from the lighter end.}$$

The center of mass of a two-dimensional thin plate is the point at which the plate can be supported as in Fig. 5.35.

Figure 5.35

If the thin plate is in a regular geometric shape, its center of mass is its geometric center because of its symmetry. Examples of four common geometric figures with each corresponding center of mass are shown in Fig. 5.36.

The center of mass of a more complex but uniform thin object can be found by subdividing the object into combinations of simpler figures. Find the center of each simpler figure. Consider the mass of each simpler figure to be concentrated at its center and proceed using the method for moments of a two-dimensional system used in Example 2 of Section 5.4.

Example 3

Find the center of mass of the uniform thin plate in Fig. 5.37.

The center of region 1 (square) is (3, 3), of region 2 (rectangle) is $(10, \frac{3}{2})$, and of region 3 (rectangle) is (16, 6). Since the plate is uniform, the mass in each region is proportional to its area. The area of region 1 is 36 units, of region 2 is 24 units, and of region 3 is 48 units. So

$$m = 36 + 24 + 48 = 108$$
$$M_y = m_1 x_1 + m_2 x_2 + m_3 x_3$$
$$= (36)(3) + (24)(10) + (48)(16)$$
$$= 1116$$
$$M_x = m_1 y_1 + m_2 y_2 + m_3 y_3$$
$$= (36)(3) + (24)(\tfrac{3}{2}) + (48)(6)$$
$$= 432$$

Then

$$\bar{x} = \frac{M_y}{m} = \frac{1116}{108} = 10\tfrac{1}{3}$$

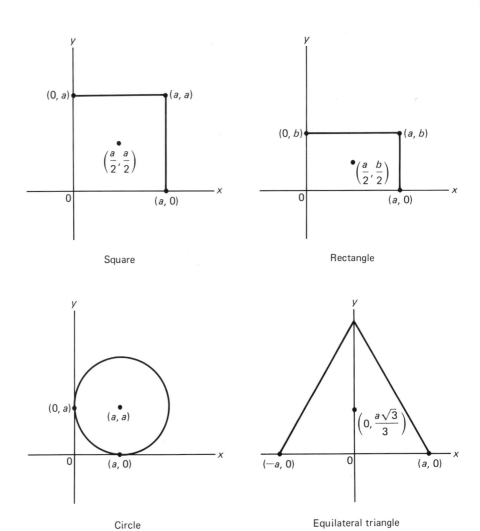

Figure 5.36

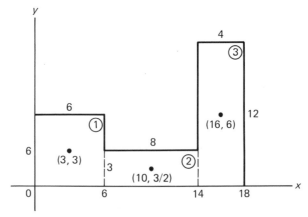

Figure 5.37

Applications of Integration   Chap. 5

$$\bar{y} = \frac{M_x}{m} = \frac{432}{108} = 4$$

The center of mass of the plate is $(10\frac{1}{3}, 4)$.

Note that in Example 3 the center of mass is not on the surface of the plate.

Next, let's find the centroid of an irregular shaped area (or thin plate) of constant density ρ between the curves shown in Fig. 5.38. First divide the area into n rectangles, each of width Δx. Let (x_i, y_i) be the center of mass of the ith rectangle. The y-value of the geometric center of the ith rectangle is

$$y_i = \frac{f(x_i) + g(x_i)}{2}$$

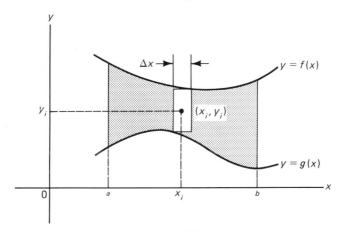

Figure 5.38

The area of the ith rectangle is $[f(x_i) - g(x_i)] \Delta x$. The mass of the ith rectangle is

$$\text{mass} = (\text{density})(\text{area})$$
$$= \rho[f(x_i) - g(x_i)] \Delta x$$

The total mass is the integral

$$m = \rho \int_a^b [f(x) - g(x)] \, dx = \rho A$$

where A is the area of the region.

Next, find the moment of the ith rectangle about the x-axis:

$$\text{moment} = (\text{mass})(\text{moment arm})$$
$$= \rho[f(x_i) - g(x_i)] \Delta x \cdot y_i$$
$$= \rho[f(x_i) - g(x_i)] \Delta x \cdot \frac{f(x_i) + g(x_i)}{2}$$
$$= \frac{\rho}{2}\{[f(x_i)]^2 - [g(x_i)]^2\} \Delta x$$

The moment about the x-axis is the integral

$$M_x = \frac{\rho}{2} \int_a^b \{[f(x)]^2 - [g(x)]^2\} \, dx$$

Similarly, the moment about the y-axis is found to be

$$M_y = \rho \int_a^b x[f(x) - g(x)] \, dx$$

Moments and Center of Mass of a Plane Area or Thin Plate

Let $g(x) \leqslant f(x)$ be continuous functions on $a \leqslant x \leqslant b$ for the area of uniform density ρ bounded by $y = f(x)$, $y = g(x)$, $x = a$, and $x = b$. The moments about the x-axis and the y-axis are

$$M_x = \frac{\rho}{2} \int_a^b \{[f(x)]^2 - [g(x)]^2\} \, dx \quad \text{and} \quad M_y = \rho \int_a^b x[f(x) - g(x)] \, dx$$

Its mass is given by

$$m = \rho \int_a^b [f(x) - g(x)] \, dx$$

and its center of mass $(\bar{x}, \bar{y})$ is

$$\bar{x} = \frac{M_y}{m} \quad \text{and} \quad \bar{y} = \frac{M_x}{m}$$

Example 4

Find the center of mass of the uniform thin plate of density ρ bounded by $y = x^2$ and $y = x + 2$.

First, graph the equations and find the points of intersection as in Fig. 5.39.

$$x^2 = x + 2$$
$$x^2 - x - 2 = 0$$
$$(x - 2)(x + 1) = 0$$
$$x = 2, -1$$

$$M_x = \frac{\rho}{2} \int_a^b \{[f(x)]^2 - [g(x)]^2\} \, dx$$

$$= \frac{\rho}{2} \int_{-1}^2 [(x + 2)^2 - (x^2)^2] \, dx$$

$$= \frac{\rho}{2} \int_{-1}^2 [x^2 + 4x + 4 - x^4] \, dx$$

$$= \frac{\rho}{2} \left(\frac{x^3}{3} + 2x^2 + 4x - \frac{x^5}{5} \right) \Big|_{-1}^2$$

$$= \frac{\rho}{2} \left[\left(\frac{8}{3} + 8 + 8 - \frac{32}{5} \right) - \left(-\frac{1}{3} + 2 - 4 + \frac{1}{5} \right) \right]$$

Applications of Integration   Chap. 5

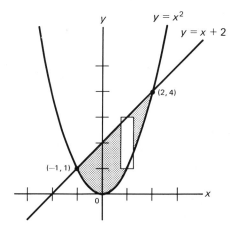

$y = x^2$

$y = x + 2$

(2, 4)

(−1, 1)

Figure 5.39

$$= \frac{\rho}{2}\left(\frac{72}{5}\right)$$

$$= \frac{36\rho}{5}$$

$$M_y = \rho \int_a^b x[f(x) - g(x)]\, dx$$

$$= \rho \int_{-1}^2 x(x + 2 - x^2)\, dx$$

$$= \rho \int_{-1}^2 (x^2 + 2x - x^3)\, dx$$

$$= \rho \left(\frac{x^3}{3} + x^2 - \frac{x^4}{4}\right)\Bigg|_{-1}^2$$

$$= \rho\left[\left(\frac{8}{3} + 4 - 4\right) - \left(-\frac{1}{3} + 1 - \frac{1}{4}\right)\right]$$

$$= \frac{9\rho}{4}$$

$$m = \rho \int_a^b [f(x) - g(x)]\, dx$$

$$= \rho \int_{-1}^2 (x + 2 - x^2)\, dx$$

$$= \rho\left(\frac{x^2}{2} + 2x - \frac{x^3}{3}\right)\Bigg|_{-1}^2$$

$$= \rho\left[\left(2 + 4 - \frac{8}{3}\right) - \left(\frac{1}{2} - 2 + \frac{1}{3}\right)\right]$$

$$= \frac{9\rho}{2}$$

Then

$$\bar{x} = \frac{M_y}{m} = \frac{9\rho/4}{9\rho/2} = \frac{1}{2}$$

$$\bar{y} = \frac{M_x}{m} = \frac{36\rho/5}{9\rho/2} = \frac{8}{5}$$

The center of mass is $(\frac{1}{2}, \frac{8}{5})$.

Note that the density ρ cancels in both $\bar{x}$ and $\bar{y}$. That is, the center of mass of a thin plate or area of uniform density depends only on its shape and not its density. Thus we may find the centroid as follows:

Centroid of a Plane Region or Thin Plate

Let $g(x) \leqslant f(x)$ be continuous functions on $a \leqslant x \leqslant b$. The centroid $(\bar{x}, \bar{y})$ of the region bounded by $y = f(x)$, $y = g(x)$, $x = a$, and $y = b$ is

$$\bar{x} = \frac{\displaystyle\int_a^b x[f(x) - g(x)]\,dx}{A}$$

and

$$\bar{y} = \frac{\displaystyle\frac{1}{2}\int_a^b \{[f(x)]^2 - [g(x)]^2\}\,dx}{A}$$

where A is the area of the region.

Example 5

Find the centroid of the region bounded by $y = x^4$ and $y = x$.

First, graph the equations and find the points of intersection. (See Fig. 5.40.)

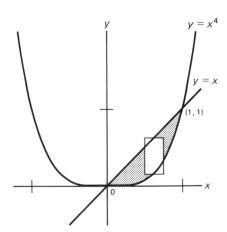

Figure 5.40

$$A = \int_0^1 (x - x^4)\, dx$$

$$= \left(\frac{x^2}{2} - \frac{x^5}{5}\right)\Big|_0^1$$

$$= \left(\frac{1}{2} - \frac{1}{5}\right) - (0) = \frac{3}{10}$$

$$\bar{x} = \frac{\int_a^b x[f(x) - g(x)]\, dx}{A}$$

$$= \frac{\int_0^1 x(x - x^4)\, dx}{3/10}$$

$$= \frac{10}{3} \int_0^1 (x^2 - x^5)\, dx$$

$$= \frac{10}{3}\left(\frac{x^3}{3} - \frac{x^6}{6}\right)\Big|_0^1$$

$$= \frac{10}{3}\left[\left(\frac{1}{3} - \frac{1}{6}\right) - (0)\right]$$

$$= \frac{10}{3}\left(\frac{1}{6}\right) = \frac{5}{9}$$

$$\bar{y} = \frac{\frac{1}{2}\int_a^b \{[f(x)]^2 - [g(x)]^2\}\, dx}{A}$$

$$= \frac{\frac{1}{2}\int_0^1 [(x)^2 - (x^4)^2]\, dx}{3/10}$$

$$= \frac{5}{3}\int_0^1 (x^2 - x^8)\, dx$$

$$= \frac{5}{3}\left(\frac{x^3}{3} - \frac{x^9}{9}\right)\Big|_0^1$$

$$= \frac{5}{3}\left[\left(\frac{1}{3} - \frac{1}{9}\right) - (0)\right]$$

$$= \frac{5}{3}\left(\frac{2}{9}\right)$$

$$= \frac{10}{27}$$

The centroid is at $(\frac{5}{9}, \frac{10}{27})$.

The most general case of finding the center of mass of a two-dimensional mass distribution requires double integration. This case is not treated in this text.

A solid of revolution of constant density has its centroid on its axis of revolution. Let the area bounded by $y = f(x)$, $x = a$, and $x = b$ be revolved about the x-axis.

We have drawn a typical disk in Fig. 5.41. Its center of mass is x units from the y-axis; therefore, the length of its moment arm is x. Its volume is $\pi y^2 \, dx$. Since the solid is of constant density ρ, its mass is proportional to its volume; that is, $m = \rho \pi y^2 \, dx$. Thus the moment about the y-axis is

$$md = (\rho \pi y^2 \, dx)x = \rho \pi x y^2 \, dx$$

The sum of the moments of all such disks may be expressed by the integral

$$M_y = \rho \pi \int_a^b x y^2 \, dx$$

The mass of the solid may be expressed by the integral

$$m = \rho \pi \int_a^b y^2 \, dx$$

Centroid of a Volume of a Solid of Revolution about the x-axis

$$\bar{x} = \frac{M_y}{m} = \frac{\displaystyle\int_a^b x y^2 \, dx}{\displaystyle\int_a^b y^2 \, dx} \quad \text{and} \quad \bar{y} = 0$$

Note: The ρ and π factors cancel.

In a similar manner, we can show:

Centroid of a Volume of a Solid of Revolution about the y-axis

$$\bar{y} = \frac{M_x}{m} = \frac{\displaystyle\int_c^d y x^2 \, dy}{\displaystyle\int_c^d x^2 \, dy} \quad \text{and} \quad \bar{x} = 0$$

Example 6

Find the centroid of the solid formed by revolving the region bounded by $y = x^2$, $x = 1$, and $y = 0$ about the x-axis. (See Fig. 5.42.)

$$\bar{x} = \frac{\displaystyle\int_a^b x y^2 \, dx}{\displaystyle\int_a^b y^2 \, dx}$$

$$= \frac{\displaystyle\int_0^1 x(x^2)^2 \, dx}{\displaystyle\int_0^1 (x^2)^2 \, dx}$$

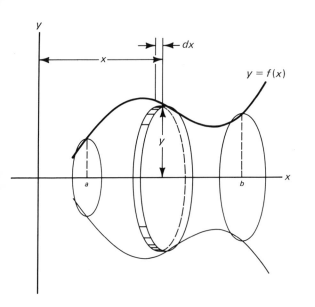

$y = f(x)$

Figure 5.41

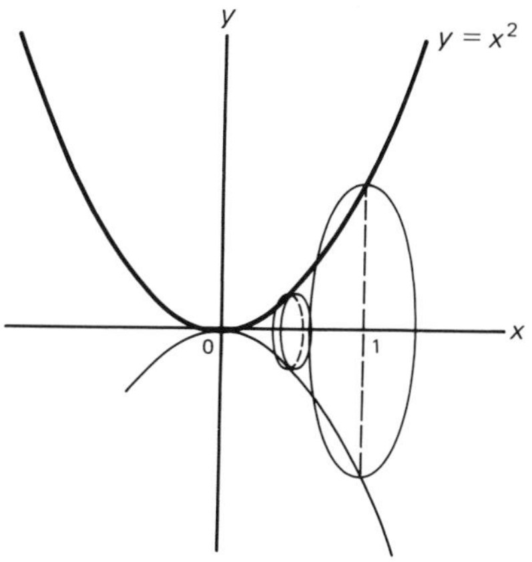

$y = x^2$

Figure 5.42

$$= \frac{\displaystyle\int_0^1 x^5\, dx}{\displaystyle\int_0^1 x^4\, dx}$$

$$= \frac{\left.\dfrac{x^6}{6}\right|_0^1}{\left.\dfrac{x^5}{5}\right|_0^1} = \frac{\dfrac{1}{6}}{\dfrac{1}{5}} = \frac{5}{6}$$

The centroid is at $(\frac{5}{6},\, 0)$.

Example 7

Find the centroid of the solid formed by revolving the region bounded by $y = 4 - x^2$, $x = 0$, and $y = 0$ about the y-axis. (See Fig. 5.43.)

$$\bar{y} = \frac{\displaystyle\int_c^d yx^2\, dy}{\displaystyle\int_c^d x^2\, dy}$$

$$= \frac{\displaystyle\int_0^4 y(4 - y)\, dy}{\displaystyle\int_0^4 (4 - y)\, dy} \qquad (Note: x^2 = 4 - y.)$$

$$= \frac{\displaystyle\int_0^4 (4y - y^2)\, dy}{\displaystyle\int_0^4 (4 - y)\, dy}$$

$$= \frac{2y^2 - \dfrac{y^3}{3}\Big|_0^4}{4y - \dfrac{y^2}{2}\Big|_0^4} = \frac{\left(32 - \dfrac{64}{3}\right) - (0)}{(16 - 8) - (0)} = \frac{32/3}{8} = \frac{4}{3}$$

The centroid is at $(0, \frac{4}{3})$.

The most general case of finding the center of mass of a three-dimensional mass distribution requires triple integration. This case is not treated in this text.

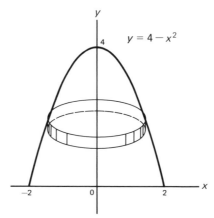

Figure 5.43

Exercises

1. Find the center of mass of a straight wire 20 cm long and of uniform density.
2. Show that the center of mass of a straight wire of length L and of uniform density is at its midpoint.
3. Find the center of mass measured from the lighter end of a straight wire 10 cm long and whose density is given by $\rho(x) = 0.1x$, where x is the distance from one end.
4. Find the center of mass measured from the lighter end of a straight wire 8 cm long and whose density is given by $\rho(x) = 0.1x^2$, where x is the distance from one end.
5. Find the center of mass measured from the lighter end of a straight wire 12 cm long and whose density is given by $\rho(x) = 4 + x^2$, where x is the distance from one end.
6. Find the center of mass measured from the lighter end of a straight wire 9 cm long and whose density is given by $\rho(x) = 3 - \sqrt{x}$, where x is the distance from one end.
7. The density of a straight wire 6 cm long is directly proportional to the distance from one end. Find its center of mass.
8. The density of a straight wire 9 cm long varies inversely as the square root of the distance from one end. Find its center of mass.

Find the center of mass of each uniform thin plate.

9.

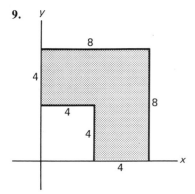

Figure 5.44

10.

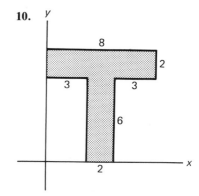

Figure 5.45

11.

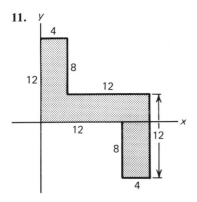

Figure 5.46

12.

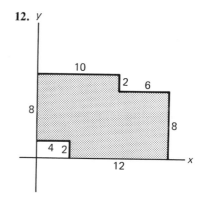

Figure 5.47

13.

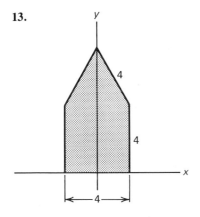

Figure 5.48

14.

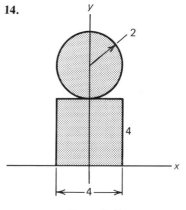

Figure 5.49

Sec. 5.5   Center of Mass of Continuous Mass Distributions

203

Find the centroid of each region bounded by the given curves.

15. $y = \sqrt{x}$, $y = 0$, and $x = 9$ **16.** $y = x^2$ and $y = 4$

17. $y = x^2 - 2x$ and $y = 0$ **18.** $y = 4 - x^2$ and $y = x^2 - 4$

19. $y = 4 - x^2$ and $y = 0$ **20.** $y = 4 - x^2$ and $y = x + 2$

21. $y = x^3$ and $y = x$ (first quadrant)   **22.** $y = x^3$, $x = 0$, and $y = 1$

23. Find the centroid of the semicircle of radius 1 with center at the origin lying in the first and second quadrants.

24. Find the centroid of the quarter circle of radius 1 with center at the origin lying in the first quadrant.

Find the centroid of the solid formed by revolving each region bounded by the given curves about the given axis.

25. $y = x^3$, $y = 0$, and $x = 1$ about x-axis

26. $y = x^3$, $y = 0$, and $x = 1$ about y-axis

27. $y = 3 - x$, $x = 0$, and $y = 0$ about x-axis

28. $y = 3 - 2x$, $x = 0$, and $y = 0$ about y-axis

29. $y = x^2$, $x = 1$, and $y = 0$ about y-axis

30. $x^2 + y^2 = 1$, $x = 0$, and $y = 0$ about x-axis

5.6 MOMENTS OF INERTIA

In Sections 5.4 and 5.5 we used moments to find centers of mass and centroids. Each moment was the product of the mass and its distance from some line. This case is called the *first moment*.

The second moment, called the *moment of inertia* about a line, is defined as the product of a mass m and the square of its distance d from a given line; that is,

$$I = md^2$$

Inertia is a property of an object that resists a change in its motion. That is, inertia is a property of an object that causes it to remain at rest if it is at rest or to continue moving with constant velocity.

Moment of Inertia of a System

Let masses $m_1, m_2, m_3, \ldots, m_n$ be at distances $d_1, d_2, d_3, \ldots, d_n$, respectively, rotating about some axis. The moment of inertia, I, of the system is

$$I = m_1 d_1^2 + m_2 d_2^2 + m_3 d_3^2 + \cdots + m_n d_n^2$$

Let m be the sum of all the masses in the system and let R be the distance of m from the axis that gives the same total moment of inertia

$$I = mR^2 = m_1 d_1^2 + m_2 d_2^2 + m_3 d_3^2 + \cdots + m_n d_n^2$$

R is called the *radius of gyration*, which corresponds to the center of mass in the preceding sections.

Example 1

Find the moment of inertia and the radius of gyration about the y-axis of the system: $m_1 = 8$ at $(2, -4)$; $m_2 = 3$ at $(-9, 8)$; and $m_3 = 6$ at $(-5, 2)$.

$$I_y = m_1 x_1^2 + m_2 x_2^2 + m_3 x_3^2$$

$$I_y = 8(2)^2 + 3(-9)^2 + 6(-5)^2 = 425$$

$$m = m_1 + m_2 + m_3 = 8 + 3 + 6 = 17$$

$$I_y = mR^2$$

$$R = \sqrt{\frac{I_y}{m}} = \sqrt{\frac{425}{17}} = 5$$

Now, let's find the moment of inertia of an area of constant density ρ about the y-axis as shown in Fig. 5.50. First, divide the area into n rectangles, each of width Δx. Let (x_i, y_i) be the center of mass of the ith rectangle. As we saw in the preceding sections, the y-value of the geometric center of the ith rectangle is

$$y_i = \frac{f(x_i) + g(x_i)}{2}$$

its area is $[f(x_i) - g(x_i)]\,\Delta x$, and its mass is $\rho[f(x_i) - g(x_i)]\,\Delta x$. The total mass is the integral

$$m = \rho \int_a^b [f(x) - g(x)]\,dx = \rho A$$

where A is the area of the region.

The distance of the center of the ith rectangle from the y-axis is x_i. Its moment of inertia, its second moment, is then

$$\text{(mass)(moment arm)}^2 = \rho[f(x_i) - g(x_i)]\,\Delta x \cdot (x_i)^2$$

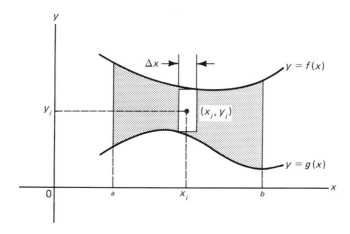

Figure 5.50

Summing the moments of all such rectangles, the moment of inertia of the region about the y-axis is the integral

$$I_y = \rho \int_a^b x^2 [f(x) - g(x)]\, dx$$

Similarly, the moment of inertia of an area about the x-axis is found to be

$$I_x = \rho \int_c^d y^2 [f(y) - g(y)]\, dy$$

The radius of gyration for each moment is found by

About y-axis	About x-axis
$I_y = mR^2$	$I_x = mR^2$
$R = \sqrt{\dfrac{I_y}{m}}$	$R = \sqrt{\dfrac{I_x}{m}}$

Moments of Inertia of an Area of Constant Density about the x- and y-axes

Let $g(x) \leqslant f(x)$ be continuous functions on $a \leqslant x \leqslant b$ for the area of constant density ρ bounded by $y = f(x)$, $y = g(x)$, $x = a$, $x = b$. The moment of inertia about the y-axis is

$$I_y = \rho \int_a^b x^2 [f(x) - g(x)]\, dx \qquad \left(R = \sqrt{\dfrac{I_y}{m}}\right)$$

Similarly, the moment of inertia about the x-axis is

$$I_x = \rho \int_c^d y^2 [f(y) - g(y)]\, dy \qquad \left(R = \sqrt{\dfrac{I_x}{m}}\right)$$

Example 2

Find the moment of inertia and the radius of gyration about the y-axis of the region bounded by $y = x^2$, $y = 0$, and $x = 3$, where the region has a constant density of 5. (See Fig. 5.51.)

$$I_y = \rho \int_a^b x^2 [f(x) - g(x)]\, dx$$

$$= 5 \int_0^3 x^2 [x^2 - 0]\, dx$$

$$= 5 \int_0^3 x^4\, dx$$

$$= 5 \frac{x^5}{5} \Big|_0^3$$

$$= 243$$

$$m = \rho \int_a^b [f(x) - g(x)]\, dx$$

$$= 5 \int_0^3 (x^2 - 0)\, dx$$

$$= 5 \int_0^3 x^2 \, dx$$

$$= 5 \cdot \frac{x^3}{3} \Big|_0^3$$

$$= 45$$

$$R = \sqrt{\frac{I_y}{m}} = \sqrt{\frac{243}{45}} = \frac{3\sqrt{15}}{5} = 2.32$$

Example 3

Find the moment of inertia and the radius of gyration about the x-axis of the region described in Example 2. (See Fig. 5.52.)

$$I_x = \rho \int_c^d y^2 [f(y) - g(y)] \, dy$$

$$= 5 \int_0^9 y^2 [3 - \sqrt{y}] \, dy \qquad (3 - x = 3 - \sqrt{y})$$

$$= 5 \int_0^9 [3y^2 - y^{5/2}] \, dy$$

$$= 5 \left[y^3 - \frac{y^{7/2}}{7/2} \right] \Big|_0^9$$

$$= 5 \left[9^3 - \frac{9^{7/2}}{7/2} \right] = 521 \qquad \text{(three significant digits)}$$

$$R = \sqrt{\frac{I_x}{m}} = \sqrt{\frac{521}{45}} = 3.40$$

To find the moment of inertia of a solid of revolution with respect to its axis of revolution, it is most convenient to use the shell method. Let the area

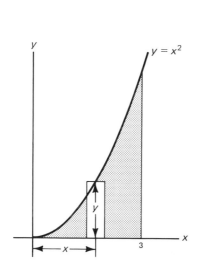

Figure 5.51

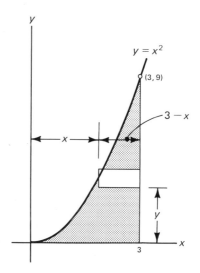

Figure 5.52

bounded by $y = f(x)$, $y = 0$, $x = a$, and $x = b$ be revolved about the y-axis. We have drawn a typical shell in Fig. 5.53. The mass of the solid is proportional to its volume $m = \rho V$:

$$m = 2\pi\rho \int_a^b x \, f(x) \, dx$$

and x^2 is the square of its distance from the y-axis. Summing all such shells gives the following results:

The moment of inertia of a solid of revolution about the y-axis is given by the integral

$$I_y = 2\pi\rho \int_a^b x^3 f(x) \, dx$$

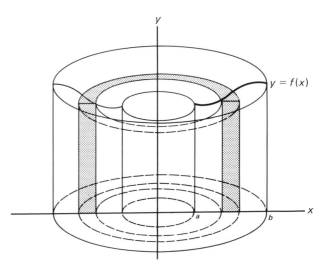

Figure 5.53

Similarly, let the area bounded by $x = f(y)$, $x = 0$, $y = c$, and $y = d$ be revolved about the x-axis as in Fig. 5.54. The moment of inertia about the x-axis is given by the integral

$$I_x = 2\pi\rho \int_c^d y^3 f(y) \, dy$$

Note: $f(x)$ and $f(y)$ correspond to the height of the shell.

Each radius of gyration is found in a similar way as that for a plane region; that is,

$$R = \sqrt{\frac{I_y}{m}} \quad \text{or} \quad R = \sqrt{\frac{I_x}{m}}$$

Example 4

Find the moment of inertia and the radius of gyration of the solid formed by revolving the region bounded by $y = 2x$, $y = 0$, and $x = 3$ about the y-axis. Assume that $\rho = 5$. (See Fig. 5.55.)

Applications of Integration   Chap. 5

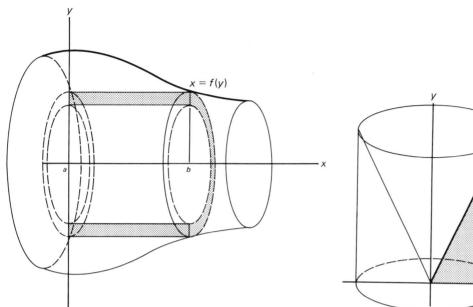

Figure 5.54   **Figure 5.55**

$$I_y = 2\pi\rho \int_a^b x^3 f(x)\, dx$$

$$= 2\pi(5) \int_0^3 x^3(2x)\, dx$$

$$= 20\pi \int_0^3 x^4\, dx$$

$$= 20\pi \cdot \frac{x^5}{5}\Big|_0^3 = 972\pi$$

$$m = 2\pi\rho \int_a^b xf(x)\, dx$$

$$= 2\pi(5) \int_0^3 x(2x)\, dx$$

$$= 20\pi \int_0^3 x^2\, dx$$

$$= 20\pi \cdot \frac{x^3}{3}\Big|_0^3 = 180\pi$$

$$R = \sqrt{\frac{I_y}{m}} = \sqrt{\frac{972\pi}{180\pi}} = 2.32$$

Example 5

Find the moment of inertia and the radius of gyration of the solid formed by revolving the region bounded by $y = x^2$, $y = 0$, and $x = 2$ about the x-axis. Assume that $\rho = 1$. (See Fig. 5.56.)

Sec. 5.6   Moments of Inertia   **209**

$$I_x = 2\pi\rho \int_c^d y^3 f(y)\, dy$$

$$= 2\pi(1) \int_0^4 y^3(2 - \sqrt{y})\, dy \qquad (2 - x = 2 - \sqrt{y})$$

$$= 2\pi \int_0^4 (2y^3 - y^{7/2})\, dy$$

$$= 2\pi \left(\frac{y^4}{2} - \frac{y^{9/2}}{9/2} \right) \Big|_0^4$$

$$= 2\pi \left(128 - \frac{1024}{9} \right) = \frac{256\pi}{9}$$

$$m = 2\pi\rho \int_c^d y\, f(y)\, dy$$

$$= 2\pi(1) \int_0^4 y(2 - \sqrt{y})\, dy$$

$$= 2\pi \int_0^4 (2y - y^{3/2})\, dy$$

$$= 2\pi \left(y^2 - \frac{y^{5/2}}{5/2} \right) \Big|_0^4$$

$$= 2\pi \left(16 - \frac{64}{5} \right) = \frac{32\pi}{5}$$

$$R = \sqrt{\frac{I_x}{m}} = \sqrt{\frac{256\pi/9}{32\pi/5}} = 2.11$$

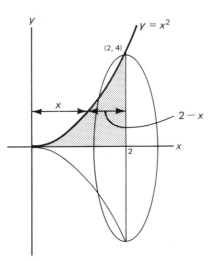

Figure 5.56

Exercises

Find the moment of inertia and the radius of gyration about the y-axis of each system.

1. $m_1 = 9$ at $(3, -2)$; $m_2 = 12$ at $(5, 4)$; $m_3 = 15$ at $(3, 7)$
2. $m_1 = 3$ at $(-6, 2)$; $m_2 = 9$ at $(-5, 8)$; $m_3 = 8$ at $(7, -2)$

Applications of Integration   Chap. 5

3. $m_1 = 15$ at $(3, -9)$; $m_2 = 10$ at $(6, -4)$; $m_3 = 18$ at $(9, 2)$; $m_4 = 12$ at $(1, 3)$

4. $m_1 = 24$ at $(-6, 3)$; $m_2 = 36$ at $(7, -9)$; $m_3 = 15$ at $(4, 4)$; $m_4 = 12$ at $(-2, -5)$; $m_5 = 18$ at $(8, 6)$

Find the moment of inertia and the radius of gyration about the x-axis of each system.

5. The system in Exercise 1.

6. $m_1 = 8$ at $(-3, 9)$; $m_2 = 16$ at $(8, -6)$; $m_3 = 14$ at $(4, 2)$

7. $m_1 = 9$ at $(-5, -9)$; $m_2 = 5$ at $(8, -2)$; $m_3 = 8$ at $(5, 0)$; $m_4 = 10$ at $(-3, 1)$

8. The system in Exercise 4.

Find the moment of inertia and the radius of gyration of each region bounded by the given curves about the given axis.

9. $y = x^2$, $x = 0$, and $y = 4$ about y-axis. $\rho = 5$.

10. Region of Exercise 9 about x-axis.

11. $y = x^2$, $y = x$, and $x > 0$ about x-axis. $\rho = 4$.

12. Region of Exercise 11 about y-axis.

13. $y = 5 - x^2$, $y = 1$, and $x = 0$ about y-axis. $\rho = 3$.

14. $x = 1 + y^2$, $x = 10$, and $y = 0$ about x-axis. $\rho = 5$.

15. $y = 1/x^2$, $y = 0$, $x = 1$, and $x = 2$ about y-axis. $\rho = 2$.

16. $y = 4x - x^2$ and $y = 0$ about y-axis. $\rho = 15$.

Find the moment of inertia and the radius of gyration of the solid formed by revolving the region bounded by the given curves about the given axis.

17. $y = 3x$, $y = 0$, and $x = 2$ about y-axis. $\rho = 15$.

18. Region of Exercise 17 about x-axis.

19. $y = 4x^2$, $y = 0$, and $x = 2$ about x-axis. $\rho = 1$.

20. Region of Exercise 19 about y-axis.

21. $y = 9 - x^2$, $y = 0$, and $x = 0$ about y-axis. $\rho = 12$.

22. $x = 4 - y^2$, $y = 0$, and $x = 0$ about x-axis. $\rho = 6$.

23. $y = 4x - x^2$ and $y = 0$ about y-axis. $\rho = 15$.

24. $y = 1/x^3$, $x = 0$, $y = 1$, and $y = 8$ about x-axis. $\rho = 2$.

5.7 WORK, FLUID PRESSURE, AND AVERAGE VALUE

Although there are still many more technical applications of the integral, we will consider only three more in this section: work, fluid pressure, and average value.

Work

When a constant force F is applied to an object, moving it through a distance s, the technical term *work* is defined to be the product $F \cdot s$. That is, work W is the product of the force and the distance through which the force acts.

Example 1

A 70-lb container is lifted 8 ft above the floor. Find the work done.

$$W = F \cdot s$$
$$= (70 \text{ lb})(8 \text{ ft})$$
$$= 560 \text{ ft-lb}$$

This formula for work is appropriate when the force remains constant. However, if the force varies as the object is moved across a given distance, then another approach must be used. If the object is moved from a to b, then we can approximate the work done by a variable force F as follows: Divide the interval from a to b into intervals, each of width Δx. Let F_k represent the value of the force acting on the object somewhere in the kth interval. (See Fig. 5.57.)

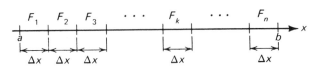

Figure 5.57

Then $\Delta W_k = F_k \cdot \Delta x$ can be considered as an approximation for the work done over the kth interval. If there are n intervals, each of width Δx, from a to b, then

$$W_{\text{approx}} = F_1 \cdot \Delta x + F_2 \cdot \Delta x + F_3 \cdot \Delta x + \cdots + F_n \cdot \Delta x$$

is an approximation for the work done moving the object from a to b. If the variable force can be expressed as a function of the distance traveled by the object, then it is possible to use integration techniques to find the actual work done.

Suppose that $F = f(x)$ expresses the force as a function of the distance traveled by the object. Then

$$W_{\text{approx}} = f(x_1) \cdot \Delta x + f(x_2) \cdot \Delta x + f(x_3) \cdot \Delta x + \cdots$$
$$+ f(x_k) \, \Delta x + \cdots + f(x_n) \cdot \Delta x$$

is an approximation for W, the actual work done, where x_k is a number in the kth interval.

The smaller we choose Δx, the better approximation we obtain for W_{approx}. In fact, if we let Δx approach 0, then W_{approx} approaches the actual work done W. But then

Work
$W = \displaystyle\int_a^b f(x)\,dx$

since this is how the definite integral of $f(x)$ from a to b was described in Chapter 4.

Example 2

Find the work done by a force F moving an object from $x = 1$ to $x = 2$ according to $F = f(x) = x^2$.

$$W = \int_1^2 x^2 \, dx$$

$$= \left. \frac{x^3}{3} \right|_1^2 = \frac{8}{3} - \frac{1}{3}$$

$$= \tfrac{7}{3}$$

Example 3

Hooke's law states that the force required to stretch a spring is directly proportional to the amount that it is stretched. Find the work done in stretching a spring 3 in. if it requires 12 lb of force to stretch it 10 in.

Let x represent the distance stretched by the force $F = f(x)$. Then by Hooke's law

$$f(x) = kx$$

At $x = 10$ we know that $f(10) = 12$, so

$$12 = k(10)$$

$$k = \tfrac{6}{5}$$

Then

$$f(x) = \tfrac{6}{5}x$$

and

$$W = \int_a^b f(x)\, dx$$

$$= \int_0^3 \tfrac{6}{5}x\, dx$$

$$= \left. \frac{6}{5}\left(\frac{x^2}{2}\right) \right|_0^3$$

$$= \tfrac{6}{5}(\tfrac{9}{2} - 0)$$

$$= \tfrac{27}{5} \text{ in.-lb}$$

Example 4

Two charged particles separated by a distance x (in metres) attract each other with a force $F = 4.65 \times 10^{-20}x^{-2}$ newton. Find the work done (in joules) in separating them over an interval from $x = 0.01$ m to $x = 0.1$ m.

$$W = \int_{0.01}^{0.1} 4.65 \times 10^{-20}x^{-2}\, dx$$

$$= 4.65 \times 10^{-20} \int_{0.01}^{0.1} x^{-2}\, dx$$

$$= 4.65 \times 10^{-20} \left(\left. \frac{x^{-1}}{-1} \right|_{0.01}^{0.1} \right)$$

$$= 4.65 \times 10^{-20} \left(\frac{-1}{0.1} - \frac{-1}{0.01} \right)$$

$$= 4.65 \times 10^{-20}(-10 + 100)$$

$$= 4.19 \times 10^{-18} \text{ N-m}$$

$$= 4.19 \times 10^{-18} \text{ J}$$

Note: The metric system unit of work is the joule (J). $1 \text{ J} = 1 \text{ N} \cdot \text{m}$.

Example 5

A chain 50 ft long and weighing 4 lb/ft is hanging from a pulley. (a) How much work is needed to pull 30 ft of the chain to the top? (b) How much work is needed to pull all of the chain to the top?

The force needed to lift the chain at any one time equals the weight of the chain hanging down at that time. If x feet of chain are hanging down and the chain weighs 4 lb/ft, the force is

$$f(x) = F = 4x$$

(a)
$$W = \int_{20}^{50} 4x \, dx$$

$$= 2x^2 \Big|_{20}^{50} = 4200 \text{ ft-lb}$$

(b)
$$W = \int_{0}^{50} 4x \, dx$$

$$= 2x^2 \Big|_{0}^{50} = 5000 \text{ ft-lb}$$

Example 6

A cylindrical tank 10 ft in diameter and 12 ft high is full of water. How much work is needed to pump all the water out over the top? The density of water $\rho = 62.4$ lb/ft^3.

First, divide the tank into n layers, each of thickness Δx as in Fig. 5.58. Let x be the distance that each layer travels as it is pumped to the top. The force is the weight of each layer of water. $F = \rho V$.

Each layer is in the shape of a cylinder, whose volume is given by

$$V = \pi r^2 h$$

So

$$F = \rho V = 62.4\pi(5^2) \, \Delta x$$

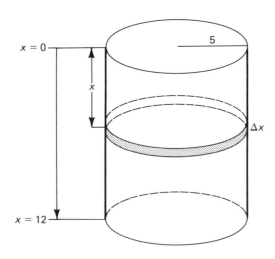

Figure 5.58

Applications of Integration   Chap. 5

is the force of each layer. Since each layer travels a distance of x feet, the work needed to move one layer to the top is

$$W = F \cdot s = 62.4\pi(5^2)\,\Delta x \cdot x$$
$$= 1560\pi x\,\Delta x$$

Summing the work of such layers gives

$$W = \int_0^{12} 1560\pi x\, dx$$
$$= 780\pi\, x^2 \,\big|_0^{12}$$
$$= 112{,}320\pi \text{ ft-lb}$$

Fluid Pressure

Hydrostatic pressure deals with the characteristics of liquids at rest and the pressure the liquid exerts on an immersed object. As you probably know, the pressure increases as you go deeper in water. Fluids are different in this respect from solids in that, where solids exert only a downward force due to gravity, the force exerted by fluids is in all directions. The pressure at any given depth in a fluid is due to its weight. This relationship may be expressed by

$$p = \rho g h$$

where p is the pressure, ρ is the mass density of the fluid, g is the force of gravity, and h is the height or depth of the fluid.

For example, if you are in a swimming pool 12 ft below the surface, the pressure you feel is

$$p = \rho g h$$
$$p = (62.4 \text{ lb/ft}^3)(12 \text{ ft}) \qquad (\rho g = 62.4 \text{ lb/ft}^3)$$
$$= 748.8 \text{ lb/ft}^2$$

The total force is given by

$$F = pA$$

Our main interest in fluid pressure is determining the total force exerted by a fluid upon the walls of its container. If the container has vertical sides, we simply calculate the total force on the *bottom* of the container; that is,

$$F = \rho g h A$$

For example, the total force on the bottom of a rectangular swimming pool 12 ft × 30 ft when the water is 8 ft deep is

$$F = (62.4 \text{ lb/ft}^3)(8 \text{ ft})(12 \text{ ft} \times 30 \text{ ft})$$
$$F = 180{,}000 \text{ lb} \quad \text{(approx.)}$$

The more difficult problem is finding the total force against the *vertical sides* of a container because the pressure is not constant. The pressure increases as the depth increases.

Let a vertical plane region be submerged into a fluid of constant density ρ as shown in Figure 5.59. We need to find the total force against this region from depth $h - a$ to $h - b$. First, divide the interval $a \le y \le b$ into n rectangles

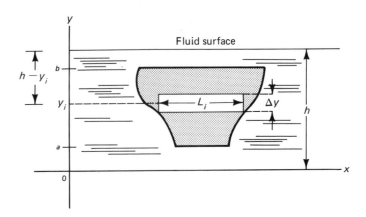

Figure 5.59

each of width Δy. The ith rectangle has length L_i, area $L_i\,\Delta y$, and depth $h - y_i$. The force on the ith rectangle is

$$\Delta F_i = \rho g(h - y_i)L_i\,\Delta y$$

Summing the forces on all such rectangles gives the following integral.

Force Exerted by a Fluid

The force F exerted by a fluid of constant mass density ρ against a submerged vertical plane region from $y = a$ to $y = b$ is given by

$$F = \rho g \int_a^b (h - y)L\,dy$$

where h is the total depth of the fluid and L is the horizontal length of the region at y.

Note: In the metric system, the mass density ρ must be known and $g = 9.80\ \text{m/s}^2$. In the English system, the weight density, ρg, must be known.

Example 7

    A vertical gate in a dam is in the shape of an isosceles trapezoid 12 ft across the top and 8 ft across the bottom, with a height of 10 ft. Find the total force against the gate if the water surface is at the top of the gate.

        The solution can be simplified if we position the trapezoid in the plane as shown in Fig. 5.60.

        The equation of the line through $(4, 0)$ and $(6, 10)$ is

$$y - 0 = 5(x - 4)$$

$$x = \frac{y + 20}{5}$$

The width of the ith rectangle is Δy; its length L may be written as

$$L = 2x = 2\left(\frac{y + 20}{5}\right) = \frac{2}{5}(y + 20)$$

Applications of Integration   Chap. 5

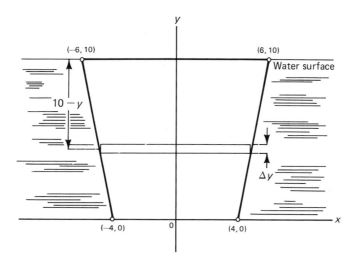

Figure 5.60

and its depth is $10 - y$. So

$$F = \rho g \int_a^b (h - y) L \, dy$$

$$F = 62.4 \int_0^{10} (10 - y)\frac{2}{5}(y + 20) \, dy$$

$$= 24.96 \int_0^{10} (10 - y)(y + 20) \, dy$$

$$= 24.96 \int_0^{10} (200 - 10y - y^2) \, dy$$

$$= 24.96 \left(200y - 5y^2 - \frac{y^3}{3} \right) \Big|_0^{10}$$

$$= 24.96 \left(2000 - 500 - \frac{1000}{3} \right)$$

$$= 29,120 \text{ lb}$$

Example 8

A vertical gate in a dam is semicircular and has a diameter of 8 m. Find the total force against the gate if the water level is 1 m from the top of the gate.

Here place the semicircle with center at the origin as in Fig. 5.61. Its equation is $x^2 + y^2 = 16$. Other positions make the integration more difficult.

$$L = 2x = 2 \sqrt{16 - y^2}$$

The depth of the rectangle is $0 - y$, or $-y$. $\rho = 1000 \text{ kg/m}^3$ and $g = 9.80 \text{ m/s}^2$.

$$F = \rho g \int_a^b (h - y) L \, dy$$

$$F = 9800 \int_{-4}^{-1} (-y) 2 \sqrt{16 - y^2} \, dy$$

$$= -19{,}600 \int_{-4}^{-1} y(16 - y^2)^{1/2} \, dy$$

$$= -19{,}600 \int yu^{1/2} \, \frac{du}{-2y}$$

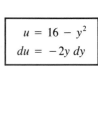

$$u = 16 - y^2$$
$$du = -2y \, dy$$

$$= 9800 \int u^{1/2} \, du$$

$$= 9800 \cdot \frac{u^{3/2}}{3/2}$$

$$= \frac{19{,}600}{3}(16 - y^2)^{3/2} \Big|_{-4}^{-1}$$

$$= \frac{19{,}600}{3}(15^{3/2} - 0)$$

$$= 379{,}600 \text{ N}$$

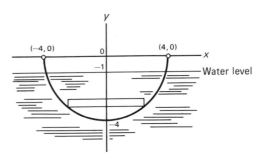

Figure 5.61

Example 9

A swimming pool is 15 ft wide and 20 ft long. The bottom is flat but sloped so that the water is 4 ft deep at one end and 12 ft deep at the other end. Find the force of the water on one 20-ft side.

First, let's position the vertical side in the plane as shown in Fig. 5.62. Notice that the right ends of the horizontal strips sometimes are on the vertical line $x = 20$ and sometimes on the line through $(0, 0)$ and $(20, 8)$, whose equation is $y = 2x/5$. Thus we need two integrals. For each integral $L = x$.

$$F = 62.4 \int_0^8 (12 - y)x \, dy + 62.4 \int_8^{12} (12 - y)x \, dy$$

$$= 62.4 \int_0^8 (12 - y)\left(\frac{5}{2}y\right) dy + 62.4 \int_8^{12} (12 - y)(20) \, dy$$

$$= 156 \int_0^8 (12y - y^2) \, dy + 1248 \int_8^{12} (12 - y) \, dy$$

$$= 156 \left(6y^2 - \frac{y^3}{3}\right)\Big|_0^8 + 1248 \left(12y - \frac{y^2}{2}\right)\Big|_8^{12}$$

$$= 33{,}280 + 9984$$

$$= 43{,}264 \text{ lb}$$

Note: The position of the vertical side in the number plane has no effect on the final result. You should repeat this problem using other positions.

Average Value

The average value of a function is another application of integration. The sum

$$A.V. = \frac{f(x_1) + f(x_2) + f(x_3) + \cdots + f(x_n)}{n}$$

is the average value of a function for the given values of x: $x_1, x_2, x_3, \ldots, x_n$. For example, if $f(x_1) = 3$, $f(x_2) = 5$, and $f(x_3) = 4$, then

$$A.V. = \frac{3 + 5 + 4}{3} = 4$$

is the average of the three given values.

Now

$$A.V. = \frac{f(x_1) + f(x_2) + f(x_3) + \cdots + f(x_n)}{n}$$

$$= \frac{[f(x_1) + f(x_2) + f(x_3) + \cdots + f(x_n)]\,\Delta x}{n\,\Delta x}$$

$$= \frac{1}{n\,\Delta x}[f(x_1)\,\Delta x + f(x_2)\,\Delta x + f(x_3)\,\Delta x + \cdots + f(x_n)\,\Delta x]$$

In Fig. 5.63 $n\,\Delta x = b - a$, so that

$$A.V. = \frac{1}{b - a}[f(x_1)\,\Delta x + f(x_2)\,\Delta x + f(x_3)\,\Delta x + \cdots + f(x_n)\,\Delta x]$$

As we let Δx approach 0, the right-hand factor approaches the integral

$$\int_a^b f(x)\,dx$$

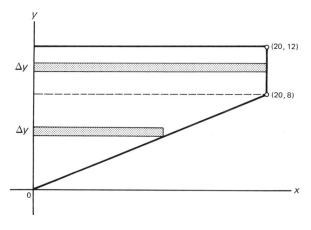

Figure 5.62

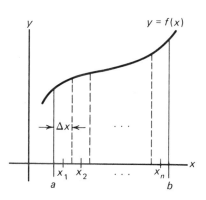

Figure 5.63

Sec. 5.7   Work, Fluid Pressure, and Average Value

219

We then consider

Average Value
The *average value* of the function $y = f(x)$ over the interval $x = a$ to $x = b$ is

$$y_{av} = \frac{1}{b-a} \int_a^b f(x)\, dx$$

A geometrical interpretation can be given to y_{av}. y_{av} is the height of a rectangle with base, $b - a$, having the same area as the area bounded by the curve $y = f(x)$, $x = a$, $x = b$, and the x-axis as shown in Figure 5.64.

Example 10

Find the average value of the curve $y = 4 - x^2$ from $x = 0$ to $x = 2$.

$$A = \int_0^2 (4 - x^2)\, dx = \left(4x - \frac{x^3}{3} \right)\bigg|_0^2 = \frac{16}{3}$$

then

$$y_{av} = \frac{1}{2-0} \int_0^2 (4 - x^2)\, dx = \left(\frac{1}{2}\right)\left(\frac{16}{3}\right) = \frac{8}{3}$$

The rectangle with height $\frac{8}{3}$ and width 2 has area also equal to A: $(\frac{8}{3})(2) = \frac{16}{3}$. (See Fig. 5.65.)

Example 11

The power developed in a resistor is given by $P = 0.5i^3$. Find the average power (in watts) as the current changes from 1 A to 3 A.

$$P_{av} = \frac{1}{3-1} \int_1^3 0.5i^3\, di$$

$$= \frac{0.5}{2} \cdot \frac{i^4}{4}\bigg|_1^3$$

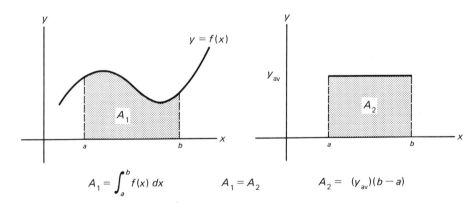

$$A_1 = \int_a^b f(x)\, dx \qquad A_1 = A_2 \qquad A_2 = (y_{av})(b-a)$$

Figure 5.64

Applications of Integration   Chap. 5

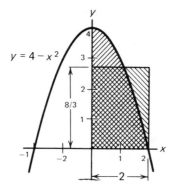

$y = 4 - x^2$

Figure 5.65

$$= \frac{0.5}{2}\left(\frac{81}{4} - \frac{1}{4}\right)$$

$$= 5.0 \text{ W}$$

Exercises

1. A force moves an object from $x = 0$ to $x = 3$ according to the equation $F = x^3 - x$. Find the work done.

2. A force moves an object from $x = 1$ to $x = 100$ according to the equation $F = \sqrt{x}$. Find the work done.

3. Find the work done in stretching a spring 5 in. if it requires a 20-lb force to stretch it 10 in.

4. Find the work done in stretching a spring 12 in. if it requires a 10-lb force to stretch it 8 in.

5. A spring with a natural length of 8 cm measures 12 cm after a 150-N weight is attached. Find the work required to stretch the spring 6 cm from its natural length.

6. A spring with a natural length of 10 cm measures 14 cm after a 60-N weight is attached. Find the work done in stretching the spring 10 cm from its natural length.

7. Two charged particles separated by a distance x attract each other with a force $F = 3.62 \times 10^{-16}x^{-2}$ N. Find the work done (in joules) in separating them over an interval from $x = 0.01$ m to $x = 0.05$ m.

8. Two charged particles separated by a distance x attract each other with a force $F = 4.52 \times 10^{-18}x^{-2}$ N. Find the work done (in joules) in separating them over an interval from $x = 0.02$ m to $x = 0.08$ m.

9. A chain 50 ft long and weighing 2 lb/ft is hanging over a pulley. How much work is needed to pull (a) 10 ft of the chain to the top? (b) half of the chain to the top? (c) all of the chain to the top?

10. Find the amount of work done in winding all of a 300-ft hanging cable that weighs 120 lb.

11. A cylindrical tank 8 ft in diameter and 12 ft high is full of water. How much work is needed to pump all the water out over the top?

12. How much work is needed in Exercise 11 to pump half the water out over the top of the tank?

13. Suppose that the tank in Exercise 11 is placed on a 10-ft platform. How much work is needed to fill the tank from the ground when the water is pumped in through a hole in the bottom of the tank?

14. How much work is needed to fill the tank in Exercise 13 if the water is pumped into the top of the tank?

15. A conical tank (inverted right circular cone) filled with water is 10 ft across the top and 12 ft high. How much work is needed to pump all the water out over the top?

16. How much work is needed in Exercise 15 to pump 4 ft of water out over the top (a) when the tank is full? (b) when the tank has 4 ft of water in it?

17. A dam contains a vertical rectangular gate 10 ft high and 8 ft wide. The top of the gate is at the water's surface. Find the force on the gate.

18. A dam contains a vertical rectangular gate 6 ft high and 8 ft wide. The top of the gate is 4 ft below the water's surface. Find the force on the gate.

19. A rectangular tank is 8 m wide and 4 m deep. If the tank is $\frac{3}{4}$ full of water, find the force against the side.

20. A cylindrical tank is lying on its side and is half-filled with water. If its diameter is 6 m, find the force against an end.

21. A cylindrical tank of oil is half full of oil ($\rho = 870$ kg/m³) and lying on its side. If the diameter of the tank is 10 m, find the force on an end of the tank.

22. A porthole on a vertical side of a ship is 1 ft square. Find the total force on the porthole if its top is 20 ft below the water's surface.

23. A trough is 12 ft long and 2 ft high. Vertical cross sections are isosceles right triangles with the hypotenuse horizontal. Find the force on one end if the trough is filled with water.

24. A trough is 12 m long and 1 m high. Vertical cross sections are equilateral triangles with the top side horizontal. Find the force on one end if the trough is filled with alcohol ($\rho = 790$ kg/m³).

25. A dam has a vertical gate in the shape of an isosceles trapezoid with upper base 10 ft, lower base 16 ft, and height 6 ft. Find the force on the gate if the upper base is 8 ft below the water's surface.

26. Find the force on the gate in Exercise 25 if the gate is inverted.

27. A swimming pool is 12 ft wide and 18 ft long. The bottom is flat but sloped so that the water is 3 ft deep at one end and 9 ft deep at the other end. Find the force on one 18-ft side.

28. A dam is in the shape of a parabola 12 ft high and 8 ft across its top. Find the force on it when the water's surface is at the top.

Find the average value of each function.

29. $y = x^2$ from $x = 1$ to $x = 3$

30. $y = \sqrt{x}$ from $x = 1$ to $x = 4$

31. $y = \dfrac{1}{\sqrt{x - 1}}$ from $x = 5$ to $x = 10$

32. $y = x^2 - 1/x^2$ from $x = 1$ to $x = 3$

33. The electric current for a certain circuit is given by $i = 6t - t^2$. Find the average value of the current (in amperes) over the interval from $t = 0.1$ s to $t = 0.5$ s.

34. The power developed in a resistor is given by $P = 0.28i^3$. Find the average power (in watts) as the current changes from 1 A to 4 A.

CHAPTER SUMMARY

1. The area between two curves $y = f(x)$ and $y = g(x)$ between $x = a$ and $x = b$ is

$$\int_a^b [f(x) - g(x)]\, dx$$

for $f(x) \geqslant g(x)$ and $a \leqslant x \leqslant b$ and the area between two curves $x = f(y)$ and $x = g(y)$ between $y = c$ and $y = d$ is

$$\int_c^d [f(y) - g(y)]\, dy$$

for $f(y) \geqslant g(y)$ and $c \leqslant y \leqslant d$.

2. *Volume of revolution: disk method*

$$V = \pi \int_a^b \overset{\substack{\text{radius} \quad \text{thickness} \\ \downarrow \qquad\quad \downarrow}}{[f(x)]^2\, dx} \qquad\qquad \text{(revolved about } x\text{-axis)}$$

$$V = \pi \int_c^d [f(y)]^2\, dy \qquad\qquad \text{(revolved about } y\text{-axis)}$$

$$V = \pi \int_a^b \{[f(x)]^2 - [g(x)]^2\}\, dx \qquad \text{(washer)}$$

3. *Volume of revolution: shell method*

$$V = 2\pi \int_a^b \overset{\substack{\text{radius} \quad \text{height} \quad \text{thickness} \\ \downarrow \qquad \downarrow \qquad\quad \downarrow}}{x \qquad\; f(x) \qquad\; dx} \qquad \text{(shells parallel to } y\text{-axis)}$$

$$V = 2\pi \int_c^d y \qquad\; f(y) \qquad\; dy \qquad \text{(shells parallel to } x\text{-axis)}$$

4. *Moment in a linear system along the x-axis:* Let $\bar{x}$ be the center of mass of a linear system along the x-axis with n masses; then

$$\bar{x} = \frac{m_1 x_1 + m_2 x_2 + m_3 x_3 + \cdots + m_n x_n}{m_1 + m_2 + m_3 + \cdots + m_n}$$

If we let M_o be the moment about the origin and m be the total mass of the system, then

$$\bar{x} = \frac{M_o}{m}$$

5. *Moments of a two-dimensional system:* Consider n masses $m_1, m_2, \ldots, m_n$ located at points $(x_1, y_1), (x_2, y_2), \ldots, (x_n, y_n)$, respectively. The moment about the y-axis, M_y, is

$$M_y = m_1 x_1 + m_2 x_2 + \cdots + m_n x_n$$

and the moment about the x-axis, M_x, is

$$M_x = m_1 y_1 + m_2 y_2 + \cdots + m_n y_n.$$

If we let m be the total mass of the system, the center of mass $(\bar{x}, \bar{y})$ is

$$\bar{x} = \frac{M_y}{m} \quad \text{and} \quad \bar{y} = \frac{M_x}{m}$$

6. The center of mass of a continuous thin mass of variable density from $x = a$ to $x = b$ is given by

$$x = \frac{M_o}{m} = \frac{\displaystyle\int_a^b \rho(x)\, x\, dx}{\displaystyle\int_a^b \rho(x)\, dx},$$

where $\rho(x)$ is the density expressed as a function of x. If the density is constant, let $\rho(x) = 1$.

7. *Moments and center of mass of a plane area or thin plate:* Let $g(x) \leqslant f(x)$ be continuous functions on $a \leqslant x \leqslant b$ for the area of uniform density ρ bounded by $y = f(x)$, $y = g(x)$, $x = a$, and $x = b$, the moments about the x-axis and the y-axis are

$$M_x = \frac{\rho}{2} \int_a^b \{[f(x)]^2 - [g(x)]^2\}\, dx \quad \text{and} \quad M_y = \rho \int_a^b x[f(x) - g(x)]\, dx$$

Its mass is given by

$$m = \rho \int_a^b [f(x) - g(x)]\, dx$$

and its center of mass $(\bar{x}, \bar{y})$ is

$$\bar{x} = \frac{M_y}{m} \quad \text{and} \quad \bar{y} = \frac{M_x}{m}$$

8. *Centroid of a plane region or thin plate:* Let $g(x) \leqslant f(x)$ be continuous functions on $a \leqslant x \leqslant b$. The centroid $(\bar{x}, \bar{y})$ of the region bounded by $y = f(x)$, $y = g(x)$, $x = a$, and $x = b$ is

$$\bar{x} = \frac{\displaystyle\int_a^b x[f(x) - g(x)]\, dx}{A} \quad \text{and} \quad \bar{y} = \frac{\dfrac{1}{2}\displaystyle\int_a^b [f(x)]^2 - [g(x)]^2\, dx}{A}$$

where A is the area of the region.

9. *Centroid of a volume of a solid of revolution*

$$\bar{x} = \frac{M_y}{m} = \frac{\displaystyle\int_a^b xy^2\, dx}{\displaystyle\int_a^b y^2\, dx} \quad \text{and} \quad \bar{y} = 0 \qquad \text{(revolved about } x\text{-axis)}$$

$$\bar{y} = \frac{M_x}{m} = \frac{\int_c^d yx^2\, dy}{\int_c^d x^2\, dy} \quad \text{and} \quad \bar{x} = 0 \qquad \text{(revolved about } y\text{-axis)}$$

10. *Moment of inertia of a system:* Let masses $m_1, m_2, m_3, \ldots, m_n$ be at distances $d_1, d_2, d_3, \ldots, d_n$, respectively, rotating about some axis. The moment of inertia, I, of the system is

$$I = m_1 d_1^2 + m_2 d_2^2 + m_3 d_3^2 + \cdots + m_n d_n^2$$

Let m be the sum of all the masses in the system and let R be the distance of m from the axis that gives the same total moment of inertia

$$I = mR^2$$

R is called the radius of gyration.

11. *Moments of inertia of an area of constant density about the x- and y-axes:* Let $g(x) \leqslant f(x)$ be continuous functions on $a \leqslant x \leqslant b$ for the area of constant density ρ bounded by $y = f(x)$, $y = g(x)$, $x = a$, and $x = b$. The moment of inertia about the y-axis is

$$I_y = \rho \int_a^b x^2[f(x) - g(x)]\, dx \quad R = \sqrt{\frac{I_y}{m}}$$

Similarly, the moment of inertia about the x-axis is

$$I_x = \rho \int_c^d y^2[f(y) - g(y)]\, dy \qquad R = \sqrt{\frac{I_x}{m}}$$

12. *Moment of inertia of a solid of revolution:* Let the area bounded by $y = f(x)$, $y = 0$, $x = a$, and $x = b$ be revolved about the y-axis. The moment of inertia about the y-axis is

$$I_y = 2\pi\rho \int_a^b x^3 f(x)\, dx$$

Its radius of gyration is $R = \sqrt{\dfrac{I_y}{m}}$, where $m = 2\pi\rho \displaystyle\int_a^b x\, f(x)\, dx$.

Similarly, let the area bounded by $x = f(y)$, $x = 0$, $y = c$, and $y = d$ be revolved about the x-axis. The moment of inertia about the x-axis is

$$I_x = 2\pi\rho \int_c^d y^3 f(y)\, dy$$

Its radius of gyration is $R = \sqrt{\dfrac{I_x}{m}}$, where $m = 2\pi\rho \displaystyle\int_c^d y\, f(y)\, dy$.

Note: $f(x)$ and $f(y)$ correspond to the height of the shell.

13. *Work of a variable force $f(x)$ acting through the distance from $x = a$ to $x = b$*

$$W = \int_a^b f(x)\, dx$$

14. *Force exerted by a fluid:* The force F exerted by a fluid of constant density ρ against a submerged vertical plane region from $y = a$ to $y = b$ is

$$F = \rho g \int_a^b (h - y)L\, dy$$

where h is the total depth of the fluid and L is the horizontal length of the region at y.

15. *Average value:* The average value of the function $y = f(x)$ over the interval $x = a$ to $x = b$ is

$$y_{av} = \frac{1}{b - a} \int_a^b f(x)\, dx$$

CHAPTER REVIEW

Find each area bounded by the curves.
1. $y = x^2 + 3$, $y = 0$, $x = 1$, and $x = 2$.
2. $y = 1 - x^2$, $y = 0$, and $x = 0$.   3. $x = y^2 - y^3$ and the y-axis.
4. $y = 3x^2 - 12x + 9$, $y = 0$, $x = 0$, and $x = 4$.
5. $x = y^4 - 2y^2$ and $x = 2y^2$.   6. $x = y^2$ and $x = 9$.

Find the volume of each solid formed by revolving the region bounded by the given curves about the given line.
7. $y = \sqrt{x}$, $y = 0$ and $x = 4$ about the x-axis (shell method).
8. $y = \sqrt{x}$, $y = 0$ and $x = 4$ about the x-axis (disk method).
9. $y = x - x^2$ and $y = 0$ about the x-axis.
10. $y = x$ and $y = 3x - x^2$ about the y-axis.
11. $y = 3x^2 - x^3$ and $y = 0$ about the y-axis.
12. $y = x^2 + 1$, $y = 0$, $x = 0$, and $x = 3$ about the y-axis.
13. $x = y^2$ and $x = 4$ about y-axis
14. $x = 4y - y^2$, $x = 0$, and $y = 3$ about x-axis
15. Find the center of mass of the linear system: $m_1 = 12$, $x_1 = -4$; $m_2 = 20$, $x_2 = 9$; $m_3 = 24$, $x_3 = 12$
16. Find the center of mass of the system: $m_1 = 24$ at $(11, -3)$; $m_2 = 36$ at $(-4, -15)$; $m_3 = 30$ at $(-7, 0)$
17. Find the center of mass of the uniform thin plate in Fig. 5.66.

Find the centroid of each region bounded by the given curves.
18. $y = 5x$ and $x = 4$ 19. $y = 6x - x^2$ and $y = 3x$
20. $y = x^2 - 6x$ and $y = 0$

Find the centroid of the solid formed by revolving each region bounded by the given curves about the given axis.
21. $y = 2x$, $x = 0$, and $y = 2$ about y-axis

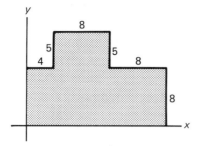

Figure 5.66

22. $y = x^2$, $x = 0$, and $y = 1$ about x-axis

23. $x = y^2 - 4y$ and $x = 0$ about y-axis

24. Find the moment of inertia and the radius of gyration about the x-axis of the system: $m_1 = 10$ at $(3, 2)$; $m_2 = 6$ at $(5, 7)$; $m_3 = 8$ at $(8, -4)$

Find the moment of inertia and the radius of gyration of each region bounded by the given curves about the given axis.

25. $y = 3x$, $x = 4$, and $y = 0$ about y-axis. $\rho = 1$.

26. Region in Exercise 25 about x-axis.

27. $x = 1 - y^2$, $y = 0$, and $x = 0$ about x-axis. $\rho = 4$.

Find the moment of inertia and the radius of gyration of the solid formed by revolving the region bounded by the given curves about the given axis.

28. $y = x^3$, $y = 0$, and $x = 1$ about x-axis. $\rho = 4$.

29. Region of Exercise 28 about y-axis.

30. $y = 1/x$, $y = 0$, $x = 1$, and $x = 4$ about y-axis. $\rho = 3$.

31. Find the work done in stretching a spring 10 in. if it requires a 16-lb force to stretch it 4 in.

32. Two charged particles separated by a distance x attract each other with a force $F = 5.24 \times 10^{-18}x^{-2}$ N. Find the work done (in joules) in separating them over an interval from $x = 0.01$ m to $x = 0.02$ m.

33. A cable weighs 4 lb/ft and has a 250-lb weight attached in a hole 200 ft below the ground. Find the amount of work needed to pull the cable and weight to ground level.

34. A dam contains a vertical rectangular gate 8 ft high and 10 ft wide. The top of the gate is 6 ft below the water's surface. Find the force on the gate.

35. A cylindrical tank, 10 m in diameter, is lying on its side and half full of water. Find the force against an end.

36. Find the average value of the voltage, V_{av}, in an electrical circuit from $t = 0$ s to $t = 3$ s if $V = t^2 + 3t + 2$.

37. Find the average value of the current, i_{av}, in an electrical circuit from $t = 4$ s to $t = 9$ s if $i = 4t^{3/2}$.

38. The power in a circuit varies according to $p = 2t^3$. Find the average power p_{av} (in watts) from $t = 1$ s to $t = 3$ s.

6

DERIVATIVES OF TRANSCENDENTAL FUNCTIONS

6.1 THE TRIGONOMETRIC FUNCTIONS

Before we begin to develop the calculus of the trigonometric functions, we briefly review some of the basics of trigonometry.

An angle is in **standard position** when its vertex is located at the origin and its initial side is lying on the positive x-axis. An angle resulting from a counter-clockwise rotation, as indicated by the direction of the arrow, is a **positive angle**. But if the rotation is clockwise, the angle is **negative**. (See Fig. 6.1.)

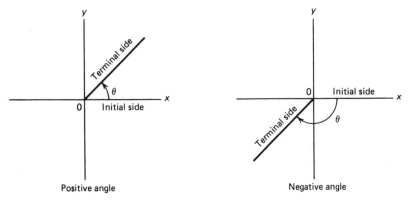

Positive angle

Negative angle

Figure 6.1

There are six trigonometric functions associated with angle θ in standard position. They can be expressed in terms of the coordinates of the point P as ratios where point P is on the terminal side of angle θ as in Fig. 6.2.

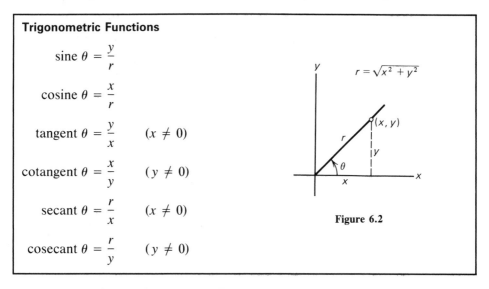

Trigonometric Functions

$$\text{sine } \theta = \frac{y}{r}$$

$$\cos\text{ine } \theta = \frac{x}{r}$$

$$\text{tangent } \theta = \frac{y}{x} \qquad (x \neq 0)$$

$$\text{cotangent } \theta = \frac{x}{y} \qquad (y \neq 0)$$

$$\text{secant } \theta = \frac{r}{x} \qquad (x \neq 0)$$

$$\text{cosecant } \theta = \frac{r}{y} \qquad (y \neq 0)$$

$r = \sqrt{x^2 + y^2}$

Figure 6.2

We must be careful to watch for angles where the point P on the terminal side has its abscissa, x, or its ordinate, y, equal to zero. Since we cannot divide by zero, $\tan \theta$ and $\sec \theta$ do not exist when $x = 0$. Likewise, $\cot \theta$ and $\csc \theta$ do not exist when $y = 0$.

Two or more angles in standard position are *coterminal* if they have the same terminal side. For example, 60°, 420°, and $-300°$ are coterminal.

A **quadrantal angle** is one which, when in standard position, has its terminal side coinciding with one of the axes.

Example 1

Find the values of $\sin \theta$, $\cos \theta$, and $\tan \theta$ if θ is in standard position and its terminal side passes through the point $(-3, 4)$. (See Fig. 6.3.)

$$r = \sqrt{x^2 + y^2} = \sqrt{9 + 16} = 5$$

$$\sin \theta = \frac{y}{r} = \frac{4}{5}$$

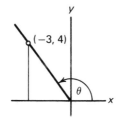

Figure 6.3

$$\cos \theta = \frac{x}{r} = \frac{-3}{5} = -\frac{3}{5}$$

$$\tan \theta = \frac{y}{x} = \frac{4}{-3} = -\frac{4}{3}$$

The **reference angle**, α, of any nonquadrantal angle, θ, in standard position is the *acute* angle between the terminal side of θ and the x-axis. Angle α is always considered to be a positive angle less than 90°.

Example 2

Find the reference angle α for each given angle θ. (See Fig. 6.4.)

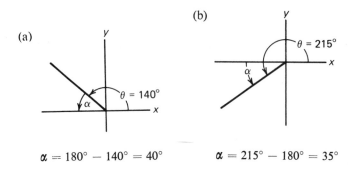

$$\alpha = 180° - 140° = 40° \qquad\qquad \alpha = 215° - 180° = 35°$$

Figure 6.4

The reference angle α is often used to determine the angle θ when the value of the trigonometric function is known.

Example 3

Given $\cos \theta = -\frac{1}{2}$, find θ for $0° \leqslant \theta < 360°$.

Using a calculator or tables, we find that $\alpha = 60°$. The cosine function is negative in quadrants II and III. The second-quadrant angle is $180° - 60° = 120°$. The third-quadrant angle is $180° + 60° = 240°$. (See Fig. 6.5.)

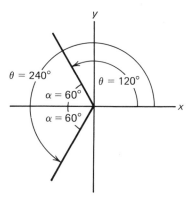

Figure 6.5

Angles are commonly given in degree measure and in radian measure. Since

$$1 \text{ revolution} = 360°$$

Derivatives of Transcendental Functions   Chap. 6

and
$$1 \text{ revolution} = 2\pi \text{ rad}$$
the basic relationship between radians and degrees is
$$\pi \text{ rad} = 180°$$

The degree and radian measure of several common angles are given in the table below.

VALUES OF COMMON ANGLES

Radians	0	$\dfrac{\pi}{6}$	$\dfrac{\pi}{4}$	$\dfrac{\pi}{3}$	$\dfrac{\pi}{2}$	$\dfrac{2\pi}{3}$	$\dfrac{3\pi}{4}$	$\dfrac{5\pi}{6}$	π	$\dfrac{3\pi}{2}$	2π
Degrees	$0°$	$30°$	$45°$	$60°$	$90°$	$120°$	$135°$	$150°$	$180°$	$270°$	$360°$
$\sin\theta$	0	$\dfrac{1}{2}$	$\dfrac{\sqrt{2}}{2}$	$\dfrac{\sqrt{3}}{2}$	1	$\dfrac{\sqrt{3}}{2}$	$\dfrac{\sqrt{2}}{2}$	$\dfrac{1}{2}$	0	-1	0
$\cos\theta$	1	$\dfrac{\sqrt{3}}{2}$	$\dfrac{\sqrt{2}}{2}$	$\dfrac{1}{2}$	0	$-\dfrac{1}{2}$	$-\dfrac{\sqrt{2}}{2}$	$-\dfrac{\sqrt{3}}{2}$	-1	0	1
$\tan\theta$	0	$\dfrac{\sqrt{3}}{3}$	1	$\sqrt{3}$	undefined	$-\sqrt{3}$	-1	$-\dfrac{\sqrt{3}}{3}$	0	undefined	0
$\cot\theta$	undefined	$\sqrt{3}$	1	$\dfrac{\sqrt{3}}{3}$	0	$-\dfrac{\sqrt{3}}{3}$	-1	$-\sqrt{3}$	undefined	0	undefined
$\sec\theta$	1	$\dfrac{2\sqrt{3}}{3}$	$\sqrt{2}$	2	undefined	-2	$-\sqrt{2}$	$-\dfrac{2\sqrt{3}}{3}$	-1	undefined	1
$\csc\theta$	undefined	2	$\sqrt{2}$	$\dfrac{2\sqrt{3}}{3}$	1	$\dfrac{2\sqrt{3}}{3}$	$\sqrt{2}$	2	undefined	-1	undefined

Table 6 of Appendix B lists the trigonometric identities commonly used in any trigonometry course, which we assume you have successfully completed as a prerequisite to this course.

In this course trigonometric identities will be used to simplify trigonometric expressions, to change a given trigonometric expression into a different but equivalent expression that is easier to differentiate or integrate, and to compare results of integration by different methods to see that they are equivalent.

Example 4

Prove $\sin^2\theta + \sin^2\theta \tan^2\theta = \tan^2\theta$.

$$\begin{aligned}
\sin^2\theta + \sin^2\theta \tan^2\theta &= \sin^2\theta\,(1 + \tan^2\theta) \\
&= \sin^2\theta \sec^2\theta \\
&= \sin^2\theta \left(\frac{1}{\cos^2\theta}\right) \\
&= \tan^2\theta
\end{aligned}$$

Therefore, $\sin^2\theta + \sin^2\theta \tan^2\theta = \tan^2\theta$.

Example 5

Prove $\dfrac{\cos x}{1 + \sin x} = \dfrac{1 - \sin x}{\cos x}$.

Multiply the numerator and the denominator of the right-hand side by $1 + \sin x$.

$$\frac{1 - \sin x}{\cos x} \cdot \frac{1 + \sin x}{1 + \sin x} = \frac{1 - \sin^2 x}{\cos x(1 + \sin x)}$$

$$= \frac{\cos^2 x}{\cos x(1 + \sin x)}$$

$$= \frac{\cos x}{1 + \sin x}$$

Therefore, $\dfrac{\cos x}{1 + \sin x} = \dfrac{1 - \sin x}{\cos x}$.

Example 6

Prove $\dfrac{1 - \cos 2x}{\sin 2x} = \tan x$.

$$\frac{1 - \cos 2x}{\sin 2x} = \frac{1 - (1 - 2\sin^2 x)}{2 \sin x \cos x}$$

$$= \frac{2 \sin^2 x}{2 \sin x \cos x}$$

$$= \frac{\sin x}{\cos x}$$

$$= \tan x$$

Therefore, $\dfrac{1 - \cos 2x}{\sin 2x} = \tan x$.

Example 7

Simplify $2 \sin 3x \cos 3x$.

Using Formula 24, Table 6,

$$2 \sin 3x \cos 3x = \sin 2(3x) = \sin 6x$$

Example 8

Simplify $1 - 2 \cos^2 5x$.

Using Formula 25b, Table 6,

$$1 - 2 \cos^2 5x = -\cos 2(5x) = -\cos 10x$$

Example 9

Simplify $\cos 2\theta \cos 3\theta - \sin 2\theta \sin 3\theta$.

By Formula 20, Table 6,

$$\cos 2\theta \cos 3\theta - \sin 2\theta \sin 3\theta = \cos (2\theta + 3\theta) = \cos 5\theta$$

Figure 6.6 shows the graphs of each of the six basic trigonometric functions. Recall that each is periodic. The *period* is the length of each cycle; that is, the period is the horizontal distance between any point on the curve and the next corresponding point in the next cycle where the graph begins to repeat itself.

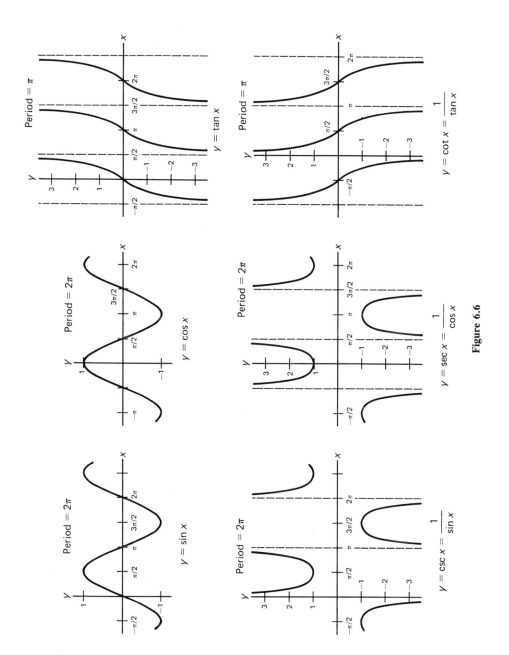

Figure 6.6

The *amplitude* of the sine and cosine functions is one-half the distance difference between the maximum and minimum values of the function.

Recall the following graphing facts and relationships for the six trigonometric functions.

$$\text{Period} = \frac{2\pi}{b} \left. \begin{cases} y = a \sin (bx + c) \\ y = a \cos (bx + c) \\ y = a \sec (bx + c) \\ y = a \csc (bx + c) \end{cases} \right\} \text{amplitude} = |a|$$

$$\text{Period} = \frac{\pi}{b} \begin{cases} y = a \tan (bx + c) \\ y = a \cot (bx + c) \end{cases}$$

Phase shift $= \dfrac{c}{b}$:

to the *left* if $\dfrac{c}{b} > 0$ and

to the *right* if $\dfrac{c}{b} < 0$

Example 10

Graph $y = 2 \cos 3x$.

The amplitude is 2 and the period is $P = \dfrac{2\pi}{b} = \dfrac{2\pi}{3}$. (See Fig. 6.7.)

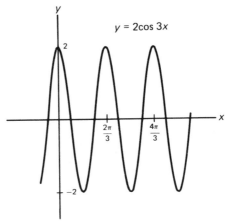

$y = 2\cos 3x$

Figure 6.7

Example 11

Graph $y = -3 \sin \tfrac{1}{2}x$.

The amplitude is 3; the period is $\dfrac{2\pi}{b} = \dfrac{2\pi}{\frac{1}{2}} = 4\pi$.

The effect of the negative sign is to flip, or invert, the curve $y = 3 \sin \tfrac{1}{2}x$. (See Fig. 6.8.)

Example 12

Graph $y = 2 \cos \left(2x + \dfrac{\pi}{4} \right)$.

The amplitude is 2. The period is $\dfrac{2\pi}{b} = \dfrac{2\pi}{2} = \pi$. The phase shift is $\dfrac{c}{b} = \dfrac{\pi/4}{2} = \dfrac{\pi}{8}$, or $\dfrac{\pi}{8}$ to the left. (See Fig. 6.9.)

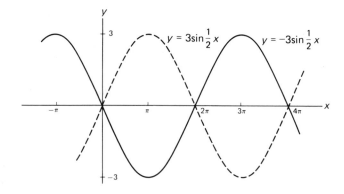

Figure 6.8

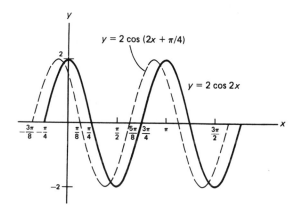

Figure 6.9

Exercises

Prove each identity.

1. $\cos x \tan x = \sin x$

2. $(1 - \cos^2 x) \csc^2 x = 1$

3. $(\cot^2 x + 1) \tan^2 x = \sec^2 x$

4. $\cos \theta (\csc \theta - \sec \theta) = \cot \theta - 1$

5. $\tan^2 \theta - \tan^2 \theta \sin^2 \theta = \sin^2 \theta$

6. $\dfrac{\sin^2 x}{1 + \cos x} = 1 - \cos x$

7. $\dfrac{\cos x \tan x + \sin x}{\tan x} = 2 \cos x$

8. $\cos^4 \theta - \sin^4 \theta = 2 \cos^2 \theta - 1$

9. $\dfrac{1 + \sec x}{\csc x} = \sin x + \tan x$

10. $\dfrac{1 + \tan^2 x}{\tan^2 x} = \csc^2 x$

11. $(\sec x - \tan x)(\csc x + 1) = \cot x$

12. $\dfrac{1 - \cos \theta}{\cos \theta \tan \theta} = \dfrac{\sin \theta}{1 + \cos \theta}$

13. $\dfrac{1 - \tan x}{1 + \tan x} = \dfrac{\cot x - 1}{\cot x + 1}$

14. $\dfrac{1 - \sin^2 x}{1 - \cos^2 x} = \cot^2 x$

15. $\sin (x + \pi) = -\sin x$

16. $\cos (x + 180°) = -\cos x$

17. $\sin (x + 2\pi) = \sin x$

18. $\tan (x + \pi) = \tan x$

19. $\tan (\pi - x) = -\tan x$

20. $\sin (90° - \theta) = \cos \theta$

21. $\cos\left(\dfrac{\pi}{2} - \theta\right) = \sin\theta$

22. $\cos\left(\dfrac{\pi}{2} + \theta\right) = -\sin\theta$

23. $\cos(x + y)\cos(x - y) = \cos^2 x - \sin^2 y$

24. $\sin(x + y)\sin(x - y) = \sin^2 x - \sin^2 y$

25. $(\sin x + \cos x)^2 = 1 + \sin 2x$

26. $\sin 2x = \dfrac{2\tan x}{1 + \tan^2 x}$

27. $\dfrac{1 - \tan^2 x}{1 + \tan^2 x} = \cos 2x$

28. $2\tan x \csc 2x = \sec^2 x$

29. $\cot 2x = \dfrac{\cot^2 x - 1}{2\cot x}$

30. $\sec 2x = \dfrac{\sec^2 x}{2 - \sec^2 x}$

31. $\tan x + \cot 2x = \csc 2x$

32. $\sin^2\dfrac{x}{2} = \dfrac{\sec x - 1}{2\sec x}$

33. $2\cos^2\dfrac{\theta}{2} = \dfrac{1 + \sec\theta}{\sec\theta}$

34. $\sec^2\dfrac{x}{2} = \dfrac{2}{1 + \cos x}$

35. $\tan\dfrac{x}{2} = \dfrac{\sin x}{1 + \cos x}$

36. $2\cos\dfrac{x}{2} = (1 + \cos x)\sec\dfrac{x}{2}$

Simplify each expression.

37. $\sin\theta\cos 3\theta + \cos\theta\sin 3\theta$

38. $\cos 2\theta\cos\theta - \sin 2\theta\sin\theta$

39. $\cos 4\theta\cos 3\theta + \sin 4\theta\sin 3\theta$

40. $\sin 2\theta\cos 3\theta - \cos 2\theta\sin 3\theta$

41. $\dfrac{\tan 3\theta + \tan 2\theta}{1 - \tan 3\theta\tan 2\theta}$

42. $\dfrac{\tan\theta - \tan 2\theta}{1 + \tan\theta\tan 2\theta}$

43. $\sin(\theta + \phi) + \sin(\theta - \phi)$

44. $\cos(\theta + \phi) + \cos(\theta - \phi)$

45. $2\sin\dfrac{x}{4}\cos\dfrac{x}{4}$

46. $20\sin^2 x\cos^2 x$

47. $1 - 2\sin^2 3x$

48. $\sqrt{\dfrac{1 - \cos 6\theta}{2}}$

49. $\sqrt{\dfrac{1 + \cos\dfrac{\theta}{4}}{2}}$

50. $\cos 2x + 2\sin^2 x$

51. $\cos^2\dfrac{x}{6} - \sin^2\dfrac{x}{6}$

52. $2\sin 3x\cos 3x$

53. $20\sin 4\theta\cos 4\theta$

54. $1 - 2\sin^2 7t$

55. $4 - 8\sin^2\theta$

56. $15\sin\dfrac{x}{6}\cos\dfrac{x}{6}$

Sketch each curve.

57. $y = 2\cos x$

58. $y = \cos 3x$

59. $y = 3\cos 6x$

60. $y = -\dfrac{1}{2}\sin\dfrac{2}{3}x$

61. $y = 2\sin 3\pi x$

62. $y = 3\sin\left(x - \dfrac{\pi}{3}\right)$

63. $y = -\sin\left(4x - \dfrac{2\pi}{3}\right)$

64. $y = -\cos(4x + \pi)$

65. $y = 3\sin\left(\dfrac{1}{2}x - \dfrac{\pi}{4}\right)$

66. $y = 5\cos\left(\dfrac{2}{3}x + \pi\right)$

67. $y = \tan 3x$

68. $y = 4\sin\left(\dfrac{\pi x}{6} - \dfrac{\pi}{3}\right)$

6.2 DERIVATIVES OF THE SINE AND COSINE FUNCTIONS

So far we have differentiated and integrated only algebraic functions; that is, functions in the form $y = f(x)$ where $f(x)$ is an algebraic expression. But many important applications involve the use of nonalgebraic functions which are called *transcendental functions*. The trigonometric and logarithmic functions are examples of transcendental functions. We will first find the derivative of $y = \sin u$.

By the definition of the derivative we have

$$\frac{d}{du}(\sin u) = \frac{dy}{du} = \lim_{\Delta u \to 0} \frac{\Delta y}{\Delta u}$$

$$= \lim_{\Delta u \to 0} \frac{\sin(u + \Delta u) - \sin(u)}{\Delta u}$$

Using the trigonometric identity (33) found in Appendix B, Table 6,

$$\sin A - \sin B = 2 \sin\left(\frac{A - B}{2}\right) \cos\left(\frac{A + B}{2}\right)$$

and letting $A = u + \Delta u$ and $B = u$, we have

$$\sin(u + \Delta u) - \sin u = 2 \sin \tfrac{1}{2}(u + \Delta u - u) \cos \tfrac{1}{2}(u + \Delta u + u)$$

$$= 2 \sin\left(\frac{\Delta u}{2}\right) \cos\left(u + \frac{\Delta u}{2}\right)$$

So now,

$$\frac{dy}{du} = \lim_{\Delta u \to 0} \frac{2 \sin\left(\dfrac{\Delta u}{2}\right) \cos\left(u + \dfrac{\Delta u}{2}\right)}{\Delta u}$$

Next, divide numerator and denominator by 2.

$$\frac{dy}{du} = \lim_{\Delta u \to 0} \frac{\sin\left(\dfrac{\Delta u}{2}\right) \cos\left(u + \dfrac{\Delta u}{2}\right)}{\dfrac{\Delta u}{2}}$$

$$= \lim_{\Delta u \to 0} \frac{\sin\left(\dfrac{\Delta u}{2}\right)}{\dfrac{\Delta u}{2}} \cdot \lim_{\Delta u \to 0} \cos\left(u + \frac{\Delta u}{2}\right)$$

The right-hand factor, $\displaystyle\lim_{\Delta u \to 0} \cos\left(u + \frac{\Delta u}{2}\right)$, is seen to be $\cos u$. We need, however, to determine

$$\lim_{\Delta u \to 0} \frac{\sin\left(\dfrac{\Delta u}{2}\right)}{\dfrac{\Delta u}{2}}.$$

For this purpose we construct the geometric figure as shown in Fig. 6.10.

Angle θ is the central angle (measured in radians) of a circle of radius $r = OA = OD$. Note that:

$$\text{area of triangle } OAD < \text{area of sector } OAD < \text{area of triangle } OAB$$

or

$$(\tfrac{1}{2}r)(r \sin \theta) < \qquad \tfrac{1}{2}r^2\theta \qquad < (\tfrac{1}{2}r)(r \tan \theta)$$

Dividing each term by $\tfrac{1}{2}r^2$, we have

$$\sin \theta < \qquad \theta \qquad < \tan \theta$$

or

$$\sin \theta < \qquad \theta \qquad < \frac{\sin \theta}{\cos \theta} \qquad \left(\tan \theta = \frac{\sin \theta}{\cos \theta}\right)$$

Dividing each term now by $\sin \theta$, we have:

$$1 < \qquad \frac{\theta}{\sin \theta} \qquad < \frac{1}{\cos \theta}$$

By taking the reciprocal of each term, we have

$$1 > \qquad \frac{\sin \theta}{\theta} \qquad > \cos \theta$$

As θ approaches 0, $\cos \theta$ approaches 1 so we have

$$\lim_{\theta \to 0} 1 = 1 \qquad \text{and} \qquad \lim_{\theta \to 0} \cos \theta = 1$$

We conclude that $\dfrac{\sin \theta}{\theta}$ must *also* approach 1 as θ approaches 0, that is,

$$\lim_{\theta \to 0} \frac{\sin \theta}{\theta} = 1$$

So

$$\frac{d}{du}(\sin u) = \lim_{\Delta u \to 0} \frac{\sin\left(\dfrac{\Delta u}{2}\right)}{\dfrac{\Delta u}{2}} \cdot \lim_{\Delta u \to 0} \cos\left(u + \frac{\Delta u}{2}\right) = 1 \cdot \cos u$$

$$= \cos u$$

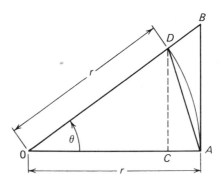

Figure 6.10

Derivatives of Transcendental Functions   Chap. 6

Thus

$$\boxed{\frac{d}{du}(\sin u) = \cos u}$$

We are still unable, however, to find the derivative of a function such as

$$\frac{d}{dx}\sin(x^2 + 3x)$$

To find the derivative of $\sin u$ with respect to x when u is itself a function of x, we use the chain rule from Section 2.7. Recall that if $y = f(u)$ where $u = g(x)$, then $y = f[g(x)]$ is a function of x. Then,

Chain Rule

$$\frac{dy}{dx} = \frac{dy}{du} \cdot \frac{du}{dx}$$

Example 1

Find the derivative of $y = \sin(x^2 + 3x)$.

Let $u = x^2 + 3x$ and $y = \sin u$ so that

$$\frac{dy}{du} = \cos u \quad \text{and} \quad \frac{du}{dx} = 2x + 3$$

and

$$\frac{dy}{dx} = \frac{dy}{du} \cdot \frac{du}{dx} = (\cos u)(2x + 3)$$

$$\frac{dy}{dx} = (2x + 3)\cos(x^2 + 3x)$$

Example 2

Find $\dfrac{d}{dx}(\sin 2x)$.

Let $u = 2x$ and $y = \sin u$.

$$\frac{dy}{dx} = \frac{dy}{du} \cdot \frac{du}{dx}$$

$$= (\cos u)(2)$$

$$= 2\cos 2x$$

Example 3

Find the derivative of $y = \sin(3x - 5)^2$.

Let $u = (3x - 5)^2$ and $y = \sin u$.

$$\frac{dy}{du} = \cos u \quad \text{and} \quad \frac{du}{dx} = 2(3x - 5)(3) = 6(3x - 5)$$

$$\frac{dy}{dx} = \frac{dy}{du} \cdot \frac{du}{dx}$$

$$= (\cos u)[6(3x - 5)]$$
$$= 6(3x - 5) \cos (3x - 5)^2$$

Example 4

Find the derivative of $y = \sin^3(2x - 3)$.

Let $u = 2x - 3$. Then $y = \sin^3(2x - 3) = \sin^3 u$ and

$$\frac{dy}{dx} = \frac{dy}{du} \cdot \frac{du}{dx} = \frac{dy}{du} \cdot 2 \qquad \left[\text{substituting } \frac{du}{dx} = \frac{d}{dx}(2x - 3) = 2\right]$$

To find $\dfrac{dy}{du}$ let $w = \sin u$. Then $y = \sin^3 u = w^3$. So

$$\frac{dy}{du} = \frac{dy}{dw} \cdot \frac{dw}{du}$$

$$= 3w^2 \frac{dw}{du} \qquad \left[\text{substituting } \frac{dw}{du} = \frac{d}{du}(\sin u) = \cos u\right]$$

$$= 3w^2(\cos u)$$

$$= (3 \sin^2 u)(\cos u) \qquad \text{(substituting } w = \sin u)$$

Finally, from above,

$$\frac{dy}{dx} = \frac{dy}{du} \cdot 2$$

$$= 2(3 \sin^2 u \cos u)$$

$$= 6 \sin^2(2x - 3) \cos (2x - 3) \qquad \text{(substituting } u = 2x - 3)$$

To find the derivative of $y = \cos u$, first recall that $\cos u = \sin\left(\dfrac{\pi}{2} - u\right)$
and let $w = \dfrac{\pi}{2} - u$. Then $y = \cos u = \sin\left(\dfrac{\pi}{2} - u\right) = \sin w$. Using the chain rule,

$$\frac{dy}{du} = \frac{dy}{dw} \cdot \frac{dw}{du}$$

and noting that

$$\frac{dy}{dw} = \frac{d}{dw}(\sin w) = \cos w \quad \text{and} \quad \frac{dw}{du} = \frac{d}{du}\left(\frac{\pi}{2} - u\right) = -1$$

then we have

$$\frac{dy}{du} = (\cos w)(-1) = -\cos w$$

$$= -\cos\left(\frac{\pi}{2} - u\right)$$

$$= -\sin u \qquad \left[\text{since } \cos\left(\frac{\pi}{2} - u\right) = \sin u\right]$$

Thus

$$\frac{d}{du}(\cos u) = -\sin u$$

Example 5

Find the derivative of $y = \cos (5x^2 + x)$.

Let $u = 5x^2 + x$ and $y = \cos u$.

$$\frac{dy}{du} = -\sin u \quad \text{and} \quad \frac{du}{dx} = 10x + 1$$

$$\frac{dy}{dx} = \frac{dy}{du} \cdot \frac{du}{dx}$$

$$= (-\sin u)(10x + 1)$$

$$= -(10x + 1) \sin (5x^2 + x)$$

Example 6

Find the derivative of $y = \cos (x^2 - 1)^5$.

$$\frac{dy}{dx} = -\sin (x^2 - 1)^5 [5(x^2 - 1)^4](2x)$$

$$= -10x(x^2 - 1)^4 \sin (x^2 - 1)^5$$

Example 7

Find the derivative of $y = \sin (7x^2 + 2) \cos 4x$.

We first need to use the product rule for differentiation.

$$\frac{dy}{dx} = \sin (7x^2 + 2) \frac{d}{dx}(\cos 4x) + \cos 4x \frac{d}{dx}[\sin (7x^2 + 2)]$$

$$= \sin (7x^2 + 2)(-\sin 4x)\frac{d}{dx}(4x) + \cos 4x \cos (7x^2 + 2)\frac{d}{dx}(7x^2 + 2)$$

$$= \sin (7x^2 + 2)(-\sin 4x)(4) + \cos 4x \cos (7x^2 + 2)(14x)$$

$$= -4 \sin (7x^2 + 2) \sin 4x + 14x \cos 4x \cos (7x^2 + 2)$$

In summary,

$$\frac{d}{dx}(\sin u) = \cos u \frac{du}{dx}$$

$$\frac{d}{dx}(\cos u) = -\sin u \frac{du}{dx}$$

Exercises

Find the derivative of each function.

1. $y = \sin 7x$
2. $y = 5 \sin 2x$
3. $y = 2 \cos 5x$
4. $y = 4 \cos 6x$
5. $y = 2 \sin x^3$
6. $y = -3 \cos x^4$

7. $y = 3 \cos 4x^2$ 8. $y = 5 \sin 8x^3$ 9. $y = 4 \sin (1 - x)$

10. $y = 6 \cos (1 - 3x)$ 11. $y = 3 \sin (x^2 + 4)$ 12. $y = \sin (x^3 + 2x^2 - 4)$

13. $y = 4 \cos (5x^2 + x)$ 14. $y = 2 \cos (x - 3)$

15. $y = \cos (x^4 - 2x^2 + 3)$ 16. $y = 3 \cos (x^2 - 4)$

17. $y = \cos^2(3x - 1)$ 18. $y = \sin^2(1 - 2x)$ 19. $y = \sin^3(2x + 3)$

20. $y = \cos^4(x^2 - 1)$ 21. $y = \sin (2x - 5)^2$ 22. $y = \cos (5x + 4)^3$

23. $y = \cos(x^3 - 4)^4$ 24. $y = 4 \sin (x^2 + x - 3)^6$

25. $y = \sin x \cos 3x$ 26. $y = \sin x^2 \cos 5x$

27. $y = \sin 5x \cos 6x$ 28. $y = \sin (3x - 1) \cos (4x + 3)$

29. $y = \cos 4x \cos 7x$ 30. $y = \sin 3x \sin 5x$

31. $y = \sin(x^2 + 2x)\cos x^3$ 32. $y = \sin (x + 2)^3 \cos (x^2 - 3)$

33. $y = (x^2 + 3x) \sin (5x - 2)$ 34. $y = \sqrt{4x - 3} \cos (x^2 + 2)$

35. $y = \dfrac{\sin 5x}{x}$ 36. $y = \dfrac{\cos 6x}{3x^2}$

37. $y = \dfrac{x^2 - 1}{\cos 3x}$ 38. $y = \dfrac{4x - 5}{\sin (2x - 1)}$

39. $y = \sin 5x + \cos 6x$ 40. $y = 2 \sin 3x - 5 \sin 6x$

41. $y = \sin (x^2 - 3x) + \cos 4x$ 42. $y = x^2 - 3 \sin^3 2x$

43. $y = \dfrac{\sin x}{\cos x}$ 44. $y = \dfrac{\cos x}{\sin x}$

45. Find $\dfrac{d^2y}{dx^2}$ for $y = \cos x$. 46. Find $\dfrac{d^2y}{dx^2}$ for $y = \sin x$.

47. Find $\dfrac{d^3y}{dx^3}$ for $y = \sin x$. 48. Find $\dfrac{d^4y}{dx^4}$ for $y = \cos x$.

49. Find $\dfrac{d^2y}{dx^2}$ for Exercise 39. 50. Find $\dfrac{d^2y}{dx^2}$ for Exercise 40.

51. Find the slope of the tangent line to $y = 4 \sin 3x$ at $x = \pi/18$.

52. Find the slope of the tangent line to $y = -4 \cos 2x$ at $x = 5\pi/12$.

53. Find the equation of the tangent line to $y = 2 \cos 5x$ at $x = \pi/10$.

54. Find the equation of the tangent line to $y = 6 \cos 4x$ at $x = 5\pi/24$.

6.3 DERIVATIVES OF OTHER TRIGONOMETRIC FUNCTIONS

Since $\tan u = \dfrac{\sin u}{\cos u}$ and using the quotient rule for differentiation, we have

$$\frac{d}{du}(\tan u) = \frac{d}{du}\left(\frac{\sin u}{\cos u}\right)$$

$$= \frac{\cos u \dfrac{d}{du}(\sin u) - \sin u \dfrac{d}{du}(\cos u)}{\cos^2 u}$$

$$= \frac{(\cos u)(\cos u) - (\sin u)(-\sin u)}{\cos^2 u}$$

Derivatives of Transcendental Functions   Chap. 6

$$= \frac{\cos^2 u + \sin^2 u}{\cos^2 u}$$

$$= \frac{1}{\cos^2 u} \qquad (\cos^2 u + \sin^2 u = 1)$$

$$= \sec^2 u \qquad \left(\frac{1}{\cos u} = \sec u \right)$$

and

$$\frac{d}{dx} (\tan u) = \sec^2 u \frac{du}{dx} \qquad \text{(using the chain rule)}$$

In a similar manner it can be shown that

$$\frac{d}{du} (\cot u) = -\csc^2 u$$

and

$$\frac{d}{dx} (\cot u) = -\csc^2 u \frac{du}{dx}$$

Since $\sec u = \dfrac{1}{\cos u}$, we have

$$\frac{d}{du} (\sec u) = \frac{d}{du} (\cos u)^{-1}$$

$$= -(\cos u)^{-2} \frac{d}{du} (\cos u)$$

$$= \frac{-1}{\cos^2 u} (-\sin u)$$

$$= \frac{\sin u}{\cos^2 u}$$

$$= \frac{1}{\cos u} \cdot \frac{\sin u}{\cos u}$$

$$= \sec u \tan u$$

and using the chain rule,

$$\frac{d}{dx} (\sec u) = \sec u \tan u \frac{du}{dx}$$

In a similar manner one can show that

$$\frac{d}{du} (\csc u) = -\csc u \cot u$$

and

$$\frac{d}{dx} (\csc u) = -\csc u \cot u \frac{du}{dx}$$

Example 1

Find the derivative of $y = \tan(x^2 + 3)$.

$$\frac{dy}{dx} = \sec^2(x^2 + 3)\frac{d}{dx}(x^2 + 3) \qquad (u = x^2 + 3)$$

$$= \sec^2(x^2 + 3)(2x)$$

$$= 2x \sec^2(x^2 + 3)$$

Example 2

Find the derivative of $y = \sec^3 5x$.

We first need to use the power rule for differentiation.

$$\frac{dy}{dx} = 3 \sec^2 5x \frac{d}{dx}(\sec 5x)$$

$$= (3 \sec^2 5x)(\sec 5x \tan 5x)\frac{d}{dx}(5x) \qquad (u = 5x)$$

$$= 15 \sec^3 5x \tan 5x$$

Example 3

Find the derivative of $y = \cot \sqrt{3x + 1}$.

$$\frac{dy}{dx} = -\csc^2 \sqrt{3x + 1} \frac{d}{dx}(3x + 1)^{1/2} \qquad (u = \sqrt{3x + 1})$$

$$= (-\csc^2 \sqrt{3x + 1})[\tfrac{1}{2}(3x + 1)^{-(1/2)}(3)]$$

$$= \frac{-3 \csc^2 \sqrt{3x + 1}}{2\sqrt{3x + 1}}$$

Example 4

Find the derivative of $y = \sec 3x \tan 4x$.

$$\frac{dy}{dx} = (\sec 3x)(\sec^2 4x)(4) + (\tan 4x)(\sec 3x \tan 3x)(3)$$

$$= 4 \sec 3x \sec^2 4x + 3 \sec 3x \tan 3x \tan 4x$$

Example 5

Find the derivative of $y = (x^2 + \cot^3 4x)^6$.

$$\frac{dy}{dx} = 6(x^2 + \cot^3 4x)^5[2x + 3(\cot^2 4x)(-\csc^2 4x)(4)]$$

$$= 6 (x^2 + \cot^3 4x)^5[2x - 12 \cot^2 4x \csc^2 4x]$$

$$= 12(x^2 + \cot^3 4x)^5(x - 6 \cot^2 4x \csc^2 4x)$$

Example 6

Find the derivative of $y = \csc(\sin 5x)$.

$$\frac{dy}{dx} = [-\csc(\sin 5x) \cot(\sin 5x)]\frac{d}{dx}(\sin 5x) \qquad (u = \sin 5x)$$

$$= -5 \csc(\sin 5x) \cot(\sin 5x) \cos 5x$$

When you prove trigonometric identities, you often have a choice of which identities to use. Similarly, here in finding the derivative of a trigonometric expression, you may use a trigonometric identity either to change the form of the original expression before finding the derivative or to simplify the result or to change the result to another form. When finding the derivative or later when finding the integral of a trigonometric expression, you will often need to use trigonometric identities to show that your answer is equivalent to the given answer.

Example 7

Find the derivative of $y = \dfrac{\sin x}{\cot^2 x}$

Method 1: First, find the derivative and then change the trigonometric functions to sine and cosine.

$$\frac{dy}{dx} = \frac{(\cot^2 x)(\cos x) - (\sin x)[(2 \cot x)(-\csc^2 x)]}{\cot^4 x}$$

$$= \frac{\cot x(\cot x \cos x + 2 \sin x \csc^2 x)}{\cot^4 x}$$

$$= \frac{\dfrac{\cos x}{\sin x} \cos x + 2 \sin x \dfrac{1}{\sin^2 x}}{\cot^3 x}$$

$$= \frac{\dfrac{\cos^2 x + 2}{\sin x}}{\dfrac{\cos^3 x}{\sin^3 x}}$$

$$= \frac{\sin^2 x(\cos^2 x + 2)}{\cos^3 x}$$

Method 2: Change the expression to sine and cosine first and then find the derivative.

$$y = \frac{\sin x}{\cot^2 x} = \frac{\sin x}{\dfrac{\cos^2 x}{\sin^2 x}} = \frac{\sin^3 x}{\cos^2 x}$$

$$\frac{dy}{dx} = \frac{(\cos^2 x)[(3 \sin^2 x)(\cos x)] - (\sin^3 x)[(2 \cos x)(-\sin x)]}{\cos^4 x}$$

$$= \frac{\cos x \sin^2 x[3 \cos^2 x + 2 \sin^2 x]}{\cos^4 x}$$

$$= \frac{\sin^2 x[\cos^2 x + 2(\cos^2 x + \sin^2 x)]}{\cos^3 x}$$

$$= \frac{\sin^2 x(\cos^2 x + 2)}{\cos^3 x}$$

In summary, the derivatives of the trigonometric functions are as follows:

$$\frac{d}{dx}(\sin u) = \cos u \frac{du}{dx}$$

$$\frac{d}{dx}(\cos u) = -\sin u \frac{du}{dx}$$

$$\frac{d}{dx}(\tan u) = \sec^2 u \frac{du}{dx}$$

$$\frac{d}{dx}(\cot u) = -\csc^2 u \frac{du}{dx}$$

$$\frac{d}{dx}(\sec u) = \sec u \tan u \frac{du}{dx}$$

$$\frac{d}{dx}(\csc u) = -\csc u \cot u \frac{du}{dx}$$

Exercises

Find the derivative of each function.

1. $y = \tan 3x$

2. $y = \cot 2x$

3. $y = \sec 7x$

4. $y = \csc 4x$

5. $y = \cot (3x^2 - 7)$

6. $y = \tan (x^3 + 4)$

7. $y = 3 \csc (3x - 4)$

8. $y = 4 \sec \left(x - \frac{\pi}{3} \right)$

9. $y = \tan^2(5x - 2)$

10. $y = \sec^4 7x$

11. $y = 4 \cot^3 2x$

12. $y = \csc^3 x^4$

13. $y = \sec \sqrt{x^2 + x}$

14. $y = \sqrt{\tan 5x}$

15. $y = \dfrac{\csc x}{3x}$

16. $y = \dfrac{\cot 2x}{x}$

17. $y = \tan 3x - \sec (x^2 + 1)$

18. $y = \tan 2x + \cot 3x$

19. $y = \sec x \tan x$

20. $y = \sin x \sec x$

21. $y = \sin^2 x \cot x$

22. $y = \tan^2 x \sin x$

23. $y = x \sec x$

24. $y = x^2 \tan^2 x$

25. $y = x^2 + x^2 \tan^2 x$

26. $y = x^2 \csc^2 x - x^2$

27. $y = \csc 3x \cot 3x$

28. $y = \cot 2x \cos 4x$

29. $y = \csc^2 3x \sin 3x$

30. $y = \cos 2x \csc^2 x$

31. $y = (\sin x - \cos x)^2$

32. $y = \sec^2 x - \tan^2 x$

33. $y = (x + \sec^2 3x)^4$

34. $y = (\sin x - \tan^2 x)^3$

35. $y = (\sec x + \tan x)^3$

36. $y = (1 + \cot^3 x)^4$

37. $y = \sin (\tan x)$

38. $y = \sec (2 \cos x)$

39. $y = \tan (\cos x)$

40. $y = \cos (\cot x)$

41. $y = \sin^2(\cos x)$

42. $y = \tan^2(\sin x)$

43. $y = \dfrac{\tan x}{\cos x}$

44. $y = \dfrac{\sec x}{\cot x}$

45. $y = \dfrac{\sin^2 x}{\tan^2 x}$ $\qquad$ **46.** $y = \dfrac{\csc^2 x}{\tan x}$ $\qquad$ **47.** $y = \dfrac{\sin x}{1 + \tan x}$

48. $y = \dfrac{\sin x}{1 + \cos x}$

Find the second derivative of each function.

49. $y = \tan 3x$ $\qquad\qquad\qquad$ **50.** $y = \sec 2x$

51. $y = x \cot x$ $\qquad\qquad\qquad$ **52.** $y = \dfrac{\tan x}{x}$

53. Find the slope of the tangent line to the curve $y = \tan x$ at $x = \pi/4$.

54. Find the slope of the tangent line to the curve $y = \sec^2 x$ at $x = \pi/4$.

55. Show that $\dfrac{d}{du}(\cot u) = -\csc^2 u.$

56. Show that $\dfrac{d}{du}(\csc u) = -\csc u \cot u.$

6.4 THE INVERSE TRIGONOMETRIC FUNCTIONS

In mathematics, a *relation* is a set of ordered pairs of the form (x, y), usually written as an equation. The *inverse* of a given equation is the resulting equation when the x and y variables are interchanged. For example,

The inverse of:	Is:
$y = x^2$	$x = y^2$
$y = 3x^2 + 4x + 7$	$x = 3y^2 + 4y + 7$
$y = \dfrac{x + 4}{3x - 1}$	$x = \dfrac{y + 4}{3y - 1}$

Likewise, each basic trigonometric equation has an inverse.

The inverse of:	Is:
$y = \sin x$	$x = \sin y$
$y = \cos x$	$x = \cos y$
$y = \tan x$	$x = \tan y$
$y = \cot x$	$x = \cot y$
$y = \sec x$	$x = \sec y$
$y = \csc x$	$x = \csc y$

We find it necessary to write the inverse trigonometric equations solved for y. There are two common forms of the inverse trigonometric equations solved for y.

		Solved for y	
The inverse of:	Is:	Is:	Is:
$y = \sin x$	$x = \sin y$	$y = \arcsin x^a$	$y = \sin^{-1} x^b$
$y = \cos x$	$x = \cos y$	$y = \arccos x$	$y = \cos^{-1} x$
$y = \tan x$	$x = \tan y$	$y = \arctan x$	$y = \tan^{-1} x$
$y = \cot x$	$x = \cot y$	$y = \text{arccot } x$	$y = \cot^{-1} x$
$y = \sec x$	$x = \sec y$	$y = \text{arcsec } x$	$y = \sec^{-1} x$
$y = \csc x$	$x = \csc y$	$y = \text{arccsc } x$	$y = \csc^{-1} x$

[a] This is read, "y equals the arc sine of x" and means that y is the angle whose sine is x.
[b] This notation will not be used here because of the confusion caused by the fact that -1 is not an exponent.

Example 1

What is the meaning of each equation?
(a) $y = \arctan x$ (a) y is the angle whose tangent is x.
(b) $y = \arccos 3x$ (b) y is the angle whose cosine is $3x$.
(c) $y = 4 \text{ arccsc } 5x$ (c) y is four times the angle whose cosecant is $5x$.

Remember that $x = \sin y$ and $y = \arcsin x$ express the same relationship. The first form expresses the relationship in terms of the function (sine) of the angle; the second form expresses the relationship in terms of the angle itself.

Example 2

Given the equation $y = \arcsin x$, find y when $x = \frac{1}{2}$. (Give the answer in radians.)
Substituting $x = \frac{1}{2}$, we have
$$y = \arcsin \tfrac{1}{2}$$

which means y is the angle whose sine is $\frac{1}{2}$. We know that $\sin \dfrac{\pi}{6} = \dfrac{1}{2}$, so $y = \dfrac{\pi}{6}$.

But we also know that

$$\sin \frac{5\pi}{6} = \frac{1}{2}, \quad \text{so} \quad y = \frac{5\pi}{6}$$

$$\sin \frac{13\pi}{6} = \frac{1}{2}, \quad \text{so} \quad y = \frac{13\pi}{6} \qquad \left(\frac{13\pi}{6} = \frac{\pi}{6} + 2\pi\right)$$

$$\sin \frac{17\pi}{6} = \frac{1}{2}, \quad \text{so} \quad y = \frac{17\pi}{6} \qquad \left(\frac{17\pi}{6} = \frac{5\pi}{6} + 2\pi\right)$$

$$\vdots \qquad\qquad \vdots$$

$$\sin\left(\frac{-7\pi}{6}\right) = \frac{1}{2}, \quad \text{so} \quad y = \frac{-7\pi}{6}$$

$$\sin\left(\frac{-11\pi}{6}\right) = \frac{1}{2}, \quad \text{so} \quad y = \frac{-11\pi}{6}$$

$$\vdots \qquad\qquad \vdots$$

Thus there are infinitely many angles whose sine is $\frac{1}{2}$. We saw this when we graphed $y = \sin x$. That is,

$$y = \frac{\pi}{6} + n \cdot 2\pi$$

$$y = \frac{5\pi}{6} + n \cdot 2\pi$$

for every integer n.

Example 3

Given the equation $y = \arctan x$, find y when $x = -1$ for $0 \leq y < 2\pi$.
Substituting $x = -1$, we have

$$y = \arctan(-1)$$

which means y is the angle whose tangent is -1. We know that

$$\tan \frac{3\pi}{4} = -1, \quad \text{so} \quad y = \frac{3\pi}{4}$$

and

$$\tan \frac{7\pi}{4} = -1, \quad \text{so} \quad y = \frac{7\pi}{4}$$

for $0 \leq y < 2\pi$.

Example 4

Find $\operatorname{arccot} \dfrac{1}{\sqrt{3}}$ for all angles in radians.

Let $y = \operatorname{arccot} \dfrac{1}{\sqrt{3}}$, which means y is the angle whose cotangent is $\dfrac{1}{\sqrt{3}}$. We know that

$$\cot \frac{\pi}{3} = \frac{1}{\sqrt{3}} \quad \text{and} \quad \cot \frac{4\pi}{3} = \frac{1}{\sqrt{3}}$$

for $0 \leq y < 2\pi$. Thus

$$\operatorname{arccot} \frac{1}{\sqrt{3}} = \begin{cases} \dfrac{\pi}{3} + 2n\pi \\[2ex] \dfrac{4\pi}{3} + 2n\pi \end{cases}$$

for every integer n.

Example 5

Solve the equation $y = \cos 2x$ for x.
The equation $y = \cos 2x$ is equivalent to

$$\arccos y = 2x$$

So

$$x = \frac{1}{2} \arccos y$$

Example 6

Solve the equation $2y = \arctan 3x$ for x.

The equation $2y = \arctan 3x$ is equivalent to

$$\tan 2y = 3x$$

So

$$x = \frac{1}{3}\tan 2y$$

Example 7

Solve the equation $y = \frac{1}{3}\operatorname{arcsec} 2x$ for x.

First, multiply both sides by 3.

$$3y = \operatorname{arcsec} 2x$$

The equation $3y = \operatorname{arcsec} 2x$ is equivalent to

$$\sec 3y = 2x$$

Thus

$$x = \frac{1}{2}\sec 3y$$

To graph the inverse trigonometric relation

$$y = \arcsin x$$

first solve for x; that is,

$$x = \sin y$$

Then, plotting ordered pairs of solutions of this equation, we have the graph in Fig. 6.11.

In a similar manner, we can graph all six inverse trigonometric relations. They are shown in Fig. 6.12.

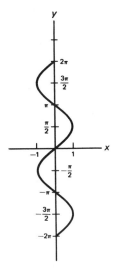

$y = \arcsin x$ **Figure 6.11**

       Derivatives of Transcendental Functions   Chap. 6

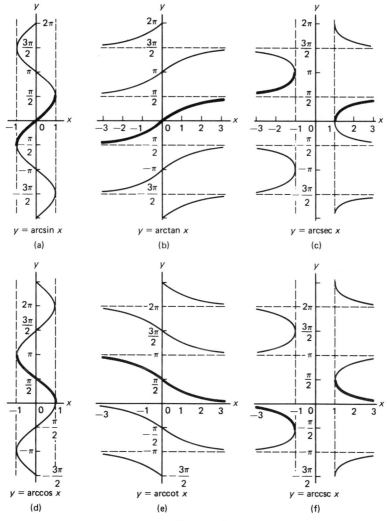

Figure 6.12

A *function* is a special relation: a set of ordered pairs in which no two distinct ordered pairs have the same first element.

As you can see from the graphs, each x-value in each domain corresponds to (infinitely) many values of y. Thus none of the inverse trigonometric relations is a function. However, if we restrict the y-values of each inverse trigonometric relation, we can define an inverse that is also a function. While this could be done in any of several ways, it is customary to restrict the y-values as follows. *Note:* The inverse trigonometric *functions* are *capitalized* to distinguish them from the inverse trigonometric relations.

Look once again at the graphs of the inverse trigonometric relations in Fig. 6.12. The extra thick lines indicate the portions of the graphs that correspond to the inverse trigonometric *functions*. These are also shown in Fig. 6.13.

Sec. 6.4   The Inverse Trigonometric Functions   **251**

Note: The three inverse trigonometric functions on pocket calculators are programmed to these same restricted ranges.

Example 8

Find $\text{Arcsin } \left(\dfrac{1}{2}\right)$.

$$\text{Arcsin } \left(\dfrac{1}{2}\right) = \dfrac{\pi}{6}$$

This is the only value in the defined range of $-\dfrac{\pi}{2} \leq y \leq \dfrac{\pi}{2}$.

Example 9

Find $\text{Arctan } (-1)$.

$$\text{Arctan } (-1) = -\dfrac{\pi}{4}$$

This is the only value in the defined range of $-\dfrac{\pi}{2} < y < \dfrac{\pi}{2}$.

Example 10

Find $\text{Arccos } \left(-\dfrac{1}{2}\right)$.

$$\text{Arccos } \left(-\dfrac{1}{2}\right) = \dfrac{2\pi}{3}$$

This is the only value in the defined range of $0 \leq y \leq \pi$.

Example 11

Find $\tan [\text{Arccos } (-1)]$.

$$\tan [\text{Arccos } (-1)] = \tan \pi = 0$$

Example 12

Find $\cos (\text{Arcsec } 2)$.

$$\cos (\text{Arcsec } 2) = \cos \dfrac{\pi}{3} = \dfrac{1}{2}$$

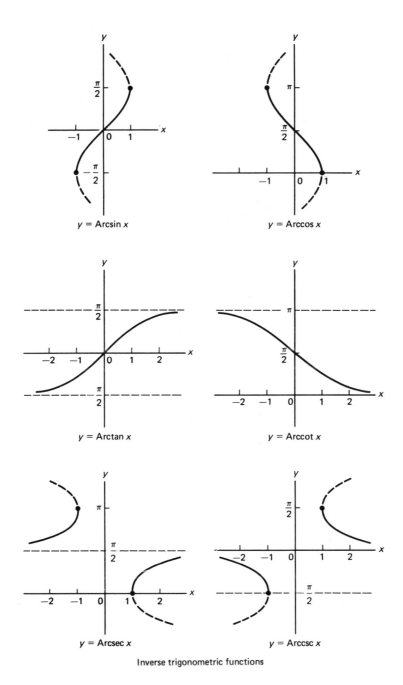

Inverse trigonometric functions

Figure 6.13

Example 13

Find $\sin\left[\operatorname{Arctan}\left(-\dfrac{1}{\sqrt{3}}\right)\right]$.

$$\sin\left[\operatorname{Arctan}\left(-\frac{1}{\sqrt{3}}\right)\right] = \sin\left(-\frac{\pi}{6}\right) = -\frac{1}{2}$$

Example 14

Find an algebraic expression for $\sin(\operatorname{Arccos} x)$.

Let $\theta = \operatorname{Arccos} x$. Then

$$\cos\theta = x = \frac{x}{1}$$

Draw a right triangle with θ as an acute angle, x as the adjacent side, and 1 as the hypotenuse as in Fig. 6.14.

Using the Pythagorean theorem,

$$c^2 = a^2 + b^2$$

$$1^2 = x^2 + (\text{side opposite } \theta)^2$$

$$\text{side opposite } \theta = \sqrt{1 - x^2}$$

Now we see that

$$\sin(\operatorname{Arccos} x) = \sin\theta$$

$$= \frac{\text{side opposite } \theta}{\text{hypotenuse}}$$

$$= \frac{\sqrt{1 - x^2}}{1}$$

$$= \sqrt{1 - x^2}$$

Example 15

Find an algebraic expression for $\sec(\operatorname{Arctan} x)$.

Let $\theta = \operatorname{Arctan} x$. Then

$$\tan\theta = x = \frac{x}{1}$$

Draw a right triangle with θ as an acute angle, x as the opposite side, and 1 as the adjacent side as in Fig. 6.15. Using the Pythagorean theorem, we find the hypotenuse is $\sqrt{x^2 + 1}$.

$$\sec(\operatorname{Arctan} x) = \sec\theta$$

$$= \frac{\text{hypotenuse}}{\text{side adjacent to } \theta}$$

$$= \frac{\sqrt{x^2 + 1}}{1}$$

$$= \sqrt{x^2 + 1}$$

Example 16

Find an algebraic expression for $\cos(2\operatorname{Arcsin} x)$.

Let $\theta = \operatorname{Arcsin} x$. Then

$$\sin \theta = \frac{x}{1}$$

Draw a right triangle with θ as an acute angle, x as the opposite side, and 1 as the hypotenuse as in Fig. 6.16. Using the Pythagorean theorem, we find the side adjacent to θ is $\sqrt{1 - x^2}$.

$$\cos (2 \text{ Arcsin } x) = \cos 2\theta$$
$$= 1 - 2 \sin^2\theta \text{ (Formula 25c in Appendix B, Table 6)}$$
$$= 1 - 2x^2$$

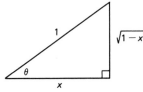

Figure 6.14

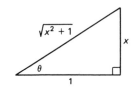

Figure 6.15

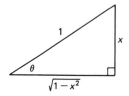

Figure 6.16

Exercises

Solve each equation for $0 \le y < 2\pi$.

1. $y = \arcsin \left(\dfrac{\sqrt{3}}{2}\right)$

2. $y = \arccos \left(\dfrac{1}{2}\right)$

3. $y = \arctan 1$

4. $y = \text{arccot } (-1)$

5. $y = \text{arcsec} \left(-\dfrac{2}{\sqrt{3}}\right)$

6. $y = \text{arccsc } \sqrt{2}$

7. $y = \arccos \left(-\dfrac{1}{2}\right)$

8. $y = \arcsin 0$

9. $y = \arccos \left(-\dfrac{1}{\sqrt{2}}\right)$

10. $y = \arctan (-\sqrt{3})$

11. $y = \text{arccot} \left(\dfrac{1}{\sqrt{3}}\right)$

12. $y = \arcsin \left(-\dfrac{1}{\sqrt{2}}\right)$

13. $y = \text{arcsec } 1$

14. $y = \arcsin (-1)$

15. $y = \arctan (1.963)$

16. $y = \arccos (-0.9063)$

Find all angles in radians for each expression.

17. $\arctan (-\sqrt{3})$

18. $\arcsin (-1)$

19. $\arccos \left(-\dfrac{\sqrt{3}}{2}\right)$

20. $\text{arcsec} \left(\dfrac{2}{\sqrt{3}}\right)$

21. $\text{arccsc } 2$

22. $\arctan 0$

23. $\text{arcsec } (-1)$

24. $\arcsin \left(\dfrac{1}{\sqrt{2}}\right)$

25. $\text{arccot } \sqrt{3}$

26. $\arcsin \left(-\dfrac{\sqrt{3}}{2}\right)$

27. $\text{arcsec} \left(\dfrac{2}{\sqrt{3}}\right)$

28. $\arctan (-\sqrt{3})$

29. $\arccos 1$

30. $\arcsin 1$

31. $\arcsin (-0.8572)$

32. $\text{arccot } (-0.8195)$

Solve each equation for x.

33. $y = \sin 3x$

34. $y = \tan 4x$

35. $y = 4 \cos x$

36. $y = 3 \sec x$

37. $y = 5 \tan \dfrac{x}{2}$

38. $y = \dfrac{1}{2} \cos 3x$

39. $y = \dfrac{3}{2} \cot \dfrac{x}{4}$

40. $y = \dfrac{5}{2} \sin \dfrac{2x}{3}$

41. $y = 3 \sin (x - 1)$

42. $y = 4 \tan (2x + 1)$

43. $y = \dfrac{1}{2} \cos (3x + 1)$

44. $y = \dfrac{1}{3} \sec (1 - 4x)$

Find the value of each expression in radians.

45. $\text{Arcsin} \left(\dfrac{\sqrt{3}}{2} \right)$

46. $\text{Arccos} \left(\dfrac{1}{2} \right)$

47. $\text{Arctan} \left(-\dfrac{1}{\sqrt{3}} \right)$

48. $\text{Arcsin} \left(-\dfrac{1}{2} \right)$

49. $\text{Arccos} \left(-\dfrac{\sqrt{3}}{2} \right)$

50. $\text{Arctan} \left(\dfrac{1}{\sqrt{3}} \right)$

51. $\text{Arcsec} (-2)$

52. $\text{Arccot} (-1)$

53. $\text{Arccsc} \sqrt{2}$

54. $\text{Arcsin} (-1)$

55. $\text{Arctan} \sqrt{3}$

56. $\text{Arccos} \, 0$

57. $\text{Arccos} \left(\dfrac{1}{\sqrt{2}} \right)$

58. $\text{Arcsin} \, 1$

59. $\text{Arcsin} \left(-\dfrac{\sqrt{3}}{2} \right)$

60. $\text{Arctan} (-\sqrt{3})$

Find the value of each expression.

61. $\cos (\text{Arctan} \sqrt{3})$

62. $\tan \left[\text{Arcsin} \left(\dfrac{1}{\sqrt{2}} \right) \right]$

63. $\sin \left[\text{Arccos} \left(-\dfrac{1}{\sqrt{2}} \right) \right]$

64. $\sin [\text{Arctan} (-1)]$

65. $\tan [\text{Arccos} (-1)]$

66. $\sec \left[\text{Arccos} \left(-\dfrac{1}{2} \right) \right]$

67. $\sin \left[\text{Arcsin} \left(\dfrac{\sqrt{3}}{2} \right) \right]$

68. $\tan [\text{Arctan} (-\sqrt{3})]$

69. $\cos \left[\text{Arcsin} \left(\dfrac{3}{5} \right) \right]$

70. $\tan \left[\text{Arcsin} \left(\dfrac{12}{13} \right) \right]$

71. $\tan [\text{Arcsin} (-0.1560)]$

72. $\sin [\text{Arccot} (1.635)]$

Find an algebraic expression for each.

73. $\cos (\text{Arcsin} \, x)$

74. $\tan (\text{Arccos} \, x)$

75. $\sin (\text{Arcsec} \, x)$

76. $\cot (\text{Arcsec} \, x)$

77. $\sec (\text{Arccos} \, x)$

78. $\sin (\text{Arctan} \, x)$

79. $\tan (\text{Arctan} \, x)$

80. $\sin (\text{Arcsin} \, x)$

81. $\cos (\text{Arcsin} \, 2x)$

82. $\tan (\text{Arccos} \, 3x)$

83. $\sin (2 \, \text{Arcsin} \, x)$

84. $\cos (2 \, \text{Arctan} \, x)$

6.5 DERIVATIVES OF INVERSE TRIGONOMETRIC FUNCTIONS

To differentiate $y = \text{Arcsin} \, u$, first differentiate its inverse—the function $u = \sin y$. Then

$$\frac{du}{dx} = \frac{d}{dx} (\sin y) = \cos y \, \frac{dy}{dx}$$

Solving for $\dfrac{dy}{dx}$,

$$\frac{dy}{dx} = \frac{1}{\cos y} \cdot \frac{du}{dx}$$

Now express $\cos y$ in terms of $\sin y$ using the identity $\sin^2 y + \cos^2 y = 1$. That is,

$$\cos y = \sqrt{1 - \sin^2 y}$$

Note that we choose $\cos y = +\sqrt{1 - \sin^2 y}$ because $-\dfrac{\pi}{2} \leqslant \text{Arcsin } u \leqslant \dfrac{\pi}{2}$ and $\cos y > 0$ in quadrants I and IV. So

$$\frac{dy}{dx} = \frac{1}{\sqrt{1 - \sin^2 y}} \cdot \frac{du}{dx}$$

$$= \frac{1}{\sqrt{1 - u^2}} \cdot \frac{du}{dx} \qquad (\text{since } u = \sin y)$$

Example 1

Find the derivative of $y = \text{Arcsin } 2x$.

Let $u = 2x$, then

$$\frac{dy}{dx} = \frac{1}{\sqrt{1 - (2x)^2}} (2) = \frac{2}{\sqrt{1 - 4x^2}}$$

To differentiate $y = \text{Arctan } u$, begin with its inverse:

$$u = \tan y$$

$$\frac{du}{dx} = \frac{d}{dx} (\tan y) = \sec^2 y \frac{dy}{dx}$$

Solving for $\dfrac{dy}{dx}$,

$$\frac{dy}{dx} = \frac{1}{\sec^2 y} \frac{du}{dx}$$

$$= \frac{1}{1 + \tan^2 y} \frac{du}{dx} \qquad (\sec^2 y = 1 + \tan^2 y)$$

$$= \frac{1}{1 + u^2} \frac{du}{dx} \qquad (u = \tan y)$$

Example 2

Find the derivative of $y = \text{Arctan } 3x^2$.

Let $u = 3x^2$, then

$$\frac{dy}{dx} = \frac{1}{1 + (3x^2)^2} (6x) = \frac{6x}{1 + 9x^4}$$

Formulas for the derivatives of the other trigonometric functions are found in a similar manner and are left as exercises. We now list the formulas for the derivatives of the six inverse trigonometric functions.

Derivatives of the Inverse Trigonometric Functions

$$\frac{d}{dx}(\text{Arcsin } u) = \frac{1}{\sqrt{1 - u^2}}\frac{du}{dx}$$

$$\frac{d}{dx}(\text{Arccos } u) = -\frac{1}{\sqrt{1 - u^2}}\frac{du}{dx}$$

$$\frac{d}{dx}(\text{Arctan } u) = \frac{1}{1 + u^2}\frac{du}{dx}$$

$$\frac{d}{dx}(\text{Arccot } u) = -\frac{1}{1 + u^2}\frac{du}{dx}$$

$$\frac{d}{dx}(\text{Arcsec } u) = \frac{1}{|u|\sqrt{u^2 - 1}}\frac{du}{dx}$$

$$\frac{d}{dx}(\text{Arccsc } u) = -\frac{1}{|u|\sqrt{u^2 - 1}}\frac{du}{dx}$$

Example 3

Find the derivative of $y = \text{Arccos}^3 6x$.

Find the derivative of a power u^3, where $u = \text{Arccos } 6x$.

$$\frac{dy}{dx} = (3 \text{ Arccos}^2 6x)\left(-\frac{1}{\sqrt{1 - (6x)^2}}\right) \quad (6)$$

$$= \frac{-18 \text{ Arccos}^2 6x}{\sqrt{1 - 36x^2}}$$

Example 4

Find the derivative of $y = x^2 \text{ Arcsec } (1 - 3x)$.

$$\frac{dy}{dx} = x^2\left(\frac{1}{|1 - 3x| \sqrt{(1 - 3x)^2 - 1}}\right)(-3) + 2x \text{ Arcsec } (1 - 3x)$$

$$= \frac{-3x^2}{|1 - 3x| \sqrt{9x^2 - 6x}} + 2x \text{ Arcsec } (1 - 3x)$$

Exercises

Find the derivative of each.

1. $y = \text{Arcsin } 5x$
2. $y = \text{Arccos } 6x$
3. $y = \text{Arctan } 3x$
4. $y = \text{Arcsec } 4x$
5. $y = \text{Arccsc } (1 - x)$
6. $y = \text{Arccot } \sqrt{x}$
7. $y = 3 \text{ Arccos } (x - 1)$
8. $y = 4 \text{ Arcsin } (1/x)$
9. $y = 2 \text{ Arccot } 3x^2$
10. $y = 3 \text{ Arctan } (x^2 - 1)$
11. $y = 5 \text{ Arcsec } x^3$
12. $y = 6 \text{ Arccsc } (x/2)$
13. $y = \text{Arcsin}^3 x$
14. $y = \text{Arctan}^2 5x$
15. $y = 2 \text{ Arccos}^2 3x$
16. $y = 4 \text{ Arcsec}^3 2x$
17. $y = 3 \text{ Arctan}^4 \sqrt{x}$
18. $y = 6 \text{ Arcsin}^2 (1 - x)$

19. $y = \text{Arcsin } x + \text{Arccos } x$

20. $y = x - \text{Arctan } x$

21. $y = \sqrt{1 - x^2} + \text{Arcsin } x$

22. $y = \sqrt{1 - x^2} + \text{Arccos } x$

23. $y = x \text{ Arcsin } 3x$

24. $y = x^2 \text{ Arccos } x$

25. $y = x \text{ Arctan } x$

26. $y = x \text{ Arcsin } x^2$

27. $y = x \text{ Arcsin } x + \sqrt{1 - x^2}$

28. $y = \dfrac{x}{\sqrt{1 - x^2}} - \text{Arcsin } x$

29. $y = \dfrac{x}{\text{Arcsin } x}$

30. $y = \dfrac{\text{Arctan } x}{x}$

Find the slope of the tangent line to each curve at the given value.

31. $y = \text{Arcsin } x$ at $x = \frac{1}{2}$

32. $y = \text{Arctan } x$ at $x = 1$

33. $y = x \text{ Arctan } x$ at $x = -1$

34. $y = \text{Arccos}^2 x$ at $x = \dfrac{\sqrt{3}}{2}$

35. Show that $\dfrac{d}{dx} \text{Arccos } u = -\dfrac{1}{\sqrt{1 - u^2}} \dfrac{du}{dx}$.

36. Show that $\dfrac{d}{dx} \text{Arccot } u = -\dfrac{1}{1 + u^2} \dfrac{du}{dx}$.

37. Show that $\dfrac{d}{dx} \text{Arcsec } u = \dfrac{1}{|u| \sqrt{(u^2 - 1)}} \dfrac{du}{dx}$.

38. Show that $\dfrac{d}{dx} \text{Arccsc } u = -\dfrac{1}{|u| \sqrt{(u^2 - 1)}} \dfrac{du}{dx}$.

6.6 EXPONENTIAL AND LOGARITHMIC FUNCTIONS

We have previously considered equations with a *constant* exponent in the form

$$y = x^n$$

These are called **power functions.** Two examples are $y = x^2$ and $y = x^3$.

Exponential Function

Equations with a variable exponent in the form

$$y = b^x$$

where $b > 0$ and $b \neq 1$ are called **exponential functions.**

Two examples are $y = 2^x$ and $y = (\frac{3}{4})^x$.

Example 1

Graph $y = 2^x$ by plotting points. (See Fig. 6.17.)

In general, for $b > 1$, $y = b^x$ is an **increasing** function. That is, as x increases, y increases.

Example 2

Graph $y = (\frac{1}{2})^x$ by plotting points. (See Figure 6.18.)

Sec. 6.6   Exponential and Logarithmic Functions

259

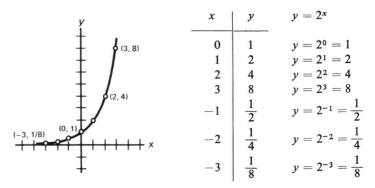

x	y	$y = 2^x$
0	1	$y = 2^0 = 1$
1	2	$y = 2^1 = 2$
2	4	$y = 2^2 = 4$
3	8	$y = 2^3 = 8$
-1	$\dfrac{1}{2}$	$y = 2^{-1} = \dfrac{1}{2}$
-2	$\dfrac{1}{4}$	$y = 2^{-2} = \dfrac{1}{4}$
-3	$\dfrac{1}{8}$	$y = 2^{-3} = \dfrac{1}{8}$

Figure 6.17

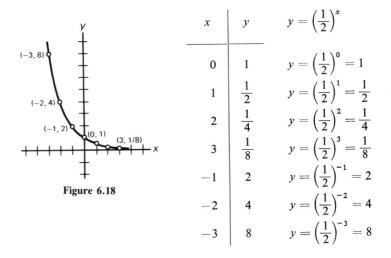

x	y	$y = \left(\dfrac{1}{2}\right)^x$
0	1	$y = \left(\dfrac{1}{2}\right)^0 = 1$
1	$\dfrac{1}{2}$	$y = \left(\dfrac{1}{2}\right)^1 = \dfrac{1}{2}$
2	$\dfrac{1}{4}$	$y = \left(\dfrac{1}{2}\right)^2 = \dfrac{1}{4}$
3	$\dfrac{1}{8}$	$y = \left(\dfrac{1}{2}\right)^3 = \dfrac{1}{8}$
-1	2	$y = \left(\dfrac{1}{2}\right)^{-1} = 2$
-2	4	$y = \left(\dfrac{1}{2}\right)^{-2} = 4$
-3	8	$y = \left(\dfrac{1}{2}\right)^{-3} = 8$

Figure 6.18

In general, for $0 < b < 1$, $y = b^x$ is a **decreasing** function. That is, as x increases, y decreases.

Some basic laws of exponents are given below.

Laws of Exponents

1. $a^m \cdot a^n = a^{m+n}$

2. $\dfrac{a^m}{a^n} = a^{m-n}$

3. $(a^m)^n = a^{mn}$

4. $(ab)^n = a^n b^n$

5. $\left(\dfrac{a}{b}\right)^n = \dfrac{a^n}{b^n}$ $(b \neq 0)$

6. $a^0 = 1$

Derivatives of Transcendental Functions   Chap. 6

When the values of x and y are interchanged in an equation, the resulting equation is called the **inverse** of the given equation. The inverse of the exponential equation, $y = b^x$, is the exponential equation, $x = b^y$. We define this inverse equation to be the **logarithmic** equation. The middle and right equations below show how to express this logarithmic equation in either exponential form or logarithmic form:

Exponential equation	Logarithmic equation in exponential form	Logarithmic equation in logarithmic form
$y = b^x$	$x = b^y$	$y = \log_b x$

That is, $x = b^y$ and $y = \log_b x$ are equivalent equations for $b > 0$ but $b \neq 1$.

The logarithm of a number is the *exponent* indicating the power to which the base must be raised to equal that number. The expression $\log_b x$ is read "the logarithm of x to the base b."

> *Remember:* A logarithm is an exponent.

Example 3

Write each equation in logarithmic form.

	Exponential Form	Logarithmic Form
(a)	$2^3 = 8$	$\log_2 8 = 3$
(b)	$5^2 = 25$	$\log_5 25 = 2$
(c)	$4^{-2} = \frac{1}{16}$	$\log_4 \left(\frac{1}{16}\right) = -2$
(d)	$36^{1/2} = 6$	$\log_{36} 6 = \frac{1}{2}$
(e)	$p^q = r$	$\log_p r = q$

Example 4

Write each equation in exponential form.

	Logarithmic Form	Exponential Form
(a)	$\log_7 49 = 2$	$7^2 = 49$
(b)	$\log_4 64 = 3$	$4^3 = 64$
(c)	$\log_{10} 0.01 = -2$	$10^{-2} = 0.01$
(d)	$\log_{27} 3 = \frac{1}{3}$	$27^{1/3} = 3$
(e)	$\log_m p = n$	$m^n = p$

Example 5

Graph $y = \log_2 x$ by plotting points.

First, change the equation from logarithmic form to exponential form. That is, $y = \log_2 x$ is equivalent to $x = 2^y$. Then choose values for y and compute values for x. (See Fig. 6.19.)

Example 6

If $\log_3 81 = x$, find x.

The exponential form of $\log_3 81 = x$ is

$$3^x = 81$$

We know that

$$3^4 = 81$$

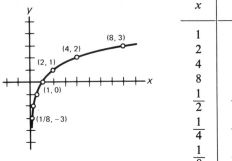

x	y	$x = 2^y$
1	0	$x = 2^0 = 1$
2	1	$x = 2^1 = 2$
4	2	$x = 2^2 = 4$
8	3	$x = 2^3 = 8$
$\dfrac{1}{2}$	-1	$x = 2^{-1} = \dfrac{1}{2}$
$\dfrac{1}{4}$	-2	$x = 2^{-2} = \dfrac{1}{4}$
$\dfrac{1}{8}$	-3	$x = 2^{-3} = \dfrac{1}{8}$

Figure 6.19

Therefore,

$$x = 4$$

Example 7

If $\log_3 x = -2$, find x.

$$\log_3 x = -2 \quad \text{or} \quad 3^{-2} = x$$

Therefore,

$$x = \frac{1}{9}$$

Example 8

If $\log_x 32 = \dfrac{5}{3}$, find x.

$$\log_x 32 = \frac{5}{3} \quad \text{or} \quad x^{5/3} = 32$$

$$x^{1/3} = 2 \qquad \text{(Take the fifth root of each side.)}$$
$$x = 8 \qquad \text{(Cube each side.)}$$

Or begin with

$$x^{5/3} = 32$$

Raise each side to the $\frac{3}{5}$ power.

$$(x^{5/3})^{3/5} = 32^{3/5}$$
$$x = 8$$

Recall the following three basic logarithmic properties.

1. *Multiplication:* If M and N are positive real numbers,

$$\boxed{\begin{array}{l} \log_a (M \cdot N) = \log_a M + \log_a N \\ \text{where } a > 0 \text{ and } a \neq 1. \end{array}}$$

That is, the logarithm of a product equals the sum of the logarithms of its factors.

2. *Division:* If M and N are positive real numbers,

$$\log_a \left(\frac{M}{N}\right) = \log_a M - \log_a N$$

where $a > 0$ and $a \neq 1$.

That is, the logarithm of a quotient equals the difference of the logarithms of its factors.

3. *Powers:* If M is a positive real number and n is any real number,

$$\log_a M^n = n \log_a M$$

where $a > 0$ and $a \neq 1$.

That is, the logarithm of a power of a number equals the product of the exponent times the logarithm of the number.

There are three special cases of the power property that are helpful.

(a) *Roots:* If M is any positive real number and n is any positive integer,

$$\log_a \sqrt[n]{M} = \frac{1}{n} \cdot \log_a M$$

Note this is a special case where $\sqrt[n]{M} = M^{1/n}$. That is, the logarithm of the root of a number equals the logarithm of the number divided by the index of the root.

(b) For $n = 0$,

$$\log_a M^0 = \log_a 1 \qquad (M^0 = 1)$$
$$\log_a M^0 = 0 \cdot \log_a M \qquad \text{(Property 3)}$$
$$= 0$$

Therefore,

$$\log_a 1 = 0$$

That is, the logarithm of one to any base is zero.

(c) For $n = -1$,

$$\log_a M^{-1} = \log_a \frac{1}{M} \qquad \left(M^{-1} = \frac{1}{M}\right)$$
$$\log_a M^{-1} = (-1)\log_a M \qquad \text{(Property 3)}$$

Therefore,

$$\log_a \frac{1}{M} = -\log_a M$$

That is, the logarithm of the reciprocal of a number is the negative of the logarithm of the number.

Example 9

Write $\log_4 2x^5y^2$ as a sum and multiple of logarithms of a single variable.

$$\log_4 2x^5y^2 = \log_4 2 + \log_4 x^5 + \log_4 y^2 \qquad \text{(Property 1)}$$
$$= \log_4 2 + 5 \log_4 x + 2 \log_4 y \qquad \text{(Property 3)}$$

Example 10

Write $\log_3 \dfrac{\sqrt{x(x-2)}}{(x+3)^2}$ as a sum, difference, or multiple of the logarithms of x, $x-2$, and $x+3$.

$$\log_3 \frac{\sqrt{x(x-2)}}{(x+3)^2} = \log_3 \frac{[x(x-2)]^{1/2}}{(x+3)^2}$$

$$= \log_3 [x(x-2)]^{1/2} - \log_3 (x+3)^2 \qquad \text{(Property 2)}$$

$$= \frac{1}{2} \log_3 [x(x-2)] - 2 \log_3 (x+3) \qquad \text{(Property 3)}$$

$$= \frac{1}{2} [\log_3 x + \log_3 (x-2)] - 2 \log_3 (x+3) \qquad \text{(Property 1)}$$

$$= \frac{1}{2} \log_3 x + \frac{1}{2} \log_3 (x-2) - 2 \log_3 (x+3)$$

Example 11

Write $3 \log_2 x + 4 \log_2 y - 2 \log_2 z$ as a single logarithmic expression.

$$3 \log_2 x + 4 \log_2 y - 2 \log_2 z = \log_2 x^3 + \log_2 y^4 - \log_2 z^2 \qquad \text{(Property 3)}$$

$$= \log_2 (x^3 y^4) - \log_2 z^2 \qquad \text{(Property 1)}$$

$$= \log_2 \frac{x^3 y^4}{z^2} \qquad \text{(Property 2)}$$

Example 12

Write $3 \log_{10} (x-1) - \dfrac{1}{3} \log_{10} x - \log_{10} (2x+3)$ as a single logarithmic expression.

$$3 \log_{10} (x-1) - \frac{1}{3} \log_{10} x - \log_{10} (2x+3)$$

$$= \log_{10} (x-1)^3 - \log_{10} x^{1/3} - \log_{10} (2x+3) \qquad \text{(Property 3)}$$

$$= \log_{10} (x-1)^3 - [\log_{10} x^{1/3} + \log_{10} (2x+3)]$$

$$= \log_{10} (x-1)^3 - [\log_{10} (x^{1/3})(2x+3)] \qquad \text{(Property 1)}$$

$$= \log_{10} \frac{(x-1)^3}{x^{1/3}(2x+3)} \qquad \text{(Property 2)}$$

Derivatives of Transcendental Functions   Chap. 6

$$= \log_{10} \frac{(x - 1)^3}{\sqrt[3]{x(2x + 3)}}$$

There are two other logarithmic properties that are useful in simplifying expressions.

$$\log_b b^x = x \quad \text{and} \quad b^{\log_b x} = x$$

To show the first one, we begin with the identity

$$(b^x) = b^x$$

Then, writing this logarithmic equation in exponential form, we have

$$b^{(\log_b x)} = x$$

To show the second one, we begin with the identity

$$\log_b x = (\log_b x)$$

Then, writing this logarithmic equation in exponential form, we have

$$b^{(\log_b x)} = x$$

Example 13

Find the value of $\log_2 16$.

$$\log_2 16 = \log_2 2^4 \quad \text{(Note that this simplification is possible}$$
$$= 4 \quad \text{because 16 is a power of 2.)}$$

Example 14

Find the value of $\log_{10} 0.01$.

$$\log_{10} 0.01 = \log_{10} 10^{-2} \quad \text{(Note that } 0.01 = \frac{1}{100} = 10^{-2}.\text{)}$$
$$= -2$$

Example 15

Find the value of $5^{\log_5 8}$.

$$5^{\log_5 8} = 8$$

One other useful law is for change of base

$$\log_b x = \frac{\log_a x}{\log_a b}$$

While all the laws of exponents and logarithms hold for any base b ($b > 0$ and $b \neq 1$), there are two bases used almost exclusively with logarithms:

1. *Common logarithms:* Base 10
   Usual notation: $\log_{10} x = \log x$

2. *Natural logarithms:* Base e
   Usual notation: $\log_e x = \ln x$

As we shall see, the most convenient base for use in calculus is base e, which is defined as

$$e = \lim_{x \to 0} (1 + x)^{1/x} \approx 2.71828$$

Recall that e is an irrational number.

Exercises

Graph each equation.

1. $y = 4^x$
2. $y = 3^x$
3. $y = (\frac{1}{3})^x$
4. $y = (\frac{1}{4})^x$
5. $y = 4^{-x}$
6. $y = (\frac{2}{3})^x$
7. $y = (\frac{4}{3})^{-x}$
8. $y = 4^{2x}$

Write each equation in logarithmic form.

9. $3^2 = 9$
10. $7^2 = 49$
11. $5^3 = 125$
12. $10^3 = 1000$
13. $9^{1/2} = 3$
14. $4^0 = 1$
15. $10^{-5} = 0.00001$
16. $d^e = f$

Write each equation in exponential form.

17. $\log_5 25 = 2$
18. $\log_8 64 = 2$
19. $\log_{25} 5 = \dfrac{1}{2}$
20. $\log_{27} 3 = \dfrac{1}{3}$
21. $\log_2 (\frac{1}{4}) = -2$
22. $\log_2 (\frac{1}{8}) = -3$
23. $\log_{10} 0.01 = -2$
24. $\log_g h = k$

Graph each equation.

25. $y = \log_4 x$
26. $y = \log_3 x$
27. $y = \log_{10} x$
28. $y = \log_5 x$
29. $y = \log_{1/4} x$
30. $y = \log_{1/3} x$

Solve for x.

31. $\log_4 x = 3$
32. $\log_2 x = -1$
33. $\log_9 3 = x$
34. $\log_6 36 = x$
35. $\log_2 8 = x$
36. $\log_3 27 = x$
37. $\log_{25} 5 = x$
38. $\log_{27} 3 = x$
39. $\log_x 25 = 2$
40. $\log_x (\frac{1}{27}) = -3$
41. $\log_{1/2} (\frac{1}{8}) = x$
42. $\log_x 3 = \frac{1}{2}$
43. $\log_{12} x = 2$
44. $\log_8 (\frac{1}{64}) = x$
45. $\log_x 9 = \frac{2}{3}$
46. $\log_x 64 = \frac{3}{2}$
47. $\log_x (\frac{1}{8}) = -\frac{3}{2}$
48. $\log_x (\frac{1}{27}) = -\frac{3}{4}$

Write each expression as a sum, difference, or multiple of single logarithms.

49. $\log_2 5x^3 y$
50. $\log_3 \dfrac{8x^2 y^3}{z^4}$
51. $\log_b \dfrac{y^3 \sqrt{x}}{z^2}$
52. $\log_b \dfrac{7xy}{\sqrt[3]{z}}$
53. $\log_b \sqrt[3]{\dfrac{x^2}{y}}$
54. $\log_5 \sqrt[4]{xy^2 z}$
55. $\log_2 \dfrac{1}{x} \sqrt{\dfrac{y}{z}}$
56. $\log_b \dfrac{1}{z^2} \sqrt[3]{\dfrac{x^2}{y}}$
57. $\log_b \dfrac{z^3 \sqrt{x}}{\sqrt[3]{y}}$

Derivatives of Transcendental Functions   Chap. 6

58. $\log_b \dfrac{\sqrt{y\sqrt{x}}}{z^2}$ **59.** $\log_b \dfrac{x^2(x + 1)}{\sqrt{x + 2}}$ **60.** $\log_b \dfrac{\sqrt{x}(x + 4)}{x^2}$

Write each as a single logarithmic expression.

61. $\log_b x + 2 \log_b y$ **62.** $2 \log_b z - 3 \log_b x$

63. $\log_b x + 2 \log_b y - 3 \log_b z$ **64.** $3 \log_7 x - 4 \log_7 y - 5 \log_7 z$

65. $\log_3 x + \frac{1}{3} \log_3 y - \frac{1}{2} \log_3 z$ **66.** $\frac{1}{2} \log_2 x - \frac{1}{3} \log_2 y - \log_2 z$

67. $2 \log_{10} x - \frac{1}{2} \log_{10} (x - 3) - \log_{10} (x + 1)$

68. $\log_3 (x + 1) + \frac{1}{2} \log_3 (x + 2) - 3 \log_3 (x - 1)$

69. $5 \log_b x + \frac{1}{3} \log_b (x - 1) - \log_b (x + 2)$

70. $\log_b (x + 1) + \frac{1}{3} \log_b (x - 7) - 2 \log_b x$

71. $\log_{10} x + 2 \log_{10} (x - 1) - \frac{1}{3}[\log_{10} (x + 2) + \log_{10} (x - 5)]$

72. $\frac{1}{2} \log_b (x + 1) - 3[\log_b x + \log_b (x - 1) + \log_b (2x - 1)]$

Find the value of each expression.

73. $\log_b b^3$ **74.** $\log_2 2^5$ **75.** $\log_3 9$

76. $\log_2 16$ **77.** $\log_5 125$ **78.** $\log_4 64$

79. $\log_2 \frac{1}{4}$ **80.** $\log_3 \frac{1}{27}$ **81.** $\log_{10} 0.001$

82. $\log_{10} 0.1$ **83.** $\log_3 1$ **84.** $\log_{10} 1$

85. $6^{\log_6 5}$ **86.** $3^{\log_3 9}$ **87.** $25^{\log_5 6}$

88. $27^{\log_3 2}$ **89.** $4^{\log_2(1/5)}$ **90.** $8^{\log_2(1/3)}$

6.7 DERIVATIVE OF THE LOGARITHMIC FUNCTION

We now find the derivative of a logarithmic function in the form $y = \log_b u$.

$$\frac{dy}{du} = \lim_{\Delta u \to 0} \frac{\Delta y}{\Delta u}$$

where

$$\frac{\Delta y}{\Delta u} = \frac{\log_b (u + \Delta u) - \log_b u}{\Delta u}$$

Recall that the difference of two logarithms is the logarithm of their quotient, so

$$\frac{\Delta y}{\Delta u} = \frac{\log_b \left(\dfrac{u + \Delta u}{u} \right)}{\Delta u}$$

$$= \frac{1}{u} \cdot \frac{u}{\Delta u} \log_b \left(\frac{u + \Delta u}{u} \right) \qquad \text{(Multiply numerator and denominator by } u.\text{)}$$

Also recall $n \log_b a = \log_b a^n$. We then have

$$\frac{\Delta y}{\Delta u} = \frac{1}{u} \log_b \left(\frac{u + \Delta u}{u} \right)^{u/\Delta u}$$

$$= \frac{1}{u} \log_b \left(1 + \frac{\Delta u}{u} \right)^{u/\Delta u}$$

and then

$$\frac{dy}{du} = \lim_{\Delta u \to 0} \frac{1}{u} \cdot \lim_{\Delta u \to 0} \log_b \left(1 + \frac{\Delta u}{u}\right)^{u/\Delta u}$$

The limit of the left-hand factor is

$$\lim_{\Delta u \to 0} \frac{1}{u} = \frac{1}{u} \qquad \text{(as } \Delta u \to 0, u \text{ itself is not affected)}$$

but it remains to determine the limit of the right-hand factor.

One can show that the function $(1 + h)^{1/h}$ approaches a limiting number as h approaches 0. We denote this number by e, that is,

$$\lim_{h \to 0} (1 + h)^{1/h} = e$$

This is an irrational number whose decimal value can only be approximated. An approximation of e to seven decimal places is 2.7182818.

Note that if we let $h = \dfrac{\Delta u}{u}$, then h approaches 0 as Δu approaches 0. Then

$$\frac{dy}{du} = \frac{1}{u} \cdot \lim_{\Delta u \to 0} \log_b \left(1 + \frac{\Delta u}{u}\right)^{(1/\Delta u)/u} \qquad \left(\frac{u}{\Delta u} = \frac{1}{\frac{\Delta u}{u}}\right)$$

$$= \frac{1}{u} \log_b e$$

Using the chain rule, we also have

$$\frac{dy}{dx} = \frac{1}{u} \log_b e \, \frac{du}{dx}$$

Thus,

$$\frac{d}{du} (\log_b u) = \frac{1}{u} \log_b e$$

and

$$\frac{d}{dx} (\log_b u) = \frac{1}{u} \log_b e \, \frac{du}{dx}$$

Since $\log_e e = 1$ and since we denote $\log_e u$ by $\ln u$ (called the natural logarithm of u), we have

$$\frac{d}{du} (\ln u) = \frac{1}{u}$$

and

$$\frac{d}{dx} (\ln u) = \frac{1}{u} \frac{du}{dx}$$

Derivatives of Transcendental Functions   Chap. 6

Note: u must always be positive since $\log_b u$ or $\ln u$ is only defined for positive values of u.

Example 1

Find the derivative of $y = \ln 5x$.

$$\frac{dy}{dx} = \frac{1}{5x}\frac{d}{dx}(5x) = \frac{5}{5x} = \frac{1}{x} \qquad (u = 5x)$$

Example 2

Find the derivative of $y = \log(3x + 1)$.

$$\frac{dy}{dx} = \left(\frac{1}{3x + 1}\right)(\log e)\left[\frac{d}{dx}(3x + 1)\right] \qquad (u = 3x + 1)$$

$$= \frac{3\log e}{3x + 1} \quad \text{or} \quad \frac{3(0.4343)}{3x + 1} = \frac{1.3029}{3x + 1}$$

Example 3

Find the derivative of $y = \ln(\sin 3x)$.

$$\frac{dy}{dx} = \frac{1}{\sin 3x}\frac{d}{dx}(\sin 3x) \qquad (u = \sin 3x)$$

$$= \frac{3\cos 3x}{\sin 3x}$$

$$= 3\cot 3x$$

You may find it helpful to make simplifications using the properties of logarithms.

Example 4

Find the derivative of $y = \ln\left(\frac{x^3}{2x - 5}\right)$.

Since

$$\ln\left(\frac{x^3}{2x - 5}\right) = \ln x^3 - \ln(2x - 5)$$

$$\ln\left(\frac{x^3}{2x - 5}\right) = 3\ln x - \ln(2x - 5)$$

Then

$$\frac{dy}{dx} = 3\left(\frac{1}{x}\right) - \frac{1}{2x - 5}\frac{d}{dx}(2x - 5)$$

$$= \frac{3}{x} - \frac{2}{2x - 5}$$

$$= \frac{4x - 15}{x(2x - 5)}$$

Example 5

Find the derivative of $y = \ln(x\sqrt{4 + x})$.

Rewriting we have

$$y = \ln x + \tfrac{1}{2}\ln(4 + x)$$

$$\frac{dy}{dx} = \frac{1}{x} + \frac{1}{2}\left(\frac{1}{4+x}\right)$$

$$= \frac{8 + 3x}{2x(4 + x)}$$

For more complicated algebraic expressions, consider the following method, called *logarithmic differentiation*.

Example 6

Find the derivative of $y = \sqrt[3]{\frac{x^2(3x + 4)}{1 - 2x^2}}$.

First, take the natural logarithm of both sides.

$$\ln y = \ln \left(\frac{x^2(3x + 4)}{1 - 2x^2}\right)^{1/3}$$

$$\ln y = \frac{1}{3} \ln \left(\frac{x^2(3x + 4)}{1 - 2x^2}\right)$$

$$\ln y = \frac{1}{3} [2 \ln x + \ln (3x + 4) - \ln (1 - 2x^2)]$$

Next, find the derivative of both sides.

$$\frac{1}{y}\frac{dy}{dx} = \frac{1}{3}\left[\frac{2}{x} + \frac{3}{3x + 4} - \frac{-4x}{1 - 2x^2}\right]$$

Solve for $\frac{dy}{dx}$:

$$\frac{dy}{dx} = \frac{y}{3}\left[\frac{2}{x} + \frac{3}{3x + 4} + \frac{4x}{1 - 2x^2}\right]$$

$$= \frac{1}{3}\sqrt[3]{\frac{x^2(3x + 4)}{1 - 2x^2}}\left[\frac{2}{x} + \frac{3}{3x + 4} + \frac{4x}{1 - 2x^2}\right]$$

Logarithmic differentiation is especially helpful for functions having both a variable base and a variable exponent.

Example 7

Find the derivative of $y = x^{4x}$.

First, take the natural logarithm of both sides.

$$\ln y = \ln x^{4x}$$

$$\ln y = 4x \ln x$$

Then find the derivative of both sides.

$$\frac{1}{y}\frac{dy}{dx} = 4x\frac{1}{x} + 4 \ln x$$

$$\frac{dy}{dx} = 4y(1 + \ln x)$$

$$= 4x^{4x} (1 + \ln x)$$

Exercises

Find the derivative of each function.

1. $y = \log (4x - 3)$
2. $y = \log (x^2 - 1)$
3. $y = \log_2 3x$
4. $y = \log_7 (5x + 2)$
5. $y = \ln (2x^3 - 3)$
6. $y = \ln (x^2 - 4)$
7. $y = \ln (\tan 3x)$
8. $y = \ln (\sec x^2)$
9. $y = \ln (x \sin x)$
10. $y = \ln (x^3 \cos 2x)$
11. $y = \ln \sqrt{3x - 2}$
12. $y = \ln (x^2 - 1)^{1/3}$
13. $y = \ln \dfrac{x^3}{x^2 + 1}$
14. $y = \ln \dfrac{5x}{1 - x^2}$
15. $y = \tan (\ln x)$
16. $y = (\ln x)(\sin x)$
17. $y = \ln (\ln x)$
18. $y = x^2 + \ln (3x - 2)$
19. $y = \text{Arctan} (\ln x^2)$
20. $y = \text{Arcsin} (\ln x)$
21. $y = \ln (\text{Arccos}^2 x)$
22. $y = \ln (\text{Arctan}^3 x)$

Find dy/dx using logarithmic differentiation.

23. $y = (3x + 2)(6x - 1)^2(x - 4)$
24. $y = \sqrt{x(2x + 1)(1 - 5x)}$
25. $y = \dfrac{(x + 1)(2x + 1)}{(3x - 4)(1 - 8x)}$
26. $y = \dfrac{x(2x - 1)^{3/2}}{\sqrt[3]{x + 1}}$
27. $y = x^x$
28. $y = x^{x+1}$
29. $y = x^{2/x}$
30. $y = (1 + x)^{1/x}$
31. $y = (\sin x)^x$
32. $y = (\tan x)^x$
33. $y = (1 + x)^{x^2}$
34. $y = x^{\sin x}$

Find the equation of the tangent line at the given value.

35. $y = \ln x$ at $x = 1$
36. $y = \ln x^2$ at $x = 1$
37. $y = \ln (\sin x)$ at $x = \pi/6$
38. $y = \tan (\ln x)$ at $x = 1$

6.8 DERIVATIVE OF THE EXPONENTIAL FUNCTION

An exponential function is a function in the form $y = b^u$ where b is a positive constant, $b \neq 1$ and u is a variable.

If $y = b^u$, then $u = \log_b y$ in logarithmic form. Then from the preceding section,

$$\frac{du}{dx} = \frac{1}{y} \cdot \log_b e \frac{dy}{dx}$$

Solving for $\dfrac{dy}{dx}$, we have

$$\frac{dy}{dx} = \frac{y}{\log_b e} \frac{du}{dx}$$

$$= \frac{b^u}{\log_b e} \frac{du}{dx} \qquad (y = b^u)$$

That is,

$$\boxed{\frac{d}{dx} (b^u) = \frac{b^u}{\log_b e} \frac{du}{dx}}$$

When the base $b = e$, the expression above simplifies to

$$\frac{d}{dx}(e^u) = e^u \frac{du}{dx} \qquad (\log_e e = 1)$$

Example 1

Find the derivative of $y = e^{3x}$.

$$\frac{dy}{dx} = e^{3x} \frac{d}{dx}(3x) = 3e^{3x} \qquad (u = 3x)$$

Example 2

Find the derivative of $y = 10^{2x}$.

$$\frac{dy}{dx} = \frac{10^{2x}}{\log e} \cdot \frac{d}{dx}(2x) \qquad (u = 2x)$$

$$= \frac{(2)10^{2x}}{\log e}$$

Example 3

Find the derivative of $y = e^{x^2+3}$.

$$\frac{dy}{dx} = e^{x^2+3}(2x) \qquad (u = x^2 + 3)$$

$$= 2xe^{x^2+3}$$

Example 4

Find the derivative of $y = e^{\tan x}$.

$$\frac{dy}{dx} = e^{\tan x}(\sec^2 x) \qquad (u = \tan x)$$

Example 5

Find the derivative of $y = x^2 e^{-5x}$.

First apply the product rule for differentiation.

$$\frac{dy}{dx} = x^2 \frac{d}{dx}(e^{-5x}) + e^{-5x} \frac{d}{dx}(x^2)$$

$$= x^2 e^{-5x}(-5) + e^{-5x}(2x)$$

$$= xe^{-5x}(2 - 5x)$$

Example 6

Find the derivative of $y = \text{Arcsin } e^{3x}$.

Let $u = e^{3x}$, then

$$\frac{dy}{dx} = \frac{1}{\sqrt{1 - (e^{3x})^2}}(3e^{3x}) = \frac{3e^{3x}}{\sqrt{1 - e^{6x}}}$$

Exercises

Find the derivative of each function.

   1. $y = e^{5x}$ **2.** $y = 2e^{3x+1}$ **3.** $y = 4e^{x^3}$

4. $y = e^{x^2 + 3x}$ **5.** $y = 10^{3x}$ **6.** $y = 4^{x^2}$

7. $y = \dfrac{1}{e^{6x}}$ **8.** $y = e^{-x^3}$ **9.** $y = e^{\sqrt{x}}$

10. $y = e^{\sqrt{3x-2}}$ **11.** $y = e^{\sin x}$ **12.** $y = 5e^{\cos x^2}$

13. $y = 6xe^{x^2-1}$ **14.** $y = xe^{x^3}$ **15.** $y = (\cos x)e^{3x^2}$

16. $y = 3e^{\tan x^2}$ **17.** $y = \ln(\cos e^{5x})$ **18.** $y = \tan e^{4x}$

19. $y = e^x - e^{-x}$ **20.** $y = \dfrac{e^x}{x}$ **21.** $y = \dfrac{2x^2}{3e^x - x}$

22. $y = \dfrac{e^{x^2}}{x-2}$ **23.** $y = 2\,\text{Arctan }e^{3x}$ **24.** $y = \text{Arcsin }e^{4x}$

25. $y = \text{Arccos}^3 e^{-2x}$ **26.** $y = \text{Arctan }e^{x^2}$ **27.** $y = xe^x - e^x$

28. $y = (e^x + e^{-x})^3$ **29.** $y = \ln e^{x^2}$ **30.** $y = \ln(x - e^x)$

6.9 APPLICATIONS

The previous calculus techniques for curve sketching, finding maximums and minimums, solving problems of motion, and so on, apply also to the transcendental functions. Some of these applications are found in this section.

Example 1

Sketch the curve $y = xe^x$.

    The x- and y-intercepts are both (0, 0). To find where the curve is above and below the x-axis:

Signs of factors x $-$ $+$
in each interval:

               e^x $+$ $+$ $\rightarrow x$

Sign of y in   0
each interval: $-$ $+$
       (below)   (above)

Note: $e^x > 0$. This factor, being always positive, does not affect the sign in any interval. So we will not list it after this.

$$\frac{dy}{dx} = xe^x + e^x = e^x(x + 1)$$

To find where the curve is increasing and decreasing:

Sign of factor
in each interval: $x + 1$ $-$ $+$
                                  $\rightarrow x$
Sign of $f'(x)$: $-$ -1 $+$
       (decreasing) (increasing)

Since $f(x)$ changes from decreasing to increasing at $x = -1$, $(-1, -1/e)$ is a relative minimum.

$$\frac{d^2y}{dx^2} = xe^x + 2e^x = e^x(x + 2)$$

To find where the curve is concave upward and concave downward:

Sign of $f''(x)$ in
each interval:

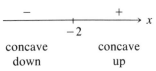

concave   concave
down   up

So $(-2, -2/e^2)$ is a point of inflection.
With this information we now sketch the curve in Fig. 6.20.

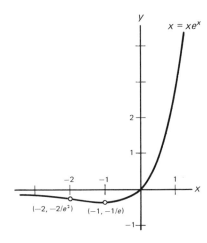

Figure 6.20

Example 2

The insulation resistance of a shielded cable is given by

$$R = \frac{\rho}{2\pi} \ln\left(\frac{r_2}{r_1}\right) \ \Omega/\text{m},$$

where ρ is the resistivity of the insulation material, r_1 is the inner radius of the insulation, and r_2 is the outer radius. Find the rate of change of the insulation resistance R with respect to the outer radius r_2.

$$\frac{dR}{dr_2} = \frac{d}{dr_2}\left[\left(\frac{\rho}{2\pi}\right) \ln\left(\frac{r_2}{r_1}\right)\right]$$

$$= \left(\frac{\rho}{2\pi}\right)\frac{d}{dr_2}\left[\ln\left(\frac{r_2}{r_1}\right)\right]$$

$$= \left(\frac{\rho}{2\pi}\right)\left(\frac{1}{r_2/r_1}\right)\frac{d}{dr_2}\left(\frac{r_2}{r_1}\right)$$

$$= \left(\frac{\rho}{2\pi}\right)\left(\frac{r_1}{r_2}\right)\left(\frac{1}{r_1}\right)$$

$$= \frac{\rho}{2\pi r_2} \ \Omega/\text{m}$$

Example 3

The current in an electrical circuit is given by $i = 10(1 - e^{-20t})$. Find the maximum value of the current.

$$\frac{di}{dt} = \frac{d}{dt}[10(1 - e^{-20t})]$$

$$= 10\frac{d}{dt}(1 - e^{-20t})$$

$$= 10(0 + 20e^{-20t})$$

$$= 200e^{-20t}$$

Since $\dfrac{di}{dt}$ is never zero and always positive, there is no maximum value. The current increases with time.

Example 4

Find the equation of the line tangent to the curve $y = \ln x^2$ at the point $(1, 0)$.

$$\frac{dy}{dx} = \frac{d}{dx}(\ln x^2) = \frac{1}{x^2}(2x) = \frac{2}{x}$$

The slope of tangent line is

$$m = \frac{dy}{dx}\bigg|_{x=1} = \frac{2}{1} = 2$$

Using the point-slope formula, we have

$$y - y_1 = m(x - x_1)$$

$$y - 0 = 2(x - 1)$$

$$y = 2x - 2$$

Exercises

Sketch each curve.

1. $y = \sin x + \cos x$ **2.** $y = x \ln x, \ x > 0$

3. $y = \dfrac{\ln x}{x}, \ x > 0$ **4.** $y = \ln (\sin x), \ 0 < x < \pi$

5. $y = e^x + e^{-x}$ **6.** $y = \ln x - x, \ x > 0$ **7.** $y = x^2 e^{-x}$

8. $y = \ln \left(\dfrac{1}{x^2 - 1}\right)$ **9.** $y = \dfrac{1}{1 - e^x}$ **10.** $y = \dfrac{x}{\ln x}, \ x > 0$

Find any relative maximums or minimums of each function.

11. $y = xe^{-2x}$ **12.** $y = x^2 e^{2x}$ **13.** $y = \dfrac{x^2}{e^{2x}}$

14. $y = \dfrac{e^x}{x}$ **15.** $y = \dfrac{\ln x}{x^2}$ **16.** $y = \dfrac{x}{\ln^2 x}$

17. Find the equation of the tangent line to $y = \sin 2x$ at the point $(\pi/2, 0)$.

18. Find the equation of the tangent line to $y = xe^x$ at the point $(1, e)$.

19. Find the equation of the tangent line to $y = x^2 + x \ln x$ at $x = 1$.

20. Find the equation of the tangent line to $y = e^{\cos x}$ at $x = \pi/2$.

Find $\dfrac{d^4y}{dx^4}$ for each function:

21. $y = e^x \sin x$ **22.** $y = e^{-x} \cos x$

23. The current in a circuit is given by $i = 50 \cos 2t$. Find the equation of the voltage V_L across a 3-henry (H) inductor $\left(V_L = L\dfrac{di}{dt}, \text{ where } L \text{ is the inductance}\right)$.

24. The current in a circuit is given by $i = 40(1 - 2e^{-20t})$. Find the equation of the voltage V_L across a 5-H inductor.

25. The voltage in a circuit is given by $V = 25e^{0.4t}$. Find $\dfrac{dV}{dt}$ at $t = 3$.

26. The current in a circuit is given by $i = 4t^2 e^{8/t}$. Find the value of t when $di/dt = 0$.

27. The apparent power p_a of a circuit is given by $p_a = p \sec \theta$ where p is the power and θ is the impedance phase angle. Find $\dfrac{dp_a}{dt}$ at $t = 1$ s if $\dfrac{d\theta}{dt} = 0.1$ rad/s and $\theta = \dfrac{\pi}{4}$. The power of the circuit, p, is 15 watts (W).

28. If work in a circuit is given by $W = 25 \sin^2 t$, find the equation for the power p $\left(p = \dfrac{dW}{dt}\right)$.

29. The charge q at any time t on a capacitor is given by $q = e^{-0.02t}(0.05 \cos 2t)$. Find the current in the circuit $\left(i = \dfrac{dq}{dt}\right)$.

30. The voltage across a capacitor at any time t is given by $V = e^{-0.001t}$. If the capacitance C is 2×10^{-5} farad (F), find the current in the circuit $\left(i = C\dfrac{dV}{dt}\right)$.

31. A particle moves along a straight line according to $s = 2e^{3t} + 5e^{-3t}$. Find expressions for its velocity and acceleration.

32. A particle moves along a straight line according to $s = \ln(t^2 + 1)$. Find expressions for its velocity and acceleration.

33. A particle moves along a straight line according to $s = e^{-t/2} \tan \dfrac{\pi t}{2}$. Find its velocity and acceleration at $t = 4$.

34. An object moves along a straight line according to $s = \sin 2t + e^t$. Find its velocity and acceleration at $t = 2$.

35. The charge of a capacitor at time t is given by $q = e^{-0.2t}(0.04 \cos 2t - 0.5 \sin 3t)$. Find the current $\left(i = \dfrac{dq}{dt}\right)$ at $t = \pi/2$.

CHAPTER SUMMARY

1. Trigonometric functions (see Fig. 6.21.)

$$\sin \theta = \frac{y}{r}$$

$$\cos \theta = \frac{x}{r}$$

$$\tan \theta = \frac{y}{x}$$

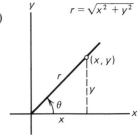

Figure 6.21

$$\cot \theta = \frac{x}{y}$$

$$\sec \theta = \frac{r}{x}$$

$$\csc \theta = \frac{r}{y}$$

2. Graphing the six trigonometric functions

$$\text{Period} = \frac{2\pi}{b} \quad \left. \begin{array}{l} y = a \sin (bx + c) \\ y = a \cos (bx + c) \end{array} \right\} \text{amplitude} = |a|$$
$$\left. \begin{array}{l} y = a \sec (bx + c) \\ y = a \csc (bx + c) \end{array} \right.$$

$$\text{Period} = \frac{\pi}{b} \quad \left\{ \begin{array}{l} y = a \tan (bx + c) \\ y = a \cot (bx + c) \end{array} \right.$$

Phase shift $= \dfrac{c}{b}$:

to the *left* if $\dfrac{c}{b} > 0$

and

to the *right* if $\dfrac{c}{b} < 0$

3. A list of trigonometric identities is given in Appendix B, Table 6.

4. Derivatives of the trigonometric functions

(a) $\dfrac{d}{dx}(\sin u) = \cos u \dfrac{du}{dx}$

(b) $\dfrac{d}{dx}(\cos u) = -\sin u \dfrac{du}{dx}$

(c) $\dfrac{d}{dx}(\tan u) = \sec^2 u \dfrac{du}{dx}$

(d) $\dfrac{d}{dx}(\cot u) = -\csc^2 u \dfrac{du}{dx}$

(e) $\dfrac{d}{dx}(\sec u) = \sec u \tan u \dfrac{du}{dx}$

(f) $\dfrac{d}{dx}(\csc u) = -\csc u \cot u \dfrac{du}{dx}$

5. Inverse trigonometric functions

$y = \text{Arcsin } x, \quad -\dfrac{\pi}{2} \leqslant y \leqslant \dfrac{\pi}{2}$

$y = \text{Arccos } x, \quad 0 \leqslant y \leqslant \pi$

$y = \text{Arctan } x, \quad -\dfrac{\pi}{2} < y < \dfrac{\pi}{2}$

$y = \text{Arccot } x, \quad 0 < y < \pi$

$y = \text{Arcsec } x, \quad 0 \leqslant y \leqslant \pi, y \neq \dfrac{\pi}{2}$

$y = \text{Arccsc } x, \quad -\dfrac{\pi}{2} \leqslant y \leqslant \dfrac{\pi}{2}, y \neq 0$

6. Derivatives of the inverse trigonometric functions

$$\frac{d}{dx}(\text{Arcsin } u) = \frac{1}{\sqrt{1 - u^2}} \frac{du}{dx}$$

$$\frac{d}{dx}(\text{Arccos } u) = -\frac{1}{\sqrt{1-u^2}}\frac{du}{dx}$$

$$\frac{d}{dx}(\text{Arctan } u) = \frac{1}{1+u^2}\frac{du}{dx}$$

$$\frac{d}{dx}(\text{Arccot } u) = -\frac{1}{1+u^2}\frac{du}{dx}$$

$$\frac{d}{dx}(\text{Arcsec } u) = \frac{1}{|u|\sqrt{u^2-1}}\frac{du}{dx}$$

$$\frac{d}{dx}(\text{Arccsc } u) = -\frac{1}{|u|\sqrt{u^2-1}}\frac{du}{dx}$$

7. Exponential function
$$y = b^x \qquad \text{where } b > 0 \quad \text{and} \quad b \neq 1$$

8. Laws of exponents
$$a^m \cdot a^n = a^{m+n}$$
$$\frac{a^m}{a^n} = a^{m-n}$$
$$(a^m)^n = a^{mn}$$
$$(ab)^n = a^n b^n$$
$$\left(\frac{a}{b}\right)^n = \frac{a^n}{b^n}$$
$$a^0 = 1$$

9. Logarithmic function

In exponential form	In logarithmic form	Is inverse of:
$x = b^y$	$y = \log_b x$	$y = b^x$

Note: $b > 0$ and $b \neq 1$.

10. Logarithmic properties
$$\log_a (M \cdot N) = \log_a M + \log_a N, \text{ where } a > 0 \text{ and } a \neq 1$$
$$\log_a \left(\frac{M}{N}\right) = \log_a M - \log_a N, \text{ where } a > 0 \text{ and } a \neq 1$$
$$\log_a M^n = n \log_a M, \text{ where } a > 0 \text{ and } a \neq 1$$
$$\log_a \sqrt[n]{M} = \frac{1}{n} \cdot \log_a M, \text{ where } a > 0 \text{ and } a \neq 1$$
$$\log_a 1 = 0, \text{ where } a > 0 \text{ and } a \neq 1$$
$$\log_a \frac{1}{M} = -\log_a M, \text{ where } a > 0 \text{ and } a \neq 1$$
$$\log_b (b^x) = x$$
$$b^{\log_b x} = x$$
$$\log_b x = \frac{\log_a x}{\log_a b}$$

11. Derivatives of the logarithmic functions
$$\frac{d}{dx}(\ln u) = \frac{1}{u}\frac{du}{dx}$$

$$\frac{d}{dx}(\log_b u) = \frac{1}{u}\log_b e \frac{du}{dx}$$

12. Derivatives of the exponential functions

$$\frac{d}{dx}(b^u) = \frac{b^u}{\log_b e}\frac{du}{dx}$$

$$\frac{d}{dx}(e^u) = e^u\frac{du}{dx}$$

CHAPTER REVIEW

Prove each identity.

1. $\sec x \cot x = \csc x$

2. $\sec^2\theta + \tan^2\theta + 1 = \dfrac{2}{\cos^2\theta}$

3. $\dfrac{\cos\theta}{\cos\theta + \sin\theta} = \dfrac{\cot\theta}{1 + \cot\theta}$

4. $\cos\left(\theta - \dfrac{3\pi}{2}\right) = -\sin\theta$

5. $\left(\sin\dfrac{1}{2}x + \cos\dfrac{1}{2}x\right)^2 = 1 + \sin x$

6. $2\cos^2\dfrac{\theta}{2} = \dfrac{1 + \sec\theta}{\sec\theta}$

7. $\dfrac{2\cot\theta}{1 + \cot^2\theta} = \sin 2\theta$

8. $\csc x - \cot x = \tan\dfrac{1}{2}x$

9. $\tan 2x = \dfrac{2\cos x}{\csc x - 2\sin x}$

10. $\tan^2\dfrac{x}{2} + 1 = 2\tan\dfrac{x}{2}\csc x$

Simplify each expression.

11. $\sin\theta\cos\theta$

12. $\cos^2 3\theta - \sin^2 3\theta$

13. $\dfrac{1 + \cos 4\theta}{2}$

14. $1 - 2\sin^2\dfrac{\theta}{3}$

15. $\cos 2x\cos 3x - \sin 2x\sin 3x$

16. $\sin 2x\cos x - \cos 2x\sin x$

Sketch each curve.

17. $y = 4\cos 6x$

18. $y = -2\sin\dfrac{1}{3}x$

19. $y = 3\sin\left(x - \dfrac{\pi}{4}\right)$

20. $y = \cos\left(2x + \dfrac{2\pi}{3}\right)$

21. $y = 4\sin\left(\pi x + \dfrac{\pi}{2}\right)$

22. $y = \tan 5x$

23. $y = -\cot 3x$

24. $y = 2\sec 4x$

Find the derivative of each function.

25. $y = \sin(x^2 + 3)$

26. $y = \cos 8x$

27. $y = \cos^3(5x - 1)$

28. $y = \sin 3x\cos 2x$

29. $y = \tan(3x - 2)$

30. $y = \sec(4x + 3)$

31. $y = \cot 6x^2$

32. $y = \csc^2(8x^2 + x)$

33. $y = \sec^2 x\sin x$

34. $y = x^2 - \csc^2 x$

35. $y = \tan(\sec x)$

36. $y = \dfrac{\cos x}{1 + \sin x}$

37. $y = (1 - \sin x)^3$

38. $y = (1 + \sec 4x)^2$

Solve each equation for $0 \le y < 2\pi$.

39. $y = \arcsin\left(\dfrac{1}{2}\right)$

40. $y = \arctan\left(-\dfrac{1}{\sqrt{3}}\right)$

41. $y = \arccos\left(\dfrac{1}{\sqrt{2}}\right)$　　　　　　　**42.** $y = \text{arcsec}\ (-2)$

Find all angles in radians for each expression.

43. $\arctan\ (-\sqrt{3})$　　　　　　　**44.** $\arcsin\ (-1)$

Solve each equation for x.

45. $y = \dfrac{1}{2}\sin\dfrac{3x}{4}$　　　　　　　**46.** $y = 5\tan\ (1 - 2x)$

Find the value of each expression in radians.

47. $\text{Arcsin}\left(\dfrac{1}{\sqrt{2}}\right)$　　　　　　　**48.** $\text{Arctan}\left(-\dfrac{1}{\sqrt{3}}\right)$

49. $\text{Arcsec}\ (-1)$　　　　　　　**50.** $\text{Arccos}\left(-\dfrac{1}{2}\right)$

Find the value of each expression.

51. $\sin\left[\text{Arccos}\left(-\dfrac{1}{2}\right)\right]$　　　　　　　**52.** $\tan\ (\text{Arctan}\ \sqrt{3})$

53. Find an algebraic expression for $\sin(\text{Arccot}\ x)$.

Find the derivative of each function.

54. $y = \text{Arcsin}\ x^3$　　　**55.** $y = \text{Arctan}\ 3x$　　　**56.** $y = 3\ \text{Arccos}\ \dfrac{1}{2x}$

57. $y = 2\ \text{Arcsec}\ 4x$　　　**58.** $y = \text{Arcsin}^2 3\sqrt{x}$　　　**59.** $y = x\ \text{Arcsin}\ x$

Graph each equation.

60. $y = 3^x$　　　　　　　**61.** $y = \log_3 x$

Solve for x.

62. $\log_9 x = 2$　　　　　**63.** $\log_x 8 = 3$　　　　　**64.** $\log_2 32 = x$

Write each expression as a sum, difference, or multiple of single logarithms.

65. $\log_4 6x^2y$　　　　　　　**66.** $\log_3 \dfrac{5x\sqrt{y}}{z^3}$

67. $\log_{10} \dfrac{x^2(x + 1)^3}{\sqrt{x - 4}}$　　　　　　　**68.** $\ln \dfrac{[x(x - 1)]^3}{\sqrt{x + 1}}$

Write each expression as a single logarithmic expression.

69. $\log_2 x + 3\log_2 y - 2\log_2 z$　　　**70.** $\dfrac{1}{2}\log_{10}(x + 1) - 3\log_{10}(x - 2)$

71. $4\ln x - 5\ln(x + 1) - \ln(x + 2)$　　　**72.** $\dfrac{1}{2}[\ln x + \ln(x + 2)] - 2\ln(x - 5)$

Simplify.

73. $\log_{10} 1000$　　　　　　　**74.** $\log_{10} 10^{x^2}$

75. $\ln e^2$　　　　　　　**76.** $\ln e^x$

Find the derivative of each function.

77. $y = \ln(2x^3 - 4)$　　　**78.** $y = \log_3(4x + 1)$　　　**79.** $y = \ln\left(\dfrac{x^2}{x^2 + 3}\right)$

80. $y = \cos(\ln x)$　　　**81.** $y = \dfrac{\sqrt{x + 1}\ (3x - 4)}{x^2(x + 2)}$　　　**82.** $y = x^{1-x}$

83. $y = e^{x^2+5}$ **84.** $y = 8^{3x}$ **85.** $y = \cot e^{2x}$

86. $y = e^{\sin x^2}$ **87.** $y = \text{Arcsin } e^{-4x}$ **88.** $y = x^3 e^{-4x}$

Sketch the graph of each equation.

89. $y = xe^{-x}$ **90.** $y = e^{-x^2}$

91. The current in a circuit is given by $i = 100 \cos 5t$. Find the equation of the voltage V_L across a 2-H inductor $\left(V_L = L\dfrac{di}{dt}, \text{ where } L \text{ is the inductance} \right)$.

92. The work done in a circuit is given by $W = 60 \cos^2 3t$. Find the equation for the power $p \left(p = \dfrac{dW}{dt} \right)$.

93. Find the equation of the tangent line to $y = \ln x^2$ at the point $(1, 0)$.

94. Find the velocity of an object moving along a straight line according to $s = e^{\sin t}$.

95. A particle moves along a straight line according to $s = 1 - 4e^{-t}$. Find its acceleration at $t = \frac{1}{2}$.

7

METHODS
OF INTEGRATION

7.1 THE GENERAL POWER FORMULA

In this chapter we develop techniques that can be used to integrate more complicated functions. The first technique is based on the use of the general power formula developed in Chapter 4:

$$\int u^n \, du = \frac{u^{n+1}}{n+1} + C, \qquad n \neq -1$$

This formula can be effectively used to integrate numerous functions involving either transcendental or algebraic functions. As seen in Chapter 4, the effective use of this formula depends on the proper choice of u and du.

Example 1

Integrate $\int \sin^5 2x \cos 2x \, dx$.

If we choose $u = \sin 2x$ then $du = 2 \cos 2x \, dx$. We can then apply the general power formula with $n = 5$.

$$\int \sin^5 2x \cos 2x \, dx = \int u^5 \frac{du}{2}$$

$$= \tfrac{1}{2} \int u^5 \, du$$

$u = \sin 2x$
$du = 2 \cos 2x \, dx$
$n = 5$

$$= \frac{1}{2} \cdot \frac{u^6}{6} + C$$

$$= \tfrac{1}{12} \sin^6 2x + C$$

Example 2

Integrate $\displaystyle\int e^{3x}(2 - e^{3x})^4\, dx$.

$$\int e^{3x}(2 - e^{3x})^4\, dx = \int u^4 \left(-\frac{du}{3} \right)$$

$$u = 2 - e^{3x}$$
$$du = -3e^{3x}\, dx$$
$$n = 4$$

$$= -\tfrac{1}{3} \int u^4\, du$$

$$= -\frac{1}{3} \cdot \frac{u^5}{5} + C$$

$$= -\frac{(2 - e^{3x})^5}{15} + C$$

Example 3

Integrate $\displaystyle\int \frac{\sec^2 x\, dx}{\sqrt{4 + \tan x}}$.

$$\int \frac{\sec^2 x\, dx}{\sqrt{4 + \tan x}} = \int (4 + \tan x)^{-(1/2)} \sec^2 x\, dx$$

$$= \int u^{-(1/2)}\, du$$

$$u = 4 + \tan x$$
$$du = \sec^2 x\, dx$$
$$n = -\tfrac{1}{2}$$

$$= \frac{u^{1/2}}{\tfrac{1}{2}} + C$$

$$= 2\sqrt{4 + \tan x} + C$$

Example 4

Evaluate $\displaystyle\int_1^e \frac{\ln x}{x}\, dx$.

$$\int \frac{\ln x}{x}\, dx = \int u\, du$$

$$u = \ln x$$
$$du = \frac{1}{x}\, dx$$
$$n = 1$$

$$= \frac{u^2}{2} + C$$

$$= \frac{(\ln x)^2}{2} + C$$

So

$$\int_1^e \frac{\ln x}{x}\, dx = \left. \frac{(\ln x)^2}{2} \right|_1^e = \frac{(\ln e)^2}{2} - \frac{(\ln 1)^2}{2} = \frac{(1)^2}{2} - \frac{(0)^2}{2}$$

$$= \tfrac{1}{2}$$

Sometimes a function may appear to be difficult to integrate. However, an appropriate change in the form of the integral can often lead to the desired solution. This is illustrated in the next example where it is desirable to multiply the function $\dfrac{\tan x}{\sec^3 x}$ by 1 $\left(\text{in the form } \dfrac{\sec x}{\sec x}\right)$.

Example 5

Integrate $\displaystyle\int \dfrac{\tan x}{\sec^3 x}\, dx$.

$$\int \frac{\tan x}{\sec^3 x} = \int \left(\frac{\sec x}{\sec x}\right)\left(\frac{\tan x}{\sec^3 x}\right) dx$$

$$= \int \frac{\sec x \tan x}{\sec^4 x}\, dx$$

$$= \int \frac{du}{u^4} = \int u^{-4}\, du$$

$$= \frac{u^{-3}}{-3} + C$$

$$= -\frac{1}{3}\frac{1}{\sec^3 x} + C$$

$$= -\tfrac{1}{3} \cos^3 x + C$$

$\boxed{\begin{aligned} u &= \sec x \\ du &= \sec x \tan x\, dx \\ n &= -4 \end{aligned}}$

Example 6

Evaluate $\displaystyle\int_{1/6}^{1/3} \dfrac{\text{Arccos}^2 3x}{\sqrt{1-9x^2}}\, dx$.

First, find the indefinite integral.

$$\int \frac{\text{Arccos}^2 3x}{\sqrt{1-9x^2}}\, dx = \int \frac{u^2}{-3}\, du$$

$$= -\frac{u^3}{9}$$

$$= -\tfrac{1}{9}\,\text{Arccos}^3 3x$$

$\boxed{\begin{aligned} u &= \text{Arccos } 3x \\ du &= \frac{-3}{\sqrt{1-9x^2}}\, dx \\ n &= 2 \end{aligned}}$

Thus

$$\int_{1/6}^{1/3} \frac{\text{Arccos}^2 3x}{\sqrt{1-9x^2}}\, dx = -\frac{1}{9}\,\text{Arccos}^3 3x \Big|_{1/6}^{1/3}$$

$$= -\frac{1}{9}\left[0 - \left(\frac{\pi}{3}\right)^3\right]$$

$$= \frac{\pi^3}{243}$$

Exercises

Integrate.

1. $\displaystyle\int \sqrt{3x + 2}\, dx$

2. $\displaystyle\int \sqrt[3]{2x - 1}\, dx$

3. $\displaystyle\int \frac{dx}{\sqrt{4 + x}}$

4. $\displaystyle\int \frac{dx}{(2x + 1)^3}$

5. $\displaystyle\int (x + 2)(x^2 + 4x)^{3/4}\, dx$

6. $\displaystyle\int x\sqrt{x^2 + 4}\, dx$

Methods of Integration   Chap. 7

7. $\displaystyle\int \cos^3 x \sin x \, dx$

8. $\displaystyle\int \frac{\cos 2x \, dx}{\sqrt{\sin 2x}}$

9. $\displaystyle\int \tan^3 4x \sec^2 4x \, dx$

10. $\displaystyle\int \cos x \sin x \, dx$

11. $\displaystyle\int \sin 4x(\cos 4x + 1) \, dx$

12. $\displaystyle\int \cos 2x(1 - \sin 2x)^2 \, dx$

13. $\displaystyle\int \sqrt{9 + \sec x} \, \sec x \tan x \, dx$

14. $\displaystyle\int \frac{\csc^2 2x \, dx}{\sqrt{8 - \cot 2x}}$

15. $\displaystyle\int \sqrt{1 + e^{2x}} \, e^{2x} \, dx$

16. $\displaystyle\int (10 - e^{-2x})^3 e^{-2x} \, dx$

17. $\displaystyle\int \frac{xe^{x^2} \, dx}{\sqrt{1 + e^{x^2}}}$

18. $\displaystyle\int \frac{e^{\tan x} \sec^2 x \, dx}{\sqrt{2 + e^{\tan x}}}$

19. $\displaystyle\int \frac{\ln(3x - 5)}{3x - 5} \, dx$

20. $\displaystyle\int \frac{\ln 4x}{x} \, dx$

21. $\displaystyle\int \frac{dx}{x \ln^2 x}$

22. $\displaystyle\int \frac{x \ln(x^2 + 1) \, dx}{x^2 + 1}$

23. $\displaystyle\int \frac{\text{Arcsin } 3x}{\sqrt{1 - 9x^2}} \, dx$

24. $\displaystyle\int \frac{\text{Arctan } 2x}{1 + 4x^2} \, dx$

25. $\displaystyle\int \frac{\cot x}{\csc^4 x} \, dx$

26. $\displaystyle\int \frac{x \sec^2 x^2 \, dx}{\sqrt{9 + \tan x^2}}$

27. $\displaystyle\int \frac{\text{Arctan}^2 x}{1 + x^2} \, dx$

28. $\displaystyle\int \frac{\ln^2 4x}{2x} \, dx$

29. $\displaystyle\int_3^5 x\sqrt{x^2 - 9} \, dx$

30. $\displaystyle\int_0^6 3x\sqrt{x^2 + 1} \, dx$

31. $\displaystyle\int_0^1 \frac{e^{3x} \, dx}{\sqrt{1 + e^{3x}}}$

32. $\displaystyle\int_0^{\pi/12} \sin^3 6x \cos 6x \, dx$

33. $\displaystyle\int_1^2 \frac{\ln(2x - 1)}{2x - 1} \, dx$

34. $\displaystyle\int_0^{\sqrt{2}/4} \frac{\text{Arcsin } 2x}{\sqrt{1 - 4x^2}} \, dx$

35. Find the area bounded by the curves $y = \sin^2 x \cos x$ from $x = 0$ to $x = \pi/2$ and $y = 0$.

36. Find the area bounded by the curves $y = \dfrac{\ln^2 x}{x}$ from $x = 1$ to $x = e$ and $y = 0$.

7.2 LOGARITHMIC AND EXPONENTIAL FORMS

Since integration is the inverse operation of differentiation and $\dfrac{d}{du}(\ln u) = \dfrac{1}{u}$, we find

$$\boxed{\int \frac{du}{u} = \ln |u| + C}$$

The absolute value of u is necessary since logarithms are only defined for positive numbers. If $u > 0$, then $\displaystyle\int \frac{du}{u} = \ln u + C$ and if $u < 0$, then $\displaystyle\int \frac{du}{u} = \int \frac{-du}{-u} = \ln(-u) + C$ so that in both cases we obtain $\ln |u| + C$. *Note:* This form is the general power formula for $n = -1$.

Example 1

Integrate $\displaystyle\int \frac{dx}{x-1}$.

$$\int \frac{dx}{x-1} = \int \frac{du}{u}$$

$$= \ln|u| + C$$

$$= \ln|x-1| + C$$

$$\boxed{\begin{array}{l} u = x - 1 \\ du = dx \end{array}}$$

Example 2

Integrate $\displaystyle\int \frac{x\,dx}{x^2+6}$.

$$\int \frac{x\,dx}{x^2+6} = \int \frac{(du/2)}{u}$$

$$= \frac{1}{2} \int \frac{du}{u}$$

$$= \tfrac{1}{2} \ln|u| + C$$

$$= \tfrac{1}{2} \ln|x^2+6| + C$$

$$\boxed{\begin{array}{l} u = x^2 + 6 \\ du = 2x\,dx \end{array}}$$

Observe that the integral formula $\displaystyle\int \frac{du}{u} = \ln|u| + C$ is applicable whenever the integrand is a quotient and the derivative of the denominator is equal to a constant multiple of the numerator.

Example 3

Integrate $\displaystyle\int \frac{\sec 3x \tan 3x\,dx}{2 + \sec 3x}$.

$$\int \frac{\sec 3x \tan 3x\,dx}{2 + \sec 3x} = \int \frac{(du/3)}{u}$$

$$= \frac{1}{3} \int \frac{du}{u}$$

$$= \tfrac{1}{3} \ln|u| + C$$

$$= \tfrac{1}{3} \ln|2 + \sec 3x| + C$$

$$\boxed{\begin{array}{l} u = 2 + \sec 3x \\ du = 3 \sec 3x \tan 3x\,dx \end{array}}$$

Example 4

Evaluate $\displaystyle\int_2^3 \frac{x^2\,dx}{1-x^3}$.

$$\int \frac{x^2\,dx}{1-x^3} = \int \frac{du/(-3)}{u}$$

$$= -\frac{1}{3} \int \frac{du}{u}$$

$$= -\tfrac{1}{3} \ln|u| + C$$

$$= -\tfrac{1}{3} \ln|1 - x^3| + C$$

$$\boxed{\begin{array}{l} u = 1 - x^3 \\ du = -3x^2\,dx \end{array}}$$

So

$$\int_2^3 \frac{x^2\,dx}{1-x^3} = -\frac{1}{3}\ln\left|1-x^3\right|\Big|_2^3$$

$$= -\tfrac{1}{3}\ln\left|-26\right| + \tfrac{1}{3}\ln\left|-7\right|$$

$$= \tfrac{1}{3}(\ln 7 - \ln 26)$$

$$= \tfrac{1}{3}\ln\left(\tfrac{7}{26}\right) \quad \text{or} \quad -0.437$$

From the derivative formula $\dfrac{d}{du}(e^u) = e^u$, we find

$$\boxed{\int e^u\,du = e^u + C}$$

Example 5

Integrate $\displaystyle\int e^{5x}\,dx$.

$$\int e^{5x}\,dx = \int e^u\left(\frac{du}{5}\right) \qquad \boxed{\begin{array}{l} u = 5x \\ du = 5\,dx \end{array}}$$

$$= \tfrac{1}{5}\int e^u\,du$$

$$= \tfrac{1}{5}e^u + C$$

$$= \tfrac{1}{5}e^{5x} + C$$

Example 6

Integrate $\displaystyle\int \frac{dx}{e^{2x}}$.

$$\int \frac{dx}{e^{2x}} = \int e^{-2x}\,dx = \int e^u\left(\frac{du}{-2}\right) \qquad \boxed{\begin{array}{l} u = -2x \\ du = -2\,dx \end{array}}$$

$$= -\tfrac{1}{2}\int e^u\,du$$

$$= -\tfrac{1}{2}e^u + C$$

$$= -\tfrac{1}{2}e^{-2x} + C$$

Example 7

Evaluate $\displaystyle\int_0^1 x^2 e^{x^3+1}\,dx$.

First,

$$\int x^2 e^{x^3+1}\,dx = \int e^u\left(\frac{du}{3}\right) \qquad \boxed{\begin{array}{l} u = x^3 + 1 \\ du = 3x^2\,dx \end{array}}$$

$$= \tfrac{1}{3}\int e^u\,du$$

$$= \tfrac{1}{3}e^u + C$$

$$= \tfrac{1}{3}e^{x^3+1} + C$$

So

$$\int_0^1 x^2 e^{x^3+1}\, dx = \tfrac{1}{3} e^{x^3+1}\Big|_0^1$$

$$= \tfrac{1}{3} e^2 - \tfrac{1}{3} e$$

$$= \tfrac{1}{3} e(e - 1)$$

For bases other than e, we have

$$\int a^u\, du = \frac{a^u}{\ln a} + C, \qquad a > 0, a \neq 1$$

Example 8

Find the area bounded by $xy = 1$, $x = 1$, $x = 2$, and $y = 0$.
The desired area is given by the definite integral (see Fig. 7.1).

$$\int_1^2 \frac{dx}{x} = \ln |x|\Big|_1^2 = \ln 2 - \ln 1 = \ln 2$$

$$= 0.6931$$

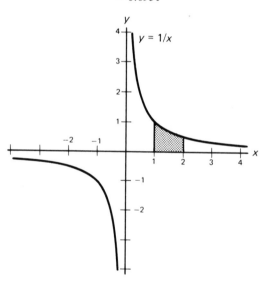

Figure 7.1

Exercises

Integrate.

1. $\displaystyle\int \frac{dx}{3x + 2}$

2. $\displaystyle\int \frac{dx}{x + 5}$

3. $\displaystyle\int \frac{dx}{1 - 4x}$

4. $\displaystyle\int \frac{dx}{4x + 2}$

5. $\displaystyle\int \frac{4x\, dx}{1 - x^2}$

6. $\displaystyle\int \frac{3x\, dx}{x^2 + 1}$

7. $\displaystyle\int \frac{x^3\, dx}{x^4 - 1}$

8. $\displaystyle\int \frac{(x + 1)\, dx}{x^2 + 2x - 3}$

9. $\displaystyle\int \frac{\csc^2 x\, dx}{\cot x}$

10. $\displaystyle\int \frac{\sin x \, dx}{\cos x}$

11. $\displaystyle\int \frac{\sec^2 3x \, dx}{1 + \tan 3x}$

12. $\displaystyle\int \frac{\cos 2x}{1 + \sin 2x} \, dx$

13. $\displaystyle\int \frac{\csc x \cot x}{1 + \csc x} \, dx$

14. $\displaystyle\int \frac{\csc^2 4x}{1 - \cot 4x} \, dx$

15. $\displaystyle\int \frac{\cos x}{1 + \sin x} \, dx$

16. $\displaystyle\int \frac{(x - \sin 2x) \, dx}{x^2 + \cos 2x}$

17. $\displaystyle\int \frac{dx}{x \ln x}$

18. $\displaystyle\int \frac{dx}{x(4 + \ln x)}$

19. $\displaystyle\int e^{2x} \, dx$

20. $\displaystyle\int e^{3x-1} \, dx$

21. $\displaystyle\int \frac{dx}{e^{4x}}$

22. $\displaystyle\int \frac{dx}{e^{2x+3}}$

23. $\displaystyle\int xe^{x^2} \, dx$

24. $\displaystyle\int x^3 e^{x^4-1} \, dx$

25. $\displaystyle\int \frac{x \, dx}{e^{x^2+9}}$

26. $\displaystyle\int xe^{-x^2} \, dx$

27. $\displaystyle\int (\sin x) \, e^{\cos x} \, dx$

28. $\displaystyle\int (\sec^2 x) \, e^{2 \tan x} \, dx$

29. $\displaystyle\int_0^2 xe^{x^2+2} \, dx$

30. $\displaystyle\int_0^{\pi/2} (\cos x) \, e^{\sin x} \, dx$

31. $\displaystyle\int 4^x \, dx$

32. $\displaystyle\int 10^{2x} \, dx$

33. $\displaystyle\int \frac{2e^x}{e^x + 4} \, dx$

34. $\displaystyle\int \frac{e^{-x}}{1 - e^{-x}} \, dx$

35. $\displaystyle\int_0^1 \frac{x \, dx}{x^2 + 1}$

36. $\displaystyle\int_{\pi/4}^{\pi/2} \frac{\cos x \, dx}{\sin x}$

37. $\displaystyle\int_1^5 \frac{4 \, dx}{2x - 1}$

38. $\displaystyle\int_0^6 \frac{dx}{5x + 4}$

39. $\displaystyle\int_0^2 e^{x/2} \, dx$

40. $\displaystyle\int_{-2}^0 e^{-4x} \, dx$

41. $\displaystyle\int_0^1 x^2 e^{x^3} \, dx$

42. $\displaystyle\int_1^2 xe^{x^2} \, dx$

43. $\displaystyle\int_0^{\pi/6} \frac{\cos x \, dx}{1 - \sin x}$

44. $\displaystyle\int_{\pi/3}^{\pi/2} \frac{\sin x}{1 + \cos x} \, dx$

45. Find the area bounded by $y = 1/(1 + 2x)$ from $x = 0$ to $x = 1$ and $y = 0$.

46. Find the area bounded by $xy = 1$, $x = 1$, $x = 3$ and $y = 0$.

47. Find the area bounded by $y = e^{2x}$, $x = 0$, $x = 4$, and $y = 0$.

48. Find the area bounded by $y = e^{-x}$, $x = 0$, $x = 5$, and $y = 0$.

7.3 BASIC TRIGONOMETRIC FORMS

We have found that

$$\frac{d}{du}(\sin u) = \cos u \qquad \text{and} \qquad \frac{d}{du}(\cos u) = -\sin u$$

Now, since integration is the inverse operation of differentiation, we have:

$$\int \cos u \, du = \sin u + C$$

and

$$\int \sin u \, du = -\cos u + C$$

Example 1

Integrate $\int \cos 3x \, dx$.

$$\int \cos 3x \, dx = \int (\cos u)\left(\frac{du}{3}\right)$$

$$\boxed{\begin{array}{l} u = 3x \\ du = 3 \, dx \end{array}}$$

$$= \tfrac{1}{3} \int \cos u \, du$$
$$= \tfrac{1}{3} \sin u + C$$
$$= \tfrac{1}{3} \sin 3x + C$$

Example 2

Integrate $\int \sin (4x + 3) \, dx$.

$$\int \sin (4x + 3) \, dx = \int (\sin u)\left(\frac{du}{4}\right)$$

$$\boxed{\begin{array}{l} u = 4x + 3 \\ du = 4 \, dx \end{array}}$$

$$= \tfrac{1}{4} \int \sin u \, du$$
$$= \tfrac{1}{4}(-\cos u) + C$$
$$= -\tfrac{1}{4} \cos (4x + 3) + C$$

Example 3

Integrate $\int x^2 \sin (x^3 + 2) \, dx$.

$$\int x^2 \sin (x^3 + 2) \, dx = \int (\sin u)\left(\frac{du}{3}\right)$$

$$\boxed{\begin{array}{l} u = x^3 + 2 \\ du = 3x^2 \, dx \end{array}}$$

$$= \tfrac{1}{3} \int \sin u \, du$$
$$= \tfrac{1}{3}(-\cos u) + C$$
$$= -\tfrac{1}{3} \cos (x^3 + 2) + C$$

Example 4

Find $\int_0^{\pi/4} \cos 2x \, dx$.

$$\int \cos 2x \, dx = \int \cos u \left(\frac{du}{2}\right)$$

$$\boxed{\begin{array}{l} u = 2x \\ du = 2 \, dx \end{array}}$$

$$= \tfrac{1}{2} \int \cos u \, du$$
$$= \tfrac{1}{2} \sin u + C$$
$$= \tfrac{1}{2} \sin 2x + C$$

So

$$\int_u^{\pi/4} \cos 2x \, dx = \frac{1}{2} \sin 2x \Big|_0^{\pi/4} = \frac{1}{2} \sin \frac{\pi}{2} - \frac{1}{2} \sin 0$$
$$= \tfrac{1}{2}(1) - \tfrac{1}{2}(0) = \tfrac{1}{2}$$

Comparing the derivatives of tan u, cot u, sec u, and csc u and noting that integration is the inverse operation of differentiation, we have:

$$\int \sec^2 u \ du = \tan u + C$$

$$\int \csc^2 u \ du = -\cot u + C$$

$$\int \sec u \tan u \ du = \sec u + C$$

$$\int \csc u \cot u \ du = -\csc u + C$$

Example 5

Integrate $\int \sec^2 3x \ dx$

$$\int \sec^2 3x \ dx = \int (\sec^2 u)\left(\frac{du}{3}\right) \qquad \boxed{\begin{aligned} u &= 3x \\ du &= 3 \ dx \end{aligned}}$$

$$= \tfrac{1}{3} \int \sec^2 u \ du$$

$$= \tfrac{1}{3} \tan u + C$$

$$= \tfrac{1}{3} \tan 3x + C$$

Example 6

Integrate $\int \sec(2x + 5) \tan (2x + 5) \ dx$.

$$\int \sec(2x + 5) \tan (2x + 5) \ dx = \int \sec u \tan u \left(\frac{du}{2}\right) \qquad \boxed{\begin{aligned} u &= 2x + 5 \\ du &= 2 \ dx \end{aligned}}$$

$$= \tfrac{1}{2} \int \sec u \tan u \ du$$

$$= \tfrac{1}{2} \sec u + C$$

$$= \tfrac{1}{2} \sec(2x + 5) + C$$

We determine $\int \tan u \ du$ as follows:

$$\int \tan u \ du = \int \left(\frac{\sin u}{\cos u}\right) du \qquad \boxed{\begin{aligned} w &= \cos u \\ dw &= -\sin u \ du \end{aligned}}$$

$$= \int \frac{1}{w}(-dw)$$

$$= -\int \frac{dw}{w}$$

$$= -\ln |w| + C$$

$$= -\ln |\cos u| + C$$

In a similar manner,

$$\int \cot u \ du = \int \frac{\cos u}{\sin u} du \qquad \boxed{\begin{aligned} w &= \sin u \\ dw &= \cos u \ du \end{aligned}}$$

$$= \int \frac{1}{w} \, dw$$

$$= \ln |w| + C$$

$$= \ln |\sin u| + C$$

For integrating sec u, we first multiply sec u by 1 in the form

$$\frac{\sec u + \tan u}{\sec u + \tan u}$$

That is,

$$\sec u = \sec u \left(\frac{\sec u + \tan u}{\sec u + \tan u} \right)$$

Then

$$\int \sec u \, du = \int \frac{\sec^2 u + \sec u \tan u}{\sec u + \tan u} \, du$$

Then note that

$$\int \sec u \, du = \int \frac{dw}{w} \qquad \boxed{\begin{array}{l} w = \sec u + \tan u \\ dw = (\sec u \tan u + \sec^2 u) \, du \end{array}}$$

$$= \ln |w| + C$$

$$= \ln |\sec u + \tan u| + C$$

The integral for csc u is obtained in a similar manner:

$$\int \csc u \, du = \ln |\csc u - \cot u| + C$$

We now can integrate all six basic trigonometric functions:

$$\boxed{\begin{array}{l} \displaystyle\int \sin u \, du = -\cos u + C \\[2mm] \displaystyle\int \cos u \, du = \sin u + C \\[2mm] \displaystyle\int \tan u \, du = -\ln |\cos u| + C \\[2mm] \displaystyle\int \cot u \, du = \ln |\sin u| + C \\[2mm] \displaystyle\int \sec u \, du = \ln |\sec u + \tan u| + C \\[2mm] \displaystyle\int \csc u \, du = \ln |\csc u - \cot u| + C \end{array}}$$

Example 7

Integrate $\displaystyle\int \tan 3x \, dx$.

$$\int \tan 3x \, dx = \int \tan u \, \frac{du}{3}$$

$$\boxed{\begin{aligned} u &= 3x \\ du &= 3\,dx \end{aligned}}$$

$$= \tfrac{1}{3} \int \tan u \, du$$

$$= \tfrac{1}{3}(-\ln |\cos u|) + C$$

$$= -\tfrac{1}{3} \ln |\cos 3x| + C$$

Example 8

Integrate $\int x \sec 3x^2 \, dx.$

$$\int x \sec 3x^2 \, dx = \int \sec u \, \frac{du}{6}$$

$$\boxed{\begin{aligned} u &= 3x^2 \\ du &= 6x\,dx \end{aligned}}$$

$$= \tfrac{1}{6} \int \sec u \, du$$

$$= \tfrac{1}{6} \ln |\sec u + \tan u| + C$$

$$= \tfrac{1}{6} \ln |\sec 3x^2 + \tan 3x^2| + C$$

Example 9

Evaluate $\int_{\pi/8}^{\pi/4} \cot 2x \, dx.$

$$\int \cot 2x \, dx = \int \cot u \, \frac{du}{2}$$

$$\boxed{\begin{aligned} u &= 2x \\ du &= 2\,dx \end{aligned}}$$

$$= \tfrac{1}{2} \int \cot u \, du$$

$$= \tfrac{1}{2} \ln |\sin u| + C$$

$$= \tfrac{1}{2} \ln |\sin 2x| + C$$

So

$$\int_{\pi/8}^{\pi/4} \cot 2x \, dx = \frac{1}{2} \ln |\sin 2x| \Big|_{\pi/8}^{\pi/4} = \frac{1}{2} \ln \left| \sin \frac{\pi}{2} \right| - \frac{1}{2} \ln \left| \sin \frac{\pi}{4} \right|$$

$$= \frac{1}{2} \ln 1 - \frac{1}{2} \ln \frac{1}{\sqrt{2}}$$

$$= 0 - \frac{1}{2} \ln 2^{-(1/2)} = \frac{1}{4} \ln 2 \quad \text{or} \quad 0.173$$

Algebraic simplifications often lead to less complicated integrals as shown below:

Example 10

Integrate $\int \dfrac{2 - \cos x}{\sin x} \, dx.$

$$\int \frac{2 - \cos x}{\sin x} \, dx = \int \frac{2 \, dx}{\sin x} - \int \frac{\cos x}{\sin x} \, dx$$

$$= 2 \int \csc x \, dx - \int \cot x \, dx$$

$$= 2 \ln |\csc x - \cot x| - \ln |\sin x| + C$$

Exercises

Integrate.

1. $\int \sin 5x \, dx$

2. $\int \cos 6x \, dx$

3. $\int \cos (3x - 1) \, dx$

4. $\int \sin (2x + 7) \, dx$

5. $\int x \sin (x^2 + 5) \, dx$

6. $\int x^3 \cos x^4 \, dx$

7. $\int (3x^2 - 2x) \cos (x^3 - x^2) \, dx$

8. $\int (x - 1) \sin (x^2 - 2x + 3) \, dx$

9. $\int \csc^2 5x \, dx$

10. $\int \sec^2 2x \, dx$

11. $\int \sec 3x \tan 3x \, dx$

12. $\int \csc 7x \cot 7x \, dx$

13. $\int \sec^2 (4x + 3) \, dx$

14. $\int \csc^2 (3x - 2) \, dx$

15. $\int \csc (2x - 3) \cot (2x - 3) \, dx$

16. $\int \sec \frac{x}{2} \tan \frac{x}{2} \, dx$

17. $\int x \sec^2 (x^2 + 3) \, dx$

18. $\int x^2 \csc^2 x^3 \, dx$

19. $\int x^2 \csc (x^3 - 1) \cot (x^3 - 1) \, dx$

20. $\int x \sec 3x^2 \tan 3x^2 \, dx$

21. $\int \tan 4x \, dx$

22. $\int x \cot x^2 \, dx$

23. $\int \sec 5x \, dx$

24. $\int x^2 \csc x^3 \, dx$

25. $\int e^x \cot e^x \, dx$

26. $\int \frac{\sec (\ln x)}{x} \, dx$

27. $\int (1 + \sec x)^2 \, dx$

28. $\int (1 + \tan 3x)^2 \, dx$

29. $\int \frac{5 + \sin x}{\cos x} \, dx$

30. $\int \frac{\tan x - \cos x}{\sin x} \, dx$

31. $\int_0^{\pi/4} \sin 2x \, dx$

32. $\int_0^{\pi/6} 2 \cos 3x \, dx$

33. $\int_0^{\pi/2} 3 \cos \left(x - \frac{\pi}{2} \right) dx$

34. $\int_0^{\pi/2} \sin \left(x - \frac{\pi}{4} \right) dx$

35. $\int_0^{\pi/4} \sec^2 x \, dx$

36. $\int_{\pi/4}^{\pi/2} \csc^2 x \, dx$

37. $\int_0^{\pi/8} \sec 2x \tan 2x \, dx$

38. $\int_{1/4}^{1/2} \csc \pi x \cot \pi x \, dx$

Find the area of each region bounded by the given curves.

39. $y = \sin x$ from $x = 0$ to $x = \pi$ and $y = 0$.

40. $y = 2 \cos x$ from $x = 0$ to $x = \pi/2$ and $y = 0$.

41. $y = \sec^2 x$ from $x = 0$ to $x = \pi/4$ and $y = 0$.

42. $y = \sec x \tan x$ from $x = 0$ to $x = \pi/4$ and $y = 0$.

43. $y = \tan x$ from $x = 0$ to $x = \pi/4$ and $y = 0$.

44. $y = \sin x + \cos x$, $x = 0$, and $y = 0$.

45. Find the volume of the solid formed by revolving the region bounded by $y = \sec x$ from $x = 0$ to $x = \pi/4$ and $y = 0$ about the x-axis.

7.4 OTHER TRIGONOMETRIC FORMS

For integrating powers of trigonometric functions it is necessary to use certain trigonometric identities. We will now demonstrate the following appropriate methods.

Methods of Integration   Chap. 7

Odd Powers of Sines or Cosines

Various forms of the trigonometric identity $\sin^2 x + \cos^2 x = 1$ are needed for this case.

Example 1

Integrate $\int \sin^5 x \, dx$.

$$\int \sin^5 x \, dx = \int \sin^4 x \sin x \, dx = \int (\sin^2 x)^2 \sin x \, dx$$

$$= \int (1 - \cos^2 x)^2 \sin x \, dx \qquad (\sin^2 x = 1 - \cos^2 x)$$

$$= \int (1 - 2\cos^2 x + \cos^4 x) \sin x \, dx$$

$$= \int \sin x \, dx - 2 \int \cos^2 x \sin x \, dx + \int \cos^4 x \sin x \, dx$$

$$= -\cos x + \frac{2 \cos^3 x}{3} - \frac{\cos^5 x}{5} + C$$

Example 2

Integrate $\int \sin^2 x \cos^3 x \, dx$.

$$\int \sin^2 x \cos^3 x \, dx = \int \sin^2 x \cos^2 x \cos x \, dx$$

$$= \int \sin^2 x (1 - \sin^2 x) \cos x \, dx \qquad (\cos^2 x = 1 - \sin^2 x)$$

$$= \int (\sin^2 x \cos x - \sin^4 x \cos x) \, dx$$

$$= \int \sin^2 x \cos x \, dx - \int \sin^4 x \cos x \, dx$$

$$= \frac{\sin^3 x}{3} - \frac{\sin^5 x}{5} + C$$

In each of the two previous examples the method used involved making a substitution so that the function to be integrated becomes a product of a power of sine (or cosine) and the first power only of cosine (or sine).

Even Powers of Sines and Cosines

For this case we use identities 30 and 31 in Appendix B, Table 6:
$$\cos^2 x = \tfrac{1}{2}(1 + \cos 2x)$$
$$\sin^2 x = \tfrac{1}{2}(1 - \cos 2x)$$

Example 3

Integrate $\int \cos^2 x \, dx$.

$$\int \cos^2 x \, dx = \int \tfrac{1}{2}(1 + \cos 2x) \, dx \qquad [\cos^2 x = \tfrac{1}{2}(1 + \cos 2x)]$$

$$= \tfrac{1}{2} \int dx + \tfrac{1}{2} \int \cos 2x \, dx$$

$$= \tfrac{1}{2} x + \tfrac{1}{4} \sin 2x + C$$

Example 4

Integrate $\int \sin^4 x \, dx$.

$$\int \sin^4 x \, dx = \int (\sin^2 x)^2 \, dx$$

$$= \int [\tfrac{1}{2}(1 - \cos 2x)]^2 \, dx \qquad [\sin^2 x = \tfrac{1}{2}(1 - \cos 2x)]$$

$$= \int \left(\frac{1}{4} - \frac{1}{2} \cos 2x + \frac{\cos^2 2x}{4} \right) dx$$

$$= \tfrac{1}{4} \int dx - \tfrac{1}{2} \int \cos 2x \, dx + \tfrac{1}{4} \int \cos^2 2x \, dx$$

$$= \frac{x}{4} - \frac{\sin 2x}{4} + \frac{1}{4} \int \cos^2 2x \, dx$$

Now find

$$\int \cos^2 2x \, dx = \int \tfrac{1}{2}(1 + \cos 4x) \, dx \qquad [\cos^2 2x = \tfrac{1}{2}(1 + \cos 4x)]$$

$$= \tfrac{1}{2} \int dx + \tfrac{1}{2} \int \cos 4x \, dx$$

$$= \frac{x}{2} + \frac{\sin 4x}{8} + C_1$$

So

$$\int \sin^4 x \, dx = \frac{x}{4} - \frac{\sin 2x}{4} + \frac{1}{4}\left(\frac{x}{2} + \frac{\sin 4x}{8} + C_1 \right)$$

$$= \frac{x}{4} - \frac{\sin 2x}{4} + \frac{x}{8} + \frac{\sin 4x}{32} + \frac{C_1}{4}$$

$$= \frac{3x}{8} - \frac{\sin 2x}{4} + \frac{\sin 4x}{32} + C$$

Other Powers

The identities $1 + \tan^2 x = \sec^2 x$ and $1 + \cot^2 x = \csc^2 x$ are useful in integrating even powers of sec x and csc x, and powers of tan x and cot x.

Example 5

Integrate $\int \sec^4 x \, dx$.

$$\int \sec^4 x \, dx = \int (1 + \tan^2 x) \sec^2 x \, dx$$

$$= \int \sec^2 x \, dx + \int \tan^2 x \sec^2 x \, dx$$

$$= \tan x + \frac{\tan^3 x}{3} + C$$

Example 6

Integrate $\int \tan^5 x \, dx$.

$$\int \tan^5 x \, dx = \int \tan^3 x \tan^2 x \, dx$$

$$= \int \tan^3 x \, (\sec^2 x - 1) \, dx \qquad (\tan^2 x = \sec^2 x - 1)$$

$$= \int \tan^3 x \sec^2 x \, dx - \int \tan^3 x \, dx$$

$$= \frac{\tan^4 x}{4} - \int \tan x \tan^2 x \, dx$$

$$= \frac{\tan^4 x}{4} - \int \tan x \, (\sec^2 x - 1) \, dx$$

$$= \frac{\tan^4 x}{4} - \int \tan x \sec^2 x \, dx + \int \tan x \, dx$$

$$= \frac{\tan^4 x}{4} - \frac{\tan^2 x}{2} - \ln |\cos x| + C$$

Example 7

Integrate $\int \csc^6 5x \, dx$.

$$\int \csc^6 5x \, dx = \int \csc^4 5x \csc^2 5x \, dx$$

$$= \int (1 + \cot^2 5x)^2 \csc^2 5x \, dx \qquad (\csc^2 5x = 1 + \cot^2 5x)$$

$$= \int (1 + u^2)^2 \frac{du}{-5} \qquad \boxed{\begin{array}{l} \text{let } u = \cot 5x \\ du = (-\csc^2 5x)\, 5 \, dx \end{array}}$$

$$= -\frac{1}{5} \int (1 + 2u^2 + u^4) \, du$$

$$= -\frac{1}{5} \left(u + \frac{2u^3}{3} + \frac{u^5}{5} \right) + C$$

$$= -\frac{1}{5} \cot 5x - \frac{2}{15} \cot^3 5x - \frac{1}{25} \cot^5 5x + C$$

Exercises

Integrate.

1. $\int \sin^3 x \, dx$
2. $\int \cos^3 x \, dx$
3. $\int \cos^5 x \, dx$

4. $\int \sin^7 x \, dx$
5. $\int \sin^2 x \cos x \, dx$
6. $\int \cos^2 x \sin x \, dx$

7. $\int \frac{\sin x \, dx}{\cos^3 x}$
8. $\int \sin^3 x \cos x \, dx$
9. $\int \sin^3 x \cos^2 x \, dx$

10. $\int \sin^2 3x \, \cos^3 3x \, dx$ **11.** $\int \sin^2 x \, dx$ **12.** $\int \cos^2 4x \, dx$

13. $\int \cos^4 3x \, dx$ **14.** $\int \sin^4 7x \, dx$ **15.** $\int \sin^2 x \, \cos^2 x \, dx$

16. $\int \sin^4 x \, \cos^2 x \, dx$ **17.** $\int \sin^2 x \, \cos^4 x \, dx$ **18.** $\int \sin^4 2x \, \cos^4 2x \, dx$

19. $\int \tan^3 x \, dx$ **20.** $\int \cot^5 4x \, dx$ **21.** $\int \cot^4 2x \, dx$

22. $\int \sec^4 7x \, dx$ **23.** $\int \sec^6 x \, dx$ **24.** $\int \csc^4 3x \, dx$

25. $\int \tan^4 2x \, dx$ **26.** $\int \cot^6 x \, dx$

27. Find the area of the region bounded by $y = \sin^2 x$ from $x = 0$ to $x = \pi$ and $y = 0$.

28. Find the volume of the solid formed by revolving the region bounded by $y = \cos^3 x$ from $x = 0$ to $x = \pi/2$ and $y = 0$ about the x-axis.

7.5 INVERSE TRIGONOMETRIC FORMS

A major use of the inverse trigonometric functions is providing solutions to certain algebraic functions. Two important integral formulas are based on the derivative of the inverse sine and inverse tangent functions:

$$\int \frac{du}{\sqrt{a^2 - u^2}} = \text{Arcsin} \frac{u}{a} + C$$

since

$$\frac{d}{du}\left(\text{Arcsin} \frac{u}{a}\right) = \frac{1}{\sqrt{1 - \left(\frac{u}{a}\right)^2}} \cdot \frac{1}{a} = \frac{a}{\sqrt{a^2 - u^2}} \frac{1}{a} = \frac{1}{\sqrt{a^2 - u^2}}$$

And

$$\int \frac{du}{a^2 + u^2} = \frac{1}{a} \text{Arctan} \frac{u}{a} + C$$

since

$$\frac{d}{du}\left(\text{Arctan} \frac{u}{a}\right) = \frac{1}{1 + \left(\frac{u}{a}\right)^2} \cdot \frac{1}{a} = \frac{a^2}{a^2 + u^2} \cdot \frac{1}{a} = \frac{a}{a^2 + u^2}$$

so that

$$\frac{d}{du}\left(\frac{1}{a} \text{Arctan} \frac{u}{a}\right) = \frac{1}{a} \cdot \frac{d}{du}\left(\text{Arctan} \frac{u}{a}\right) = \frac{1}{a^2 + u^2}$$

Example 1

Integrate $\displaystyle\int \frac{dx}{\sqrt{16 - x^2}}$.

$$\int \frac{dx}{\sqrt{16 - x^2}} = \int \frac{du}{\sqrt{a^2 - u^2}}$$

$$= \text{Arcsin}\,\frac{u}{a} + C$$

$$= \text{Arcsin}\,\frac{x}{4} + C$$

$$\begin{array}{|c|} \hline u = x \\ du = dx \\ a = 4 \\ \hline \end{array}$$

Example 2

Integrate $\displaystyle\int \frac{dx}{4x^2 + 9}$.

$$\int \frac{dx}{4x^2 + 9} = \int \frac{\left(\dfrac{du}{2}\right)}{a^2 + u^2} = \frac{1}{2} \int \frac{du}{a^2 + u^2}$$

$$= \frac{1}{2}\left(\frac{1}{a}\,\text{Arctan}\,\frac{u}{a}\right) + C$$

$$= \frac{1}{2}\left(\frac{1}{3}\,\text{Arctan}\,\frac{2x}{3}\right) + C$$

$$= \frac{1}{6}\,\text{Arctan}\,\frac{2x}{3} + C$$

$$\begin{array}{|c|} \hline u = 2x \\ dx = 2\,dx \\ a = 3 \\ \hline \end{array}$$

Example 3

Integrate $\displaystyle\int \frac{dx}{x^2 + 2x + 5}$.

It is not clear that this can be integrated using the techniques established so far. However, by using the method of completing the square in the denominator $x^2 + 2x + 5 = (x + 1)^2 + 4 = (x + 1)^2 + (2)^2$.

Then if we let $u = x + 1$ and $a = 2$, we have

$$\int \frac{dx}{x^2 + 2x + 5} = \int \frac{dx}{(x + 1)^2 + (2)^2} = \int \frac{du}{u^2 + a^2}$$

$$= \frac{1}{a}\,\text{Arctan}\,\frac{u}{a} + C$$

$$= \frac{1}{2}\,\text{Arctan}\,\left(\frac{x + 1}{2}\right) + C$$

$$\begin{array}{|c|} \hline u = x + 1 \\ du = dx \\ a = 2 \\ \hline \end{array}$$

Example 4

Evaluate $\displaystyle\int_1^2 \frac{dx}{\sqrt{25 - 4x^2}}$.

$$\int \frac{dx}{\sqrt{25 - 4x^2}} = \int \frac{\left(\dfrac{du}{2}\right)}{\sqrt{a^2 - u^2}}$$

$$\begin{array}{|c|} \hline u = 2x \\ du = 2\,dx \\ a = 5 \\ \hline \end{array}$$

$$= \frac{1}{2} \int \frac{du}{\sqrt{a^2 - u^2}}$$

$$= \frac{1}{2} \text{Arcsin} \frac{u}{a} + C$$

$$= \frac{1}{2} \text{Arcsin} \frac{2x}{5} + C$$

So

$$\int_1^2 \frac{dx}{\sqrt{25 - 4x^2}} = \frac{1}{2} \text{Arcsin} \frac{2x}{5}\Big|_1^2 = \frac{1}{2}\left(\text{Arcsin} \frac{4}{5} - \text{Arcsin} \frac{2}{5}\right)$$

$$= \tfrac{1}{2}(0.93 - 0.41) = 0.26 \quad \text{(radian measure)}$$

Exercises

Integrate.

1. $\displaystyle\int \frac{dx}{\sqrt{1 - 9x^2}}$

2. $\displaystyle\int \frac{dx}{\sqrt{1 - x^2}}$

3. $\displaystyle\int \frac{dx}{\sqrt{9 - x^2}}$

4. $\displaystyle\int \frac{dx}{\sqrt{144 - 25x^2}}$

5. $\displaystyle\int \frac{dx}{x^2 + 25}$

6. $\displaystyle\int \frac{dx}{x^2 + 4}$

7. $\displaystyle\int \frac{dx}{9x^2 + 4}$

8. $\displaystyle\int \frac{dx}{16 + 25x^2}$

9. $\displaystyle\int \frac{dx}{\sqrt{36 - 25x^2}}$

10. $\displaystyle\int \frac{dx}{\sqrt{5 - 6x^2}}$

11. $\displaystyle\int \frac{dx}{\sqrt{3 - 12x^2}}$

12. $\displaystyle\int \frac{dx}{\sqrt{3 - 4x^2}}$

13. $\displaystyle\int \frac{dx}{4 + (x - 1)^2}$

14. $\displaystyle\int \frac{dx}{16 + (x - 3)^2}$

15. $\displaystyle\int \frac{dx}{x^2 + 6x + 25}$

16. $\displaystyle\int \frac{dx}{x^2 + 4x + 13}$

17. $\displaystyle\int \frac{e^x \, dx}{\sqrt{1 - e^{2x}}}$

18. $\displaystyle\int \frac{\sec x \tan x \, dx}{1 + \sec^2 x}$

19. $\displaystyle\int \frac{\sin x \, dx}{1 + \cos^2 x}$

20. $\displaystyle\int \frac{\cos x \, dx}{\sqrt{4 - \sin^2 x}}$

21. $\displaystyle\int_0^1 \frac{dx}{1 + x^2}$

22. $\displaystyle\int_0^2 \frac{dx}{\sqrt{4 - x^2}}$

23. $\displaystyle\int_0^1 \frac{dx}{\sqrt{25 - 9x^2}}$

24. $\displaystyle\int_0^1 \frac{dx}{25x^2 + 9}$

25. A force is acting on an object according to the equation

$$F = \frac{100}{1 + 4x^2}$$

where x is measured in metres and F is measured in newtons. Find the work done in moving the object from $x = 1$ m to $x = 2$ m.

26. Find an equation for the distance traveled by an object moving along a straight line if the velocity at time t is given by

$$v = \frac{1}{\sqrt{9 - t^2}}$$

The object was 10 m from the point of reference at $t = 0$ s.

300   Methods of Integration   Chap. 7

7.6 PARTIAL FRACTIONS

In algebra we add two fractions such as

$$\frac{5}{x+1} + \frac{6}{x-2} = \frac{5(x-2)}{(x+1)(x-2)} + \frac{6(x+1)}{(x+1)(x-2)}$$

$$= \frac{5x - 10 + 6x + 6}{(x+1)(x-2)}$$

$$= \frac{11x - 4}{(x+1)(x-2)}$$

At times, we find it necessary to express a fraction as the sum of two or more others that are simpler than the original; that is, we reverse the operation. Such simpler fractions are called *partial fractions,* whose numerators are of lower degree than their denominators.

We will separate our study of partial fractions into four cases. In each case we will assume that the given fraction is expressed in lowest terms and the degree of its numerator is less than the degree of its denominator.

Case 1 Nonrepeated Linear Denominator Factors

For every nonrepeated factor $ax + b$ of the denominator of the given fraction, there corresponds the partial fraction $\dfrac{A}{ax+b}$, where A is a constant.

Example 1

Find the partial fractions of $\dfrac{11x - 4}{(x+1)(x-2)}$.

The possible partial fractons are $\dfrac{A}{x+1}$ and $\dfrac{B}{x-2}$. So we have

$$\frac{11x - 4}{(x+1)(x-2)} = \frac{A}{x+1} + \frac{B}{x-2}$$

Then multiply each side of this equation by the LCD: $(x+1)(x-2)$.

$$11x - 4 = A(x-2) + B(x+1)$$

Removing parentheses and rearranging terms, we have

$$11x - 4 = Ax - 2A + Bx + B$$
$$11x - 4 = Ax + Bx - 2A + B$$
$$11x - 4 = (A+B)x - 2A + B$$

Next, the coefficients of x must be equal and the constant terms must be equal. This gives the following system of linear equations.

$$A + B = 11$$
$$-2A + B = -4$$

Subtracting the two equations gives

$$3A = 15$$
$$A = 5$$

Substituting $A = 5$ into either equation above gives

$$B = 6$$

Then

$$\frac{11x - 4}{(x + 1)(x - 2)} = \frac{5}{x + 1} + \frac{6}{x - 2}$$

Example 2

Find the partial fractions of $\dfrac{3x^2 - 27x - 12}{x(2x + 1)(x - 4)}$.

The possible partial fractions are

$$\frac{A}{x}, \quad \frac{B}{2x + 1}, \quad \text{and} \quad \frac{C}{x - 4}$$

So we have

$$\frac{3x^2 - 27x - 12}{x(2x + 1)(x - 4)} = \frac{A}{x} + \frac{B}{2x + 1} + \frac{C}{x - 4}$$

Now, multiply each side of this equation by the LCD: $x(2x + 1)(x - 4)$.

$$3x^2 - 27x - 12 = A(2x + 1)(x - 4) + Bx(x - 4) + Cx(2x + 1)$$

Removing parentheses and rearranging terms, we have

$$3x^2 - 27x - 12 = 2Ax^2 - 7Ax - 4A + Bx^2 - 4Bx + 2Cx^2 + Cx$$
$$3x^2 - 27x - 12 = (2A + B + 2C)x^2 + (-7A - 4B + C)x - 4A$$

Then the coefficients of x^2 must be equal, the coefficients of x must be equal, and the constant terms must be equal. This gives the following system of linear equations.

$$2A + B + 2C = 3$$
$$-7A - 4B + C = -27$$
$$-4A = -12$$

Note that $A = 3$ from the third equation. Substituting $A = 3$ into the first two equations gives

$$6 + B + 2C = 3$$
$$-21 - 4B + C = -27$$

or

$$B + 2C = -3$$
$$-4B + C = -6$$

Multiplying the second equation by 2 gives

$$B + 2C = -3$$
$$-8B + 2C = -12$$

Subtracting these two equations gives

$$9B = 9$$
$$B = 1$$

Then, $C = -2$ and

$$\frac{3x^2 - 27x - 12}{x(2x + 1)(x - 4)} = \frac{3}{x} + \frac{1}{2x + 1} - \frac{2}{x - 4}$$

Case 2 Repeated Linear Denominator Factors

For every factor $(ax + b)^k$ of the denominator of the given fraction, there correspond the possible partial fractions

$$\frac{A_1}{ax + b}, \frac{A_2}{(ax + b)^2}, \frac{A_3}{(ax + b)^3}, \cdots, \frac{A_k}{(ax + b)^k}$$

where $A_1, A_2, A_3, \ldots, A_k$ are constants.

Example 3

Find the partial fractions of $\dfrac{-x^2 - 8x + 27}{x(x - 3)^2}$.

The possible partial fractions are

$$\frac{A}{x}, \quad \frac{B}{x - 3}, \quad \text{and} \quad \frac{C}{(x - 3)^2}.$$

So we have

$$\frac{-x^2 - 8x + 27}{x(x - 3)^2} = \frac{A}{x} + \frac{B}{x - 3} + \frac{C}{(x - 3)^2}$$

Then multiply each side of this equation by the LCD: $x(x - 3)^2$.

$$-x^2 - 8x + 27 = A(x - 3)^2 + Bx(x - 3) + Cx$$

Removing parentheses and rearranging terms, we have

$$-x^2 - 8x + 27 = Ax^2 - 6Ax + 9A + Bx^2 - 3Bx + Cx$$
$$-x^2 - 8x + 27 = (A + B)x^2 + (-6A - 3B + C)x + 9A$$

Equating coefficients, we have

$$A + B = -1$$
$$-6A - 3B + C = -8$$
$$9A = 27$$

From the third equation, we have $A = 3$. Substituting $A = 3$ into the first equation gives $B = -4$. Then, substituting $A = 3$ and $B = -4$ into the second equation gives $C = -2$. Then

$$\frac{-x^2 - 8x + 27}{x(x - 3)^2} = \frac{3}{x} - \frac{4}{x - 3} - \frac{2}{(x - 3)^2}$$

Example 4

Find the partial fractions of $\dfrac{3x^2 - 12x + 17}{(x - 2)^3}$.

Since $x - 2$ is repeated as a linear factor three times, the possible partial fractions are

$$\frac{A}{x - 2} + \frac{B}{(x - 2)^2} + \frac{C}{(x - 2)^3}$$

So we have

$$\frac{3x^2 - 12x + 17}{(x - 2)^3} = \frac{A}{x - 2} + \frac{B}{(x - 2)^2} + \frac{C}{(x - 2)^3}$$

Then, multiply each side of this equation by the LCD: $(x - 2)^3$.

$$3x^2 - 12x + 17 = A(x - 2)^2 + B(x - 2) + C$$

Removing parentheses and rearranging terms, we have

$$3x^2 - 12x + 17 = Ax^2 - 4Ax + 4A + Bx - 2B + C$$

$$3x^2 - 12x + 17 = Ax^2 + (-4A + B)x + 4A - 2B + C$$

Equating coefficients, we have the system

$$A = 3$$
$$-4A + B = -12$$
$$4A - 2B + C = 17$$

Substituting $A = 3$ into the second equation, we have $B = 0$. Then, substituting $A = 3$ and $B = 0$ into the third equation, we have $C = 5$. Then

$$\frac{3x^2 - 12x + 17}{(x - 2)^3} = \frac{3}{x - 2} + \frac{5}{(x - 2)^3}$$

Case 3 Nonrepeated Quadratic Denominator Factors

For every nonrepeated factor $ax^2 + bx + c$ of the denominator of the given fraction, there corresponds the partial fraction

$$\frac{Ax + B}{ax^2 + bx + c}, \qquad \text{where } A \text{ and } B \text{ are constants}$$

Example 5

Find the partial fractions of $\dfrac{11x^2 + 8x - 12}{(2x^2 + x - 4)(x + 1)}$.

The possible partial fractions are

$$\frac{Ax + B}{2x^2 + x - 4} \quad \text{and} \quad \frac{C}{x + 1}.$$

So we have

$$\frac{11x^2 + 8x - 12}{(2x^2 + x - 4)(x + 1)} = \frac{Ax + B}{2x^2 + x - 4} + \frac{C}{x + 1}$$

Then multiply each side of this equation by the LCD: $(2x^2 + x - 4)(x + 1)$.

$$11x^2 + 8x - 12 = (Ax + B)(x + 1) + C(2x^2 + x - 4)$$

Removing parentheses and rearranging terms, we have

$$11x^2 + 8x - 12 = Ax^2 + Ax + Bx + B + 2Cx^2 + Cx - 4C$$

$$11x^2 + 8x - 12 = (A + 2C)x^2 + (A + B + C)x + B - 4C$$

Equating coefficients, we have

$$A + 2C = 11$$
$$A + B + C = 8$$
$$B - 4C = -12$$

The solution of this system of linear equations is $A = 5$, $B = 0$, $C = 3$.

Then

$$\frac{11x^2 + 8x - 12}{(2x^2 + x - 4)(x + 1)} = \frac{5x}{2x^2 + x - 4} + \frac{3}{x + 1}$$

Example 6

Find the partial fractions of $\dfrac{2x^3 - 2x^2 + 8x + 7}{(x^2 + 1)(x^2 + 4)}$.

The possible partial fractions are

$$\frac{Ax + B}{x^2 + 1} \quad \text{and} \quad \frac{Cx + D}{x^2 + 4}$$

So we have

$$\frac{2x^3 - 2x^2 + 8x + 7}{(x^2 + 1)(x^2 + 4)} = \frac{Ax + B}{x^2 + 1} + \frac{Cx + D}{x^2 + 4}$$

Then multiply each side of this equation by the LCD: $(x^2 + 1)(x^2 + 4)$.

$$2x^3 - 2x^2 + 8x + 7 = (Ax + B)(x^2 + 4) + (Cx + D)(x^2 + 1)$$

Removing parentheses and rearranging terms, we have

$$2x^3 - 2x^2 + 8x + 7 = Ax^3 + Bx^2 + 4Ax + 4B + Cx^3 + Dx^2 + Cx + D$$

$$2x^3 - 2x^2 + 8x + 7 = (A + C)x^3 + (B + D)x^2 + (4A + C)x + 4B + D$$

Equating coefficients, we have

$$A + C = 2$$
$$B + D = -2$$
$$4A + C = 8$$
$$4B + D = 7$$

The solution of this system of linear equations is $A = 2, B = 3, C = 0, D = -5$. Then

$$\frac{2x^3 - 2x^2 + 8x + 7}{(x^2 + 1)(x^2 + 4)} = \frac{2x + 3}{x^2 + 1} - \frac{5}{x^2 + 4}$$

Case 4 Repeated Quadratic Denominator Factors

For every factor $(ax^2 + bx + c)^k$ of the denominator of the given fraction, there corresponds the possible partial fractions

$$\frac{A_1x + B_1}{ax^2 + bx + c}, \quad \frac{A_2x + B_2}{(ax^2 + bx + c)^2}, \quad \frac{A_3x + B_3}{(ax^2 + bx + c)^3}, \quad \cdots, \quad \frac{A_kx + B_k}{(ax^2 + bx + c)^k}$$

where $A_1, A_2, A_3, \ldots, A_k, B_1, B_2, B_3, \ldots, B_k$ are constants.

Example 7

Find the partial fractions of $\dfrac{5x^4 - x^3 + 44x^2 - 5x + 75}{x(x^2 + 5)^2}$.

The possible partial fractions are

$$\frac{A}{x}, \quad \frac{Bx + C}{x^2 + 5}, \quad \text{and} \quad \frac{Dx + E}{(x^2 + 5)^2}$$

So we have

$$\frac{5x^4 - x^3 + 44x^2 - 5x + 75}{x(x^2 + 5)^2} = \frac{A}{x} + \frac{Bx + C}{x^2 + 5} + \frac{Dx + E}{(x^2 + 5)^2}$$

Then multiply each side of this equation by the LCD: $x(x^2 + 5)^2$.

$$5x^4 - x^3 + 44x^2 - 5x + 75 = A(x^2 + 5)^2 + (Bx + C)(x^2 + 5)(x) + (Dx + E)x$$

Removing parentheses and rearranging terms, we have

$$5x^4 - x^3 + 44x^2 - 5x + 75 = Ax^4 + 10Ax^2 + 25A + Bx^4 + Cx^3$$
$$+ 5Bx^2 + 5Cx + Dx^2 + Ex$$

$$5x^4 - x^3 + 44x^2 - 5x + 75 = (A + B)x^4 + Cx^3 + (10A + 5B + D)x^2$$
$$+ (5C + E)x + 25A$$

Equating coefficients, we have

$$A + B = 5$$
$$C = -1$$
$$10A + 5B + D = 44$$
$$5C + E = -5$$
$$25A = 75$$

The solution of this system of linear equations is $A = 3$, $B = 2$, $C = -1$, $D = 4$, $E = 0$. Then

$$\frac{5x^4 - x^3 + 44x^2 - 5x + 75}{x(x^2 + 5)^2} = \frac{3}{x} + \frac{2x - 1}{x^2 + 5} + \frac{4x}{(x^2 + 5)^2}$$

If the degree of the numerator is greater than or equal to the degree of the denominator of the original fraction, you must first divide the numerator by the denominator using long division. Then find the partial fractions of the resulting remainder.

Example 8

Find the partial fractions of $\dfrac{x^3 + 3x^2 + 7x + 4}{x^2 + 2x}$.

Since the degree of the numerator is greater than the degree of the denominator, divide as follows:

$$
\begin{array}{r}
x + 1 \\
x^2 + 2x \enclose{longdiv}{x^3 + 3x^2 + 7x + 4} \\
\underline{x^3 + 2x^2} \\
x^2 + 7x \\
\underline{x^2 + 2x} \\
5x + 4
\end{array}
$$

or

$$\frac{x^3 + 3x^2 + 7x + 4}{x^2 + 2x} = x + 1 + \frac{5x + 4}{x^2 + 2x}$$

Now find the partial fractions of the remainder.

$$\frac{5x + 4}{x(x + 2)} = \frac{A}{x} + \frac{B}{x + 2}$$

$$5x + 4 = A(x + 2) + Bx$$

$$5x + 4 = Ax + 2A + Bx$$

$$5x + 4 = (A + B)x + 2A$$

Then

$$A + B = 5$$

$$2A = 4$$

So $A = 2$ and $B = 3$ and

$$\frac{x^3 + 3x^2 + 7x + 4}{x^2 + 2x} = x + 1 + \frac{2}{x} + \frac{3}{x + 2}$$

Exercises

Find the partial fractions of each expression.

1. $\dfrac{8x - 29}{(x + 2)(x - 7)}$

2. $\dfrac{10x - 34}{(x - 4)(x - 2)}$

3. $\dfrac{-x - 18}{2x^2 - 5x - 12}$

4. $\dfrac{17x - 18}{3x^2 + x - 2}$

5. $\dfrac{61x^2 - 53x - 28}{x(3x - 4)(2x + 1)}$

6. $\dfrac{11x^2 - 7x - 42}{(2x + 3)(x^2 - 2x - 3)}$

7. $\dfrac{x^2 + 7x + 10}{(x + 1)(x + 3)^2}$

8. $\dfrac{3x^2 - 18x + 9}{(2x - 1)(x - 1)^2}$

9. $\dfrac{48x^2 - 20x - 5}{(4x - 1)^3}$

10. $\dfrac{x^2 + 8x}{(x + 4)^3}$

11. $\dfrac{11x^2 - 18x + 3}{x(x - 1)^2}$

12. $\dfrac{6x^2 + 4x + 4}{x^3 + 2x^2}$

13. $\dfrac{-x^2 - 4x + 3}{(x^2 + 1)(x^2 - 3)}$

14. $\dfrac{-6x^3 + 2x^2 - 3x + 10}{(2x^2 + 1)(x^2 + 5)}$

15. $\dfrac{4x^3 - 21x - 6}{(x^2 + x + 1)(x^2 - 5)}$

16. $\dfrac{x^3 + 6x^2 + 2x - 2}{(3x^2 - x - 1)(x^2 + 4)}$

17. $\dfrac{4x^3 - 16x^2 - 93x - 9}{(x^2 + 5x + 3)(x^2 - 9)}$

18. $\dfrac{12x^2 + 8x - 72}{(x^2 + x - 1)(x^2 - 16)}$

19. $\dfrac{8x^4 - x^3 + 13x^2 - 6x + 5}{x(x^2 + 1)^2}$

20. $\dfrac{-4x^4 + 6x^3 + 8x^2 - 19x + 17}{(x - 1)(x^2 - 3)^2}$

21. $\dfrac{x^5 - 2x^4 - 8x^2 + 4x - 8}{x^2(x^2 + 2)^2}$

22. $\dfrac{3x^5 + x^4 + 24x^3 + 10x^2 + 48x + 16}{x^2(x^2 + 4)^2}$

23. $\dfrac{6x^2 + 108x + 54}{x^4 - 81}$

24. $\dfrac{x^6 + 2x^4 + 3x^2 + 1}{x^2(x^2 + 1)^3}$

25. $\dfrac{x^3}{x^2 - 1}$

26. $\dfrac{x^4 + x^2}{(x + 1)(x - 2)}$

27. $\dfrac{x^3 - x^2 + 8}{x^2 - 4}$

28. $\dfrac{2x^3 - 2x^2 + 8x - 3}{x(x - 1)}$

29. $\dfrac{3x^4 - 2x^3 - 2x + 5}{x(x^2 + 1)}$

30. $\dfrac{x^5 - x^4 - 3x^3 + 7x^2 + 3x + 20}{(x + 2)(x^2 + 2)}$

7.7 INTEGRATION USING PARTIAL FRACTIONS

Integrals of the form $\displaystyle\int \frac{P(x)}{Q(x)}\, dx$, where $P(x)$ and $Q(x)$ are polynomials, may be integrated by first writing the rational expression as a sum of partial fractions. Then integrate each term.

Example 1

Integrate $\displaystyle\int \frac{3x + 1}{x^2 - x - 6}\, dx$.

First, write $\dfrac{3x + 1}{x^2 - x - 6}$ as a sum of partial fractions.

$$\frac{3x + 1}{(x - 3)(x + 2)} = \frac{A}{x - 3} + \frac{B}{x + 2}$$

Multiply each side of this equation by the LCD: $(x - 3)(x + 2)$.

$$3x + 1 = A(x + 2) + B(x - 3)$$

Removing parentheses and rearranging terms, we have

$$3x + 1 = Ax + 2A + Bx - 3B$$
$$3x + 1 = Ax + Bx + 2A - 3B$$
$$3x + 1 = (A + B)x + 2A - 3B$$

Next, equate the coefficients of x and the constant terms, which gives the following system of linear equations.

$$3 = A + B$$
$$1 = 2A - 3B$$

Solving this system, we have

$$A = 2,\, B = 1$$

So we write

$$\frac{3x + 1}{x^2 - x - 6} = \frac{2}{x - 3} + \frac{1}{x + 2}$$

So

$$\int \frac{3x + 1}{x^2 - x - 6}\, dx = \int \left(\frac{2}{x - 3} + \frac{1}{x + 2} \right) dx$$

$$= 2 \int \frac{dx}{x - 3} + \int \frac{dx}{x + 2}$$

$$= 2 \ln |x - 3| + \ln |x + 2| + C$$

Using the properties of logarithms, we can write the result above as

$$\int \frac{3x + 1}{x^2 - x - 6}\, dx = \ln |(x - 3)^2(x + 2)| + C$$

We thus see how a complicated integral can be written as a sum of less complicated integrals by using partial fractions.

Example 2

Integrate $\displaystyle\int \frac{x^3 + 3x^2 + 7x + 4}{x^2 + 2x}\, dx$.

This rational expression was written as a sum of partial fractions in Example 8 in the preceding section.

$$\int \frac{x^3 + 3x^2 + 7x + 4}{x^2 + 2x}\, dx = \int \left(x + 1 + \frac{2}{x} + \frac{3}{x + 2} \right) dx$$

$$= \int x \, dx + \int dx + 2 \int \frac{dx}{x} + 3 \int \frac{dx}{x+2}$$

$$= \frac{x^2}{2} + x + 2 \ln |x| + 3 \ln |x + 2| + C$$

$$= \frac{x^2}{2} + x + \ln |x^2(x + 2)^3| + C$$

Since the method of writing a rational expression as a sum of partial fractions was presented in detail in the preceding section, the next examples will concentrate on the integration. The process of finding the partial fractions is left to the student.

Example 3

Integrate $\int \frac{x^2 - x + 2}{x^3 - 2x^2 + x} \, dx$.

$$\int \frac{x^2 - x + 2}{x^3 - 2x^2 + x} \, dx = \int \left[\frac{2}{x} - \frac{1}{x - 1} + \frac{2}{(x - 1)^2} \right] dx$$

$$= 2 \int \frac{dx}{x} - \int \frac{dx}{x - 1} + 2 \int (x - 1)^{-2} \, dx$$

$$= 2 \ln |x| - \ln |x - 1| - \frac{2}{x - 1} + C$$

$$= \ln \left| \frac{x^2}{x - 1} \right| - \frac{2}{x - 1} + C$$

Example 4

Integrate $\int \frac{x^2 + x - 1}{x^3 + x} \, dx$.

$$\int \frac{x^2 + x - 1}{x^3 + x} \, dx = \int \left(\frac{2x + 1}{x^2 + 1} - \frac{1}{x} \right) dx$$

$$= \int \frac{2x + 1}{x^2 + 1} \, dx - \int \frac{dx}{x}$$

$$= \int \frac{2x}{x^2 + 1} \, dx + \int \frac{dx}{x^2 + 1} - \int \frac{dx}{x}$$

$$= \ln |x^2 + 1| + \text{Arctan } x - \ln |x| + C$$

$$= \ln \left| \frac{x^2 + 1}{x} \right| + \text{Arctan } x + C$$

Example 5

Integrate $\int \frac{4x^2 - 3x + 2}{x^3 - x^2 - 2x} \, dx$.

$$\int \frac{4x^2 - 3x + 2}{x^3 - x^2 - 2x} \, dx = \int \left(-\frac{1}{x} + \frac{2}{x - 2} + \frac{3}{x + 1} \right) dx$$

$$= -\int \frac{dx}{x} + 2 \int \frac{dx}{x-2} + 3 \int \frac{dx}{x+1}$$

$$= -\ln|x| + 2 \ln|x-2| + 3 \ln|x+1| + C$$

$$= \ln \left| \frac{(x-2)^2(x+1)^3}{x} \right| + C$$

Exercises

Integrate.

1. $\displaystyle\int \frac{dx}{1-x^2}$

2. $\displaystyle\int \frac{dx}{x^2 - 5x + 6}$

3. $\displaystyle\int \frac{dx}{x^2 + 2x - 8}$

4. $\displaystyle\int \frac{dx}{x^2 + x}$

5. $\displaystyle\int \frac{x\,dx}{x^2 - 3x + 2}$

6. $\displaystyle\int \frac{x\,dx}{x^3 - 3x^2 + 2x}$

7. $\displaystyle\int \frac{x+1}{x^2 + 4x - 5}\,dx$

8. $\displaystyle\int \frac{3x-4}{2-x-x^2}\,dx$

9. $\displaystyle\int \frac{dx}{x(x+1)^2}$

10. $\displaystyle\int \frac{7x-4}{(x-1)^2(x+2)}\,dx$

11. $\displaystyle\int \frac{2x^2 + x + 3}{x^2(x+3)}\,dx$

12. $\displaystyle\int \frac{x^2 - x + 2}{x^2(x+2)}\,dx$

13. $\displaystyle\int \frac{x^3\,dx}{x^2 + 3x + 2}$

14. $\displaystyle\int \frac{x^2\,dx}{x^2 + 2x + 1}$

15. $\displaystyle\int \frac{x^2 - 2}{(x^2 + 1)x}\,dx$

16. $\displaystyle\int \frac{5x^2 - x + 11}{(x^2 + 4)(x-1)}\,dx$

17. $\displaystyle\int \frac{x^3 + 2x^2 - 9}{x^2(x^2 + 9)}\,dx$

18. $\displaystyle\int \frac{2x^3 + 3x}{(x^2 + 1)(x^2 + 2)}\,dx$

19. $\displaystyle\int \frac{x^3\,dx}{(x^2 + 1)^2}$

20. $\displaystyle\int \frac{x^2 - 2x + 1}{(x^2 + 1)^2}\,dx$

21. $\displaystyle\int_2^3 \frac{3\,dx}{1 - x^2}$

22. $\displaystyle\int_2^3 \frac{5x+1}{x^2 + x - 2}\,dx$

23. $\displaystyle\int_2^4 \frac{x\,dx}{x^2 + 4x - 5}$

24. $\displaystyle\int_1^3 \frac{x-2}{x^3 + x^2}\,dx$

25. Find the area of the region bounded by $y = \dfrac{4x}{x^2 + 2x - 3}$ from $x = 2$ to $x = 4$ and $y = 0$.

26. Find the area of the region bounded by $y = \dfrac{x+1}{(x+2)^2}$ from $x = 0$ to $x = 1$ and $y = 0$.

7.8 INTEGRATION BY PARTS

Integration by parts is another method of transforming integrals into a form that can be integrated by using familiar integration formulas. This method is based on the chain rule formula for differentiating the product of two functions u and v:

$$\frac{d}{dx}(u \cdot v) = u\frac{dv}{dx} + v\frac{du}{dx}$$

The differential form is then

$$d(u \cdot v) = u\,dv + v\,du$$

Integrating each side, we have:

$$\int d(u \cdot v) = \int (u\,dv + v\,du)$$

$$u \cdot v = \int u \, dv + \int v \, du$$

Solving for $\int u \, dv$, we can write this equation as

> **Integration by parts**
>
> $$\int u \, dv = u \cdot v - \int v \, du$$

We will now demonstrate the method of integration by parts.

Example 1

Integrate $\int xe^{2x} \, dx$.

Let $u = x$ and $dv = e^{2x} \, dx$, then

$$\int (x)(e^{2x} \, dx) = \int u \, dv$$

which is the left-hand side of the formula for integration by parts. Note that what we choose to call dv must contain the factor dx.

If $u = x$, then $du = dx$ and if $dv = e^{2x} \, dx$, then $v = \int dv = \int e^{2x} \, dx = \dfrac{e^{2x}}{2} + C_1$. Since

$$\int u \, dv = \qquad uv \qquad - \int v \, du,$$

we have

$$\int (x)(e^{2x} \, dx) = (x)\left(\frac{e^{2x}}{2} + C_1\right) - \int \left(\frac{e^{2x}}{2} + C_1\right) dx$$

$$= \frac{xe^{2x}}{2} + C_1 x - \int \frac{e^{2x}}{2} \, dx - C_1 \int dx$$

$$= \frac{xe^{2x}}{2} + C_1 x - \frac{e^{2x}}{4} - C_1 x + C$$

$$= \frac{xe^{2x}}{2} - \frac{e^{2x}}{4} + C$$

Note: The constant of integration C_1 that arose from integrating dv does not appear in the final result. This will *always* occur when using the method of integration by parts. We will then ignore C_1 when integrating dv.

The decision as to what to choose for u and dv is not always clear. A trial and error approach may be necessary. For example, we could have chosen $u = e^{2x}$ and $dv = x \, dx$ in Example 1. We would then have had $du = 2e^{2x} \, dx$ and $v = x^2/2$. So,

$$\int e^{2x}(x \, dx) = \frac{x^2 e^{2x}}{2} - \int x^2 e^{2x} \, dx$$

and the right-hand integral is more complicated than the original integral. When this occurs, one should try another choice for u and dv.

Example 2

Integrate $\int x \sin x \, dx$.

$$\int u \, dv \qquad\qquad \int v \, du$$

$$\boxed{\begin{array}{l} u = x \\ dv = \sin x \, dx \end{array}} \quad\longrightarrow\quad \boxed{\begin{array}{l} du = dx \\ v = -\cos x \end{array}}$$

$$\int u \, dv = \quad uv \quad - \int v \, du$$

$$\int x \sin x \, dx = -x \cos x - \int (-\cos x) \, dx$$

$$= -x \cos x + \int \cos x \, dx$$

$$= -x \cos x + \sin x + C$$

Example 3

Integrate $\int x^3 \ln x \, dx$.

$$\int u \, dv \qquad\qquad \int v \, du$$

$$\boxed{\begin{array}{l} u = \ln x \\ dv = x^3 \, dx \end{array}} \quad\longrightarrow\quad \boxed{\begin{array}{l} du = \dfrac{1}{x} \, dx \\ v = \dfrac{x^4}{4} \end{array}}$$

$$\int u \, dv = \quad uv \quad - \int v \, du$$

$$\int (\ln x)(x^3 \, dx) = (\ln x)\left(\frac{x^4}{4}\right) - \int \left(\frac{x^4}{4}\right)\left(\frac{1}{x} \, dx\right)$$

$$= \frac{x^4 \ln x}{4} - \frac{1}{4} \int x^3 \, dx$$

$$= \frac{x^4 \ln x}{4} - \frac{x^4}{16} + C$$

Example 4

Integrate $\int \text{Arcsin } x \, dx$.

$$\int u \, dv \qquad\qquad \int v \, du$$

$$\boxed{\begin{array}{l} u = \text{Arcsin } x \\ dv = dx \end{array}} \quad\longrightarrow\quad \boxed{\begin{array}{l} du = \dfrac{dx}{\sqrt{1 - x^2}} \\ v = x \end{array}}$$

$$\int u\, dv = \quad uv \quad - \int v\, du$$

$$\int (\text{Arcsin } x)(dx) = (\text{Arcsin } x)(x) - \int (x)\left(\frac{dx}{\sqrt{1 - x^2}}\right)$$

$$= x \text{ Arcsin } x - \int (1 - x^2)^{-(1/2)}\, x\, dx$$

$$= x \text{ Arcsin } x + \sqrt{1 - x^2} + C$$

Example 5

Integrate $\int \sec^3 x\, dx$.

$$\int u\, dv \qquad\qquad \int v\, du$$

$u = \sec x$
$dv = \sec^2 x\, dx$

$\longrightarrow$

$du = \sec x \tan x\, dx$
$v = \tan x$

$$\int u\, dv = \quad uv \quad - \int v\, du$$

$$\int (\sec x)(\sec^2 x\, dx) = (\sec x)(\tan x) - \int (\tan x)(\sec x \tan x\, dx)$$

$$= \sec x \tan x - \int \sec x \tan^2 x\, dx$$

$$= \sec x \tan x - \int \sec x(\sec^2 x - 1)\, dx$$

$$= \sec x \tan x - \int \sec^3 x\, dx + \int \sec x\, dx$$

or

$$\int \sec^3 x\, dx = \sec x \tan x - \int \sec^3 x\, dx + \ln |\sec x + \tan x| + C$$

If we add $\int \sec^3 x\, dx$ to each side, we have

$$2 \int \sec^3 x\, dx = \sec x \tan x + \ln |\sec x + \tan x| + C$$

$$\int \sec^3 x\, dx = \tfrac{1}{2}[\sec x \tan x + \ln | \sec x + \tan x|] + C$$

Example 5 demonstrates a possible use of integration by parts that should not be overlooked. Whenever a multiple (not equal to 1) of the original integral appears on the right-hand side, it may be combined with the left-hand integral to complete the integration process.

Example 6

Evaluate $\int_0^{\pi/2} e^{2x} \sin x\, dx$.

$$\int u\, dv \qquad\qquad\qquad \int v\, du$$

$$\boxed{\begin{array}{l} u = e^{2x} \\ dv = \sin x\, dx \end{array}} \quad\longrightarrow\quad \boxed{\begin{array}{l} du = 2e^{2x}\, dx \\ v = -\cos x \end{array}}$$

$$\int u\, dv = \quad uv \quad - \int v\, du$$

$$\int (e^{2x})(\sin x\, dx) = (e^{2x})(-\cos x) - \int (-\cos x)(2e^{2x}\, dx)$$

$$= -e^{2x}\cos x + 2\int e^{2x}\cos x\, dx \qquad\qquad (1)$$

In this example, we need to repeat the integration by parts process for the integral $\int e^{2x}\cos x\, dx$.

$$\int u'\, dv' \qquad\qquad\qquad \int v'\, du'$$

$$\boxed{\begin{array}{l} u' = e^{2x} \\ dv' = \cos x\, dx \end{array}} \quad\longrightarrow\quad \boxed{\begin{array}{l} du' = 2e^{2x}\, dx \\ v' = \sin x \end{array}}$$

$$\int u'\, dv' = u'v' \quad - \int v'\, du'$$

$$\int (e^{2x})(\cos x\, dx) = (e^{2x})(\sin x) - \int (\sin x)(2e^{2x}\, dx)$$

$$= e^{2x}\sin x - 2\int e^{2x}\sin x\, dx$$

So, substituting this result in equation (1), we have

$$\int e^{2x}\sin x\, dx = -e^{2x}\cos x + 2\left(e^{2x}\sin x - 2\int e^{2x}\sin x\, dx\right)$$

$$\int e^{2x}\sin x\, dx = -e^{2x}\cos x + 2e^{2x}\sin x - 4\int e^{2x}\sin x\, dx$$

Adding $4\int e^{2x}\sin x\, dx$ to each side and including a constant of integration, we have

$$5\int e^{2x}\sin x\, dx = -e^{2x}\cos x + 2e^{2x}\sin x + C_1$$

$$\int e^{2x}\sin x\, dx = \frac{e^{2x}}{5}(2\sin x - \cos x) + C$$

$$\int_0^{\pi/2} e^{2x}\sin x\, dx = \frac{e^{2x}}{5}(2\sin x - \cos x)\Big|_0^{\pi/2}$$

$$= \frac{e^{\pi}}{5}(2 - 0) - \frac{1}{5}(0 - 1)$$

$$= \tfrac{1}{5}(2e^{\pi} + 1)$$

Exercises

Integrate.

1. $\int \ln x \; dx$ **2.** $\int x \cos x \; dx$ **3.** $\int xe^x \; dx$

4. $\int xe^{-x} \; dx$ **5.** $\int \sqrt{x} \ln x \; dx$ **6.** $\int x^2 \ln x \; dx$

7. $\int \ln x^2 \; dx$ **8.** $\int \dfrac{\ln x}{\sqrt{x}} \; dx$ **9.** $\int \text{Arccos } x \; dx$

10. $\int \text{Arctan } x \; dx$ **11.** $\int e^x \cos x \; dx$ **12.** $\int x^2 \sin x \; dx$

13. $\int x^2 \cos x \; dx$ **14.** $\int x^3 \ln x \; dx$ **15.** $\int x \sec^2 x \; dx$

16. $\int x^2 e^{2x} \; dx$ **17.** $\int (\ln x)^2 \; dx$ **18.** $\int x \text{ Arctan } x \; dx$

19. $\int x \sec x \tan x \; dx$ **20.** $\int x^3 e^{3x} \; dx$ **21.** $\int_0^1 xe^{3x} \; dx$

22. $\int_0^1 \text{Arcsin } x \; dx$ **23.** $\int_1^2 x\sqrt{x - 1} \; dx$ **24.** $\int_0^3 x\sqrt{x + 1} \; dx$

25. $\int_1^2 \ln (x + 1) \; dx$ **26.** $\int_0^{\pi/2} x^2 \cos x \; dx$

27. Find the area of the region bounded by $y = \ln 2x$ from $x = \frac{1}{2}$ to $x = 1$ and $y = 0$.
28. Find the area of the region bounded by $y = x^2 e^x$ from $x = 0$ to $x = 2$ and $y = 0$.
29. Find the volume of the solid generated by revolving the region bounded by $y = e^x$, $y = 0$, $x = 0$, and $x = 1$ about the y-axis.
30. Find the volume of the solid generated by revolving the region bounded by $y = \sin x$, $y = 0$, $x = 0$, and $x = \pi$ about the y-axis.

7.9 INTEGRATION BY TRIGONOMETRIC SUBSTITUTION

Certain algebraic functions cannot yet be integrated by the methods presented so far. Appropriate trigonometric substitutions often lead to an integration solution. Algebraic functions involving the expressions $\sqrt{a^2 - u^2}$, $\sqrt{u^2 - a^2}$, or $\sqrt{u^2 + a^2}$ can often be integrated by substitutions based on the diagrams in Fig. 7.2:

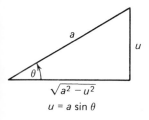

$u = a \sin \theta$

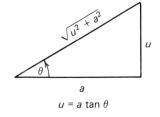

$u = a \tan \theta$

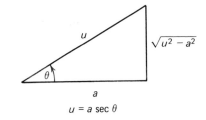

$u = a \sec \theta$

Figure 7.2

The following examples illustrate the use of these substitutions.

Example 1

Integrate $\int \dfrac{x^2\, dx}{\sqrt{9 - x^2}}$. (See Fig. 7.3.)

Since $\sqrt{9 - x^2}$ appears in the integral, we will use the substitution

$$u = a \sin \theta \qquad (u = x,\ a = 3)$$
$$x = 3 \sin \theta$$
$$dx = 3 \cos \theta\, d\theta$$
$$9 - x^2 = 9 - (3 \sin \theta)^2$$
$$= 9 - 9 \sin^2 \theta$$
$$= 9(1 - \sin^2 \theta)$$
$$= 9 \cos^2 \theta$$

So

$$\sqrt{9 - x^2} = 3 \cos \theta$$

Then

$$\int \frac{x^2\, dx}{\sqrt{9 - x^2}} = \int \frac{(3 \sin \theta)^2 (3 \cos \theta\, d\theta)}{3 \cos \theta}$$

$$= 9 \int \sin^2 \theta\, d\theta$$

$$= 9 \int \tfrac{1}{2}(1 - \cos 2\theta)\, d\theta \qquad [\sin^2 \theta = \tfrac{1}{2}(1 - \cos 2\theta)]$$

$$= \tfrac{9}{2} \int d\theta - \tfrac{9}{2} \int \cos 2\theta\, d\theta$$

$$= \tfrac{9}{2}\theta - \tfrac{9}{4} \sin 2\theta + C$$

$$= \tfrac{9}{2}\theta - \tfrac{9}{2} \sin \theta \cos \theta + C \qquad (\sin 2\theta = 2 \sin \theta \cos \theta)$$

From Fig. 7.3

$$\sin \theta = \frac{u}{a} = \frac{x}{3}$$

$$\cos \theta = \frac{\sqrt{a^2 - u^2}}{a} = \frac{\sqrt{9 - x^2}}{3}$$

and

$$\theta = \text{Arcsin}\,\frac{u}{a} = \text{Arcsin}\,\frac{x}{3}$$

So making these substitutions, we have

$$\int \frac{x^2\, dx}{\sqrt{9 - x^2}} = \frac{9}{2} \text{Arcsin}\,\frac{x}{3} - \frac{x\sqrt{9 - x^2}}{2} + C$$

Example 2

Integrate $\int \dfrac{dx}{\sqrt{x^2 + 4}}$. (See Fig. 7.4)

Since $\sqrt{x^2 + 4}$ appears in the integral, we use the substitutions:

$$u = a \tan \theta \qquad (u = x, a = 2)$$
$$x = 2 \tan \theta$$
$$dx = 2 \sec^2 \theta \, d\theta$$
$$x^2 + 4 = (2 \tan \theta)^2 + 4$$
$$= 4 \tan^2 \theta + 4$$
$$= 4(\tan^2 \theta + 1)$$
$$= 4 \sec^2 \theta$$

So

$$\sqrt{x^2 + 4} = 2 \sec \theta$$

Then

$$\int \frac{dx}{\sqrt{x^2 + 4}} = \int \frac{2 \sec^2 \theta \, d\theta}{2 \sec \theta}$$
$$= \int \sec \theta \, d\theta$$
$$= \ln |\sec \theta + \tan \theta| + C$$

From Fig. 7.4 we have

$$\sec \theta = \frac{\sqrt{u^2 + a^2}}{a} = \frac{\sqrt{x^2 + 4}}{2}$$

$$\tan \theta = \frac{u}{a} = \frac{x}{2}$$

So making these substitutions, we have

$$\int \frac{dx}{\sqrt{x^2 + 4}} = \ln \left| \frac{\sqrt{x^2 + 4}}{2} + \frac{x}{2} \right| + C$$

$$= \ln \left| \frac{\sqrt{x^2 + 4} + x}{2} \right| + C$$

$$= \ln |\sqrt{x^2 + 4} + x| - \ln 2 + C$$

$$= \ln |\sqrt{x^2 + 4} + x| + C'$$

Example 3

Integrate $\int \frac{dx}{\sqrt{(4x^2 - 25)^3}}$. (See Fig. 7.5)

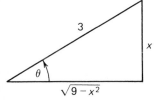

Figure 7.3

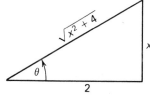

Figure 7.4

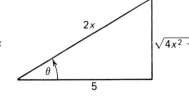

Figure 7.5

Sec. 7.9   Integration by Trigonometric Substitution

The denominator can be written as $\sqrt{(4x^2 - 25)^3} = (4x^2 - 25)\sqrt{4x^2 - 25}$. This suggests the substitutions:

$$u = a \sec\theta \qquad (u = 2x,\ a = 5)$$
$$2x = 5 \sec\theta$$
$$x = \tfrac{5}{2} \sec\theta$$
$$dx = \tfrac{5}{2} \sec\theta \tan\theta\, d\theta$$
$$4x^2 - 25 = 4\left(\frac{25}{4}\sec^2\theta\right) - 25$$
$$= 25 \sec^2\theta - 25$$
$$= 25(\sec^2\theta - 1)$$
$$= 25 \tan^2\theta$$

so

$$\sqrt{4x^2 - 25} = 5 \tan\theta$$

Then

$$\int \frac{dx}{\sqrt{(4x^2 - 25)^3}} = \frac{\tfrac{5}{2} \sec\theta \tan\theta\, d\theta}{(5 \tan\theta)^3}$$

$$= \frac{1}{50} \int \frac{\sec\theta \tan\theta\, d\theta}{\tan^3\theta}$$

$$= \frac{1}{50} \int \frac{\sec\theta}{\tan^2\theta}\, d\theta$$

$$= \frac{1}{50} \int \frac{\dfrac{1}{\cos\theta}}{\dfrac{\sin^2\theta}{\cos^2\theta}}\, d\theta$$

$$= \frac{1}{50} \int \frac{\cos\theta}{\sin^2\theta}\, d\theta$$

$$= \frac{1}{50}\left(-\frac{1}{\sin\theta}\right) + C$$

$$= -\frac{1}{50} \csc\theta + C \qquad \left(\csc\theta = \frac{2x}{\sqrt{4x^2 - 25}} \text{ from Fig. 7.5}\right)$$

$$= \frac{-x}{25\sqrt{4x^2 - 25}} + C$$

Exercises

Integrate.

1. $\displaystyle\int \frac{dx}{\sqrt{9 + 4x^2}}$

2. $\displaystyle\int \frac{dx}{\sqrt{9 - 16x^2}}$

3. $\displaystyle\int \frac{x^2}{\sqrt{4 - 9x^2}}\, dx$

4. $\displaystyle\int \frac{x^2}{\sqrt{1 - 16x^2}}\, dx$

5. $\displaystyle\int_0^1 \frac{dx}{\sqrt{9 - x^2}}$

6. $\displaystyle\int_0^2 \frac{dx}{\sqrt{16 + x^2}}$

Methods of Integration   Chap. 7

7. $\displaystyle\int_0^1 \frac{dx}{\sqrt{(4 - x^2)^3}}$

8. $\displaystyle\int_2^3 \frac{dx}{\sqrt{4x^2 - 9}}$

9. $\displaystyle\int \frac{dx}{x\sqrt{x^2 + 4}}$

10. $\displaystyle\int \frac{dx}{x\sqrt{9 - x^2}}$

11. $\displaystyle\int \frac{\sqrt{x^2 - 9}}{x^2}\, dx$

12. $\displaystyle\int \frac{dx}{\sqrt{(x^2 + 4)^3}}$

13. $\displaystyle\int \frac{dx}{x^2\sqrt{16 - x^2}}$

14. $\displaystyle\int \frac{dx}{x\sqrt{x^2 + 9}}$

15. $\displaystyle\int \frac{\sqrt{9 + x^2}}{x}\, dx$

16. $\displaystyle\int \frac{\sqrt{4 + 9x^2}}{x}\, dx$

17. $\displaystyle\int \frac{dx}{(25 - x^2)^{3/2}}$

18. $\displaystyle\int \frac{dx}{(x^2 - 5)^{3/2}}$

19. $\displaystyle\int \frac{dx}{\sqrt{x^2 - 9}}$

20. $\displaystyle\int \frac{dx}{x\sqrt{1 - x^2}}$

21. $\displaystyle\int \frac{x^3\, dx}{\sqrt{9x^2 + 4}}$

22. $\displaystyle\int \frac{x^2\, dx}{(4 + x^2)^2}$

23. $\displaystyle\int \frac{dx}{\sqrt{x^2 - 6x + 8}}$ *Hint:* Complete the square under the radical.

24. $\displaystyle\int \frac{dx}{\sqrt{-9x^2 + 18x - 5}}$

25. $\displaystyle\int \frac{dx}{(x^2 + 8x + 15)^{3/2}}$

26. $\displaystyle\int \frac{dx}{\sqrt{9x^2 + 36x + 52}}$

27. Find the area of the region bounded by the curves $y = \dfrac{1}{\sqrt{x^2 + 4}}$ from $x = 0$ to $x = 2$ and $y = 0$.

28. Find the area of the region bounded by the curves $y = \sqrt{4 - x^2}$ from $x = 0$ to $x = 2$ and $y = 0$.

7.10 INTEGRATION USING TABLES

Table 10 of Appendix B lists some standard integration formulas for integrating selected functions. These formulas have been developed using the techniques of this chapter and other methods of integrating various complicated functions. A more extensive list of integration formulas (usually more than 400) can be found in most standard handbooks of mathematical tables.

We now illustrate how to use these tables. In the table of integrals (Table 10) u represents a function of x.

Example 1

Integrate $\displaystyle\int x\sqrt{3 + 4x}\, dx$.

If we let $u = x$, $a = 3$, and $b = 4$, this integral is in the form of Formula 11 in Table 3. Substituting these values of u, a and b in this formula, we have

$$\int x\sqrt{3 + 4x} = \frac{-2[2(3) - 3(4)(x)][(3) + (4)(x)]^{3/2}}{15(4)^2} + C$$

$$= \frac{(2x - 1)(3 + 4x)^{3/2}}{20} + C$$

Example 2

Integrate $\displaystyle\int \frac{dx}{x\sqrt{6 + 5x}}$.

If we let $u = x$, $a = 6$ and $b = 5$, then the integral is in the form of Formula 15. Substituting these values of u, a and b in this formula, we have

$$\int \frac{dx}{x\sqrt{6 + 5x}} = \frac{1}{\sqrt{6}} \ln \left| \frac{\sqrt{6 + 5x} - \sqrt{6}}{\sqrt{6 + 5x} + \sqrt{6}} \right| + C$$

Example 3

Integrate $\int x^{100} \ln x \, dx$.

If we let $u = x$ and $n = 100$, then this integral is in the form of Formula 57.

$$\int x^{100} \ln x \, dx = \frac{x^{101} \ln |x|}{101} - \frac{x^{101}}{(101)^2} + C$$

Example 4

Integrate $\int 2^{\sin x} \cos x \, dx$.

If we let $u = \sin x$ and $a = 2$, then since $du = \cos x \, dx$, this integral is in the form of Formula 52.

$$\int 2^{\sin x} \cos x \, dx = \int 2^u \, du$$

$$= \frac{2^u}{\ln 2} + C$$

$$= \frac{2^{\sin x}}{\ln 2} + C$$

$$\boxed{\begin{array}{l} u = \sin x \\ du = \cos x \, dx \end{array}}$$

Example 5

Integrate $\int \sin^5 3x \, dx$.

If we let $u = 3x$ and $n = 5$, then the integral is in the form of Formula 83.

$$\int \sin^5 3x \, dx = \int \sin^5 u \, \frac{du}{3}$$

$$= \tfrac{1}{3} \int \sin^5 u \, du$$

$$= \tfrac{1}{3} \left(-\tfrac{1}{5} \sin^{5-1} u \cos u + \tfrac{4}{5} \int \sin^3 u \, du \right)$$

$$= -\tfrac{1}{15} \sin^4 u \cos u + \tfrac{4}{15} \int \sin^3 u \, du$$

$$\boxed{\begin{array}{l} u = 3x \\ du = 3 \, dx \end{array}}$$

The solution is not complete. We still need to integrate $\int \sin^3 u \, du$, which can be integrated by using Formula 83 again.

$$\int \sin^3 u \, du = -\tfrac{1}{3} \sin^2 u \cos u + \tfrac{2}{3} \int \sin u \, du$$

$$= -\tfrac{1}{3} \sin^2 u \cos u - \tfrac{2}{3} \cos u$$

So

$$\int \sin^5 3x \, dx = -\tfrac{1}{15} \sin^4 u \cos u + \tfrac{4}{15}(-\tfrac{1}{3} \sin^2 u \cos u - \tfrac{2}{3} \cos u) + C$$

$$= -\tfrac{1}{15} \sin^4 3x \cos 3x - \tfrac{4}{45} \sin^2 3x \cos 3x - \tfrac{8}{45} \cos 3x + C$$

Exercises

Integrate using Table 10. Give the number of the formula used.

1. $\displaystyle \int \frac{dx}{x\sqrt{x+5}}$

2. $\displaystyle \int \frac{x\,dx}{2+7x}$

3. $\displaystyle \int \frac{dx}{\sqrt{x^2-4}}$

4. $\displaystyle \int \frac{dx}{\sqrt{6+x^2}}$

5. $\displaystyle \int \frac{x\,dx}{\sqrt{2x+3}}$

6. $\displaystyle \int x\sqrt{4x+7}\,dx$

7. $\displaystyle \int \sin 7x \sin 3x\,dx$

8. $\displaystyle \int \cos 5x \cos 2x\,dx$

9. $\displaystyle \int \frac{x^2\,dx}{\sqrt{9-x^2}}$

10. $\displaystyle \int \frac{dx}{x\sqrt{16-x^2}}$

11. $\displaystyle \int \frac{dx}{x(1+9x)^2}$

12. $\displaystyle \int x^2 e^{4x}\,dx$

13. $\displaystyle \int \frac{dx}{x^2-25}$

14. $\displaystyle \int \frac{dx}{x\sqrt{9-16x^2}}$

15. $\displaystyle \int \sqrt{x^2+4}\,dx$

16. $\displaystyle \int \frac{dx}{x(2+5x)^2}$

17. $\displaystyle \int \frac{x\,dx}{(3+4x)^2}$

18. $\displaystyle \int \frac{5x\,dx}{\sqrt{2+4x}}$

19. $\displaystyle \int \frac{dx}{x\sqrt{9x^2-16}}$

20. $\displaystyle \int xe^{5x}\,dx$

21. $\displaystyle \int e^{3x}\sin 4x\,dx$

22. $\displaystyle \int e^{2x}\cos 5x\,dx$

23. $\displaystyle \int (2x-3)\sin(2x-3)\,dx$

24. $\displaystyle \int x^3 \cos x^2\,dx$

25. $\displaystyle \int \sin^4 x\,dx$

26. $\displaystyle \int \cos^5 x\,dx$

27. $\displaystyle \int \frac{\sqrt{9x^2-16}}{x}\,dx$

28. $\displaystyle \int \frac{dx}{(25-4x^2)^{3/2}}$

7.11 NUMERICAL METHODS OF INTEGRATION

Despite the numerous integration techniques and formulas available, there are still many functions which are difficult to integrate. In fact the value of some integrals cannot be exactly determined by any known method of integration. However, several numerical techniques have been developed for approximating the value of an integral. These numerical methods can easily be used with the help of a calculator or a computer.

The *trapezoidal rule* can be demonstrated by considering $\displaystyle \int_a^b f(x)\,dx$ as representing the area bounded by the curves $y = f(x)$, $x = a$, $x = b$, and the x-axis. The trapezoidal rule is based on approximating this area by the sum of the areas of selected trapezoids.

The line segment from a to b is divided into n intervals, each of width $\Delta x = \dfrac{b-a}{n}$ as shown in Fig. 7.6. $\left(\text{Here } n = 4, \text{ so } \Delta x = \dfrac{b-a}{4}.\right)$ This process determines $(n+1)$ numbers on the x-axis: $x_0 = a$, $x_1 = a + \Delta x$, $x_2 = a + 2\,\Delta x$, $x_3 = a + 3\,\Delta x$, $\ldots$, $x_n = a + n\,\Delta x = b$. Then forming the $n = 4$ trapezoids as shown in Fig. 7.6, we have:

$\frac{1}{2}[f(a) + f(x_1)]\Delta x$ as the area of the first trapezoid

$\frac{1}{2}[f(x_1) + f(x_2)]\Delta x$ as the area of the second trapezoid

$\frac{1}{2}[f(x_2) + f(x_3)]\Delta x$ as the area of the third trapezoid

$\frac{1}{2}[f(x_3) + f(b)]\Delta x$ as the area of the fourth trapezoid

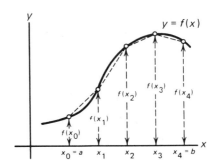

Figure 7.6

The sum of these four areas is then used as an approximation for the area $\int_a^b f(x)\,dx$. In summing these four areas, we have

$$\frac{\Delta x}{2}[f(a) + 2f(x_1) + 2f(x_2) + 2f(x_3) + f(b)]$$

In general,

Trapezoidal Rule

$$\int_a^b f(x)\,dx \doteq \frac{\Delta x}{2}[f(a) + 2f(x_1) + 2f(x_2) + 2f(x_3) + \cdots + 2f(x_{n-1}) + f(b)]$$

The symbol $\doteq$ means "approximately equals."

Note: The pattern of coefficients is 1, 2, 2, 2, . . . , 2, 2, 1.

Example 1

Use the trapezoidal rule with $n = 4$ to find the approximate value of $\int_1^2 \frac{dx}{x}$. (See Fig. 7.7.)

Since $n = 4$, $\Delta x = \dfrac{b-a}{n} = \dfrac{2-1}{4} = \dfrac{1}{4}$ and

$$x_0 = a = 1 \qquad\qquad f(a) = \tfrac{1}{1} = 1$$

$$x_1 = 1 + \tfrac{1}{4} = \tfrac{5}{4} \qquad\qquad f(x_1) = \frac{1}{\tfrac{5}{4}} = \frac{4}{5}$$

$$x_2 = 1 + 2(\tfrac{1}{4}) = \tfrac{3}{2} \qquad\qquad f(x_2) = \frac{1}{\tfrac{3}{2}} = \frac{2}{3}$$

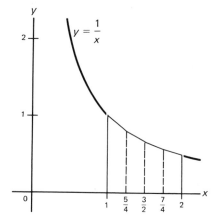

Figure 7.7

$$x_3 = 1 + 3(\tfrac{1}{4}) = \tfrac{7}{4} \qquad\qquad f(x_3) = \frac{1}{\frac{7}{4}} = \frac{4}{7}$$

$$x_4 = b = 1 + 4(\tfrac{1}{4}) = 2 \qquad f(b) = \tfrac{1}{2}$$

$$\int_1^2 \frac{dx}{x} \doteq \frac{\Delta x}{2}[f(a) + 2f(x_1) + 2f(x_2) + 2f(x_3) + f(b)]$$

$$= \frac{\frac{1}{4}}{2}[1 + 2(\tfrac{4}{5}) + 2(\tfrac{2}{3}) + 2(\tfrac{4}{7}) + \tfrac{1}{2}]$$

$$= \tfrac{1}{8}(1 + \tfrac{8}{5} + \tfrac{4}{3} + \tfrac{8}{7} + \tfrac{1}{2}) = 0.6970$$

We also know that $\displaystyle\int_1^2 \frac{dx}{x} = \ln x\big|_1^2 = \ln 2 - \ln 1 = \ln 2 - 0 = \ln 2 = 0.6931.$

Example 2

Use the trapezoidal rule with $n = 8$ to find the approximate value of

$$\int_1^5 x\sqrt{x - 1}\; dx. \quad \text{(See Fig. 7.8.)}$$

$$\text{First, } \Delta x = \frac{b - a}{n} = \frac{5 - 1}{8} = \frac{4}{8} = \frac{1}{2} = 0.5$$

$$
\begin{aligned}
x_0 = a = 1 &\qquad f(a) = f(1) = 0 \\
x_1 = 1.5 &\qquad f(x_1) = f(1.5) = 1.061 \\
x_2 = 2 &\qquad f(x_2) = f(2) = 2 \\
x_3 = 2.5 &\qquad f(x_3) = f(2.5) = 3.062 \\
x_4 = 3 &\qquad f(x_4) = f(3) = 4.243 \\
x_5 = 3.5 &\qquad f(x_5) = f(3.5) = 5.534 \\
x_6 = 4 &\qquad f(x_6) = f(4) = 6.928 \\
x_7 = 4.5 &\qquad f(x_7) = f(4.5) = 8.419 \\
x_8 = b = 5 &\qquad f(b) = f(5) = 10
\end{aligned}
$$

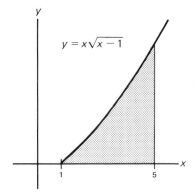

$$y = x\sqrt{x - 1}$$

Figure 7.8

So

$$\int_1^5 x\sqrt{x - 1}\ dx \doteq \frac{\Delta x}{2}[f(a) + 2f(x_1) + 2f(x_2) + 2f(x_3) + 2f(x_4) + 2f(x_5)$$

$$+ 2f(x_6) + 2f(x_7) + f(b)]$$

$$= \frac{0.5}{2}[0 + 2(1.061) + 2(2) + 2(3.062) + 2(4.243)$$

$$+ 2(5.534) + 2(6.928) + 2(8.419) + 10]$$

$$= 18.124$$

Note: Slight variations in the results may occur due to rounding.

Numerical methods of integration are especially helpful when the function to be integrated is not completely known.

Example 3

Find the work done in moving an object along a straight line for 10 ft if the following measurements were made.

Distance moved (ft)	0	2	4	6	8	10
Force (lb)	12	9	7	5	4	2

The formula for work W done in moving an object from a to b is $W = \int_a^b f(x)\ dx$, where $f(x)$ is the force exerted on the object at x.

In this example we do not know what the function $f(x)$ looks like, but we do know its values at 2-ft intervals ($\Delta x = 2$). Here $x_0 = a = 0$, $x_1 = 2$, $x_2 = 4$, $x_3 = 6$, $x_4 = 8$ and $x_5 = b = 10$. So,

$$W = \int_0^{10} f(x)\ dx \doteq \frac{\Delta x}{2}[f(a) + 2f(x_1) + 2f(x_2) + 2f(x_3) + 2f(x_4) + f(b)]$$

$$= \tfrac{2}{2}[12 + 2(9) + 2(7) + 2(5) + 2(4) + 2]$$

$$= 64 \text{ foot-pounds (ft-lb)}$$

Methods of Integration   Chap. 7

A second method of approximating the value of an integral is called Simpson's rule, which uses parabolic intervals instead of trapezoidal intervals and is often more accurate. As before, divide the line segment from a to b into n intervals, each of width $\Delta x = \dfrac{b - a}{n}$ (n must be even here). Then fit parabolas to adjacent triples of points as shown in Fig. 7.9.

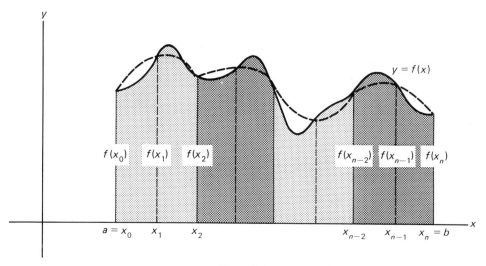

Figure 7.9

The areas of the parabolic segments are then summed. The development of this formula is rather difficult and can be found in more advanced texts.

Simpson's Rule

$$\int_a^b f(x)\,dx \doteq \frac{\Delta x}{3}[f(a) + 4f(x_1) + 2f(x_2)$$

$$+ 4f(x_3) + 2f(x_4) + \cdots + 2f(x_{n-2}) + 4f(x_{n-1}) + f(b)]$$

(n must be an even integer)

Note: The pattern of coefficients is 1, 4, 2, 4, 2, 4, 2, . . . , 2, 4, 1.

We now use Simpson's rule to obtain another approximation for $\int_1^2 \dfrac{dx}{x}$. (See Example 1.)

$$\int_1^2 \frac{dx}{x} \doteq \frac{\frac{1}{4}}{3}[1 + 4(\tfrac{4}{5}) + 2(\tfrac{2}{3}) + 4(\tfrac{4}{7}) + \tfrac{1}{2}]$$

$$= \tfrac{1}{12}(1 + \tfrac{16}{5} + \tfrac{4}{3} + \tfrac{16}{7} + \tfrac{1}{2}) = 0.6933$$

Example 4

Use Simpson's rule with $n = 6$ to find the approximate value of $\int_1^4 \dfrac{dx}{x + 2}$.

First, $\Delta x = \dfrac{b - a}{n} = \dfrac{4 - 1}{6} = \dfrac{3}{6} = \dfrac{1}{2} = 0.5$.

$$x_0 = a = 1 \qquad f(a) = f(1) = \frac{1}{3}$$

$$x_1 = 1.5 \qquad f(x_1) = f(1.5) = \frac{1}{3.5}$$

$$x_2 = 2 \qquad f(x_2) = f(2) = \frac{1}{4}$$

$$x_3 = 2.5 \qquad f(x_3) = f(2.5) = \frac{1}{4.5}$$

$$x_4 = 3 \qquad f(x_4) = f(3) = \frac{1}{5}$$

$$x_5 = 3.5 \qquad f(x_5) = f(3.5) = \frac{1}{5.5}$$

$$x_6 = b = 4 \qquad f(b) = f(4) = \frac{1}{6}$$

Then,

$$\int_1^4 \frac{dx}{x + 2} \doteq \frac{\Delta x}{3}[f(a) + 4f(x_1) + 2f(x_2) + 4f(x_3) + 2f(x_4) + 4f(x_5) + f(b)]$$

$$= \frac{0.5}{3}\left[\frac{1}{3} + 4\left(\frac{1}{3.5}\right) + 2\left(\frac{1}{4}\right) + 4\left(\frac{1}{4.5}\right) + 2\left(\frac{1}{5}\right) + 4\left(\frac{1}{5.5}\right) + \frac{1}{6}\right]$$

$$= 0.693$$

Computer Examples: Trapezoidal Rule, Simpson's Rule

The following examples show the use of the trapezoidal rule and Simpson's rule in approximating the definite integral of a function. The first example shows both rules applied to the function $f(x) = x^3 - 6x^2 + 8x + 4$ with antiderivative $F(x) = x^4/4 - 2x^3 + 4x^2 + 4x + C$. The approximations are shown as the number of intervals, N, is set at 4, 8, 16, 32, and 64. The second example shows both rules applied to the function $f(x) = \sin x$ with antiderivative $F(x) = -\cos x + C$.

A combined flowchart showing the trapezoidal rule and the changes for Simpson's rule is also shown.

Exercises

Use the trapezoidal rule to approximate the value of each integral.

1. $\int_1^3 \dfrac{dx}{x}, \ n = 4$.

2. $\int_1^2 \dfrac{dx}{x}, \ n = 10$

3. $\int_0^1 \dfrac{dx}{1 + x^2}, \ n = 4$

4. $\displaystyle\int_0^2 \frac{dx}{4+x}, \ n = 6$ **5.** $\displaystyle\int_0^1 \sqrt{4-x^2}\ dx, \ n = 10$ **6.** $\displaystyle\int_0^2 2^x\ dx, \ n = 4$

7. $\displaystyle\int_0^3 \sqrt{1+x^3}\ dx, \ n = 6$ **8.** $\displaystyle\int_0^2 \frac{dx}{\sqrt{1+x^3}}, \ n = 4$ **9.** $\displaystyle\int_0^{\pi/3} \cos x^2\ dx, \ n = 8$

10. $\displaystyle\int_0^{\pi/4} \tan x^2\ dx, \ n = 4$

11. Use the trapezoidal rule to find the work done in moving an object along a straight line for 14 ft if the following measurements were made:

Distance moved (ft)	0	2	4	6	8	10	12	14
Force (lb)	24	21	18	17	15	12	10	9

12. Use the trapezoidal rule to find the distance traveled in 6 s by an object moving along a straight line if the following data were recorded:

Time (s)	0	1	2	3	4	5	6
Velocity (m/s)	20	30	50	60	40	30	10

Use the trapezoidal rule to find the approximate area under the curve through each set of points.

13.

x	3	6	9	12	15	18	21
y	12	11	18	25	19	6	10

14.

x	-4	0	4	8	12	16	20
y	11	9	3	10	21	30	40

Use Simpson's rule to approximate the value of each integral.

15. $\displaystyle\int_0^2 \frac{dx}{\sqrt{1+x^2}}, \ n = 4$ **16.** $\displaystyle\int_0^2 \sqrt[3]{8-x^2}\ dx, \ n = 4$ **17.** $\displaystyle\int_2^6 \frac{dx}{1+x^3}, \ n = 8$

18. $\displaystyle\int_1^3 x^x\ dx, \ n = 8$ **19.** $\displaystyle\int_0^1 e^{x^2}\ dx, \ n = 4$ **20.** $\displaystyle\int_0^1 e^{-x^2}\ dx, \ n = 4$

21. $\displaystyle\int_0^{\pi/4} x \tan x\ dx, \ n = 4$ **22.** $\displaystyle\int_1^2 \sqrt{x} \sin x\ dx, \ n = 4$

23. $\displaystyle\int_0^{\pi/2} \sqrt{1+\cos^2 x}\ dx, \ n = 6$ **24.** $\displaystyle\int_{\pi/6}^{\pi/3} \csc x\ dx, \ n = 4$

Approximate the value of each integral by (a) using the trapezoidal rule and (b) using Simpson's rule.

25. $\displaystyle\int_0^3 \sqrt{9-x^2}\ dx, \ n = 6$ **26.** $\displaystyle\int_0^\pi \frac{\sin x}{1+x}\ dx, \ n = 6$

27. $\displaystyle\int_0^1 xe^x\ dx, \ n = 4$ **28.** $\displaystyle\int_0^1 xe^{x^2}\ dx, \ n = 4$

29. $\displaystyle\int_0^{\pi/2} \cos \sqrt{x}\ dx, \ n = 6$ **30.** $\displaystyle\int_0^{\pi/2} \cos^2 x\ dx, \ n = 6$

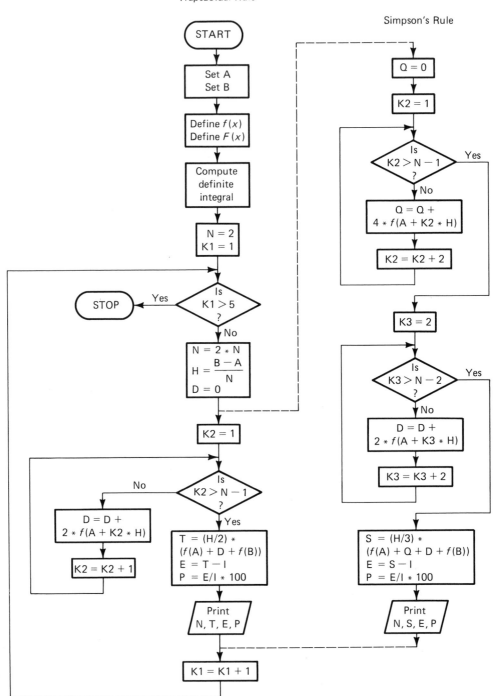

Figure 7.10

Trapezoidal rule

```
10 CLS
20 F1$ = " ## ##.#### #.##### #.#####"
30 A = 1
40 B = 4
50 DEF FN F(X) = X*X*X - 6*X*X + 8*X + 4
60 DEF FN F0(X) = X*X*X*X/4 - 2*X*X*X + 4*X*X + 4*X
70 I = FN F0 (B) - FN F0(A)
80 PRINT " N   TRAP   ERROR   % ERROR"
90 N = 2
100 FOR K1 = 1 TO 5
110 N = 2*N
120 H = (B-A)/N
130 D = 0
140 FOR K2 =1 TO N-1
150 D = D + 2*FN F(A+K2*H)
160 NEXT K2
170 T = (H/2)*(FN F(A) + D +FN F(B))
180 E = T-I
190 P = E/I*100
200 PRINT USING F1$; N, T, E, P
210 NEXT K1
220 END
```

N	TRAP	ERROR	% ERROR
4	10.1719	0.42188	4.32692
8	9.8555	0.10547	1.08173
16	9.7764	0.02637	0.27043
32	9.7566	0.00659	0.06761
64	9.7517	0.00165	0.01692

```
10 CLS
20 F1$ = " ## ##.#### ##.##### ##.#####"
30 A = 0
40 B = 3.14159
50 DEF FN F(X) = SIN(X)
60 DEF FN F0(X) = -COS(X)
70 I = FN F0 (B) - FN F0(A)
80 PRINT " N   TRAP   ERROR   % ERROR"
90 N = 2
100 FOR K1 = 1 TO 5
110 N = 2*N
120 H = (B-A)/N
130 D = 0
140 FOR K2 =1 TO N-1
150 D = D + 2*FN F(A+K2*H)
160 NEXT K2
170 T = (H/2)*(FN F(A) + D +FN F(B))
180 E = T-I
190 P = E/I*100
200 PRINT USING F1$; N, T, E, P
210 NEXT K1
220 END
```

N	TRAP	ERROR	% ERROR
4	1.8961	-0.10388	-5.19406
8	1.9742	-0.02577	-1.28842
16	1.9936	-0.00643	-0.32149
32	1.9984	-0.00161	-0.08034
64	1.9996	-0.00040	-0.02009

Simpson's rule

```
10 CLS
20 F1$ = " ## ##.#### #.##### #.####"
30 A = 1
40 B = 4
50 DEF FN F(X) = X*X*X - 6*X*X + 8*X + 4
60 DEF FN F0(X) = X*X*X*X/4 - 2*X*X*X + 4*X*X + 4*X
70 I = FN F0 (B) - FN F0(A)
80 PRINT " N   SIMP   ERROR   % ERROR"
90 N = 2
100 FOR K1 = 1 TO 5
110 N = 2*N
120 H = (B-A)/N
130 Q = 0
140 D = 0
150 FOR K2 = 1 TO N-1 STEP 2
160 Q = Q + 4*FN F(A+K2*H)
170 NEXT K2
180 FOR K3 = 2 TO N-2 STEP 2
190 D = D + 2*FN F(A+K3*H)
200 NEXT K3
210 S = (H/3)*(FN F(A) + Q + D +FN F(B))
220 E = S - I
230 P = E/I*100
240 PRINT USING F1$; N, S, E, P
250 NEXT K1
260 END
```

N	SIMP	ERROR	% ERROR
4	9.7500	0.00000	0.0000
8	9.7500	0.00000	0.0000
16	9.7500	0.00000	0.0000
32	9.7500	0.00000	0.0000
64	9.7500	0.00000	0.0000

```
10 CLS
20 F1$ = " ## ##.#### #.##### #.#####"
30 A = 0
40 B = 3.14159
50 DEF FN F(X) = SIN(X)
60 DEF FN F0(X) = -COS(X)
70 I = FN F0 (B) - FN F0(A)
80 PRINT " N   SIMP   ERROR   % ERROR"
90 N = 2
100 FOR K1 = 1 TO 5
110 N = 2*N
120 H = (B-A)/N
130 Q = 0
140 D = 0
150 FOR K2 = 1 TO N-1 STEP 2
160 Q = Q + 4*FN F(A+K2*H)
170 NEXT K2
180 FOR K3 = 2 TO N-2 STEP 2
190 D = D + 2*FN F(A+K3*H)
200 NEXT K3
210 S = (H/3)*(FN F(A) + Q + D +FN F(B))
220 E = S - I
230 P = E/I*100
240 PRINT USING F1$; N, S, E, P
250 NEXT K1
260 END
```

N	SIMP	ERROR	% ERROR
4	2.0046	0.00456	0.22799
8	2.0003	0.00027	0.01347
16	2.0000	0.00002	0.00083
32	2.0000	0.00000	0.00006
64	2.0000	0.00000	0.00000

31. Find the approximate area of the field in Fig. 7.11 (a) using the trapezoidal rule and (b) using Simpson's rule.

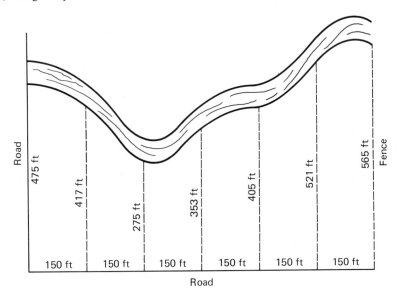

Figure 7.11

CHAPTER SUMMARY

1. *Basic integration formulas*

(a) $\displaystyle\int u^n\, du = \frac{u^{n+1}}{n+1} + C, \qquad n \neq -1$

(b) $\displaystyle\int \frac{du}{u} = \ln |u| + C$

(c) $\displaystyle\int e^u\, du = e^u + C$

(d) $\displaystyle\int a^u\, du = \frac{a^u}{\ln a} + C$

(e) $\displaystyle\int \ln u\, du = u \ln |u| - u + C$

(f) $\displaystyle\int \sin u\, du = -\cos u + C$

(g) $\displaystyle\int \cos u\, du = \sin u + C$

(h) $\displaystyle\int \tan u\, du = -\ln |\cos u| + C$

(i) $\displaystyle\int \cot u\, du = \ln |\sin u| + C$

(j) $\displaystyle\int \sec u \, du = \ln |\sec u + \tan u| + C$

(k) $\displaystyle\int \csc u \, du = \ln |\csc u - \cot u| + C$

(l) $\displaystyle\int \sec^2 u \, du = \tan u + C$

(m) $\displaystyle\int \csc^2 u \, du = -\cot u + C$

(n) $\displaystyle\int \sec u \tan u \, du = \sec u + C$

(o) $\displaystyle\int \csc u \cot u \, du = -\csc u + C$

(p) $\displaystyle\int \frac{du}{\sqrt{a^2 - u^2}} = \text{Arcsin } \frac{u}{a} + C$

(q) $\displaystyle\int \frac{du}{a^2 + u^2} = \frac{1}{a} \text{Arctan } \frac{u}{a} + C$

(r) $\displaystyle\int u \, dv = uv - \int v \, du$ (integration by parts)

Other integrals can often be written in terms of these basic integrals by using the methods of partial fractions and trigonometric substitution.

2. *Partial fractions*

(a) *Case 1 Nonrepeated linear denominator factors:* For every nonrepeated factor $ax + b$ of the denominator of the given fraction, there corresponds the partial fraction $\dfrac{A}{ax + b}$, where A is a constant.

(b) *Case 2 Repeated linear denominator factors:* For every factor $(ax + b)^k$ of the denominator of the given fraction, there correspond the possible partial fractions

$$\frac{A_1}{ax + b}, \quad \frac{A_2}{(ax + b)^2}, \quad \frac{A_3}{(ax + b)^3}, \quad \cdots \cdots, \quad \frac{A_k}{(ax + b)^k}$$

where $A_1, A_2, A_3, \ldots, A_k$ are constants.

(c) *Case 3 Nonrepeated quadratic denominator factors:* For every non-repeated factor $ax^2 + bx + c$ of the denominator of the given fraction, there corresponds the partial fraction

$$\frac{Ax + B}{ax^2 + bx + c}, \qquad \text{where } A \text{ and } B \text{ are constants}$$

(d) *Case 4 Repeated quadratic denominator factors:* For every factor $(ax^2 + bx + c)^k$ of the denominator of the given fraction, there correspond the possible partial fractions

$$\frac{A_1 x + B_1}{ax^2 + bx + c}, \quad \frac{A_2 x + B_2}{(ax^2 + bx + c)^2}, \quad \frac{A_3 x + B_3}{(ax^2 + bx + c)^3}, \quad \cdots \cdots, \quad \frac{A_k x + B_k}{(ax^2 + bx + c)^k}$$

where $A_1, A_2, A_3, \ldots, A_k, B_1, B_2, B_3, \ldots, B_k$ are constants.

3. Trigonometric substitutions

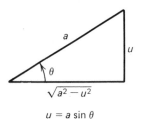

$u = a \sin \theta$

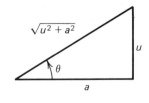

$u = a \tan \theta$

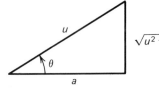

$u = a \sec \theta$

4. Trapezoidal rule for approximating the value of an integral

$$\int_a^b f(x)\, dx \doteq \frac{\Delta x}{2}[f(a) + 2f(x_1) + 2f(x_2) + 2f(x_3)$$
$$+ \cdots + 2f(x_{n-1}) + f(b)]$$

5. Simpson's rule for approximating the value of an integral

$$\int_a^b f(x)\, dx \doteq \frac{\Delta x}{3}[f(a) + 4f(x_1) + 2f(x_2) + 4f(x_3)$$
$$+ 2f(x_4) + \cdots + 2f(x_{n-2}) + 4f(x_{n-1})$$
$$+ f(b)], \qquad n \text{ must be an even integer}$$

CHAPTER REVIEW

Integrate without using a table of integrals.

1. $\displaystyle\int \frac{\cos 3x\, dx}{\sqrt{2 + \sin 3x}}$ **2.** $\displaystyle\int (5 + \tan 2x)^3 \sec^2 2x\, dx$

3. $\displaystyle\int \cos 3x\, dx$ **4.** $\displaystyle\int \frac{x\, dx}{x^2 - 5}$ **5.** $\displaystyle\int xe^{3x^2}\, dx$

6. $\displaystyle\int \frac{dx}{9 + 4x^2}$ **7.** $\displaystyle\int \frac{dx}{\sqrt{16 - x^2}}$ **8.** $\displaystyle\int \sec^2 (7x + 2)\, dx$

9. $\displaystyle\int \frac{\sec^2 x\, dx}{3 + 5\tan x}$ **10.** $\displaystyle\int x^2 \sin (x^3 + 4)\, dx$ **11.** $\displaystyle\int_0^1 \frac{dx}{16 + 9x^2}$

12. $\displaystyle\int_0^{1/2} \sin \pi x\, dx$ **13.** $\displaystyle\int_0^{\pi/4} \frac{\sin x\, dx}{\cos x}$ **14.** $\displaystyle\int_0^1 \frac{dx}{\sqrt{9 - 4x^2}}$

15. $\displaystyle\int \frac{dx}{x\sqrt{16x^2 - 9}}$ **16.** $\displaystyle\int \frac{\tan 3x}{\sec^4 3x}\, dx$ **17.** $\displaystyle\int \frac{\text{Arctan } 3x}{1 + 9x^2}\, dx$

18. $\displaystyle\int \sec 5x\, dx$ **19.** $\displaystyle\int x \tan x^2\, dx$ **20.** $\displaystyle\int \sin^5 2x \cos^2 2x\, dx$

21. $\int \cos^4 3x \, \sin^2 3x \, dx$

22. $\int \dfrac{dx}{x^2 + 2x - 3}$

23. $\int e^{3x} \cos 4x \, dx$

24. $\int \sin x \ln (\cos x) \, dx$

25. $\int \tan^4 x \, dx$

26. $\int \cos^2 5x \, dx$

27. $\int \dfrac{x - 3}{6x^2 - x - 1} \, dx$

28. $\int \sqrt{x} \, \ln x \, dx$

29. $\int e^{-x} (\cos^2 e^{-x}) (\sin e^{-x}) \, dx$

30. $\int \dfrac{4 \, dx}{x^2 - 4}$

31. $\int x^2 \sin x \, dx$

32. $\int \dfrac{\text{Arcsin } 5x}{\sqrt{1 - 25x^2}} \, dx$

33. $\int_2^3 \dfrac{dx}{x\sqrt{x^2 - 1}}$

34. $\int_0^1 x e^{4x} \, dx$

35. $\int_0^{\pi/8} 4 \tan 2x \, dx$

36. $\int_0^1 \dfrac{8 \, dx}{4 - x^2}$

37. $\int \dfrac{3x^2 - 11x + 12}{x^3 - 4x^2 + 4x} \, dx$

38. $\int \dfrac{dx}{x\sqrt{16 - 9x^2}}$

Find the area bounded by the given curves.

39. $y(x - 1) = 1$, $x = 2$, $x = 4$, and $y = 0$.

40. $y = e^{x+2}$, $x = 1$, $x = 3$, and $y = 0$.

41. $y = e^{2x}$, $x = 0$, $x = 1$, and $y = 0$.

42. $y(1 + x^2) = 1$, $x = 0$, $x = 1$, and $y = 0$.

43. $y^2(4 - x^2) = 1$, $x = 0$, $x = 1$, and $y = 0$.

44. $xy = 1$, $y = x$, $y = 0$, and $x = 2$.

45. $y = \sec^2 x$, $y = 0$, $x = 0$, and $x = \pi/4$.

46. $y(4 + x^2) = 1$, $x = 0$, $x = 2$, and $y = 0$.

47. $y = \tan x$ from $x = 0$ to $x = \pi/4$ and $y = 0$.

48. Find the volume of the solid formed by revolving the region bounded by $y = \ln x$ from $x = 1$ to $x = 2$ and $y = 0$ about the x-axis.

49. Find the volume of the solid formed by revolving the region bounded by $y = e^x$, $x = 0$, $x = 1$ and $y = 0$ about the x-axis.

50. The current i in an electric circuit varies according to the equation $i = 4t \sin 3t$ amp. Find an equation for the charge q (in coulombs, C) transferred as a function of time ($q = \int i \, dt$).

51. The current i in an electric circuit varies according to the equation $i = \dfrac{5t + 1}{t^2 + t - 2}$.

Find an equation for the charge q (in coulombs) transferred as a function of time.

52. A force is acting on an object according to the equation $F = xe^{x^2}$ where x is measured in metres and F is measured in newtons. Find the work done in moving the object from $x = 1$ m to $x = 2$ m.

Use the trapezoidal rule to approximate the value of each integral.

53. $\int_1^4 \dfrac{dx}{2x - 1}$, $n = 6$

54. $\int_1^3 \dfrac{dx}{9 + x^2}$, $n = 4$

55. $\int_0^4 \sqrt{16 - x^2} \, dx$, $n = 8$

56. $\int_1^4 \sqrt[3]{18 - x^2} \, dx$, $n = 6$

Use the trapezoidal rule to find the approximate area under the curve through each set of points.

57.

x	1	3	5	7	9
y	2.3	2.8	3.4	2.7	2.1

58.

x	1.6	1.8	2.0	2.2	2.4	2.6
y	0.9	0.7	0.8	1.1	1.3	0.9

Use Simpson's rule to approximate the value of each integral.

59. $\int_0^4 \sqrt{16 + x^2} \, dx, \ n = 8$

60. $\int_0^{12} \dfrac{4}{1 + x^2} \, dx, \ n = 6$

61. $\int_0^{\pi/2} \dfrac{dx}{2 + \sin x}, \ n = 4$

62. $\int_0^8 \dfrac{x^3}{\sqrt{1 + x^3}} \, dx, \ n = 8$

8

POLAR COORDINATES

8.1 POLAR COORDINATES

Each point in the number plane has been associated with an ordered pair of real numbers (x, y), which are called rectangular or Cartesian coordinates. Point $P(x, y)$ is shown in Fig. 8.1. Point P can also be located by specifying an angle θ from the positive x-axis and a directed distance r from the origin, and described by the ordered pair (r, θ) called *polar coordinates*. The polar coordinate system has a fixed point in the number plane called the *pole* or *origin*. From the pole draw a horizontal ray directed to the right, which is called the *polar axis*. (See Fig. 8.2.)

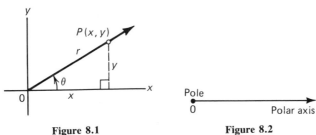

Figure 8.1   Figure 8.2

Angle θ is a directed angle: $\theta > 0$ is measured counterclockwise; $\theta < 0$ is measured clockwise. Angle θ is commonly expressed in either degrees or radians. Distance r is a directed distance: $r > 0$ is measured in the direction of the ray (terminal side of θ); $r < 0$ is measured in the direction opposite the direction of the ray.

Example 1

Graph each point whose polar coordinates are given: (a) (2, 120°), (b) (4, 4π/3), (c) (4, −2π/3), (d) (−5, 135°), (e) (−2, −60°), (f) (−3, 570°) (see Figs. 8.3–8.8.)

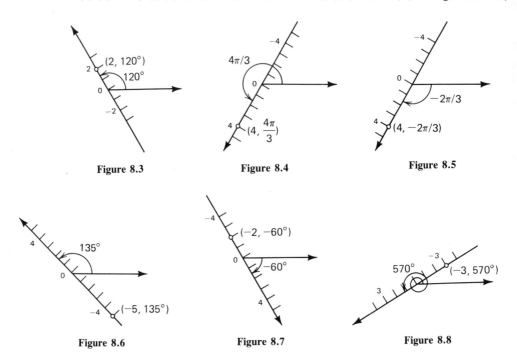

Figure 8.3   Figure 8.4   Figure 8.5

Figure 8.6   Figure 8.7   Figure 8.8

From the results of Example 1, you can see that there is a major difference between the rectangular coordinate system and the polar coordinate system. In the rectangular system there is a one-to-one correspondence between points in the plane and ordered pairs of real numbers. That is, each point is named by exactly one ordered pair, and each ordered pair corresponds to exactly one point. This one-to-one correspondence is not a property of the polar coordinate system. In Example 1 parts (a) and (e) describe the same point in the plane; and parts (b) and (c) describe the same point. In fact, each point may be named by infinitely many polar coordinates. In general, the point $P(r, \theta)$ may be represented by

$$(r, \theta + k \cdot 360°) \quad \text{or} \quad (r, \theta + k2\pi)$$

where k is any integer. $P(r, \theta)$ may also be represented by

$$(-r, \theta + k \cdot 180°) \quad \text{or} \quad (-r, \theta + k\pi)$$

where k is any odd integer.

Example 2

Name an ordered pair of polar coordinates that corresponds to the pole or origin.

Any set of coordinates in the form (0, θ), where θ is any angle, corresponds to the pole. For example, (0, 64°), (0, 2π/3), and (0, −π/6) name the pole.

Polar Coordinates   Chap. 8

Polar graph paper is available for working with polar coordinates. Fig. 8.9 shows graph paper in both degrees and radians.

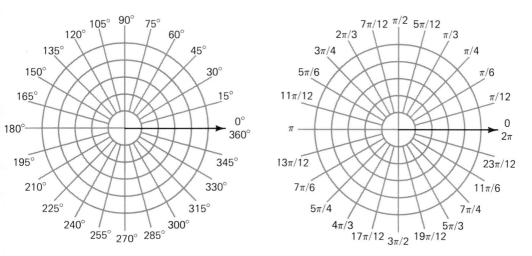

Figure 8.9

Example 3

Plot each point whose polar coordinates are given. Use polar graph paper in degrees. (See Fig. 8.10.)

$$A(5, 60°), B(4, 270°), C(3, -210°),$$
$$D(-6, 45°), E(-2, -150°), F(4, 480°)$$

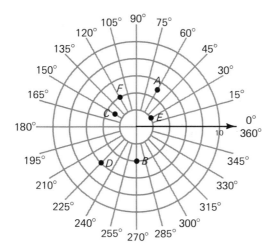

Figure 8.10

Example 4

Plot each point whose polar coordinates are given. Use polar graph paper in radians. (See Fig. 8.11.)

$$A\left(3, \frac{3\pi}{4}\right), B\left(5, \frac{11\pi}{6}\right), C\left(6, -\frac{\pi}{4}\right),$$

$$D\left(-2, \frac{\pi}{3}\right), E(-4, -\pi), F\left(7, \frac{13\pi}{2}\right)$$

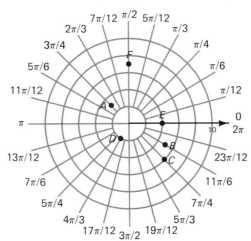

Figure 8.11

Example 5

Given the point $P(4, 150°)$, name three other sets of polar coordinates for P such that $-360° \leqslant \theta \leqslant 360°$.

$$\text{For } r > 0 \text{ and } \theta < 0: (4, -210°)$$
$$\text{For } r < 0 \text{ and } \theta > 0: (-4, 330°)$$
$$\text{For } r < 0 \text{ and } \theta < 0: (-4, -30°)$$

Example 6

Graph $r = 10 \cos \theta$ by plotting points. Assign θ values of 0°, 30°, 45°, 60°, and so on until you have a smooth curve.

Make a table for the ordered pairs as follows. [*Note:* Although r and θ are given in the same order as the ordered pair (r, θ), θ is actually the independent variable in the following tables.]

r	θ	$r = 10 \cos \theta$
10	0°	$r = 10 \cos 0° = 10$
8.7	30°	$r = 10 \cos 30° = 8.7$
7.1	45°	$r = 10 \cos 45° = 7.1$
5	60°	$r = 10 \cos 60° = 5$
0	90°	$r = 10 \cos 90° = 0$
−5	120°	$r = 10 \cos 120° = -5$
−7.1	135°	$r = 10 \cos 135° = -7.1$
−8.7	150°	$r = 10 \cos 150° = -8.7$
−10	180°	$r = 10 \cos 180° = -10$

Polar Coordinates   Chap. 8

Then plot the points as shown in Fig. 8.12.

Note: You should plot values of θ from 0° to 360°, since the period of the cosine function is 360°. In this case choosing values of θ between 180° and 360° will give ordered pairs that duplicate those in Fig. 8.12.

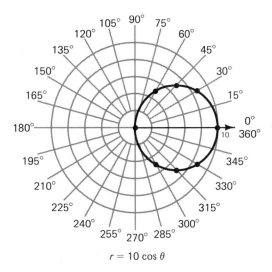

$$r = 10 \cos \theta$$

Figure 8.12

Let point $P(x, y)$ be any point in the rectangular plane. Let the polar plane coincide with the rectangular plane so that $P(x, y)$ and $P(r, \theta)$ represent the same point, as shown in Fig. 8.13.

Note the following relationships:

Polar-Rectangular Relationships

1. $\cos \theta = \dfrac{x}{r}$ or $x = r \cos \theta$

2. $\sin \theta = \dfrac{y}{r}$ or $y = r \sin \theta$

3. $\tan \theta = \dfrac{y}{x}$ or $\theta = \arctan \dfrac{y}{x}$

4. $x^2 + y^2 = r^2$

5. $\cos \theta = \dfrac{x}{\sqrt{x^2 + y^2}}$

6. $\sin \theta = \dfrac{y}{\sqrt{x^2 + y^2}}$

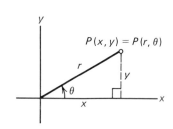

Figure 8.13

Suppose that we wish to change coordinates from one system to the other.

Example 7

Change $A(4, 60°)$ and $B(8, 7\pi/6)$ to rectangular coordinates.

For point A:

$$x = r \cos \theta \qquad\qquad y = r \sin \theta$$

$$x = 4 \cos 60° \qquad\qquad y = 4 \sin 60°$$

$$= 4\left(\frac{1}{2}\right) \qquad\qquad = 4\left(\frac{\sqrt{3}}{2}\right)$$

$$= 2 \qquad\qquad\qquad = 2\sqrt{3}$$

Thus $A(4, 60°) = (2, 2\sqrt{3})$.

For point B:

$$x = r \cos \theta \qquad\qquad y = r \sin \theta$$

$$x = 8 \cos \frac{7\pi}{6} \qquad\qquad y = 8 \sin \frac{7\pi}{6}$$

$$= 8\left(-\frac{\sqrt{3}}{2}\right) \qquad\qquad = 8\left(-\frac{1}{2}\right)$$

$$= -4\sqrt{3} \qquad\qquad\qquad = -4$$

Thus $B(8, 7\pi/6) = (-4\sqrt{3}, -4)$.

Example 8

Find polar coordinates for each point: $C(2\sqrt{3}, 2)$ in degrees $0° \leq \theta < 360°$ and $D(6, -6)$ in radians $0 \leq \theta < 2\pi$.

Note: The signs of x and y determine the quadrant for θ. That is, the signs of x and y determine in which quadrant the point lies and hence the quadrant in which θ must lie.

For point C:

$$r^2 = x^2 + y^2 \qquad\qquad\qquad \theta = \arctan \frac{y}{x}$$

$$r^2 = (2\sqrt{3})^2 + 2^2 = 16 \qquad\qquad \theta = \arctan \frac{2}{2\sqrt{3}}$$

$$r = 4 \qquad\qquad\qquad\qquad \theta = 30°$$

Thus $C(2\sqrt{3}, 2) = (4, 30°)$.

For point D:

$$r^2 = x^2 + y^2 \qquad\qquad\qquad \theta = \arctan \frac{y}{x}$$

$$r^2 = 6^2 + (-6)^2 = 72 \qquad\qquad \theta = \arctan\left(\frac{-6}{6}\right) = \arctan(-1)$$

$$r = 6\sqrt{2} \qquad\qquad\qquad\qquad \theta = \frac{7\pi}{4} \left(Note: \alpha = \frac{\pi}{4}.\right)$$

Thus $D(6, -6) = (6\sqrt{2}, 7\pi/4)$.

Some curves are most simply expressed and easiest to work with in rectangular coordinates; others are most simply expressed and easiest to work with in polar coordinates. As a result you must be able to change a polar equation to a rectangular equation and to change a rectangular equation to a polar equation.

Example 9

Change $x^2 + y^2 - 4x = 0$ to polar form.

Substituting $x^2 + y^2 = r^2$ and $x = r \cos \theta$, we have

$$r^2 - 4r \cos \theta = 0$$

Factoring, we have

$$r(r - 4 \cos \theta) = 0$$

So

$$r = 0 \quad \text{or} \quad r - 4 \cos \theta = 0$$

But $r = 0$ (the pole) is a point that is included in the graph of the equation $r - 4 \cos \theta = 0$. Note that $(0, \pi/2)$ is an ordered pair that satisfies the second equation and names the pole. Thus the simplest polar equation is

$$r = 4 \cos \theta$$

Example 10

Change $r = 4 \sin \theta$ to rectangular form.

Multiply both sides of the given equation by r:

$$r^2 = 4r \sin \theta$$

Substituting $r^2 = x^2 + y^2$ and $r \sin \theta = y$, we have

$$x^2 + y^2 = 4y$$

Note that by multiplying both sides of the given equation by r, we added the root $r = 0$. But the point represented by that root is already included in the original equation. So no new points are added to those represented by the original equation.

Example 11

Change $r \cos^2 \theta = 6 \sin \theta$ to rectangular form.

First multiply both sides by r:

$$r^2 \cos^2 \theta = 6r \sin \theta$$

$$(r \cos \theta)^2 = 6r \sin \theta$$

Substituting $r \cos \theta = x$ and $r \sin \theta = y$, we have

$$x^2 = 6y$$

Example 12

Change $r = \dfrac{2}{1 - \cos \theta}$ to rectangular form.

$$r = \frac{2}{1 - \cos \theta} \tag{1}$$

First multiply both sides by $1 - \cos \theta$:

$$r(1 - \cos \theta) = 2$$

$$r - r \cos \theta = 2$$

$$r = 2 + r \cos \theta \tag{2}$$

Substituting $r = \pm\sqrt{x^2 + y^2}$ and $r \cos \theta = x$, we have

$$\pm\sqrt{x^2 + y^2} = 2 + x$$

Squaring both sides, we have

$$x^2 + y^2 = 4 + 4x + x^2$$
$$y^2 = 4x + 4$$

Note that squaring both sides was a risky operation because we introduced the possible extraneous solutions

$$r = -(2 + r \cos \theta) \tag{3}$$

However, in this case both Equations (2) and (3) have the same graph. To show this solve Equation (3) for r:

$$r = \frac{-2}{1 + \cos \theta} \tag{4}$$

Recall that the ordered pairs (r, θ) and $(-r, \theta + \pi)$ represent the same point. Let us replace (r, θ) by $(-r, \theta + \pi)$ in Equation (4):

$$-r = \frac{-2}{1 + \cos (\theta + \pi)}$$

$$r = \frac{2}{1 - \cos \theta} \qquad \text{[Recall } \cos (\theta + \pi) = -\cos \theta.\text{]}$$

Equations (2) and (3) and thus Equations (1) and (4) have the same graph, and no extraneous solutions were introduced when we squared both sides. So our result $y^2 = 4x + 4$ is correct.

Computer Examples: Coordinate Conversions

The following examples show how a computer can be used to perform conversions between rectangular and polar coordinates. The first example shows how to convert rectangular coordinates to polar coordinates (angle in degrees); the second shows how to convert polar coordinates to rectangular coordinates.

Rectangular to Polar. You have seen that rectangular coordinates can be converted to polar coordinates using the equations

$$r = \sqrt{x^2 + y^2} \quad \text{and} \quad \tan \theta = \frac{y}{x}$$

You must be careful in finding θ, since Arctan y/x is not always equal to θ. You must take into account the quadrant in which the point lies.

The goal of this exercise is to write a program that will automatically find correct values for r and θ given x and y. The first step in designing the program is to analyze the possible cases. Figure 8.14 shows that there are nine separate cases to be considered in this problem.

Case 1: $x = 0$, $y = 0$, $\theta = 0°$
Case 2: $x > 0$, $y = 0$, $\theta = 0°$
Case 3: $x > 0$, $y > 0$, $\theta = $ Arctan y/x
Case 4: $x = 0$, $y > 0$, $\theta = 90°$

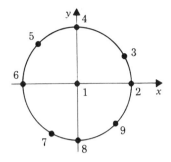

Figure 8.14

Case 5: $x < 0$, $y > 0$, $\theta = \text{Arctan } y/x + 180°*$
Case 6: $x < 0$, $y = 0$, $\theta = 180°$
Case 7: $x < 0$, $y < 0$, $\theta = \text{Arctan } y/x + 180°$
Case 8: $x = 0$, $y < 0$, $\theta = 270°$
Case 9: $x > 0$, $y < 0$, $\theta = \text{Arctan } y/x + 360°*$

The flowchart in Fig. 8.15 shows a method of solution. Note that cases 2 and 3 both correspond to $\theta = \text{Arctan } y/x$, whereas cases 5, 6, and 7 all yield $\theta = 180° + \text{Arctan } y/x$. The symbol A will be used to represent the angle θ in both the flowchart and the BASIC program, since most computer keyboards do not have a key for the symbol θ.

A BASIC language version of a program using the flowchart is given in Fig. 8.15. (*Note:* "↑" means exponentiation.)

Polar to Rectangular. The process of converting polar coordinates to rectangular coordinates is much less complicated. The flowchart in Fig. 8.16 and sample BASIC program below allow the user to enter the angle in either radians or degrees.

Exercises

Plot each point whose polar coordinates are given.
1. $A(3, 150°)$, $B(7, -45°)$, $C(2, -120°)$, $D(-4, 225°)$
2. $A(5, -90°)$, $B(2, -210°)$, $C(6, -270°)$, $D(-5, 30°)$
3. $A\left(4, \dfrac{\pi}{3}\right)$, $B\left(5, -\dfrac{\pi}{4}\right)$, $C\left(3, -\dfrac{7\pi}{6}\right)$, $D\left(-6, \dfrac{11\pi}{6}\right)$
4. $A\left(4, \dfrac{5\pi}{3}\right)$, $B\left(5, -\dfrac{3\pi}{2}\right)$, $C\left(3, -\dfrac{19\pi}{12}\right)$, $D\left(-6, -\dfrac{2\pi}{3}\right)$

* Note that Arctan $y/x < 0$ when $x < 0$ and $y > 0$, or when $x > 0$ and $y < 0$.

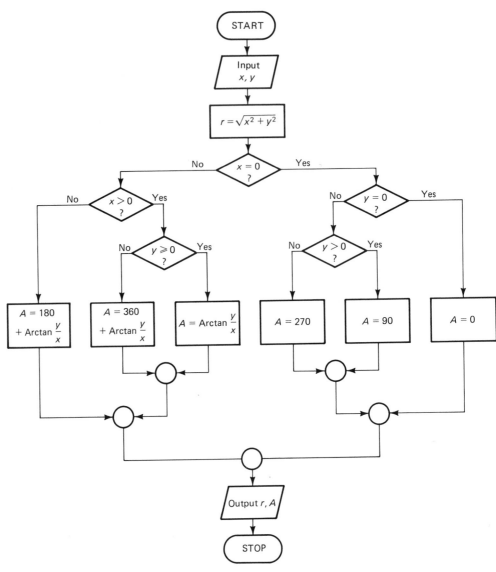

```
  5 PRINT "ENTER X AND Y"          110 GOTO 190
 10 INPUT X, Y                     120 IF Y = 0 THEN 180
 20 R = (X*X + Y*Y)↑.5             130 IF Y > 0 THEN 160
 30 IF X = 0 THEN 120              140 A = 270
 40 IF X > 0 THEN 70               150 GOTO 190
 50 A = 180 + ATN(Y/X)*180/3.1416  160 A = 90
 60 GOTO 190                       170 GOTO 190
 70 IF Y >= 0 THEN 100             180 A = 0
 80 A = 360 + ATN(Y/X)*180/3.1416  190 PRINT"R ="; R; "ANGLE ="; A
 90 GOTO 190                       200 END
100 A = ATN(Y/X)*180/3.1416
```

Figure 8.15

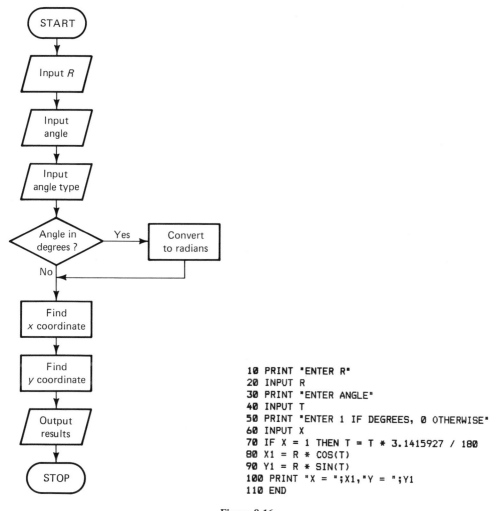

```
10 PRINT "ENTER R"
20 INPUT R
30 PRINT "ENTER ANGLE"
40 INPUT T
50 PRINT "ENTER 1 IF DEGREES, 0 OTHERWISE"
60 INPUT X
70 IF X = 1 THEN T = T * 3.1415927 / 180
80 X1 = R * COS(T)
90 Y1 = R * SIN(T)
100 PRINT "X = ";X1,"Y = ";Y1
110 END
```

Figure 8.16

For each given point name three other sets of polar coordinates such that $-360° \le \theta \le 360°$.

5. $(3, 60°)$ **6.** $(2, 240°)$ **7.** $(-5, 315°)$

8. $(-6, 90°)$ **9.** $(4, -135°)$ **10.** $(-1, -180°)$

For each given point name three other sets of polar coordinates such that $-2\pi \le \theta \le 2\pi$.

11. $\left(3, \dfrac{\pi}{6}\right)$ **12.** $\left(-7, \dfrac{\pi}{2}\right)$ **13.** $\left(-9, \dfrac{2\pi}{3}\right)$

14. $\left(-2, -\dfrac{5\pi}{6}\right)$ **15.** $\left(-4, -\dfrac{7\pi}{4}\right)$ **16.** $\left(5, -\dfrac{5\pi}{3}\right)$

Graph each equation by plotting points. Assign θ values of 0°, 30°, 45°, 60°, and so on, until you have a smooth curve.

Sec. 8.1   Polar Coordinates

17. $r = 10 \sin \theta$ **18.** $r = -10 \sin \theta$ **19.** $r = 4 + 4 \cos \theta$

20. $r = 4 + 4 \sin \theta$ **21.** $r \cos \theta = 4$ **22.** $r \sin \theta = -4$

Graph each equation by plotting points. Assign θ values of 0, $\pi/6$, $\pi/4$, $\pi/3$, and so on, until you have a smooth curve.

23. $r = -10 \cos \theta$ **24.** $r = 6 \sin \theta$ **25.** $r = 4 - 4 \sin \theta$

26. $r = 4 - 4 \cos \theta$ **27.** $r = \theta, \ 0 \leqslant \theta \leqslant 4\pi$ **28.** $r = 2\theta, \ 0 \leqslant \theta \leqslant 2\pi$

Change each set of polar coordinates to rectangular coordinates.

29. $(3, 30°)$ **30.** $(2, 180°)$ **31.** $\left(2, \dfrac{\pi}{3}\right)$

32. $\left(7, \dfrac{5\pi}{6}\right)$ **33.** $(-4, 150°)$ **34.** $(1, 420°)$

35. $\left(-6, \dfrac{3\pi}{2}\right)$ **36.** $(3, -\pi)$ **37.** $(-5, -240°)$

38. $(2, -120°)$ **39.** $\left(2, -\dfrac{7\pi}{4}\right)$ **40.** $\left(-1, -\dfrac{5\pi}{3}\right)$

Change each set of rectangular coordinates to polar coordinates in degrees $0° \leqslant \theta < 360°$.

41. $(5, 5)$ **42.** $(-\sqrt{3}, 1)$ **43.** $(0, 4)$

44. $(-3, 0)$ **45.** $(-2, -2\sqrt{3})$ **46.** $(-1, 1)$

Change each set of rectangular coordinates to polar coordinates in radians $0 \leqslant \theta < 2\pi$.

47. $(-4, 4)$ **48.** $(-1, -\sqrt{3})$ **49.** $(-\sqrt{6}, \sqrt{2})$

50. $(5\sqrt{2}, -5\sqrt{2})$ **51.** $(0, -4)$ **52.** $(0, 0)$

Change each equation to polar form.

53. $x = 3$ **54.** $y = 5$

55. $x^2 + y^2 = 36$ **56.** $y^2 = 5x$

57. $x^2 + y^2 + 2x + 5y = 0$ **58.** $2x + 3y = 6$

59. $4x - 3y = 12$ **60.** $ax + by = c$ **61.** $9x^2 + 4y^2 = 36$

62. $4x^2 - 9y^2 = 36$ **63.** $x^3 = 4y^2$ **64.** $x^4 - 2x^2y^2 + y^4 = 0$

Change each equation to rectangular form.

65. $r \sin \theta = -3$ **66.** $r \cos \theta = 7$ **67.** $r = 5$

68. $r = 3 \sec \theta$ **69.** $\theta = \dfrac{\pi}{4}$ **70.** $\theta = -\dfrac{2\pi}{3}$

71. $r = 5 \cos \theta$ **72.** $r = 6 \sin \theta$ **73.** $r = 6 \cos\left(\theta + \dfrac{\pi}{3}\right)$

74. $r = 4 \sin\left(\theta - \dfrac{\pi}{4}\right)$ **75.** $r \sin^2 \theta = 3 \cos \theta$ **76.** $r^2 = \tan^2 \theta$

77. $r^2 \sin 2\theta = 2$ **78.** $r^2 \cos 2\theta = 6$ **79.** $r^2 = \sin 2\theta$

80. $r^2 = \cos 2\theta$ **81.** $r = \tan \theta$ **82.** $r = 4 \tan \theta \sec \theta$

83. $r = \dfrac{3}{1 + \sin \theta}$ **84.** $r = \dfrac{-4}{1 + \cos \theta}$ **85.** $r = 4 \sin 3\theta$

86. $r = 4 \cos 2\theta$ **87.** $r = 2 + 4 \sin \theta$ **88.** $r = 1 - \cos \theta$

89. Find the distance between the points whose polar coordinates are $(3, 60°)$ and $(2, 330°)$.

90. Find the distance between the points whose polar coordinates are $(5, \pi/2)$ and $(1, 7\pi/6)$.

91. Find a formula for the distance between two points whose polar coordinates are $P_1(r_1, \theta_1)$ and $P_2(r_2, \theta_2)$.

8.2 GRAPHS OF POLAR EQUATIONS

As you undoubtedly know, a graph of any equation may be made by finding and plotting "enough" ordered pairs that satisfy the equation and connecting them with a curve. As you also undoubtedly know, this is often tedious and time consuming at best. We need a method for sketching the graph of a polar equation that minimizes the number of ordered pairs that must be found and plotted. One such method involves symmetry. We shall present tests for three kinds of symmetry:

Symmetry with Respect to the:

1. *Horizontal axis:* Replace θ by $-\theta$ in the original equation. If the resulting equation is equivalent to the original equation, then the graph of the original equation is symmetric with respect to the *horizontal* axis.
2. *Vertical axis:* Replace θ by $\pi - \theta$ in the original equation. If the resulting equation is equivalent to the original equation, then the graph of the original equation is symmetric with respect to the *vertical* axis.
3. *Pole*
   (a) Replace r by $-r$ in the original equation. If the resulting equation is equivalent to the original equation, then the graph of the original equation is symmetric with respect to the *pole*.
   (b) Replace θ by $\pi + \theta$ in the original equation. If the resulting equation is equivalent to the original equation, then the graph of the original equation is symmetric with respect to the *pole*.

You should note that these tests for symmetry are sufficient conditions for symmetry; that is, they are sufficient to assure symmetry. You should also note that these are not necessary conditions for symmetry; that is, symmetry may exist even though the test fails.

If either test 3(a) or 3(b) is satisfied, then the graph is symmetric with respect to the pole. It is also true that if any two of the three kinds of symmetry hold, then the remaining third symmetry automatically holds. Can you explain why?

In order to help you quickly test for symmetry, the following identities are listed for your convenience.

Polar Coordinate Identities for Testing Symmetry
$$\sin(-\theta) = -\sin\theta$$
$$\cos(-\theta) = \cos\theta$$

$$\tan(-\theta) = -\tan\theta$$
$$\sin(\pi - \theta) = \sin\theta$$
$$\cos(\pi - \theta) = -\cos\theta$$
$$\tan(\pi - \theta) = -\tan\theta$$
$$\sin(\pi + \theta) = -\sin\theta$$
$$\cos(\pi + \theta) = -\cos\theta$$
$$\tan(\pi + \theta) = \tan\theta$$

Example 1

Graph $r = 4 + 2\cos\theta$.

Replacing θ by $-\theta$, we see that the graph is symmetric with respect to the horizontal axis. The other tests fail. Thus we need to make a table as follows (note that because of symmetry with respect to the horizontal axis, we need only generate ordered pairs for $0° \le \theta \le 180°$):

r	θ	$r = 4 + 2\cos\theta$
6	0°	$r = 4 + 2\cos 0° = 6$
5.7	30°	$r = 4 + 2\cos 30° = 5.7$
5	60°	$r = 4 + 2\cos 60° = 5$
4	90°	$r = 4 + 2\cos 90° = 4$
3	120°	$r = 4 + 2\cos 120° = 3$
2.3	150°	$r = 4 + 2\cos 150° = 2.3$
2	180°	$r = 4 + 2\cos 180° = 2$

Plot the points as shown in Fig. 8.17. Because of the symmetry with respect to the horizontal axis, plot the corresponding mirror-image points below the horizontal axis, (Fig. 8.18).

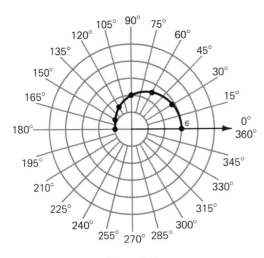

Figure 8.17

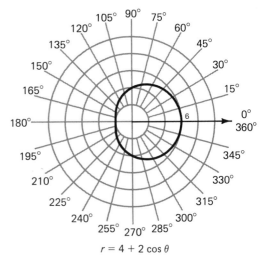

$r = 4 + 2\cos\theta$

Figure 8.18

Polar Coordinates   Chap. 8

Example 2

Graph $r = 4 + 4 \sin \theta$.

Replacing θ by $\pi - \theta$, we see that the graph is symmetric with respect to the vertical axis. The other tests fail. Thus, make a table as follows (note that because of symmetry with respect to the vertical axis, we need only generate ordered pairs for $-\pi/2 \leq \theta \leq \pi/2$).

r	θ	$r = 4 + 4 \sin \theta$
4	0	$r = 4 + 4 \sin 0 = 4$
6	$\pi/6$	$r = 4 + 4 \sin \pi/6 = 6$
7.5	$\pi/3$	$r = 4 + 4 \sin \pi/3 = 7.5$
8	$\pi/2$	$r = 4 + 4 \sin \pi/2 = 8$
2	$-\pi/6$	$r = 4 + 4 \sin (-\pi/6) = 2$
0.54	$-\pi/3$	$r = 4 + 4 \sin (-\pi/3) = 0.54$
0	$-\pi/2$	$r = 4 + 4 \sin (-\pi/2) = 0$

Plot the points as shown in Fig. 8.19. Because of the symmetry with respect to the vertical axis, plot the corresponding mirror-image points to the left of the vertical axis (Fig. 8.20).

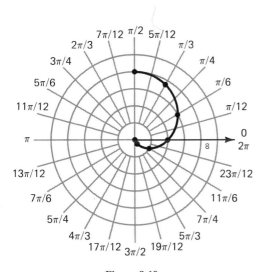

Figure 8.19

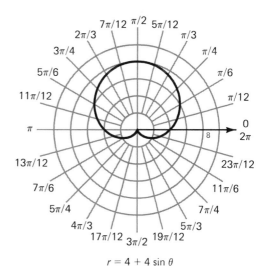

$r = 4 + 4 \sin \theta$

Figure 8.20

Example 3

Graph $r^2 = 16 \sin \theta$.

Replacing r by $-r$, we see that the graph is symmetric with respect to the pole. And, replacing θ by $\pi - \theta$, we see that the graph is also symmetric with respect to the vertical axis. Since two of the three kinds of symmetry hold, the graph is also symmetric with respect to the horizontal axis. (*Note:* Replacing θ by $-\theta$ gives the resulting equation $r^2 = -16 \sin \theta$, which is different from the original equation. However, its solutions when graphed give the same curve.)

Sec. 8.2 Graphs of Polar Equations

r	θ	$r^2 = 16 \sin \theta$
0	0°	$r^2 = 16 \sin 0° = 0; r = 0$
2.8	30°	$r^2 = 16 \sin 30° = 8; r = 2.8$
3.7	60°	$r^2 = 16 \sin 60° = 13.9; r = 3.7$
4	90°	$r^2 = 16 \sin 90° = 16; r = 4$

Plot the points as shown in Fig. 8.21. Because of the symmetry with respect to the horizontal and vertical axes, plot the corresponding mirror-image points below the horizontal axis. Then plot the mirror image points of all resulting points to the left of the vertical axis (See Fig. 8.22.)

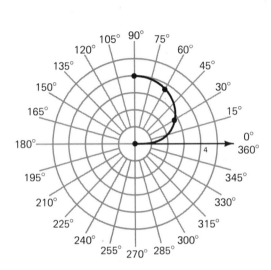

Figure 8.21

$r^2 = 16 \sin \theta$

Figure 8.22

Example 4

Graph $r^2 = 25 \cos 2\theta$.

By replacing θ by $-\theta$ and r by $-r$, we have symmetry with respect to the horizontal axis and the pole, respectively. Thus we also have symmetry with respect to the vertical axis. Working in the first quadrant, we have

r	θ	$r^2 = 25 \cos 2\theta$
5	0	$r^2 = 25 \cos 2(0) = 25; r = 5$
4.7	$\pi/12$	$r^2 = 25 \cos 2(\pi/12) = 21.7; r = 4.7$
3.5	$\pi/6$	$r^2 = 25 \cos 2(\pi/6) = 12.5; r = 3.5$
0	$\pi/4$	$r^2 = 25 \cos 2(\pi/4) = 0; r = 0$

Note: For the interval $\pi/4 < \theta \leq \pi/2$, $r^2 < 0$ and r is undefined.

Plot the points as shown in Fig. 8.23. Because of the symmetry with respect to the horizontal and vertical axes, plot the corresponding mirror-image points below the horizontal axis and to the left of the vertical axis (See Fig. 8.24.)

Polar Coordinates   Chap. 8

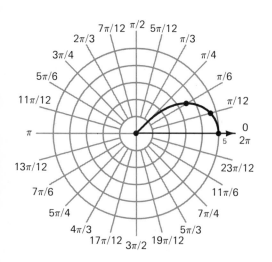

Figure 8.23

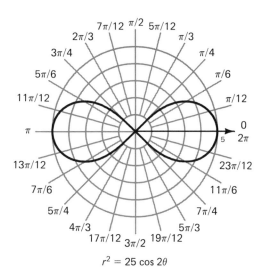

$r^2 = 25 \cos 2\theta$

Figure 8.24

There are various general polar equations whose graphs may be classified as shown in Fig. 8.25. What do you think the graphs of the various forms of $r = a + b \sin \theta$ are like?

Equations in the form

$$r = a \sin n\theta \qquad \text{or} \qquad r = a \cos n\theta$$

where n is a positive integer, are called *petal* or *rose curves*. The number of petals is equal to n if n is an *odd* integer, and is equal to $2n$ if n is an *even* integer. This is because the graph "retraces" itself as θ goes from 0° to 360° when n is odd; so there are only half as many distinct petals. (For $n = 1$ there is one circular petal. See Example 6, Section 8.1). The value of a corresponds to the length of each petal.

The tests for symmetry may be used to graph petal curves. However, we shall illustrate a somewhat different, as well as easier and quicker, method for graphing petal curves.

Example 5

Graph $r = 6 \cos 2\theta$.

First, note that $n = 2$, which is even. Therefore, we have four petals. The petals are always uniform; each petal occupies 360°/4, or 90°, of the polar coordinate system. Next, find the tip of a petal; this occurs when r is maximum or when

$$\cos 2\theta = 1$$
$$2\theta = 0°$$
$$\theta = 0°$$

That is, $r = 6$ when $\theta = 0°$.

Finally, sketch four petals, each having a maximum length of six and occupying 90°. (See Fig. 8.26.) For more accuracy, you may graph the ordered pairs corresponding to a "half petal" ($0° \leqslant \theta \leqslant 45°$ in this case).

Limacons $(r = a + b \cos \theta)$

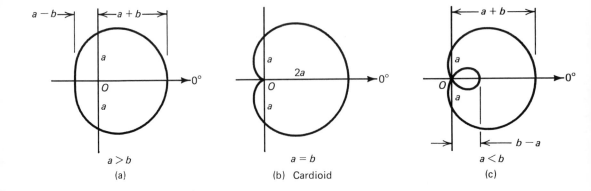

$a > b$	$a = b$	$a < b$
(a)	(b) Cardioid	(c)

Lemniscates

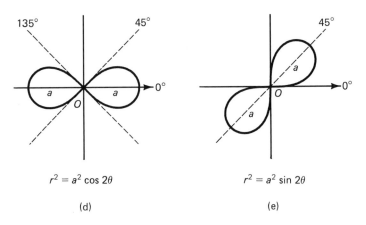

$$r^2 = a^2 \cos 2\theta \qquad\qquad r^2 = a^2 \sin 2\theta$$

(d) (e)

Figure 8.25

Polar coordinates are especially useful for the study and graphing of *spirals*. The *spiral of Archimedes* has an equation in the form

$$r = a\theta.$$

Its graph is shown in Fig. 8.27. (The dashed portion of the graph corresponds to $\theta < 0$.)

The *logarithmic spiral* has an equation of the form

$$\log_b r = \log_b a + k\theta \qquad \text{or} \qquad r = a \cdot b^{k\theta}$$

Its graph is shown in Fig. 8.28.

Polar Coordinates   Chap. 8

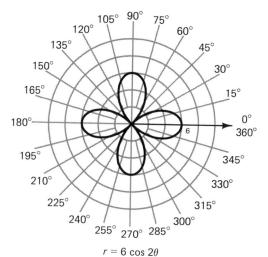

$r = 6 \cos 2\theta$

Figure 8.26

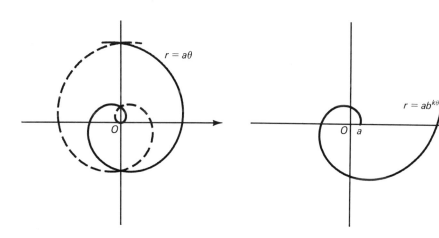

$r = a\theta$

Figure 8.27

$r = ab^{k\theta}$

Figure 8.28

Exercises

Graph each equation.

1. $r = 6$ **2.** $r = 3$ **3.** $r = -2$

4. $r = -4$ **5.** $\theta = 30°$ **6.** $\theta = -120°$

7. $\theta = -\dfrac{\pi}{3}$ **8.** $\theta = \dfrac{7\pi}{6}$ **9.** $r = 5 \sin \theta$

10. $r = 8 \cos \theta$ **11.** $r = 6 \cos \left(\theta + \dfrac{\pi}{3} \right)$ **12.** $r = 4 \sin \left(\theta - \dfrac{\pi}{4} \right)$

13. $r = 4 + 2 \sin \theta$ **14.** $r = 8 + 2 \cos \theta$ **15.** $r = 4 - 2 \cos \theta$

16. $r = 4 - 2 \sin \theta$ **17.** $r = 3 + 3 \cos \theta$ **18.** $r = 5 + 5 \sin \theta$

Sec. 8.2   Graphs of Polar Equations   **353**

19. $r = 2 + 4 \sin \theta$ **20.** $r = 2 + 8 \cos \theta$ **21.** $r = 2 - 4 \cos \theta$

22. $r = 2 - 4 \sin \theta$ **23.** $r = 3 - 3 \cos \theta$ **24.** $r = 5 - 5 \sin \theta$

25. $r \cos \theta = 6$ **26.** $r \sin \theta = -4$ **27.** $r^2 = 25 \cos \theta$

28. $r^2 = -9 \sin \theta$ **29.** $r^2 = 9 \sin 2\theta$ **30.** $r^2 = 16 \cos 2\theta$

31. $r^2 = -36 \cos 2\theta$ **32.** $r = -36 \sin 2\theta$ **33.** $r = 5 \sin 3\theta$

34. $r = 4 \cos 5\theta$ **35.** $r = 3 \cos 2\theta$ **36.** $r = 6 \sin 4\theta$

37. $r = 9 \sin^2 \theta$ **38.** $r = 16 \cos^2 \theta$ **39.** $r = 4 \cos \dfrac{\theta}{2}$

40. $r = 5 \sin^2 \dfrac{\theta}{2}$ **41.** $r = \tan \theta$ **42.** $r = 2 \csc \theta$

43. $r = 3\theta, \ \theta > 0$ **44.** $r = \dfrac{3}{\theta}, \ \theta > 0$ **45.** $r = 2^{3\theta}$

46. $r = 2 \cdot 3^{2\theta}$ **47.** $r = \dfrac{4}{\sin \theta + \cos \theta}$ **48.** $r = \dfrac{-2}{\sin \theta + \cos \theta}$

49. $r(1 + \cos \theta) = 4$ **50.** $r(1 + 2 \sin \theta) = -4$

8.3 AREAS IN POLAR COORDINATES

Finding areas in polar coordinates is similar to finding the area of a region in rectangular coordinates. Suppose that we have a region bounded by $r = f(\theta)$ and the terminal sides of angles α and β in standard position, where $\alpha < \beta$. (See Fig. 8.29.) Now let's subdivide the angular region into n subintervals of circular sectors, instead of rectangles as in rectangular coordinates, as follows:

$$\alpha = \theta_0, \theta_1, \theta_2, \ldots, \theta_n = \beta$$

And let r_1 be any ray within the ith sector. Then $\Delta\theta_i$ is the central angle of the ith sector and $r_i = f(\theta_i)$ is the radius of the ith sector.

Recall that the area of a sector of a circle whose central angle is θ (in radians) and radius r is

$$A = \frac{1}{2} r^2 \theta$$

The area of the ith sector in Fig. 8.30 is

$$\frac{1}{2} r_i^2 \Delta\theta_i = \frac{1}{2} [f(\theta_i)]^2 \Delta\theta_i$$

The total area is

$$\frac{1}{2} r_1^2 \Delta\theta_1 + \frac{1}{2} r_2^2 \Delta\theta_2 + \frac{1}{2} r_3^2 \Delta\theta_3 + \cdots + \frac{1}{2} r_n^2 \Delta\theta_n$$

which can be represented by the integral

Area of a Region in Polar Coordinates

$$A = \frac{1}{2} \int_\alpha^\beta r^2 \, d\theta = \frac{1}{2} \int_\alpha^\beta [f(\theta)]^2 \, d\theta$$

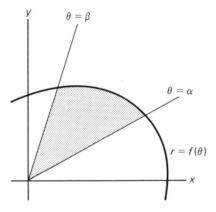

Figure 8.29

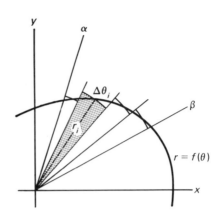

Figure 8.30

Due to the squaring of r, you should review Section 7.4 and the integration of even powers of sine and cosine functions. Recall that the identities used are

$$\sin^2 x = \frac{1}{2}(1 - \cos 2x)$$

$$\cos^2 x = \frac{1}{2}(1 + \cos 2x)$$

Example 1

Find the area of the region inside $r^2 = 4 \sin 2\theta$.

First, graph $r^2 = 4 \sin 2\theta$. (See Fig. 8.31.)

Due to symmetry, the area may be expressed by

$$A = 2\left[\frac{1}{2}\int_0^{\pi/2} r^2\, d\theta\right]$$

$$= \int_0^{\pi/2} 4 \sin 2\theta\, d\theta$$

$$= 4\left(-\frac{1}{2}\cos 2\theta\right)\Big|_0^{\pi/2}$$

$$= -2[-1 - 1] = 4$$

Example 2

Find the area of the region in the first quadrant within $r = 4 + 4 \sin \theta$.

First, graph $r = 4 + 4 \sin \theta$. (See Fig. 8.32.)

$$A = \frac{1}{2}\int_\alpha^\beta r^2\, d\theta$$

$$= \frac{1}{2}\int_0^{\pi/2} (4 + 4 \sin \theta)^2\, d\theta$$

$$= 8\int_0^{\pi/2} (1 + \sin \theta)^2\, d\theta$$

Sec. 8.3   Areas In Polar Coordinates

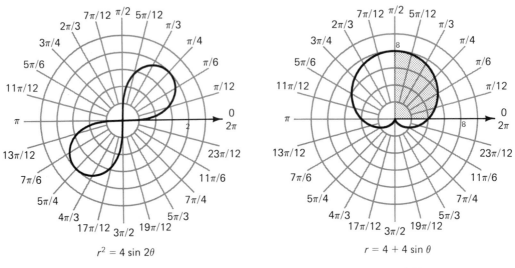

$r^2 = 4 \sin 2\theta$

Figure 8.31

$r = 4 + 4 \sin \theta$

Figure 8.32

$$= 8 \int_0^{\pi/2} (1 + 2 \sin \theta + \sin^2\theta) \, d\theta$$

$$= 8 \int_0^{\pi/2} [1 + 2 \sin \theta + \frac{1}{2}(1 - \cos 2\theta)] \, d\theta$$

$$= 8 \int_0^{\pi/2} \left[\frac{3}{2} + 2 \sin \theta - \frac{1}{2} \cos 2\theta \right] d\theta$$

$$= 8 \left(\frac{3\theta}{2} - 2 \cos \theta - \frac{1}{4} \sin 2\theta \right) \Bigg|_0^{\pi/2}$$

$$= 8 \left[\left(\frac{3}{2} \cdot \frac{\pi}{2} - 2 \cdot 0 - \frac{1}{4} \cdot 0 \right) - \left(\frac{3}{2} \cdot 0 - 2 \cdot 1 - \frac{1}{4} \cdot 0 \right) \right]$$

$$= 8 \left(\frac{3\pi}{4} + 2 \right) = 6\pi + 16$$

Example 3

Find the area within the inner loop of $r = 2 + 4 \cos \theta$.

First, graph $r = 2 + 4 \cos \theta$. (See Fig. 8.33.) The endpoints of the inner loop are found by noting that $r = 0$ at the endpoints.

$$r = 2 + 4 \cos \theta$$

$$0 = 2 + 4 \cos \theta$$

$$\cos \theta = -\frac{1}{2}$$

$$\theta = \frac{2\pi}{3}, \frac{4\pi}{3}$$

So the inner loop is determined by $\dfrac{2\pi}{3} \leq \theta \leq \dfrac{4\pi}{3}$.

$$A = \frac{1}{2} \int_{2\pi/3}^{4\pi/3} r^2 \, d\theta$$

$$= \frac{1}{2} \int_{2\pi/3}^{4\pi/3} (2 + 4 \cos \theta)^2 \, d\theta$$

$$= 2 \int_{2\pi/3}^{4\pi/3} (1 + 4 \cos \theta + 4 \cos^2 \theta) \, d\theta$$

$$= 2 \int_{2\pi/3}^{4\pi/3} [1 + 4 \cos \theta + 4 \cdot \frac{1}{2}(1 + \cos 2\theta)] \, d\theta$$

$$= 2 \int_{2\pi/3}^{4\pi/3} (3 + 4 \cos \theta + 2 \cos 2\theta) \, d\theta$$

$$= 2[3\theta + 4 \sin \theta + \sin 2\theta] \Big|_{2\pi/3}^{4\pi/3}$$

$$= 2 \left[\left(3 \cdot \frac{4\pi}{3} + 4 \cdot \left(\frac{-\sqrt{3}}{2} \right) + \frac{\sqrt{3}}{2} \right) - \left(3 \cdot \frac{2\pi}{3} + 4 \cdot \frac{\sqrt{3}}{2} + \left(-\frac{\sqrt{3}}{2} \right) \right) \right]$$

$$= 2 \left[4\pi - 2\sqrt{3} + \frac{\sqrt{3}}{2} - 2\pi - 2\sqrt{3} + \frac{\sqrt{3}}{2} \right]$$

$$= 2[2\pi - 3\sqrt{3}]$$

$$= 4\pi - 6\sqrt{3}$$

Example 4

Find the area of the region inside $r = 1$ and outside $r = 1 + \sin \theta$.

First, graph the two equations. (See Fig. 8.34.)

Use the substitution method of solving equations simultaneously to find the points of intersection.

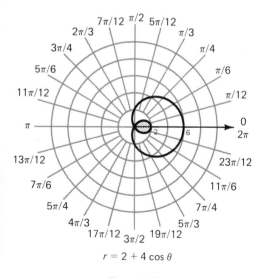

$r = 2 + 4 \cos \theta$

Figure 8.33

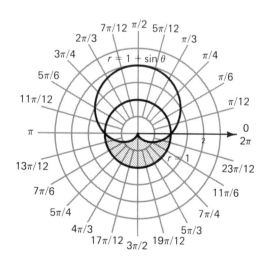

Figure 8.34

$$1 + \sin \theta = 1$$
$$\sin \theta = 0$$
$$\theta = 0, \pi$$

Due to symmetry with respect to the y-axis, the area may be expressed by

$$A = 2 \left[\frac{1}{2} \int_\pi^{3\pi/2} [1^2 - (1 + \sin \theta)^2] \, d\theta \right]$$

$$= \int_\pi^{3\pi/2} (1 - 1 - 2 \sin \theta - \sin^2\theta) \, d\theta$$

$$= \int_\pi^{3\pi/2} [-2 \sin \theta - \frac{1}{2}(1 - \cos 2\theta)] \, d\theta$$

$$= \int_\pi^{3\pi/2} (-2 \sin \theta - \frac{1}{2} + \frac{1}{2} \cos 2\theta) \, d\theta$$

$$= (2 \cos \theta - \frac{\theta}{2} + \frac{1}{4} \sin 2\theta) \Big|_\pi^{3\pi/2}$$

$$= \left(2 \cdot 0 - \frac{3\pi/2}{2} + \frac{1}{4} \cdot 0 \right) - \left(2(-1) - \frac{\pi}{2} + \frac{1}{4} \cdot 0 \right)$$

$$= -\frac{3\pi}{4} + 2 + \frac{\pi}{2}$$

$$= 2 - \frac{\pi}{4}$$

Exercises

Find the area of each region.

1. $r = 2, 0 \le \theta \le \frac{2\pi}{3}$
2. $r = 4, \frac{\pi}{3} \le \theta \le \frac{7\pi}{6}$
3. Inside $r = 3 \cos \theta$
4. Inside $r = 6 \sin \theta$
5. Inside $r^2 = 9 \sin 2\theta$
6. Inside $r^2 = \cos 2\theta$
7. Inside $r = 1 + \cos \theta$
8. Inside $r = 2 - \sin \theta$
9. Inside $r = 4 \sin 2\theta$
10. Inside $r = 2 \sin 3\theta$
11. Inside $r^2 = 16 \cos^2\theta$
12. Inside $r^2 = 4 \sin^2\theta$
13. Inside $r = 4 + 3 \cos \theta$
14. Inside $r = 3 + 3 \sin \theta$
15. $r = e^\theta, 0 \le \theta \le \pi$
16. $r = 2^\theta, 0 \le \theta \le \pi$
17. Inside the inner loop of $r = 1 - 2 \cos \theta$
18. Inside the outer loop and outside the inner loop of $r = 2 - 4 \sin \theta$
19. Inside $r = 2 \sin \theta + 2 \cos \theta$
20. Inside $r = 2 \sin \theta - 2 \cos \theta$
21. Inside $r = \sin \theta$ and $r = \sin 2\theta$
22. Inside $r = \sin \theta$ and $r = \cos \theta$
23. Inside $r = 2 \cos 2\theta$ and outside $r = 1$
24. Inside $r^2 = 8 \cos 2\theta$ and outside $r = 2$
25. Inside $r = 2 + \sin \theta$ and outside $r = 5 \sin \theta$
26. Inside $r = -6 \cos \theta$ and outside $r = 2 - 2 \cos \theta$
27. Inside $r = 3 + 3 \cos \theta$ and outside $r = 3 + 3 \sin \theta$
28. Inside the resulting inner loops of $r = 3 + 3 \sin \theta$ and $r = 3 - 3 \sin \theta$

CHAPTER SUMMARY

1. Each point $P(x, y)$ in the rectangular coordinate system may be described by the ordered pair $P(r, \theta)$ in the polar coordinate system.

2. In the rectangular coordinate system, there is a one-to-one correspondence between points in the plane and ordered pairs of real numbers. This one-to-one correspondence is not a property of the polar coordinate system. In general the point $P(r, \theta)$ may be represented by

$$(r, \theta + k \cdot 360°) \quad \text{or} \quad (r, \theta + k2\pi)$$

where k is any integer. $P(r, \theta)$ may also be represented by

$$(-r, \theta + k \cdot 180°) \quad \text{or} \quad (-r, \theta + k\pi)$$

where k is any odd integer.

3. The relationships between the rectangular and polar coordinate systems are

(a) $x = r \cos \theta$ 　　　　　　　　　　(b) $y = r \sin \theta$

(c) $\tan \theta = \dfrac{y}{x}$ or $\theta = \arctan \dfrac{y}{x}$ 　　(d) $x^2 + y^2 = r^2$

(e) $\cos \theta = \dfrac{x}{\sqrt{x^2 + y^2}}$ 　　　　(f) $\sin \theta = \dfrac{y}{\sqrt{x^2 + y^2}}$

4. *Symmetry tests for graphing polar equations*

   (a) *Horizontal axis:* Replace θ by $-\theta$ in the original equation. If the resulting equation is equivalent to the original equation, then the graph of the original equation is symmetric with respect to the *horizontal* axis.

   (b) *Vertical axis:* Replace θ by $\pi - \theta$ in the original equation. If the resulting equation is equivalent to the original equation, then the graph of the original equation is symmetric with respect to the *vertical* axis.

   (c) *Pole*

   (i) Replace r by $-r$ in the original equation. If the resulting equation is equivalent to the original equation, then the graph of the original equation is symmetric with respect to the *pole*.

   (ii) Replace θ by $\pi + \theta$ in the original equation. If the resulting equation is equivalent to the original equation, then the graph of the original equation is symmetric with respect to the *pole*.

5. *Area of a region in polar coordinates*

$$A = \frac{1}{2} \int_\alpha^\beta r^2 \, d\theta = \frac{1}{2} \int_\alpha^\beta [f(\theta)]^2 \, d\theta$$

CHAPTER REVIEW

Plot each point whose polar coordinates are given.

1. $A(6, 60°)$, $B(3, -210°)$, $C(-2, -270°)$, $D(-4, 750°)$

2. $A\left(5, \dfrac{\pi}{6}\right)$, $B\left(2, -\dfrac{5\pi}{4}\right)$, $C\left(-3, -\dfrac{\pi}{2}\right)$, $D\left(-5, \dfrac{19\pi}{2}\right)$

3. For the point $A(5, 135°)$, name three other sets of polar coordinates such that $-360° \leqslant \theta \leqslant 360°$.

4. For the point $B(-2, 7\pi/6)$, name three other sets of polar coordinates such that $-2\pi \leqslant \theta \leqslant 2\pi$.

Change each set of polar coordinates to rectangular coordinates.

5. $(3, 210°)$ **6.** $(2, -120°)$ **7.** $\left(-5, \dfrac{11\pi}{6}\right)$ **8.** $\left(-6, -\dfrac{\pi}{2}\right)$

Change each set of rectangular coordinates to polar coordinates in degrees for $0° \leqslant \theta < 360°$.

9. $(-3, 3)$ **10.** $(0, -6)$ **11.** $(-1, \sqrt{3})$

Change each set of rectangular coordinates to polar coordinates in radians for $0 \leqslant \theta < 2\pi$.

12. $(-5, 0)$ **13.** $(-6\sqrt{3}, 6)$ **14.** $(1, -1)$

Change each equation to polar form.

15. $x^2 + y^2 = 49$ **16.** $y^2 = 9x$ **17.** $5x + 2y = 8$

18. $x^2 - 4y^2 = 12$ **19.** $y^3 = 6x^2$ **20.** $y(x^2 + y^2) = x^2$

Change each equation to rectangular form.

21. $r \cos \theta = 12$ **22.** $r = 9$ **23.** $\theta = \dfrac{2\pi}{3}$

24. $r = 8 \cos \theta$ **25.** $r \sin^2\theta = 5 \cos \theta$ **26.** $r^2 \sin 2\theta = 8$

27. $r^2 = 4 \cos 2\theta$ **28.** $r = \csc \theta$ **29.** $r = 1 + \sin \theta$

30. $r = \dfrac{2}{1 - \sin \theta}$

Graph each equation.

31. $r = 7$ **32.** $\theta = -\dfrac{\pi}{4}$ **33.** $r = 5 \cos \theta$

34. $r = 6 + 3 \sin \theta$ **35.** $r = 6 - 3 \sin \theta$ **36.** $r = 4 + 4 \cos \theta$

37. $r = 3 - 6 \cos \theta$ **38.** $r \sin \theta = 5$ **39.** $r^2 = 36 \cos \theta$

40. $r = 6 \sin 5\theta$ **41.** $r^2 = 25 \sin 2\theta$ **42.** $r(1 - \sin \theta) = 6$

Find the area of each region.

43. Inside $r^2 = 4 \cos 2\theta$ **44.** Inside $r = 1 - \sin \theta$

45. Inside $r = 3 - 2 \sin \theta$ **46.** Inside $r = \frac{1}{2} \cos 3\theta$

47. Inside $r = \frac{1}{2} (\theta + \pi)$ and outside $r = \theta$ for $0 \leqslant \theta \leqslant \pi$

48. Inside $r = 2 \cos \theta$ and outside $r = 1$

49. Inside the inner loop of $r = 2 - 4 \cos \theta$

50. Inside $r = 1$ and outside $r = 1 + \sin \theta$

51. Inside $r^2 = 8 \sin 2\theta$ and outside $r = 2$

52. Inside both $r^2 = \cos 2\theta$ and $r^2 = \sin 2\theta$

9

THREE-SPACE

Partial Derivatives
and Double Integrals

9.1 FUNCTIONS IN THREE-SPACE

We have been discussing functions of only one variable whose graphs lie in a plane, which is often called the Euclidean plane or two-space. Now, we need to extend our discussion to functions of two variables whose graphs require three dimensions, sometimes called three-space.

To begin, consider three mutually perpendicular number lines (x-, y-, and z-axes) with their zero points intersecting at point O, called the origin. Although these axes may be oriented any way one wishes, we will use the orientation as shown in Fig. 9.1, which is often called the right-handed system.

Think of the y- and z-axes lying in the plane of the paper with the positive y-direction to the right and the positive z-direction upward. The x-axis is perpendicular to the paper and its positive direction is towards us. This is called a right-handed system because if the fingers on the right hand are cupped so they curve from the positive x-axis to the positive y-axis, the thumb points along the positive z-axis as shown in Fig. 9.2.

The three axes determine three planes (xy-, xz-, and yz-planes), which divide three-space into octants as shown in Fig. 9.3.

Each point in three-space is represented by an ordered triple of numbers in the form (x, y, z), which indicate the directed distances from the three planes as shown in Fig. 9.4.

Example 1

Plot the points $(2, -4, 5)$ and $(-3, 5, -4)$. See Fig. 9.5.

The graph of an equation expressed in three variables is normally a surface. We shall now discuss the graphs of some surfaces on a case-by-case basis.

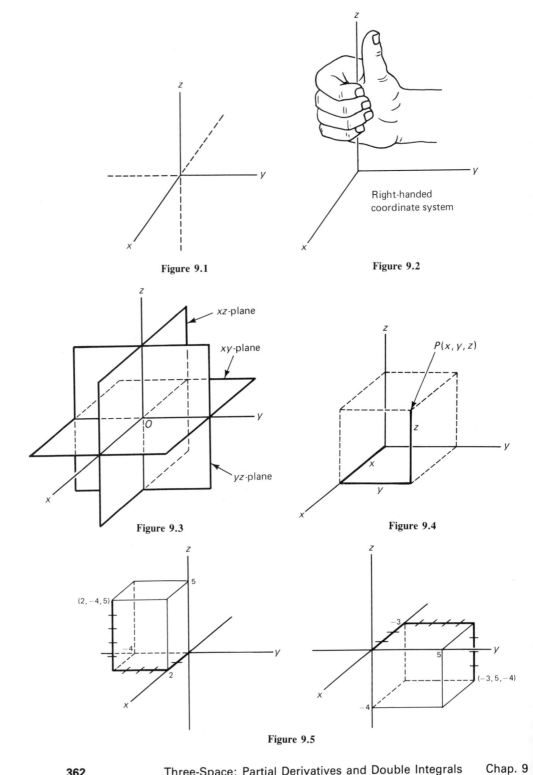

Figure 9.1

Figure 9.2

Right-handed
coordinate system

Figure 9.3

xz-plane

xy-plane

yz-plane

Figure 9.4

$P(x, y, z)$

Figure 9.5

$(2, -4, 5)$

$(-3, 5, -4)$

Three-Space: Partial Derivatives and Double Integrals   Chap. 9

> **Plane**
>
> The graph of an equation in the form
>
> $$Ax + By + Cz + D = 0, \qquad \text{where } A^2 + B^2 + C^2 \neq 0$$
>
> is a plane.

Example 2

Sketch $3x + 6y + 2z = 12$.

   If the plane does not pass through the origin, its sketch may be found by graphing the intercepts. To find the x-intercept, set y and z equal to zero and solve for x: $x = 4$, which corresponds to the point $(4, 0, 0)$. Similarly, the y- and z-intercepts are $(0, 2, 0)$ and $(0, 0, 6)$, respectively. These three points determine the plane. The lines through pairs of these points are in the coordinate planes and are called *traces,* as shown in Fig. 9.6.

Example 3

Sketch $3x + 2y = 12$ in three-space.

   The x- and y-intercepts are $(4, 0, 0)$ and $(0, 6, 0)$, respectively, which determine the trace in the xy-plane. Note that the plane never touches or crosses the z-axis because both x and y cannot be zero. Thus the plane is parallel to the z-axis, as shown in Fig. 9.7

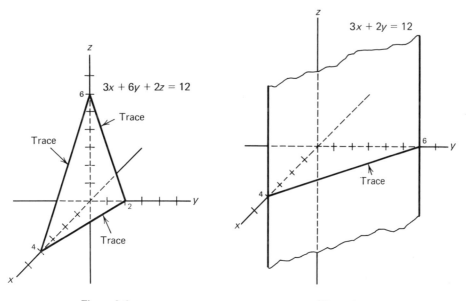

Figure 9.6   Figure 9.7

Example 4

Sketch $x = 3$ in three-space.

   Its x-intercept is $(3, 0, 0)$. The plane is parallel to both the y- and z-axes, as shown in Fig. 9.8.

The formula for the distance between two points in three-space is given in Exercise 41. Using this formula, you should see that the equation of a sphere is given below.

Sphere

A sphere with center at (h, k, l) and radius r (see Fig. 9.9) is given by the equation

$$(x - h)^2 + (y - k)^2 + (z - l)^2 = r^2$$

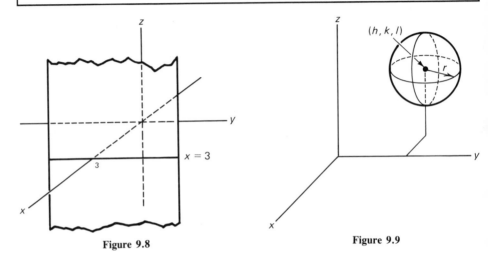

Figure 9.8   Figure 9.9

Example 5

Find the center and radius of the sphere whose equation is

$$x^2 + y^2 + z^2 - 6x + 10y - 16z + 82 = 0$$

and sketch its graph.

First, complete the square on each variable.

$$(x^2 - 6x \quad) + (y^2 + 10y \quad) + (z^2 - 16z \quad) = -82$$
$$(x^2 - 6x + 9) + (y^2 + 10y + 25) + (z^2 - 16z + 64) = -82 + 9 + 25 + 64$$
$$(x - 3)^2 + (y + 5)^2 + (z - 8)^2 = 16$$

The center is at $(3, -5, 8)$ and the radius is 4. (See Fig. 9.10.)

In general, graphing more complex surfaces can be quite complicated. One helpful technique involves graphing the intersections of the surface with the coordinate planes. Such intersections of the surfaces and the coordinate planes are called *traces*.

Cylindrical Surface

The graph in three-space of an equation expressed in only two variables is a cylindrical surface. The surface is parallel to the coordinate axis of the missing variable.

Example 6

Sketch the graph of $y^2 = 4x$ in three-space.

First, graph the trace $y^2 = 4x$ in the xy-plane. Then extend the cylindrical surface parallel to the z-axis as in Fig. 9.11.

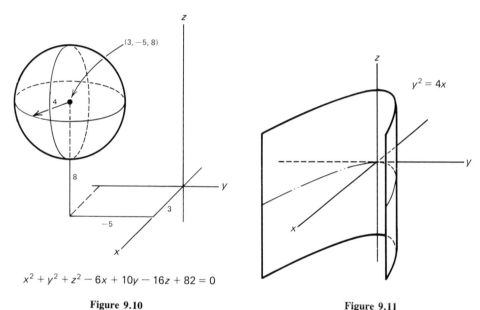

$$x^2 + y^2 + z^2 - 6x + 10y - 16z + 82 = 0$$

Figure 9.10

Figure 9.11

Example 7

Sketch the graph of $z = \sin y$ in three-space.

First, graph the trace $z = \sin y$ in the yz-plane. Then extend the cylindrical surface parallel to the x-axis as in Fig. 9.12.

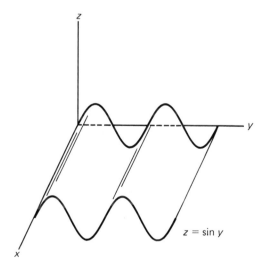

$z = \sin y$

Figure 9.12

Surface	Equation	Traces
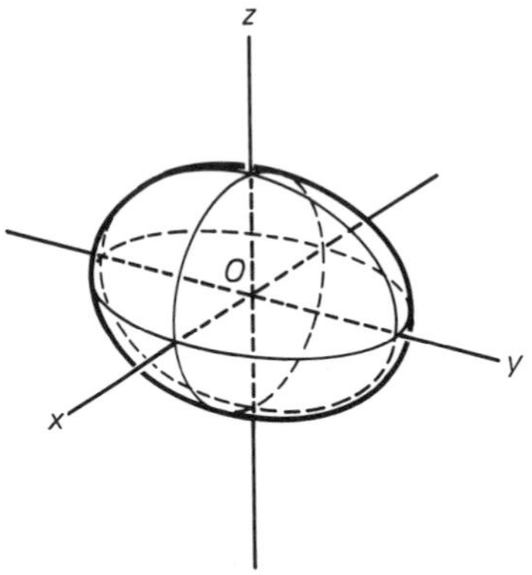 Ellipsoid	$\dfrac{x^2}{a^2} + \dfrac{y^2}{b^2} + \dfrac{z^2}{c^2} = 1$ Note: If $a = b = c$, the surface is a sphere.	xy-plane: ellipse xz-plane: ellipse yz-plane: ellipse All traces in planes parallel to the coordinate planes are ellipses.
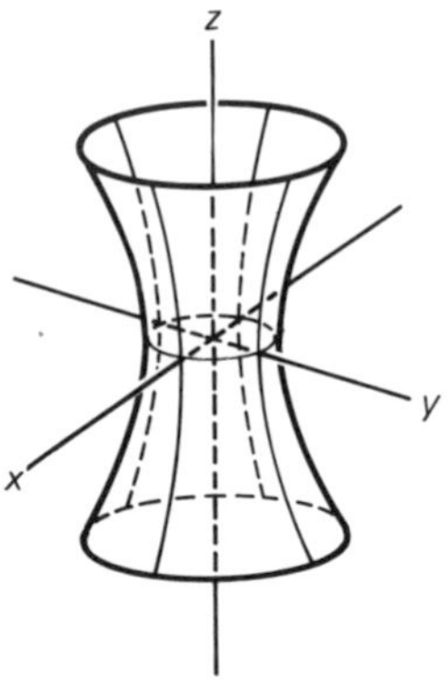 Hyperboloid of one sheet	$\dfrac{x^2}{a^2} + \dfrac{y^2}{b^2} - \dfrac{z^2}{c^2} = 1$ Note: If $a = b$, the surface may be generated by rotating a hyperbola about its conjugate axis.	xy-plane: ellipse xz-plane: hyperbola yz-plane: hyperbola All traces parallel to the xy-plane are ellipses.
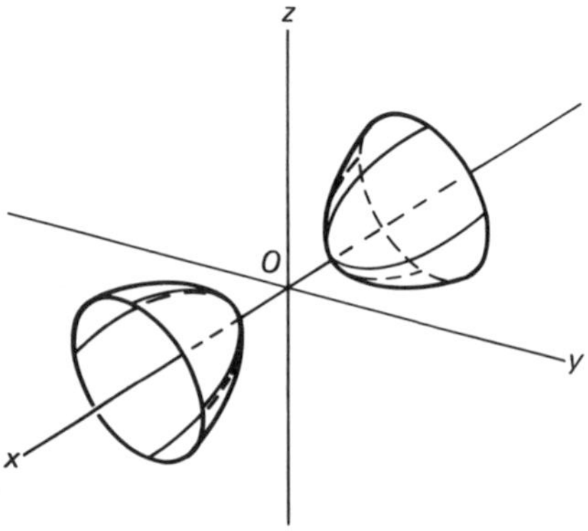 Hyperboloid of two sheets	$\dfrac{x^2}{a^2} - \dfrac{y^2}{b^2} - \dfrac{z^2}{c^2} = 1$ Note: If $b = c$, the surface may be generated by rotating a hyperbola about its transverse axis.	xy-plane: hyperbola xz-plane: hyperbola yz-plane: no trace All traces parallel to the yz-plane that intersect the surface are ellipses.

Figure 9.13

Surface	Equation	Traces
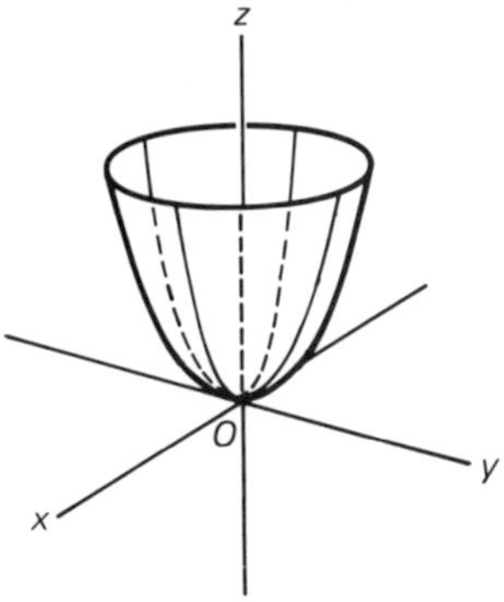 Elliptic paraboloid	$\dfrac{x^2}{a^2} + \dfrac{y^2}{b^2} = \dfrac{z}{c}$ Note: If $a = b$, the surface may be generated by rotating a parabola about its axis.	xy-plane: point xz-plane: parabola yz-plane: parabola If $c > 0$, all traces parallel to and above the xy-plane are ellipses; below the xy-plane, there is no intersection. If $c < 0$, the situation is reversed.
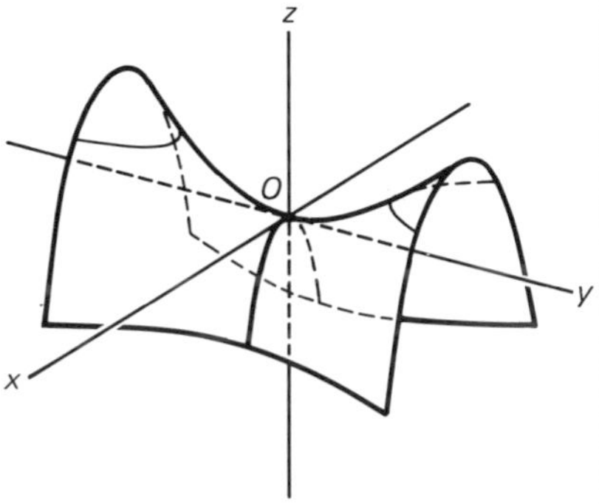 Hyperbolic paraboloid	$\dfrac{x^2}{a^2} - \dfrac{y^2}{b^2} = \dfrac{z}{c}$ Note: The origin is called a saddle point because this surface looks like a saddle.	xy-plane: pair of intersecting lines xz-plane: parabola ⎱ one opens up, yz-plane: parabola ⎰ one opens down Traces parallel to the xy-plane are hyperbolas.
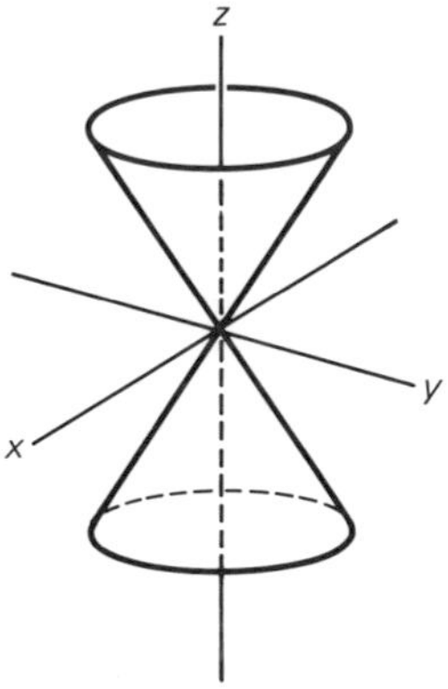 Elliptic cone	$\dfrac{x^2}{a^2} + \dfrac{y^2}{b^2} - \dfrac{z^2}{c^2} = 0$	xy-plane: point xz-plane: pair of intersecting lines yz-plane: pair of intersecting lines Traces parallel to the xy-plane are ellipses. Traces parallel to the xz- and yz-planes are hyperbolas.

Figure 9.13 Continued

Quadric Surface

The graph in three-space of a second-degree equation is a quadric surface. Plane sections, intersections, or slices of the given surface with planes are conics.

The second-degree equation in three-space has the form

$$Ax^2 + By^2 + Cz^2 + Dxy + Exz + Fyz + Gx + Hy + Iz + J = 0$$

One can show that the equations of the basic or central quadrics after translations and/or rotations can be expressed in one of the two forms

$$Ax^2 + By^2 + Cz^2 + J = 0$$

or

$$Ax^2 + By^2 + Iz = 0$$

Figure 9.13 shows six basic types of quadric surfaces.

Example 8

Name and sketch the graph of $\dfrac{x^2}{16} + \dfrac{y^2}{9} - \dfrac{z^2}{36} = 1$.

First, find the traces in the three coordinate planes:

xy-plane: Set $z = 0$, $\dfrac{x^2}{16} + \dfrac{y^2}{9} = 1$, an ellipse

xz-plane: Set $y = 0$, $\dfrac{x^2}{16} - \dfrac{z^2}{36} = 1$, a hyperbola

yz-plane: Set $x = 0$, $\dfrac{y^2}{9} - \dfrac{z^2}{36} = 1$, a hyperbola

This is a hyperboloid of one sheet. (See Fig. 9.14.)

Example 9

Name and sketch the graph of $2x^2 + 2y^2 = 8z$.
First, divide both sides by 2.

$$x^2 + y^2 = 4z$$

Then, find the traces in the three coordinate planes:

xy-plane: Set $z = 0$, $x^2 + y^2 = 0$, the origin (degenerate circle)
xz-plane: Set $y = 0$, $x^2 = 4z$, a parabola
yz-plane: Set $x = 0$, $y^2 = 4z$, a parabola (See Fig. 9.15.)

Exercises

Plot each point.

1. $(2, 4, 5)$ 2. $(3, 5, 2)$ 3. $(3, -2, 4)$
4. $(-2, 0, 3)$ 5. $(-1, 4, -3)$ 6. $(3, -2, -4)$

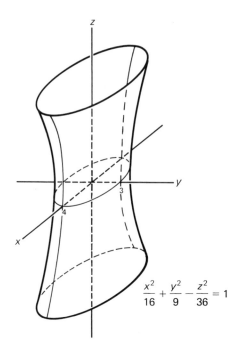

$$\frac{x^2}{16} + \frac{y^2}{9} - \frac{z^2}{36} = 1$$

$$2x^2 + 2y^2 = 8z$$

Figure 9.14

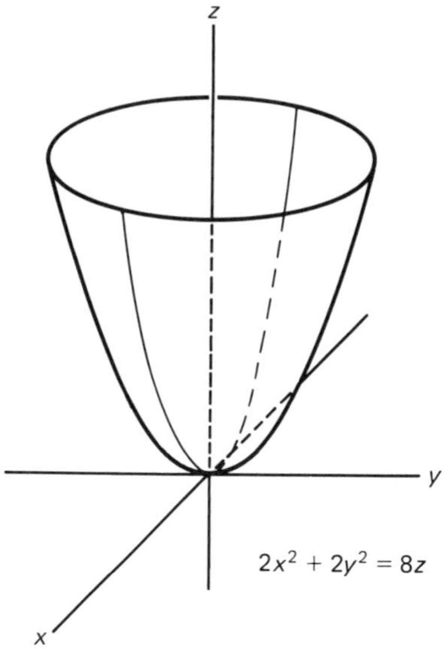

Figure 9.15

Name and sketch the graph of each equation in three-space. Sketch the traces in the coordinate planes when appropriate.

7. $2x + 3y + 8z = 24$ **8.** $6x - 3y + 9z = 18$ **9.** $x + 3y - z = 9$

10. $-4x + 8y + 3z = 12$ **11.** $2x + 5y = 20$ **12.** $4x - 3y = 12$

13. $y = 4$ **14.** $z = -5$

15. $x^2 + y^2 + z^2 = 25$ **16.** $(x - 3)^2 + y^2 + (z - 2)^2 = 36$

17. $x^2 + y^2 + z^2 - 8x + 6y - 4z + 4 = 0$

18. $x^2 + y^2 + z^2 + 6x - 6y + 9 = 0$

19. $z^2 = 8y$ **20.** $x^2 + z^2 = 16$ **21.** $9x^2 + 4y^2 = 36$

22. $y^2 - z^2 = 4$ **23.** $4x^2 + 9y^2 + z^2 = 36$ **24.** $x^2 - y^2 = 16z$

25. $x^2 + y^2 - 4z = 0$ **26.** $y = e^x$

27. $9x^2 - 36y^2 - 4z^2 = 36$ **28.** $16x^2 + 9y^2 - 36z^2 = 144$

29. $y = \cos x$ **30.** $4x^2 + 9y^2 + 9z^2 = 36$

31. $y^2 - z^2 = 8x$ **32.** $y^2 + z^2 = 4x$

33. $4x^2 + 9y^2 - 9z^2 = 0$ **34.** $9x^2 - 16y^2 - 16z^2 = 144$

35. $81y^2 + 36z^2 - 4x^2 = 324$ **36.** $x^2 + y^2 - z^2 = 0$

37. Find the equation of the sphere with center at $(3, -2, 4)$ and radius 6.

38. Find the equation of the sphere with center at $(3, -2, 4)$ and tangent to the xz-plane.

39. Find the equation of the sphere with center at $(3, -2, 4)$ and tangent to the xy-plane.

40. Find the equation of the sphere of radius 4 that is tangent to the three coordinate planes and whose center is in the first octant.

Sec. 9.1   Functions In Three-Space

41. Given two points $P_1(x_1, y_1, z_1)$ and $P_2(x_2, y_2, z_2)$, show that the distance between them is given by

$$d = P_1P_2 = \sqrt{(x_2 - x_1)^2 + (y_2 - y_1)^2 + (z_2 - z_1)^2}$$

Find the distance between each pair of points.

42. (3, 6, 8) and (7, −2, 7)

43. (5, −3, 2) and (3, 1, −2)

44. (−4, 0, 6) and (2, 5, −2)

45. Show that (5, 6, 3), (2, 8, 4), and (3, 5, 6) are vertices of an equilateral triangle.

46. Show that (1, 1, 2), (3, 1, 0), and (5, 3, 2) are vertices of a right triangle. (*Hint:* Use the Pythagorean theorem.)

9.2 PARTIAL DERIVATIVES

Let $z = f(x, y)$ represent a function z with independent variables x and y. That is, there exists a unique value of z for every pair of x and y values.

Partial Derivatives

If $z = f(x, y)$ is a function of two variables, the partial derivatives of z with respect to x (y is treated as a constant) is defined by

$$\frac{\partial z}{\partial x} = \lim_{\Delta x \to 0} \frac{f(x + \Delta x, y) - f(x, y)}{\Delta x}$$

provided that this limit exists.

If $z = f(x, y)$ is a function of two variables, the partial derivative of z with respect to y (x is treated as a constant) is defined by

$$\frac{\partial z}{\partial y} = \lim_{\Delta y \to 0} \frac{f(x, y + \Delta y) - f(x, y)}{\Delta y}$$

provided that this limit exists.

Other common notations for partial derivatives include

$$\frac{\partial z}{\partial x} \quad \text{or} \quad \frac{\partial f}{\partial x} \quad \text{or} \quad f_x \quad \text{or} \quad f_x(x, y)$$

$$\frac{\partial z}{\partial y} \quad \text{or} \quad \frac{\partial f}{\partial y} \quad \text{or} \quad f_y \quad \text{or} \quad f_y(x, y)$$

Note that in the definition of $\frac{\partial z}{\partial x}$, y is treated as a constant (held fixed) and only x is allowed to vary. Think of z as a function of only one variable x and find the usual derivative with respect to x treating y as a constant. For $\frac{\partial z}{\partial y}$, find the usual derivative with respect to y treating x as a constant.

Example 1

If $z = 5x^2 - 4x^2y + y^3$, find $\frac{\partial z}{\partial x}$ and $\frac{\partial z}{\partial y}$.

$$\frac{\partial z}{\partial x} = 10x - 8xy$$

$$\frac{\partial z}{\partial y} = -4x^2 + 3y^2$$

Example 2

If $z = x^2 \ln y - e^{xy}$, find $\dfrac{\partial z}{\partial x}$ and $\dfrac{\partial z}{\partial y}$.

$$\frac{\partial z}{\partial x} = 2x \ln y - ye^{xy}$$

$$\frac{\partial z}{\partial y} = \frac{x^2}{y} - xe^{xy}$$

Example 3

If $z = e^{-x} \sin y + \ln (x^2 + y^2)$, find $\dfrac{\partial z}{\partial x}$ and $\dfrac{\partial z}{\partial y}$.

$$\frac{\partial z}{\partial x} = -e^{-x} \sin y + \frac{2x}{x^2 + y^2}$$

$$\frac{\partial z}{\partial y} = e^{-x} \cos y + \frac{2y}{x^2 + y^2}$$

Now, let's consider a graphical interpretation of partial derivatives. Let $z = f(x, y)$ be a function of x and y. Suppose that we are considering the partial derivative of z with respect to x at the point (x_0, y_0, z_0). We treat y as a constant and take the usual derivative with respect to x; this is graphically equivalent to finding the slope at $x = x_0$ of the curve, which is determined by the intersection of the plane $y = y_0$ and the surface $z = f(x, y)$. (See Fig. 9.16a.)

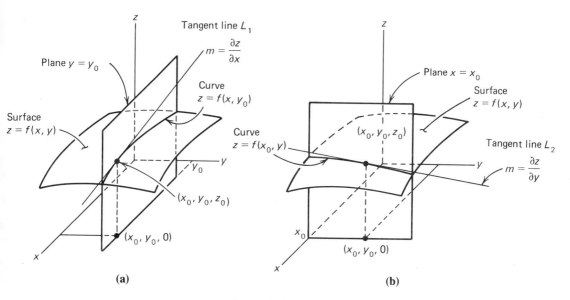

Figure 9.16

Likewise for the partial derivative of z with respect to y; this is graphically equivalent to finding the slope at $y = y_0$ of the curve, which is determined by the intersection of the plane $x = x_0$ and the surface $z = f(x, y)$. (See Fig. 9.16b).

Example 4

Given the surface $z = 4x^2 + y^2$, find the slope of the tangent lines parallel to the xz- and yz-planes and through the point $(2, -1, 9)$.

First,

$$\frac{\partial z}{\partial x} = 8x \quad \text{and} \quad \frac{\partial z}{\partial y} = 2y$$

The slope of each tangent line is the value of each derivative evaluated at the point $(2, -1, 9)$. Thus

$$\frac{\partial z}{\partial x}\bigg|_{(2,-1,9)} = 16 \quad \text{and} \quad \frac{\partial z}{\partial y}\bigg|_{(2,-1,9)} = -2$$

So the slope of the tangent line parallel to the xz-plane is 16 and the slope of the tangent line parallel to the yz-plane is -2.

Example 5

In an RC-circuit, the current I may be given by

$$I = \frac{E}{R} e^{-t/RC}$$

Assume that all quantities are constant except I and R. Find $\dfrac{\partial I}{\partial R}$.

$$\frac{\partial I}{\partial R} = \frac{E}{R} e^{-t/RC} \left(\frac{-t}{C} \cdot \frac{-1}{R^2} \right) + e^{-t/RC} \left(\frac{-E}{R^2} \right)$$

$$= \frac{tE}{CR^3} e^{-t/RC} - \frac{E}{R^2} e^{-t/RC}$$

$$= \frac{E}{R^2} e^{-t/RC} \left(\frac{t}{RC} - 1 \right)$$

Exercises

Find (a) $\dfrac{\partial z}{\partial x}$ and (b) $\dfrac{\partial z}{\partial y}$ for each function.

1. $z = 4x^3y^2$
2. $z = 8\pi x^2 \sqrt{y}$
3. $z = 6x^2y^4 + 2xy^2$

4. $z = 3xy^5 - 4x^2y^2$
5. $z = \sqrt{x^2 + y^2}$
6. $z = \ln \sqrt{x^2 + y^2}$

7. $z = \dfrac{x^2 - y^2}{2xy}$
8. $z = \frac{1}{3}\pi x^2 y$
9. $z = axy$

10. $z = e^{xy} - e^{-x}$
11. $z = \ln \dfrac{x}{y}$
12. $z = (x^2 - y^2)^{-3}$

13. $z = \tan(x - y)$
14. $z = \sin(x - y)$
15. $z = e^{3x} \sin xy$

16. $z = \tan \dfrac{x}{y}$
17. $z = \dfrac{e^{xy}}{x \sin y}$
18. $z = ye^{\sin x}$

19. $z = 2 \sin y \cos x$
20. $z = x^2 \cos 2y$
21. $z = xy \tan xy$

22. $z = x^2y \cos y$

The following formulas are used in electronics and physics. Find the indicated partial derivative.

23. $P = I^2R; \dfrac{\partial P}{\partial I}$

24. $P = \dfrac{V^2}{R}; \dfrac{\partial P}{\partial R}$

25. $I = \dfrac{E}{R + r}; \dfrac{\partial I}{\partial R}$

26. $R = \dfrac{R_1R_2}{R_1 + R_2}; \dfrac{\partial R}{\partial R_1}$

27. $Z = \sqrt{R^2 + X_L^2}; \dfrac{\partial Z}{\partial R}$

28. $Z = \sqrt{R^2 + (X_L - X_C)^2}; \dfrac{\partial Z}{\partial X_L}$

29. $e = E \sin 2\pi ft; \dfrac{\partial e}{\partial t}$

30. $E = I_2R_2 + I_3R_3; \dfrac{\partial E}{\partial I_2}$

31. $E = I_2R_2 + I_2R_3; \dfrac{\partial E}{\partial I_2}$

32. $I = \dfrac{E_2R_1 + E_2R_3 - E_1R_3}{R_1R_2 + R_1R_3 + R_2R_3}; \dfrac{\partial I}{\partial R_1}$

33. $I = \dfrac{E}{R} e^{-t/RC}; \dfrac{\partial I}{\partial t}$

34. $I = \dfrac{E}{R} e^{-t/RC}; \dfrac{\partial I}{\partial C}$

35. $q = CEe^{-t/RC}; \dfrac{\partial q}{\partial C}$

36. $q = CE(1 - e^{-t/RC}); \dfrac{\partial q}{\partial C}$

37. $\tan \phi = \dfrac{X_L}{R}; \dfrac{\partial \phi}{\partial R}$

38. $\tan \phi = \dfrac{X_L - X_C}{R}; \dfrac{\partial \phi}{\partial X_L}$

For each surface, find the slope of the tangent lines parallel to (a) the xz-plane and (b) the yz-plane and through the given point.

39. $z = 9x^2 + 4y^2; (1, -2, 25)$

40. $z = 8y^2 - x^2; (-2, 1, 4)$

41. $z = \sqrt{25x^2 + 36y^2 + 164}; (1, -1, 15)$

42. $z = 6x^2 + 2xy + 4y^2 - 6x + 8y - 2; (1, -2, -6)$

43. The ideal gas law is $PV = nRT$, where P is the pressure, V is the volume, T is the absolute temperature, n is the number of moles of gas, and R is a constant. Show that

$$\frac{\partial P}{\partial V} \cdot \frac{\partial V}{\partial T} \cdot \frac{\partial T}{\partial P} = -1$$

44. There is only one point on the surface $z = x^2 + 3xy + 4y^2 - 10x - 8y + 4 = 0$ at which its tangent plane is horizontal. Find it.

45. The volume of a right circular cylinder is given by $V = \pi r^2 h$. If the height h is fixed at 8 cm, find the rate of change of the volume V with respect to the radius r when $r = 6$ cm.

46. Given that $z = \dfrac{e^{x+y}}{e^x + e^y}$, show that $\dfrac{\partial z}{\partial x} + \dfrac{\partial z}{\partial y} = z$.

47. Given $w = t^2 + \tan te^{1/s}$, show that $s^2 \dfrac{\partial w}{\partial s} + t \dfrac{\partial w}{\partial t} = 2t^2$.

9.3 APPLICATIONS OF PARTIAL DERIVATIVES

Earlier applications of the derivative of a function of one variable may be extended to functions of two variables by using partial derivatives. In Section 3.5 we defined the differential of a function of one variable as

$$dy = f'(x)\, dx$$

Similarly, we define the total differential of a function of two variables as

Total Differential of $z = f(x, y)$
$$dz = \frac{\partial f}{\partial x}\, dx + \frac{\partial f}{\partial y}\, dy$$

The same type of definition may be extended to functions of three or more variables.

Example 1

Find dz for $z = x^3 + 5x^2y - 4xy^2$.

$$dz = \frac{\partial z}{\partial x}\, dx + \frac{\partial z}{\partial y}\, dy$$

$$= (3x^2 + 10xy - 4y^2)\, dx + (5x^2 - 8xy)\, dy$$

In Section 3.5 we used differential approximations to find changes in the function as small changes were made in the variable. Now we can use the total differential to find changes in a function of two variables as small changes are made in one or both variables.

Example 2

The height of a right circular cylinder measures 20.00 cm with a possible error of 0.10 cm, while its radius measures 8.00 cm with a possible error of 0.05 cm. Find the maximum error in its volume.

$$V = \pi r^2 h$$

$$dV = \frac{\partial V}{\partial r}\, dr + \frac{\partial V}{\partial h}\, dh$$

$$= 2\pi rh\, dr + \pi r^2\, dh$$

$$= 2\pi(8.00 \text{ cm})(20.00 \text{ cm})(0.05 \text{ cm}) + \pi(8.00 \text{ cm})^2(0.10 \text{ cm})$$

$$= 70.4 \text{ cm}^3$$

Example 3

The angle of elevation to the top of a monument when measured 150 ft (± 0.5 ft) from its base is $31.00°$ with a maximum error of $0.05°$. Find the maximum error in measuring its height. (See Fig. 9.17.)

$$\tan \theta = \frac{y}{x}$$

$$y = x \tan \theta$$

$$dy = \frac{\partial y}{\partial x}\, dx + \frac{\partial y}{\partial \theta}\, d\theta$$

$$= \tan \theta\, dx + x \sec^2\theta\, d\theta$$

$$= (\tan 31.00°)(0.5 \text{ ft}) + (150 \text{ ft})(\sec 31.00°)^2 \left(0.05° \times \frac{\pi \text{ rad}}{180°}\right)$$

$$= 0.479 \text{ ft}$$

One principal application of the derivative of a function in one variable is finding relative maximums and minimums. This application may also be extended to functions of two variables by using partial derivatives. Geometrically, let's start this discussion with a surface $z = f(x, y)$ as shown in Fig. 9.18. If point P is a relative maximum (or minimum), then every tangent line through P must be horizontal to the xy-plane. That is, $\dfrac{\partial z}{\partial x} = 0$ and $\dfrac{\partial z}{\partial y} = 0$.

Note: *Both* partial derivatives must equal zero. However, this result gives only critical points, that is, only possible relative maximum or minimum values. Unfortunately, analysis of this application is rather sophisticated and is usually found in advanced calculus texts. Here we have included only functions that can be analyzed using the first derivative test along both the x- and y-axes.

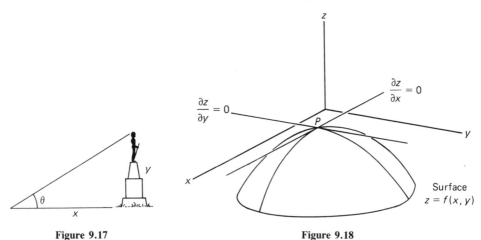

Figure 9.17   Figure 9.18

Example 4

Find any relative maximum or minimum points for $z = x^2 + xy + y^2 - 3x$.
First, find the partial derivatives.

$$\frac{\partial z}{\partial x} = 2x + y - 3 \quad \text{and} \quad \frac{\partial z}{\partial y} = x + 2y$$

Set each expression equal to zero and solve the resulting system of equations.

$$2x + y - 3 = 0$$
$$x + 2y = 0$$

Multiply the second equation by 2 and subtract.

$$2x + y\ = 3$$
$$\underline{2x + 4y = 0}$$
$$-3y = 3$$
$$y = -1 \quad \text{and} \quad x = 2$$

Then

$$z = x^2 + xy + y^2 - 3x$$
$$z = 2^2 + 2(-1) + (-1)^2 - 3(2) = -3$$

Next we need to determine if the point $(2, -1, -3)$ is a maximum or a minimum.

$$\text{For } y = -1, \frac{\partial z}{\partial x} = 2x + y - 3 = 2x + (-1) - 3 = 2x - 4$$

$$\text{For } x < 2, \frac{\partial z}{\partial x} < 0$$

$$\text{For } x > 2, \frac{\partial z}{\partial y} > 0$$

That is, in the plane $y = -1$, the trace changes from decreasing to increasing at $(2, -1, -3)$.

$$\text{For } x = 2, \frac{\partial z}{\partial y} = x + 2y = 2 + 2y$$

$$\text{For } y < -1, \frac{\partial z}{\partial y} < 0$$

$$\text{For } y > -1, \frac{\partial z}{\partial y} > 0$$

Thus, in the plane $x = 2$, the trace changes from decreasing to increasing at the point $(2, -1, -3)$. Since the trace in both planes changes from decreasing to increasing, the point $(2, -1, -3)$ is a relative minimum.

In summary, to find a relative maximum or minimum of a function in two variables in the form $z = f(x, y)$, set both

$$\frac{\partial z}{\partial x} = 0 \quad \text{and} \quad \frac{\partial z}{\partial y} = 0$$

which determine the critical points. Using the first derivative test for both $\frac{\partial z}{\partial x}$ and $\frac{\partial z}{\partial y}$, check each critical point as follows:

(a) If both partial derivatives indicate a change from decreasing to increasing at the critical point, the critical point is a relative minimum.
(b) If both partial derivatives indicate a change from increasing to decreasing at the critical point, the critical point is a relative maximum.
(c) If one partial derivative indicates a change from decreasing to increasing and the other partial derivative indicates a change from increasing to decreasing, the critical point is a saddle point.

Example 5

Find any relative maximum or minimum points for $z = x^2 - y^2 - 6x + 4y$.
First, find the partial derivatives.

$$\frac{\partial z}{\partial x} = 2x - 6 \quad \text{and} \quad \frac{\partial z}{\partial y} = -2y + 4$$

Setting each expression equal to zero, we have

$$\begin{array}{cc} 2x - 6 = 0 & -2y + 4 = 0 \\ x = 3 & y = 2 \end{array}$$

Then,

$$z = x^2 - y^2 - 6x + 4y$$
$$= 3^2 - 2^2 - 6(3) + 4(2) = -5$$

Thus the critical point is $(3, 2, -5)$.

$$\text{For } y = 2, \frac{\partial z}{\partial x} = 2x - 6$$

$$\text{For } x < 3, \frac{\partial z}{\partial x} < 0$$

$$\text{For } x > 3, \frac{\partial z}{\partial x} > 0$$

That is, in the plane $y = 2$, the trace changes from decreasing to increasing at $(3, 2, -5)$.

$$\text{For } x = 3, \frac{\partial z}{\partial y} = -2y + 4$$

$$\text{For } y < 2, \frac{\partial z}{\partial y} > 0$$

$$\text{For } y > 2, \frac{\partial z}{\partial y} < 0$$

That is, in the plane $x = 3$, the trace changes from increasing to decreasing at $(3, 2, -5)$.

Since the trace in one plane changes from decreasing to increasing while the trace in the other plane changes from increasing to decreasing, the point $(3, 2, -5)$ is neither a relative maximum nor a relative minimum. The point $(3, 2, -5)$ is a saddle point.

Example 6

Find the dimensions of a rectangular box, with no top, having a volume of 32 m³ and using the least amount of material.

The area or amount of the material is $A = lw + 2wh + 2lh$. The volume of the box is $V = lwh = 32$. (See Fig. 9.19.) Next, solve this volume equation for h, $h = \frac{32}{lw}$, and substitute into the area equation.

$$A = lw + 2w\left(\frac{32}{lw}\right) + 2l\left(\frac{32}{lw}\right)$$

$$A = lw + \frac{64}{l} + \frac{64}{w}$$

which expresses A as a function of two variables. Now find the partial derivatives.

$$\frac{\partial A}{\partial l} = w - \frac{64}{l^2} \quad \text{and} \quad \frac{\partial A}{\partial w} = l - \frac{64}{w^2}$$

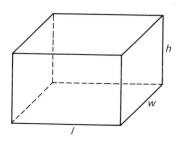

h

w

l

Figure 9.19

Setting each partial derivative equal to zero, we have

$$w - \frac{64}{l^2} = 0 \qquad l - \frac{64}{w^2} = 0$$

$$w = \frac{64}{l^2}$$

Substitute $w = 64/l^2$ into the second equation.

$$l - \frac{64}{w^2} = 0$$

$$l - \frac{64}{(64/l^2)^2} = 0$$

$$l - \frac{l^4}{64} = 0$$

$$l\left(1 - \frac{l^3}{64}\right) = 0$$

Then $l = 0$ or $l = 4$. (*Note:* $l = 0$ is impossible.) When $l = 4$,

$$w = \frac{64}{l^2} = \frac{64}{16} = 4$$

and

$$h = \frac{32}{lw} = \frac{32}{4 \cdot 4} = 2$$

So the dimensions of the box are 4 m × 4 m × 2 m.

Exercises

Find the total differential for each function.

1. $z = 3x^2 + 4xy + y^3$

2. $z = x^2y + 4x^2y^2 - 5x^3y^3$

3. $z = x^2 \cos y$

4. $z = \frac{y}{x} + \ln x$

5. $z = \frac{x - y}{xy}$

6. $z = \text{Arctan } xy$

7. $z = \ln \sqrt{1 + xy}$

8. $z = e^{xy} + \sin xy$

9. The sides of a box are measured to be 26.00 cm × 26.00 cm × 12.00 cm with a maximum error of 0.15 cm on each side. Use a differential to approximate the change in volume.

10. A 6.00-in.-radius cylindrical rod is 2 ft long. Use a differential to approximate how much nickel (in in³) is needed to coat the entire rod with a thickness of 0.12 in.

11. The total resistance R of two resistors R_1 and R_2, connected in parallel, is $R = \frac{R_1 R_2}{R_1 + R_2}$. R_1 measures 400 Ω, with a maximum error of 25 Ω. R_2 measures 600 Ω with a maximum error of 50 Ω. Use a differential to approximate the change in R.

12. The angle of elevation to the top of a hill is 18.00°. The distance to the top of the hill is 450 ± 15 ft. If the maximum error in measuring the angle is 0.5°, find the maximum error in calculating the height of the hill.

13. The height of a right circular cone increases from 21.00 cm to 21.10 cm while the radius decreases from 12.00 cm to 11.85 cm. Use a total differential to approximate the change in volume.

14. The electric current in a circuit containing a variable resistor R is given by $i = 25(1 - e^{-Rt/25})$. Use a total differential to approximate the change in current as R changes from 8.00 Ω to 8.25 Ω and t changes from 4.0 s to 4.1 s.

Find any relative maximum or minimum points or saddle points for each function.

15. $z = 9 + 6x - 8y - 3x^2 - 2y^2$

16. $z = 4x^2 + 3y^2 - 16x - 24y + 5$

17. $z = \dfrac{1}{x} + \dfrac{1}{y} + xy$

18. $z = 60x + 60y - xy - x^2 - y^2$

19. $z = x^2 - y^2 - 2x - 4y - 4$

20. $z = x^2 + y^2 + 2x + 3y + 3$

21. $z = x^3 + y^3 - 3xy + 4$

22. $z = x^3 + y^3 + 3xy + 4$

23. $z = 4x^3 + y^2 - 12x^2 - 36x - 2y$

24. $z = x^3 + y^3 - 3x^2 - 9y^2 - 24x$

25. Find the dimensions of a rectangular box, with no top, having a volume of 500 cm³ and using the least amount of material.

26. A rectangular box with no top and a volume of 6 ft³ needs to be built from material that costs $6/ft² for the bottom, $2/ft² for the front and back, and $1/ft² for the sides. Find the dimensions of the box in order to minimize the cost.

27. Find three positive numbers whose sum is 30 and whose product is a maximum.

28. Find three positive numbers whose sum is 30 and whose sum of squares is a minimum.

9.4 DOUBLE INTEGRALS

In Section 9.2 on partial differentiation, we showed how to differentiate functions of two variables by differentiating with respect to one variable at a time while holding the other variable constant. Next, we need to consider the inverse process of partial differentiation of a function of two variables, that is, if $z = f(x, y)$ and we integrate by first holding either variable x or y constant and integrate with respect to the other. Then we integrate with respect to the variable first held constant. This is called double integration or a double integral.

Double Integral

$$\int_a^b \left[\int_{g(x)}^{G(x)} f(x, y)\, dy \right] dx = \int_a^b \int_{g(x)}^{G(x)} f(x, y)\, dy\, dx$$

$$\int_c^d \left[\int_{h(y)}^{H(y)} f(x, y)\, dx \right] dy = \int_c^d \int_{h(y)}^{H(y)} f(x, y)\, dx\, dy$$

Note: The brackets are normally not included when writing a double integral. To integrate the double integral, $\displaystyle\int_a^b \int_{g(x)}^{G(x)} f(x, y)\, dy\, dx$:

1. Integrate $f(x, y)$ with respect to y holding x constant.
2. Evaluate this integral by substituting the limits on the inner or right integral. These limits are either numbers or functions of x.
3. Integrate the result from step 2 with respect to x.

4. Evaluate this integral by substituting the remaining limits (those on the outer or left integral), which are numerical values or limits of x.

Example 1

Evaluate $\displaystyle\int_0^1 \int_x^{3x^2} (6xy - x)\, dy\, dx$.

First, integrate with respect to y holding x constant.

$$\int_0^1 \int_x^{3x^2} (6xy - x)\, dy\, dx = \int_0^1 (3xy^2 - xy)\Big|_x^{3x^2} dx$$

$$= \int_0^1 \{[3x(3x^2)^2 - x(3x^2)] - [3x(x^2) - x(x)]\}\, dx$$

$$= \int_0^1 [27x^5 - 3x^3 - 3x^3 + x^2]\, dx$$

$$= \int_0^1 [27x^5 - 6x^3 + x^2]\, dx$$

$$= \frac{9x^6}{2} - \frac{3x^4}{2} + \frac{x^3}{3}\Big|_0^1$$

$$= \left(\frac{9}{2} - \frac{3}{2} + \frac{1}{3}\right) - (0)$$

$$= \frac{10}{3}$$

Example 2

Evaluate $\displaystyle\int_0^4 \int_{-y}^{3y} \sqrt{x + y}\, dx\, dy$.

First, integrate with respect to x holding y constant.

$$\int_0^4 \int_{-y}^{3y} \sqrt{x + y}\, dx\, dy = \int_0^4 \frac{2}{3}(x + y)^{3/2}\Big|_{-y}^{3y} dy$$

$$= \int_0^4 \left[\frac{2}{3}(3y + y)^{3/2} - \frac{2}{3}(-y + y)^{3/2}\right] dy$$

$$= \int_0^4 \frac{2}{3}(4y)^{3/2}\, dy$$

$$= \int_0^4 \frac{16}{3} y^{3/2}\, dy \qquad (4^{3/2} = 8)$$

$$= \frac{16}{3}\frac{y^{5/2}}{5/2}\Big|_0^4$$

$$= \frac{16}{3} \cdot \frac{2}{5}(4^{5/2} - 0)$$

$$= \frac{16}{3} \cdot \frac{2}{5} \cdot 32$$

$$= \frac{1024}{15}$$

Three-Space: Partial Derivatives and Double Integrals   Chap. 9

Geometrically, a double integral may be defined in terms of the volume under a curve. Let $z = f(x, y)$ be a continuous function whose surface is shown in Fig. 9.20. Let R be a region in the xy-plane bounded by $x = a$, $x = b$, $y = G(x)$, and $y = g(x)$.

First, divide region R into n rectangles, each of which has dimensions Δx and Δy or dx and dy. Note we have chosen to draw a typical rectangle. Next, construct a box with the rectangle as the base and z or $f(x, y)$ as its height. The volume of this typical volume element can be approximated by

$$z \, \Delta x \, \Delta y$$

Now let $n \to \infty$, where n is the number of such rectangles and the number of volume elements. Then we have

$$V = \int_a^b \int_{g(x)}^{G(x)} f(x, y) \, dy \, dx$$

Volume of a Solid

Let R be a region in the xy-plane bounded by $x = a$, $x = b$, $y = G(x)$, and $y = g(x)$. The volume of the solid between R and the surface $z = f(x, y)$ is

$$\int_a^b \int_{g(x)}^{G(x)} f(x, y) \, dy \, dx$$

Similarly, let R be a region in the xy-plane bounded by $y = c$, $y = d$, $x = H(y)$, and $x = h(y)$. The volume of the solid between R and the surface $z = f(x, y)$ is

$$\int_c^d \int_{h(y)}^{H(y)} f(x, y) \, dx \, dy$$

Example 3

Find the volume of the solid bounded by the plane $2x + 3y + z = 6$ and the coordinate planes.

First, make a sketch as in Fig. 9.21.

The region R is bounded in the xy-plane by $x = 0$, $y = 0$, and $2x + 3y = 6$ or

$$y = \frac{6 - 2x}{3}.$$

Next, you need to determine the variable and constant limits for the double integral.

$$\text{constant limits for } x: \quad 0 \leqslant x \leqslant 3$$

$$\text{variable limits for } y: \quad 0 \leqslant y \leqslant \frac{6 - 2x}{3}$$

The volume is then

$$V = \int_0^3 \int_0^{(6-2x)/3} (6 - 2x - 3y) \, dy \, dx$$

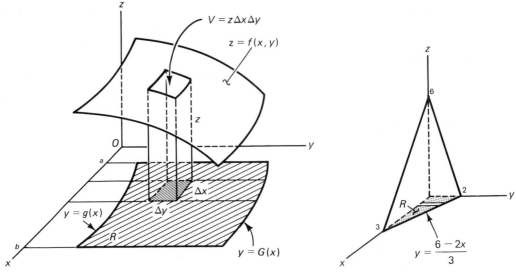

Figure 9.20   **Figure 9.21**

$$= \int_0^3 \left(6y - 2xy - \frac{3y^2}{2} \right) \Bigg|_0^{(6-2x)/3} dx$$

$$= \int_0^3 \left\{ \left[6\left(\frac{6-2x}{3} \right) - 2x\left(\frac{6-2x}{3} \right) - \frac{3}{2}\left(\frac{6-2x}{3} \right)^2 \right] - [0] \right\} dx$$

$$= \int_0^3 \left(12 - 4x - 4x + \frac{4}{3}x^2 - 6 + 4x - \frac{2}{3}x^2 \right) dx$$

$$= \int_0^3 \left(6 - 4x + \frac{2}{3}x^2 \right) dx$$

$$= \left(6x - 2x^2 + \frac{2}{9}x^3 \right) \Bigg|_0^3$$

$$= (18 - 18 + 6) - (0)$$

$$= 6$$

Note: Since we have a pyramid, we can check this result using the formula

$$V = \frac{1}{3} Bh, \qquad \text{where } B \text{ is the area of the base and } h \text{ is the height of the pyramid}$$

$$B = \frac{1}{2}(3)(2) = 3$$

$$V = \frac{1}{3}(3)(6) = 6$$

What happens if you reverse the order of the integration in Example 3?

constant limits for y: $0 \leqslant y \leqslant 2$

variable limits for x: $0 \leqslant x \leqslant \dfrac{6 - 3y}{2}$

Then the double integral is

$$V = \int_0^2 \int_0^{(6-3y)/2} (6 - 2x - 3y)\, dx\, dy$$

Show that this double integral also gives 6 as a result.

Example 4

Find the volume of the solid bounded by the cylinder $x^2 + y^2 = 4$, below the plane $y = z$, and above the xy-plane.

Region R as in Fig. 9.22 is bounded in the xy-plane by $x^2 + y^2 = 4$ and $y = 0$.

$$\text{constant limits for } x: \quad -2 \leqslant x \leqslant 2$$
$$\text{variable limits for } y: \quad 0 \leqslant y \leqslant \sqrt{4 - x^2}$$

Note that the height of a typical volume element is $z = y$.

The volume is then

$$V = \int_{-2}^2 \int_0^{\sqrt{4-x^2}} y\, dy\, dx$$

$$= \int_{-2}^2 \frac{y^2}{2}\bigg|_0^{\sqrt{4-x^2}} dx$$

$$= \int_{-2}^2 \left(\frac{4 - x^2}{2}\right) dx$$

$$= \frac{1}{2}\left(4x - \frac{x^3}{3}\right)\bigg|_{-2}^2$$

$$= \frac{1}{2}\left(8 - \frac{8}{3}\right) - \frac{1}{2}\left(-8 + \frac{8}{3}\right)$$

$$= \frac{8}{3} + \frac{8}{3} = \frac{16}{3}$$

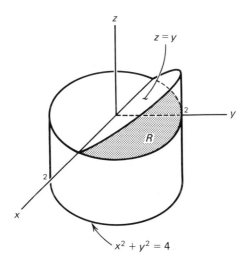

Figure 9.22

Exercises

Evaluate each double integral.

1. $\int_0^2 \int_0^{3x} (x + y) \, dy \, dx$

2. $\int_0^3 \int_0^{1-y} (x + y) \, dx \, dy$

3. $\int_0^1 \int_y^{2y} (3x^2 + xy) \, dx \, dy$

4. $\int_0^2 \int_0^x (4x^2 - 3xy^2 + 8y^3) \, dy \, dx$

5. $\int_{-1}^1 \int_x^{x^2} (4xy + 9x^2 + 6y) \, dy \, dx$

6. $\int_0^1 \int_{x^2}^{3x+1} x \, dy \, dx$

7. $\int_0^1 \int_0^{\sqrt{1-y^2}} (x + y) \, dx \, dy$

8. $\int_0^2 \int_0^{\sqrt{4-y^2}} \frac{2}{\sqrt{4 - y^2}} \, dx \, dy$

9. $\int_0^1 \int_0^x e^{x+y} \, dy \, dx$

10. $\int_0^1 \int_0^{x^3} e^{y/x} \, dy \, dx$

11. $\int_0^3 \int_0^{4y} \sqrt{y^2 + 16} \, dx \, dy$

12. $\int_0^1 \int_0^{4x} \frac{1}{x^2 + 1} \, dy \, dx$

13. $\int_0^{\pi/2} \int_0^x \cos x \sin y \, dy \, dx$

14. $\int_0^\pi \int_0^x x \sin y \, dy \, dx$

Find the volume of each solid bounded by the given surfaces.

15. $3x + y + 6z = 12$ and the coordinate planes

16. $3x + 4y + 2z = 12$ and the coordinate planes

17. $z = xy$, $y = x$, $x = 2$ (first octant)

18. $x^2 + y^2 = 4$, $z = x + y$ (first octant)

19. $x^2 + y^2 = 4$, $z = 2x + 3y$, and above xy-plane

20. $x^2 + y^2 = 1$ and $x^2 + z^2 = 1$

CHAPTER SUMMARY

1. The graph of an equation expressed in three variables or in the form $z = f(x, y)$ is normally a surface.

2. The graph of an equation in the form
$$Ax + By + Cz + D = 0, \qquad \text{where } A^2 + B^2 + C^2 \neq 0$$
is a plane.

3. A sphere with center at (h, k, l) and radius r is given by the equation
$$(x - h)^2 + (y - k)^2 + (z - l)^2 = r^2$$

4. The graph in three-space of an equation expressed in only two variables is a cylindrical surface. The surface is parallel to the coordinate axis of the missing variable.

5. The graph in three-space of a second-degree equation is a quadric surface. Plane sections, intersections, or slices of the given surface with planes are

conics. The graphs, equations, and trace information of six basic quadric surfaces are given on pages 366 and 367.

6. The distance between two points $P_1(x_1, y_1, z_1)$ and $P_2(x_2, y_2, z_2)$ is given by

$$d = P_1P_2 = \sqrt{(x_2 - x_1)^2 + (y_2 - y_1)^2 + (z_2 - z_1)^2}$$

7. *Partial derivatives:* If $z = f(x, y)$ is a function of two variables, the partial derivative of z with respect to x (y is treated as a constant) is defined by

$$\frac{\partial z}{\partial x} = \lim_{\Delta x \to 0} \frac{f(x + \Delta x, y) - f(x, y)}{\Delta x}$$

provided that this limit exists.

If $z = f(x, y)$ is a function of two variables, the partial derivative of z with respect to y (x is treated as a constant) is defined by

$$\frac{\partial z}{\partial y} = \lim_{\Delta y \to 0} \frac{f(x, y + \Delta y) - f(x, y)}{\Delta y}$$

provided that this limit exists.

8. *Total differential of $z = f(x, y)$*

$$dz = \frac{\partial f}{\partial x} dx + \frac{\partial f}{\partial y} dy$$

9. To find a relative maximum or minimum of a function in two variables in the form $z = f(x, y)$, set both

$$\frac{\partial z}{\partial x} = 0 \quad \text{and} \quad \frac{\partial z}{\partial y} = 0$$

which determine the critical points. Using the first derivative test for both $\frac{\partial z}{\partial x}$ and $\frac{\partial z}{\partial y}$, check each critical point as follows:

(a) If both partial derivatives indicate a change from decreasing to increasing at the critical point, the critical point is a relative minimum.
(b) If both partial derivatives indicate a change from increasing to decreasing at the critical point, the critical point is a relative maximum.
(c) If one partial derivative indicates a change from decreasing to increasing and the other partial derivative indicates a change from increasing to decreasing, the critical point is a saddle point.

10. To integrate the double integral $\int_a^b \int_{g(x)}^{G(x)} f(x, y) \, dy \, dx$:

(a) Integrate $f(x, y)$ with respect to y holding x constant.
(b) Evaluate this integral by substituting the limits on the inner or right integral. These limits are either numbers or functions of x.
(c) Integrate the result from step (b) with respect to x.
(d) Evaluate this integral by substituting the remaining limits (those on the outer or left integral), which are numerical values or limits of x.

11. *Volume of a solid:* Let R be a region in the xy-plane bounded by $x = a$,

$x = b$, $y = G(x)$, and $y = g(x)$. The volume of the solid between R and the surface $z = f(x, y)$ is

$$\int_a^b \int_{g(x)}^{G(x)} f(x, y) \, dy \, dx$$

Similarly, let R be a region in the xy-plane bounded by $y = c$, $y = d$, $x = H(y)$, and $x = h(y)$. The volume of the solid between R and the surface $z = f(x, y)$ is

$$\int_c^d \int_{h(y)}^{H(y)} f(x, y) \, dx \, dy$$

CHAPTER REVIEW

Name and sketch the graph of each equation in three-space. Sketch the traces in the coordinate planes when appropriate.

1. $3x + 6y + 4z = 36$
2. $9x^2 + 9y^2 = 3z$
3. $36y^2 + 9z^2 - 16x^2 = 144$
4. $16x^2 + 9y^2 + 36z^2 = 144$
5. $x^2 = -12y$
6. $9x^2 - 9y^2 = 3z$
7. $9y^2 - 36x^2 - 16z^2 = 144$
8. $9x^2 + 36y^2 - 16z^2 = 0$
9. Find the equation of the sphere with the center at $(-3, 2, 1)$ and tangent to the yz-plane.
10. Find the distance between the points $(1, 2, 9)$ and $(4, -2, 4)$.

Find (a) $\dfrac{\partial z}{\partial x}$ and (b) $\dfrac{\partial z}{\partial y}$ for each function.

11. $z = x^3 + 3x^2y + 2y^2$
12. $z = 3x^2 e^{2y}$
13. $z = \ln (3x^2 y)$
14. $z = \sin 3x \sin 3y$
15. $z = \dfrac{y \, e^{x^2}}{x \ln y}$
16. $z = \dfrac{e^x \sin y}{y \sin x}$

Find the indicated partial derivative.

17. $v = \sqrt{\dfrac{P}{w}}; \ \dfrac{\partial v}{\partial w}$
18. $U = \dfrac{1}{2} LI^2; \ \dfrac{\partial U}{\partial I}$
19. $Z = \sqrt{R^2 + X_C^2}; \ \dfrac{\partial Z}{\partial X_C}$
20. $V = \dfrac{RE}{R + r}; \ \dfrac{\partial V}{\partial R}$
21. Find the slope of the tangent lines to $z = y^2 + xy + 3x^2 - 4x$ parallel to (a) the xz-plane and (b) the yz-plane and through the point $(2, 0, 4)$.

Find the total differential for each function.

22. $z = \ln \sqrt{x^2 + xy}$
23. $z = \dfrac{x - y}{x + y}$
24. A cylindrical tank has a radius of 2.000 m and a length of 10.000 m. Use a total differential to approximate the amount of paint (in litres) needed to paint one coat a thickness of 3 mm over its total exterior.
25. An electric current varies according to $i = 50(1 - e^{-t/50R})$. Use a total differential to approximate the change in current as R changes from 150 Ω to 160 Ω and t changes from 6.0 s to 6.5 s.

Find any relative maximum or minimum points or saddle points for each function.

26. $z = x^2 + 2xy - y^2 - 14x - 6y + 8$ **27.** $z = y^2 - xy - x^2 + 4x - 3y - 6$

28. $z = -x^2 + xy - y^2 + 4x - 8y + 9$

29. A rectangular box with no top needs to be built to hold 8 m³. The bottom costs twice as much as the material for the sides. Find the dimensions of the box in order to minimize the cost.

Evaluate each double integral.

30. $\displaystyle\int_0^1 \int_0^{x^2} (x - 2y)\, dy\, dx$

31. $\displaystyle\int_0^2 \int_y^{4y} (4xy + 6x^2 - 9y^2)\, dx\, dy$

32. $\displaystyle\int_1^3 \int_0^{\ln y} ye^x\, dx\, dy$

33. $\displaystyle\int_0^{\pi/4} \int_0^x \sec^2 y\, dy\, dx$

Find the volume of each solid bounded by the given surfaces.

34. $ax + by + cz = d$, $a > 0$, $b > 0$, $c > 0$, and the coordinate planes.

35. $x^2 + y^2 = 1$, $z = 2x$ (first octant)

36. $z = 1 - y - x^2$ (first octant)

10

SERIES

10.1 ARITHMETIC AND GEOMETRIC PROGRESSIONS

Many technical applications make use of infinite series. One in particular, the Fourier series, is very useful in the field of electronics. An infinite series is the summation of an infinite sequence of numbers. We will first look, however, at sums of finite sequences.

A *finite sequence* consists of a succession of quantities $a_1, a_2, a_3, \ldots,$ a_n, where the three dots represent the quantities or terms between a_3 and a_n. The term a_n is called the nth term or last term. An *infinite sequence* is a succession of quantities which continue indefinitely and may be written

$$a_1, a_2, a_3, \ldots, a_n, \ldots \text{ or } a_1, a_2, a_3, \ldots.$$

A basic example of an infinite sequence is an arithmetic progression. An *arithmetic progression* is a sequence of terms where each term differs from the immediately preceding term by a fixed number, d, which is called the *common difference* of the progression. For example, the sequence 5, 10, 15, 20, 25, 30, ... is an arithmetic progression with a common difference of 5, and the sequence $1, \frac{1}{3}, -\frac{1}{3}, -1, -1\frac{2}{3}, \ldots$ is an arithmetic progression with a common difference of $-\frac{2}{3}$.

In any arithmetic progression if a is the first term and d is the common difference, then the second term would be $a + d$. Likewise, the third term would be $a + 2d$ and the fourth term would be $a + 3d$. Continuing in this manner we could express the arithmetic progression as follows:

$$a, a + d, a + 2d, a + 3d, \ldots$$

Note that the first n terms of an arithmetic progression, an example of a finite sequence, can be written as

$$a, a + d, a + 2d, a + 3d, \ldots, a + (n - 1)d$$

The nth or last term l of such a finite arithmetic progression is then given by

$$\boxed{l = a + (n - 1)d}$$

Example 1

Find the 6th term of the arithmetic progression 5, 9, 13, ...
The common difference is the difference between *any* term and the preceding term, that is,

$$d = 9 - 5 = 13 - 9 = 4, \qquad a = 5 \quad \text{and} \quad n = 6$$

We then have

$$l = a + (n - 1)d$$
$$= 5 + (6 - 1)(4)$$
$$= 25$$

Example 2

Find the 22nd term of the arithmetic progression $1, \frac{1}{3}, -\frac{1}{3}, \ldots$.
Since $d = \frac{1}{3} - 1 = -\frac{2}{3}$, $a = 1$ and $n = 22$, we have

$$l = a + (n - 1)d$$
$$= 1 + (22 - 1)(-\tfrac{2}{3})$$
$$= -13$$

To find the sum S of the first n terms of an arithmetic progression, we note that the first n terms can be written as an expression involving l instead of a:

$$a, \ldots, l - 2d, l - d, l$$

In fact, we can indicate the sum S of these terms by

$$S = a + (a + d) + (a + 2d) + \cdots + (l - 2d) + (l - d) + l$$

or

$$S = l + (l - d) + (l - 2d) + \cdots + (a + 2d) + (a + d) + a$$

where the terms of the last equation are written in reverse order.
If we add these two equations for S, we have

$$2S = (a + l) + (a + l) + (a + l) + \cdots + (a + l) + (a + l) + (a + l)$$

since, term by term, the multiples of d add to zero. Note that we obtain a sum of n terms of the form $(a + l)$. That is,

$$2S = n(a + l)$$

or

$$\boxed{S = \frac{n}{2}(a + l)}$$

Example 3

Find the sum of the first 12 terms of the arithmetic progression 6, 11, 16, ...

Since $d = 11 - 6 = 5$, $a = 6$, and $n = 12$, we have

$$l = a + (n - 1)d$$
$$l = 6 + (12 - 1)(5)$$
$$= 61$$

$$S = \frac{n}{2}(a + l)$$
$$= \tfrac{12}{2}(6 + 61)$$
$$= 402$$

Example 4

Find the sum of the first 500 positive integers.

Since $a = 1$, $d = 1$, $n = 500$, and $l = 500$, we have

$$S = \frac{n}{2}(a + l)$$
$$= (\tfrac{500}{2})(1 + 500)$$
$$= 125{,}250$$

A geometric progression is another example of a sequence. A *geometric progression* is a sequence of terms each of which can be obtained by multiplying the preceding term by a fixed number r, which is called the *common ratio*. For example, 1, $\tfrac{1}{2}$, $\tfrac{1}{4}$, $\tfrac{1}{8}$, . . . is a geometric progression with a common ratio of $\tfrac{1}{2}$ and -6, -18, -54, -162, . . . is a geometric progression with a common ratio of 3.

For any geometric progression if a is the first term and r is the common ratio, then ar would be the second term. Likewise, $(ar)r = ar^2$ would be the third term and $(ar^2)r = ar^3$ would be the fourth term. Continuing in this manner we could express a geometric progression as follows:

$$a, ar, ar^2, ar^3, \ldots ar^{n-1}, \ldots$$

where ar^{n-1} is the nth or last term, l, of the progression. That is,

$$\boxed{l = ar^{n-1}}$$

Example 5

Find the 8th term of the geometric progression 1, $\tfrac{1}{2}$, $\tfrac{1}{4}$, $\tfrac{1}{8}$, ...

The common ratio is found by dividing *any* term by the preceding term. So

$$r = \frac{\tfrac{1}{2}}{1} = \frac{\tfrac{1}{4}}{\tfrac{1}{2}} = \frac{\tfrac{1}{8}}{\tfrac{1}{4}} = \tfrac{1}{2}$$

Since $a = 1$ and $n = 8$, we have

$$l = ar^{n-1}$$
$$= (1)(\tfrac{1}{2})^{8-1}$$
$$= \tfrac{1}{128}$$

Example 6

Find the 10th term of the geometric progression $3, -6, 12, -24, \ldots$

In this example $a = 3$, $r = -6/3 = -2$, and $n = 10$.

$$l = ar^{n-1}$$
$$= (3)(-2)^{10-1}$$
$$= -1536$$

To find the sum S of the first n terms of a geometric progression note that

$$S = a + ar + ar^2 + ar^3 + \cdots + ar^{n-2} + ar^{n-1}$$

Multiplying each side by r, we have

$$rS = \quad ar + ar^2 + ar^3 + ar^4 + \cdots + \quad ar^{n-1} + ar^n$$

If we subtract the last equation from the first, we obtain

$$S - rS = a - ar^n$$

Solving for S, we have

$$S(1 - r) = a(1 - r^n)$$

or

$$\boxed{S = \frac{a(1 - r^n)}{1 - r}}$$

Example 7

Find the sum of the first eight terms of the geometric progression $1, \frac{1}{2}, \frac{1}{4}, \frac{1}{8}, \ldots$

Since $r = \frac{1}{2}$, $a = 1$, and $n = 8$, we have

$$S = \frac{a(1 - r^n)}{1 - r}$$
$$= \frac{(1)[1 - (\frac{1}{2})^8]}{1 - \frac{1}{2}}$$
$$= \frac{1 - \frac{1}{256}}{\frac{1}{2}} = \frac{255}{128}$$

Example 8

Find the sum of the first five terms of the geometric progression $2, -\frac{2}{3}, \frac{2}{9}, -\frac{2}{27}, \ldots$

In this example $a = 2$, $r = -\frac{1}{3}$, and $n = 5$.

$$S = \frac{a(1 - r^n)}{1 - r}$$
$$= \frac{(2)[1 - (-\frac{1}{3})^5]}{1 + \frac{1}{3}}$$
$$= \frac{(2)(1 + \frac{1}{243})}{\frac{4}{3}} = \frac{122}{81}$$

Example 9

If $3000 is deposited annually in a savings account at 8% interest compounded annually, find the total amount in this account after 4 years.

The total amount in the account is the sum of a geometric progression:

$$(3000)(1.08) + (3000)(1.08)^2 + (3000)(1.08)^3 + (3000)(1.08)^4$$

since the value of each dollar in the account increases by 8% each year. Note that the first term $(3000)(1.08) = \$3240$ represents the amount in the account after one year. Thus $a = 3240$, $r = 1.08$, and $n = 4$.

$$S = \frac{a(1 - r^n)}{1 - r}$$

$$= \frac{(3240)(1 - 1.08^4)}{1 - 1.08}$$

$$= \$14{,}600$$

In Example 7 we found that the sum of the first eight terms of the geometric progression $1, \frac{1}{2}, \frac{1}{4}, \frac{1}{8}, \ldots$ is $\frac{255}{128}$. The sum of the first nine terms can be shown to be $\frac{511}{256}$ and the sum of the first ten terms is $\frac{1023}{512}$. This last sum is close to the value 2. In fact, the sum of the first 50 terms is given by

$$S = \frac{(1)[1 - (\frac{1}{2})^{50}]}{1 - \frac{1}{2}} = \frac{1 - (\frac{1}{2})^{50}}{\frac{1}{2}} = 2[1 - (\frac{1}{2})^{50}]$$

But since $(\frac{1}{2})^{50} = \frac{1}{1,125,899,906,842,624}$, which is practically zero, we conclude that the sum S is very close to the value 2. The sum S gets closer and closer to 2 as n is given a larger and larger value.

If we denote the sum of the first n terms of a geometric progression by S_n, then we write $\lim_{n \to \infty} S_n = 2$ to express the result above that the sum of the first n terms of the geometric progression $1, \frac{1}{2}, \frac{1}{4}, \frac{1}{8}, \ldots$ approaches 2 as a limiting value as the number of terms n approaches infinity (∞).

The term *series* is used to denote the sum of a sequence of terms. Each S_n is thus a finite series. The methods of computing the sums S_n of these finite arithmetic and geometric series have already been shown.

An infinite series is the indicated sum of an infinite sequence of terms. For example, $1 + \frac{1}{2} + \frac{1}{4} + \frac{1}{8} + \cdots$ is an infinite series. Since it is the infinite summation of the terms of a geometric sequence, it is called an *infinite geometric series*.

We have seen that for this geometric series

$$\lim_{n \to \infty} S_n = 2$$

That is, the sum of $1 + \frac{1}{2} + \frac{1}{4} + \frac{1}{8} + \cdots = 2$.

Not every infinite series has a finite sum. In general, we call S the sum of an infinite series $a_1 + a_2 + a_3 + \cdots + a_n + \cdots$ if

$$S = \lim_{n \to \infty} S_n$$

where $S_n = a_1 + a_2 + a_3 + \cdots + a_n$, the sum of the first n terms. When the sum exists, the series is said to *converge*. When the limit does not exist, the series is said to *diverge*.

Example 10

Find the sum of the infinite geometric series

$$2 + \tfrac{2}{3} + \tfrac{2}{9} + \cdots + 2(\tfrac{1}{3})^{n-1} + \cdots$$

The common ratio is $r = \tfrac{1}{3}$ and $a = 2$. Then

$$S_n = 2 + \tfrac{2}{3} + \tfrac{2}{9} + \cdots + 2(\tfrac{1}{3})^{n-1}$$

$$= \frac{2[1 - (\tfrac{1}{3})^n]}{1 - \tfrac{1}{3}}$$

$$= \frac{2[1 - (\tfrac{1}{3})^n]}{\tfrac{2}{3}}$$

$$= 3[1 - (\tfrac{1}{3})^n]$$

As n approaches ∞, $(\tfrac{1}{3})^n = \dfrac{1}{3^n}$ approaches 0 and $3[1 - (\tfrac{1}{3})^n]$ approaches $3(1 - 0)$

$$= 3.$$

So $\lim\limits_{n \to \infty} S_n = 3$ and we write

$$2 + \tfrac{2}{3} + \tfrac{2}{9} + \cdots + 2(\tfrac{1}{3})^{n-1} + \cdots = 3$$

Example 11

Find, if possible, the sum of the infinite geometric series

$$1 + 2 + (2)^2 + (2)^3 + \cdots + (2)^{n-1} + \cdots$$

Since $r = 2$ and $a = 1$, we have

$$S_n = \frac{1 - (2)^n}{1 - 2}$$

$$= \frac{1 - (2)^n}{-1}$$

$$= (2)^n - 1$$

As n approaches ∞, 2^n becomes large without limit. That is, 2^n approaches ∞. Thus $\lim\limits_{n \to \infty} S_n$ does not exist and the series has no sum.

In general, if $|r| < 1$, the infinite geometric series

$$a + ar + ar^2 + \cdots + ar^{n-1} + \cdots$$

has sum

$$S = \lim_{n \to \infty} S_n = \lim_{n \to \infty} \frac{a(1 - r^n)}{1 - r} = \frac{a}{1 - r} \qquad \begin{array}{l}(r^n \text{ approaches 0 as } n \\ \text{approaches } \infty)\end{array}$$

If $|r| \geq 1$, the infinite geometric series has no sum as illustrated in Example 11.

Example 12

Find the sum of the infinite geometric series $3 + \tfrac{3}{5} + \tfrac{3}{25} + \cdots + 3(\tfrac{1}{5})^{n-1} + \cdots$
Since $r = \tfrac{1}{5}$ and $a = 3$, we have

$$S = \frac{a}{1 - r}$$

$$= \frac{3}{1 - \tfrac{1}{5}} = \tfrac{15}{4}$$

Example 13

Find a fraction which is equivalent to the decimal $0.232323\ldots$.

We can write this decimal as the infinite series

$$0.23 + 0.0023 + 0.000023 + \cdots$$

Then $a = 0.23$ and $r = 0.01$. Thus

$$S = \frac{a}{1 - r} = \frac{0.23}{1 - 0.01} = \frac{0.23}{0.99} = \frac{23}{99}$$

Note that any repeating decimal can be expressed as a fraction. The fractional form can be found by using the method shown in Example 13.

Exercises

Find the nth term of each arithmetic progression.

1. $2, 5, 8, \ldots,\quad n = 6$
2. $-3, -7, -11, \ldots,\quad n = 7$
3. $3, 4\frac{1}{2}, 6, \ldots,\quad n = 15$
4. $-2, \frac{1}{5}, 2\frac{2}{5}, \ldots,\quad n = 8$
5. $4, -5, -14, \ldots,\quad n = 12$
6. $10, 50, 90, \ldots,\quad n = 9$

7–12. Find the sum of the first n terms of the progressions in Exercises 1–6.

Write the first five terms of each arithmetic progression whose first term is a and whose common difference is d.

13. $a = 2, d = -3$
14. $a = -4, d = 2$
15. $a = 5, d = \frac{2}{3}$
16. $a = 3, d = -\frac{1}{2}$

17. Find the first term of an arithmetic progression whose 10th term is 12 and whose sum of the first ten terms is 80.

18. Find the common difference of an arithmetic progression whose first term is 7 and whose 8th term is 16.

19. Find the sum of the first 1000 odd positive integers.

20. Find the sum of the first 500 even positive integers.

21. A man is employed at an initial salary of $24,000. If he receives an annual raise of $800, what is his salary for the tenth year?

22. Equipment purchased at an original value of $1360 is depreciated $120 per year for ten years. Find the depreciated value after four years. Find the scrap value (depreciated value after ten years).

Find the nth term of each geometric progression.

23. $20, \frac{20}{3}, \frac{20}{9}, \ldots,\quad n = 8$
24. $\frac{1}{8}, -\frac{1}{4}, \frac{1}{2}, \ldots,\quad n = 7$
25. $\sqrt{2}, 2, 2\sqrt{2}, \ldots,\quad n = 6$
26. $6, 3, \frac{3}{2}, \ldots,\quad n = 8$
27. $8, -4, 2, \ldots,\quad n = 10$
28. $3, 12, 48, \ldots,\quad n = 5$

29–34. Find the sum of the first n terms of the progressions in Exercises 23–28.

Write the first five terms of each geometric progression whose first term is a and whose common ratio is r.

35. $a = 3, r = \frac{1}{2}$
36. $a = -6, r = \frac{1}{3}$
37. $a = 5, r = -\frac{1}{4}$
38. $a = 2, r = -\frac{3}{2}$
39. $a = -4, r = 3$
40. $a = -5, r = -2$

41. Find the common ratio of a geometric progression whose first term is 6 and whose fourth term is $\frac{3}{4}$.

42. Find the first term of a geometric progression whose common ratio is $\frac{1}{3}$ and whose sum of the first three terms is 13.

43. If $1000 is deposited annually in an account at 10% interest compounded annually, find the total amount in the account after ten years.

44. If $200 is deposited quarterly in an account at 9% interest compounded quarterly, find the total amount in the account after ten years.

45. A ball is dropped from a height of 12 ft. After each bounce it rebounds to $\frac{1}{2}$ the height of the previous height from which it fell. Find the distance the ball rises after the fifth bounce.

46. The half-life of chlorine isotope, ^{38}Cl, used in radio isotope therapy is 37 min. This means that half of a given amount will disintegrate in 37 min. This means also that after 74 min three-fourths will have disintegrated. Find how much will have disintegrated in 148 min.

47. A salt solution is being cooled such that the temperature decreases 20% each minute. Find the temperature of the solution after 8 min if the original temperature was 90°C.

48. A tank contains 400 gallons of acid. Then 100 gal is drained out and refilled with water. Then 100 gal of the mixture is drained out and refilled with water. Assuming that this process continues, how much acid remains in the tank after five quantities of 100-gal units are drained out?

Find the sum, when possible, for each infinite geometric series.

49. $4 + \frac{4}{7} + \frac{4}{49} + \cdots + 4(\frac{1}{7})^{n-1} + \cdots$

50. $6 + \frac{6}{11} + \frac{6}{121} + \cdots + 6(\frac{1}{11})^{n-1} + \cdots$

51. $3 - \frac{3}{8} + \frac{3}{64} - \cdots + 3(-\frac{1}{8})^{n-1} + \cdots$

52. $1 - \frac{1}{9} + \frac{1}{81} - \cdots + (-\frac{1}{9})^{n-1} + \cdots$

53. $4 + 12 + 36 + \cdots + 4(3)^{n-1} + \cdots$

54. $-5 - \frac{5}{2} - \frac{5}{4} - \cdots - 5(\frac{1}{2})^{n-1} - \cdots$

55. $3 + 3.1 + 3.01 + \cdots + [3 + (0.1)^{n-1}] + \cdots$

56. $2 + 2 + 2 + \cdots + 2(1)^{n-1} + \cdots$

Find the fraction which is equivalent to each decimal.

57. $0.3333 \ldots$ 58. $0.135135135 \ldots$ 59. $0.0121212 \ldots$

60. $0.6252525 \ldots$ 61. $0.86666 \ldots$ 62. $0.365365365 \ldots$

10.2 SERIES AND CONVERGENCE

We now begin a more general study of series. First, we introduce the sigma notation for writing series. The Greek letter Σ (sigma) is used to indicate that the given expression is a sum. The term following Σ represents the general form of each term. For example,

$$\sum_{n=1}^{6} n^2 = 1^2 + 2^2 + 3^2 + 4^2 + 5^2 + 6^2$$

where the general form of each term is n^2. The numbers below and above the Σ must be integers and indicate the values of n to be used for the first and last terms, respectively, of the series. The other terms are found by replacing n by the consecutive integers between 1 and 6. The sum

$$1 + 2 + 3 + \cdots + n \quad \text{can be represented by} \quad \sum_{n=1}^{n} n$$

while the infinite sum

$$\frac{2}{3} + \frac{2}{9} + \frac{2}{27} + \cdots \text{ can be represented by } \sum_{n=1}^{\infty} 2\left(\frac{1}{3}\right)^n$$

The following example illustrates the use of sigma notation.

Example 1

$\sum$ notation	Expanded form of sum
(a) $\sum_{n=1}^{6} (2n)$	$2 + 4 + 6 + 8 + 10 + 12$
(b) $\sum_{n=5}^{9} (3n - 4)$	$11 + 14 + 17 + 20 + 23$
(c) $\sum_{n=1}^{n} (n - 2)$	$-1 + 0 + 1 + 2 + 3 + \cdots + (n - 2)$
(d) $\sum_{n=0}^{n} 2^n$	$1 + 2 + 2^2 + 2^3 + \cdots + 2^n$
(e) $\sum_{n=1}^{n} A_n$	$A_1 + A_2 + A_3 + \cdots + A_n$

Example 2

Write the expanded form of $\sum_{n=1}^{5} \dfrac{4}{2n + 1}$.

$$\sum_{n=1}^{5} \frac{4}{2n + 1} = \frac{4}{2 \cdot 1 + 1} + \frac{4}{2 \cdot 2 + 1} + \frac{4}{2 \cdot 3 + 1} + \frac{4}{2 \cdot 4 + 1} + \frac{4}{2 \cdot 5 + 1}$$

$$= \frac{4}{3} + \frac{4}{5} + \frac{4}{7} + \frac{4}{9} + \frac{4}{11}$$

Example 3

Write the sum $\dfrac{6}{9} + \dfrac{6}{16} + \dfrac{6}{25} + \cdots + \dfrac{6}{121}$ using sigma notation.

Build the general form of each term. Note that 6 appears in the numerator of each term while the denominator changes in a pattern of perfect squares beginning with 3^2. Do you see that this series has nine terms?

$$\sum_{n=3}^{11} \frac{6}{n^2} \quad \text{or} \quad \sum_{n=1}^{9} \frac{6}{(n + 2)^2} \quad \text{are acceptable forms.}$$

Next, we need to study an important application of series, which is to find the value of infinite summations. To begin this process, consider the series

Definition of Convergence and Divergence

$$\sum_{n=1}^{\infty} a_n = a_1 + a_2 + a_3 + \cdots$$

Then

$$S_1 = a_1$$

$$S_2 = a_1 + a_2$$

$$S_3 = a_1 + a_2 + a_3$$

$$\vdots$$

$$S_n = a_1 + a_2 + a_3 + \cdots + a_n$$

$$\vdots$$

where S_1 is the sum of the first term, S_2 is the sum of the first two terms, S_3 is the sum of the first three terms, . . ., S_n is the sum of the first n terms (or sometimes called the nth partial sum). . . .

(a) If $\lim_{n \to \infty} S_n = S$ (where S is finite), the series $\sum_{n=1}^{\infty} a_n$ converges and S is the *sum of the infinite series.*

(b) If $\lim_{n \to \infty} S_n$ does not exist, the series $\sum_{n=1}^{\infty} a_n$ *diverges.*

The geometric series $1 + \dfrac{1}{2} + \dfrac{1}{4} + \dfrac{1}{8} + \cdots$ from Section 10.1 converges because $\lim_{n \to \infty} S_n = 2$. The arithmetic series $1 + 2 + 3 + 4 + \cdots$ diverges because $\lim_{n \to \infty} S_n = \infty$.

In the remainder of this section and Sections 10.3 to 10.5, we seek answers to the following two questions regarding infinite series:

1. Does a given series converge or diverge?
2. If the series converges, to what value does it converge?

The answers to these questions are not always easy, especially the second one. There is no general method for finding the sum of a convergent infinite series. We will begin the study of these two questions with a simple test for divergence.

Test for Divergence of a Series

If $\lim_{n \to \infty} a_n \neq 0$, then the series *diverges.*

Example 4

Show that each series diverges by using the test for divergence.

(a) $\sum_{n=1}^{\infty} 2^n$ (b) $\sum_{n=1}^{\infty} \dfrac{n}{n+1}$ (c) $1 - 1 + 1 - 1 + 1 - 1 + \cdots$

The series in parts (a) and (b) diverge because

(a) $\lim_{n \to \infty} 2^n = \infty$ and (b) $\lim_{n \to \infty} \dfrac{n}{n+1} = \lim_{n \to \infty} \dfrac{1}{1 + \dfrac{1}{n}} = 1.$

The series in part (c) also diverges because $\lim_{n \to \infty} a_n$ clearly is not zero.

---

nth Term Test for Convergence of a Series

If the series $\displaystyle\sum_{n=1}^{\infty} a_n$ converges, then $\displaystyle\lim_{n\to\infty} a_n = 0$.

Note of warning: *The converse of the nth Term Test is not true.* If $\displaystyle\lim_{n\to\infty} a_n = 0$, this is no guarantee that the series converges. For example, the series

$$\sum_{n=1}^{\infty} \frac{1}{n} = 1 + \frac{1}{2} + \frac{1}{3} + \frac{1}{4} + \cdots \qquad \text{(called the harmonic series)}$$

diverges even though $\displaystyle\lim_{n\to\infty} \frac{1}{n} = 0$.

We now present some tests to determine whether a given series of positive terms converges or diverges. The next special series that we need to consider is the *p*-series. We can show that the conditions for convergence and divergence of the *p*-series are the following:

Convergence and Divergence of a *p*-series

Any series in the form

$$\sum_{n=1}^{\infty} \frac{1}{n^p} = \frac{1}{1^p} + \frac{1}{2^p} + \frac{1}{3^p} + \cdots$$

where *p* is a real number, is called a *p*-series. The *p*-series
(a) converges for $p > 1$, and
(b) diverges for $p \leq 1$.

Example 5

Determine whether each *p*-series converges or diverges.

(a) $1 + \dfrac{1}{2^3} + \dfrac{1}{3^3} + \dfrac{1}{4^3} + \cdots$ (b) $1 + \dfrac{1}{2} + \dfrac{1}{3} + \dfrac{1}{4} + \cdots$

(c) $1 + \dfrac{1}{\sqrt{2}} + \dfrac{1}{\sqrt{3}} + \cdots$

(a) This *p*-series ($p = 3$) converges.
(b) This *p*-series ($p = 1$) diverges. Recall this series is called the harmonic series.
(c) This *p*-series ($p = 1/2$) diverges.

The next test for convergence is called the comparison test.

Comparison Test for Convergence and Divergence

Let N be a positive integer, $\displaystyle\sum_{n=1}^{\infty} a_n$ and $\displaystyle\sum_{n=1}^{\infty} b_n$ be series of positive terms,

and $0 \leq a_n \leq b_n$ for all $n > N$; then:

(a) If $\displaystyle\sum_{n=1}^{\infty} b_n$ converges, then $\displaystyle\sum_{n=1}^{\infty} a_n$ also converges.

(b) If $\displaystyle\sum_{n=1}^{\infty} a_n$ diverges, then $\displaystyle\sum_{n=1}^{\infty} b_n$ also diverges.

In other words, the comparison test says:

1. A series of positive terms that is term by term smaller than a known convergent series must also converge.
2. A series of positive terms that is term by term larger than a known divergent series must also diverge.

Example 6

Use the comparison test to determine whether each series converges or diverges.

(a) $\displaystyle\sum_{n=1}^{\infty} \frac{1}{2^n + 1}$ (b) $\displaystyle\sum_{n=2}^{\infty} \frac{1}{\sqrt{n} - 1}$

(a) We know that the geometric series $\displaystyle\sum_{n=1}^{\infty} \frac{1}{2^n}$ converges. And

$$\frac{1}{2^n + 1} \leq \frac{1}{2^n} \qquad \text{for } n \geq 1$$

Then by the comparison test, $\displaystyle\sum_{n=1}^{\infty} \frac{1}{2^n + 1}$ also converges.

(b) We know that the p-series $\displaystyle\sum_{n=2}^{\infty} \frac{1}{\sqrt{n}}$ $(p = \frac{1}{2})$ diverges. And

$$\frac{1}{\sqrt{n} - 1} \geq \frac{1}{\sqrt{n}} \qquad \text{for } n \geq 2$$

Then by the comparison test, $\displaystyle\sum_{n=2}^{\infty} \frac{1}{\sqrt{n} - 1}$ also diverges.

A test that is easier to apply than the comparison test is the limit comparison test. First, a definition:

Let $\displaystyle\sum_{n=1}^{\infty} a_n$ and $\displaystyle\sum_{n=1}^{\infty} b_n$ be two series of positive terms.

(a) Then $\displaystyle\sum_{n=1}^{\infty} a_n$ and $\displaystyle\sum_{n=1}^{\infty} b_n$ have the *same order of magnitude* if

$$\lim_{n \to \infty} \frac{a_n}{b_n} = L, \text{ where } L \text{ is a real number and } L > 0.$$

(b) The series $\displaystyle\sum_{n=1}^{\infty} a_n$ has a *lesser order of magnitude* than $\displaystyle\sum_{n=1}^{\infty} b_n$ if

$$\lim_{n \to \infty} \frac{a_n}{b_n} = 0.$$

(c) The series $\sum\limits_{n=1}^{\infty} a_n$ has a *greater order of magnitude* than $\sum\limits_{n=1}^{\infty} b_n$ if

$$\lim_{n\to\infty} \frac{a_n}{b_n} = \infty.$$

Example 7

Compare the orders of magnitude of each pair of series.

(a) $\sum\limits_{n=1}^{\infty} (2n)$ and $\sum\limits_{n=1}^{\infty} (n - 4)$

(b) $\sum\limits_{n=1}^{\infty} (3n)$ and $\sum\limits_{n=1}^{\infty} (n^2 + 1)$

(c) $\sum\limits_{n=1}^{\infty} \left(\frac{1}{n}\right)$ and $\sum\limits_{n=1}^{\infty} \frac{1}{n^2 + 1}$

(a) $\lim\limits_{n\to\infty} \dfrac{2n}{n - 4} = \lim\limits_{n\to\infty} \dfrac{2}{1 - \dfrac{4}{n}} = 2$. These two series have the same order of

magnitude.

(b) $\lim\limits_{n\to\infty} \dfrac{3n}{n^2 + 1} = \lim\limits_{n\to\infty} \dfrac{3}{n + \dfrac{1}{n}} = 0$. So $\sum\limits_{n=1}^{\infty} (3n)$ has a lesser order of magnitude

than $\sum\limits_{n=1}^{\infty} (n^2 + 1)$.

(c) $\lim\limits_{n\to\infty} \dfrac{\dfrac{1}{n}}{\dfrac{1}{n^2 + 1}} = \lim\limits_{n\to\infty} \dfrac{n^2 + 1}{n} = \lim\limits_{n\to\infty} \left(n + \dfrac{1}{n}\right) = \infty$. Then $\sum\limits_{n=1}^{\infty} \left(\dfrac{1}{n}\right)$ has a larger

order of magnitude than $\sum\limits_{n=1}^{\infty} \dfrac{1}{n^2 + 1}$.

We can now give the limit comparison test:

Limit Comparison Test

Let $\sum\limits_{n=1}^{\infty} a_n$ and $\sum\limits_{n=1}^{\infty} b_n$ be series of positive terms.

(a) If both series have the same order of magnitude, then either both series converge or both series diverge.

(b) If the series $\sum\limits_{n=1}^{\infty} a_n$ has a lesser order of magnitude than $\sum\limits_{n=1}^{\infty} b_n$ and

$\sum\limits_{n=1}^{\infty} b_n$ is known to converge, then $\sum\limits_{n=1}^{\infty} a_n$ also converges.

(c) If $\sum\limits_{n=1}^{\infty} a_n$ has a greater order of magnitude than $\sum\limits_{n=1}^{\infty} b_n$ and $\sum\limits_{n=1}^{\infty} b_n$ is known to diverge, then $\sum\limits_{n=1}^{\infty} a_n$ also diverges.

Example 8

Use the limit comparison test to determine whether each series converges or diverges.

(a) $\sum\limits_{n=1}^{\infty} \dfrac{1}{n(3n+1)}$ (b) $\sum\limits_{n=1}^{\infty} \dfrac{\ln n}{n}$

(a) Let's compare with $\sum\limits_{n=1}^{\infty} \dfrac{1}{n^2}$ (p-series, $p = 2$), which converges.

$$\lim_{n\to\infty} \frac{\dfrac{1}{n(3n+1)}}{\dfrac{1}{n^2}} = \lim_{n\to\infty} \frac{n^2}{3n^2 + n} = \lim_{n\to\infty} \frac{1}{3 + \dfrac{1}{n}} = \frac{1}{3}$$

Since $\sum\limits_{n=1}^{\infty} \dfrac{1}{n^2}$ converges and both series have the same order of magnitude, the series $\sum\limits_{n=1}^{\infty} \dfrac{1}{n(3n+1)}$ also converges.

(b) Let's compare with $\sum\limits_{n=1}^{\infty} \dfrac{1}{n}$ (harmonic series), which diverges.

$$\lim_{n\to\infty} \frac{\dfrac{\ln n}{n}}{\dfrac{1}{n}} = \lim_{n\to\infty} \ln n = \infty$$

Since $\sum\limits_{n=1}^{\infty} \dfrac{1}{n}$ diverges and the series $\sum\limits_{n=1}^{\infty} \dfrac{\ln n}{n}$ has a greater order of magnitude, then the series $\sum\limits_{n=1}^{\infty} \dfrac{\ln n}{n}$ also diverges.

Exercises

Write the expanded form of each series.

1. $\sum\limits_{n=1}^{6} (4n + 1)$

2. $\sum\limits_{n=1}^{5} (1 - n^2)$

3. $\sum\limits_{n=3}^{8} (n^2 + 1)$

4. $\sum\limits_{n=1}^{6} (n^2 - 4n)$

5. $\sum\limits_{n=1}^{n} \dfrac{n^2}{n + 1}$

6. $\sum\limits_{n=1}^{n} \dfrac{4}{n(n + 1)}$

7. $\sum\limits_{n=1}^{\infty} (-1)^n \dfrac{1}{n^2}$

8. $\sum\limits_{n=1}^{\infty} (-1)^{n+1} \dfrac{n}{n + 1}$

Write each sum using sigma notation.

9. $1 + 2 + 3 + \cdots + 12$

10. $1 + 3 + 5 + \cdots + 43$

11. $2 + 4 + 6 + \cdots + 100$

12. $1^3 + 2^3 + 3^3 + \cdots + (n - 1)^3$

13. $1 + 3 + 5 + \cdots + (2n - 1)$

14. $\dfrac{1}{2} + \dfrac{1}{4} + \dfrac{1}{8} + \cdots + \dfrac{1}{2^n}$

15. $10 + 17 + 26 + 37 + \cdots + (n^2 + 1)$

16. $\dfrac{1}{2} + \dfrac{2}{3} + \dfrac{3}{4} + \cdots + \dfrac{n + 1}{n + 2}$

Determine whether each series converges or diverges.

17. $3 + 9 + 27 + \cdots + 3^n + \cdots$

18. $1 + 2 + 3 + \cdots + n + \cdots$

19. $\displaystyle\sum_{n=2}^{\infty} \dfrac{2n}{n - 1}$

20. $\displaystyle\sum_{n=1}^{\infty} \dfrac{3n + 1}{2n - 1}$

21. $1 + \dfrac{1}{\sqrt[4]{2}} + \dfrac{1}{\sqrt[4]{3}} + \dfrac{1}{\sqrt[4]{4}} + \cdots + \dfrac{1}{\sqrt[4]{n}} + \cdots$

22. $1 + \dfrac{1}{2^4} + \dfrac{1}{3^4} + \dfrac{1}{4^4} + \cdots + \dfrac{1}{n^4} + \cdots$

23. $\displaystyle\sum_{n=1}^{\infty} \dfrac{1}{n^2}$

24. $\displaystyle\sum_{n=1}^{\infty} \dfrac{1}{\sqrt[3]{n}}$

25. $\displaystyle\sum_{n=1}^{\infty} \dfrac{1}{(n + 1)^2}$

26. $\displaystyle\sum_{n=4}^{\infty} \dfrac{1}{n - 3}$

27. $\dfrac{1}{1 \cdot 2} + \dfrac{1}{2 \cdot 3} + \dfrac{1}{3 \cdot 4} + \cdots + \dfrac{1}{n(n + 1)} + \cdots$

28. $1 + \dfrac{1}{3} + \dfrac{1}{5} + \cdots + \dfrac{1}{2n - 1} + \cdots$

29. $\dfrac{1}{2} + \dfrac{1}{4} + \dfrac{1}{6} + \dfrac{1}{8} + \cdots + \dfrac{1}{2n} + \cdots$

30. $\displaystyle\sum_{n=1}^{\infty} \dfrac{1}{n^2 + 3n}$

31. $\displaystyle\sum_{n=1}^{\infty} \dfrac{1}{(2n - 1)^2}$

32. $\displaystyle\sum_{n=1}^{\infty} \dfrac{1}{n^2 + 1}$

33. $\displaystyle\sum_{n=1}^{\infty} \dfrac{1}{\sqrt{n^2 + 1}}$

34. $\displaystyle\sum_{n=2}^{\infty} \dfrac{1}{n\sqrt{n^2 - 1}}$

35. $\displaystyle\sum_{n=1}^{\infty} \dfrac{1}{\sqrt{n}(n + 1)}$

36. $\displaystyle\sum_{n=1}^{\infty} \dfrac{1}{\sqrt[3]{n^2 + 1}}$

37. $\displaystyle\sum_{n=1}^{\infty} \dfrac{1}{2^n + 2n}$

38. $\displaystyle\sum_{n=3}^{\infty} \dfrac{1}{n^2 - 4}$

39. $\displaystyle\sum_{n=2}^{\infty} \dfrac{1}{\ln n}$

40. $\displaystyle\sum_{n=3}^{\infty} \dfrac{n^2}{n^2 - 4}$

41. $\displaystyle\sum_{n=1}^{\infty} \dfrac{1 + \sin n\pi}{n^2}$

42. $\displaystyle\sum_{n=1}^{\infty} \dfrac{3n}{(n + 1)(n + 2)}$

43. $\displaystyle\sum_{n=1}^{\infty} \dfrac{1}{\sqrt{n(n + 1)}}$

44. $\displaystyle\sum_{n=1}^{\infty} \dfrac{1}{n^n}$

10.3 RATIO AND INTEGRAL TESTS

The tests for convergence and divergence in the preceding section work for some series but not others. As a result, we offer two additional tests in this section.

Ratio Test for Convergence and Divergence

Let $\displaystyle\sum_{n=1}^{\infty} a_n$ be a series of positive terms and

---

$$r = \lim_{n \to \infty} \frac{a_{n+1}}{a_n}$$

(a) If $r < 1$, the series converges.
(b) If $r > 1$ (including $r = \infty$), the series diverges.
(c) If $r = 1$, the test fails. Some other test must be used.

Example 1

Determine whether the series $\sum_{n=1}^{\infty} \dfrac{2n}{3^n}$ converges or diverges.

$$r = \lim_{n \to \infty} \frac{a_{n+1}}{a_n} = \lim_{n \to \infty} \frac{\dfrac{2(n+1)}{3^{n+1}}}{\dfrac{2n}{3^n}} = \lim_{n \to \infty} \frac{3^n \, 2(n+1)}{3^{n+1} \, (2n)} = \lim_{n \to \infty} \frac{n+1}{3n}$$

$$= \lim_{n \to \infty} \frac{1 + \dfrac{1}{n}}{3} = \frac{1}{3} < 1$$

Since $r < 1$, the given series converges.

The ratio test is especially helpful in testing expressions involving factorials. Recall that

$5! = 5 \cdot 4 \cdot 3 \cdot 2 \cdot 1$

$10! = 10 \cdot 9 \cdot 8 \cdot 7 \cdot 6 \cdot 5 \cdot 4 \cdot 3 \cdot 2 \cdot 1$

$n! = n(n - 1)(n - 2) \cdots 4 \cdot 3 \cdot 2 \cdot 1$ (n must be a positive integer)

Example 2

Determine whether the series $\sum_{n=1}^{\infty} \dfrac{2^n}{n!}$ converges or diverges.

$$r = \lim_{n \to \infty} \frac{a_{n+1}}{a_n} = \lim_{n \to \infty} \frac{\dfrac{2^{n+1}}{(n+1)!}}{\dfrac{2^n}{n!}} = \lim_{n \to \infty} \frac{n! \, 2^{n+1}}{(n+1)! \, 2^n}$$

$$= \lim_{n \to \infty} \frac{2}{n+1} = 0 < 1$$

Since $r < 1$, the given series converges.

Example 3

Determine whether the series $\sum_{n=1}^{\infty} \dfrac{2^n}{n^2}$ converges or diverges.

$$r = \lim_{n \to \infty} \frac{a_{n+1}}{a_n} = \lim_{n \to \infty} \frac{\dfrac{2^{n+1}}{(n+1)^2}}{\dfrac{2^n}{n^2}} = \lim_{n \to \infty} \frac{n^2 \, 2^{n+1}}{(n+1)^2 \, 2^n}$$

$$= \lim_{n \to \infty} \frac{1(2)}{\left(1 + \frac{1}{n}\right)^2} = 2 > 1$$

Since $r > 1$, the given series diverges.

Integral Test for Convergence and Divergence

Let $\displaystyle\sum_{n=1}^{\infty} a_n$ be a series of positive terms, $f(x)$ be a continuous, decreasing function for $x \geq 1$ such that $f(n) = a_n$ for all positive integers n. Then either $\displaystyle\sum_{n=1}^{\infty} a_n$ and $\displaystyle\int_1^{\infty} f(x)\, dx$ both converge or both diverge.

Note: Before using the integral test, you must be certain that the function $f(x)$ is decreasing and continuous.

Example 4

Determine whether the series $\displaystyle\sum_{n=2}^{\infty} \frac{1}{n \ln n}$ converges or diverges.

Note: This series begins with $n = 2$ because $\ln 1 = 0$.

Let $f(x) = \dfrac{1}{x \ln x}$ for $x \geq 2$; $f(x)$ is continuous and decreasing for $x \geq 2$.

$$\int_2^{\infty} f(x)\, dx = \int_2^{\infty} \frac{dx}{x \ln x} = \lim_{b \to \infty} \int_2^b \frac{dx}{x \ln x}$$

To integrate $\displaystyle\int \frac{dx}{x \ln x}$, make the following substitution.

Then

$$\int \frac{dx}{x \ln x} = \int \frac{du}{u}$$

$$= \ln u$$

$$= \ln \ln x$$

$$\boxed{\begin{aligned} u &= \ln x \\ du &= \frac{1}{x}\, dx \end{aligned}}$$

Then

$$\lim_{b \to \infty} \int_2^b \frac{dx}{x \ln x} = \lim_{b \to \infty} \ln \ln x \Big|_2^b = \lim_{b \to \infty} (\ln \ln b - \ln \ln 2) = \infty$$

Since $\displaystyle\int_2^{\infty} f(x)\, dx$ diverges, the given series also diverges.

Example 5

Determine whether the series $\displaystyle\sum_{n=1}^{\infty} \frac{n}{e^n}$ converges or diverges.

Let $f(x) = xe^{-x}$ for $x \geq 1$; $f(x)$ is continuous and decreasing for $x \geq 1$.

$$\int_1^{\infty} f(x)\, dx = \int_1^{\infty} xe^{-x}\, dx = \lim_{b \to \infty} \int_1^b xe^{-x}\, dx$$

To integrate $\int xe^{-x}\,dx$, we use integration by parts.

$$\int u\,dv \qquad\qquad \int v\,du$$

$u = x$
$dv = e^{-x}\,dx$

$du = dx$
$v = -e^{-x}$

$$\int u\,dv = uv - \int v\,du$$

$$\int xe^{-x}\,dx = -xe^{-x} - \int(-e^{-x})\,dx$$

$$= -xe^{-x} - e^{-x}$$

$$= e^{-x}(-x - 1)$$

Then

$$\lim_{b\to\infty}\int_1^b xe^{-x}\,dx = \lim_{b\to\infty}\left[e^{-x}(-x-1)\right]\Big|_1^b$$

$$= \lim_{b\to\infty}\left[e^{-b}(-b-1) - e^{-1}(-1-1)\right]$$

$$= \lim_{b\to\infty}\left[\frac{-b-1}{e^b} + \frac{2}{e}\right] = \frac{2}{e}$$

Since $\int_1^\infty f(x)\,dx$ converges, the given series also converges.

The integral test can be used to find the values of p for which the p-series converges. Let's begin by considering the p-series

$$\sum_{n=1}^{\infty}\frac{1}{n^p} \qquad \text{for } p > 0$$

Note: For $p \le 0$, $\lim\limits_{n\to\infty}\dfrac{1}{n^p} \ne 0$, so the p-series diverges for $p \le 0$.

Let $f(x) = \dfrac{1}{x^p}$ for $p > 0$; $f(x)$ is continuous and decreasing for $x > 0$. We need two cases:

For $p \ne 1$,

$$\int_1^\infty \frac{dx}{x^p} = \lim_{b\to\infty}\int_1^\infty x^{-p}\,dx$$

$$= \lim_{b\to\infty}\frac{x^{1-p}}{1-p}\Big|_1^b$$

$$= \lim_{b\to\infty}\left(\frac{b^{1-p}}{1-p} - \frac{1}{1-p}\right) = \begin{cases} \dfrac{1}{p-1} & \text{if } p > 1 \\[2mm] \infty & \text{if } p < 1 \end{cases}$$

For $p = 1$,

$$\int_1^\infty \frac{dx}{x} = \lim_{b\to\infty} \int_1^b \frac{dx}{x} = \lim_{b\to\infty} \ln x \Big|_1^b = \lim_{b\to\infty} (\ln b - \ln 1) = \infty$$

So the p-series converges for $p > 1$ and diverges for $p \leqslant 1$.

Exercises

Use either the ratio test or the integral test to determine whether each series converges or diverges.

1. $\displaystyle\sum_{n=1}^\infty \frac{n+1}{n \cdot 3^n}$

2. $\displaystyle\sum_{n=1}^\infty \frac{2^{n+1}}{3^{n-1}}$

3. $\displaystyle\sum_{n=1}^\infty \frac{1}{n!}$

4. $\displaystyle\sum_{n=1}^\infty \frac{n+2}{n!}$

5. $\displaystyle\sum_{n=1}^\infty \frac{n^2}{n!}$

6. $\displaystyle\sum_{n=1}^\infty \frac{n^2}{2^n}$

7. $\displaystyle\sum_{n=1}^\infty \frac{3^n}{n \cdot 2^n}$

8. $\displaystyle\sum_{n=1}^\infty \frac{n!}{10^n}$

9. $\displaystyle\sum_{n=1}^\infty \frac{2n+3}{2^n}$

10. $\displaystyle\sum_{n=1}^\infty \frac{2^n}{n^2 + 1}$

11. $\displaystyle\sum_{n=1}^\infty \frac{1}{2n+1}$

12. $\displaystyle\sum_{n=1}^\infty \frac{1}{n\sqrt{n}}$

13. $\displaystyle\sum_{n=2}^\infty \frac{1}{n\sqrt{\ln n}}$

14. $\displaystyle\sum_{n=1}^\infty \frac{1}{\sqrt[3]{n}}$

15. $1 + \dfrac{1}{3} + \dfrac{1}{5} + \dfrac{1}{7} + \cdots$

16. $\dfrac{1}{2} + \dfrac{1}{4} + \dfrac{1}{6} + \dfrac{1}{8} + \cdots$

17. $\displaystyle\sum_{n=1}^\infty \frac{n}{n^2 + 1}$

18. $\displaystyle\sum_{n=2}^\infty \frac{\ln n}{n}$

19. $\displaystyle\sum_{n=1}^\infty \frac{n^2}{e^n}$

20. $\displaystyle\sum_{n=1}^\infty \frac{n}{\sqrt{n^2 + 1}}$

10.4 ALTERNATING SERIES AND CONDITIONAL CONVERGENCE

Up to now we have considered only series with all positive terms. Next, we need to study series having both positive and negative terms.

An *alternating series* is a series whose terms are alternately positive and negative. Examples of alternating series are

$$\sum_{n=1}^\infty (-1)^{n+1} 2^n = 2 - 4 + 8 - 16 + \cdots$$

$$\sum_{n=1}^\infty (-1)^n \frac{3}{n} = -3 + \frac{3}{2} - \frac{3}{3} + \frac{3}{4} - \frac{3}{5} + \cdots$$

$$\sum_{n=1}^\infty (-1)^{n+1} a_n = a_1 - a_2 + a_3 - a_4 + \cdots \qquad (a_n > 0 \text{ for each } n)$$

We have a relatively simple test for convergence of alternating series.

The alternating series

$$\sum_{n=1}^{\infty} (-1)^{n+1} a_n = a_1 - a_2 + a_3 - a_4 + \cdots \qquad (a_n > 0 \text{ for each } n)$$

converges provided that both conditions are fulfilled:
(a) $0 < a_{n+1} \leq a_n$ for $n \geq 1$ and
(b) $\lim_{n \to \infty} a_n = 0$

In addition, if S is the sum of the infinite series and S_n is the nth partial sum, then

$$|S - S_n| \leq a_{n+1}$$

Note: Condition (a) states in other words that the absolute value of each term is less than its preceding term.

Example 1

Determine whether the alternating series $\sum_{n=1}^{\infty} \dfrac{(-1)^{n+1}}{n} = 1 - \dfrac{1}{2} + \dfrac{1}{3} - \dfrac{1}{4} + \cdots$ converges or diverges.

Since $a_{n+1} = \dfrac{1}{n+1} < \dfrac{1}{n} = a_n$ and $\lim_{n \to \infty} \dfrac{1}{n} = 0$, this alternating series converges.

We have already shown that the harmonic series

$$\sum_{n=1}^{\infty} \frac{1}{n} = 1 + \frac{1}{2} + \frac{1}{3} + \frac{1}{4} + \cdots$$

diverges, even though $a_{n+1} < a_n$ and $\lim_{n \to \infty} a_n = 0$. This alternating series is sometimes called the alternating harmonic series.

Example 2

Determine whether the series $\sum_{n=2}^{\infty} \dfrac{(-1)^{n+1}}{n \ln n}$ converges or diverges.

Since

$$a_{n+1} = \frac{1}{(n+1) \ln (n+1)} < \frac{1}{n \ln n} = a_n \quad \text{and} \quad \lim_{n \to \infty} \frac{1}{n \ln n} = 0$$

this alternating series converges.

We have shown that the series of positive terms $\sum_{n=2}^{\infty} \dfrac{1}{n \ln n}$ diverges in Example 4 of Section 10.3.

Absolute and Conditional Convergence

Suppose that $\sum_{n=1}^{\infty} a_n$ converges.

(a) If $\sum\limits_{n=1}^{\infty} |a_n|$ converges, then $\sum\limits_{n=1}^{\infty} a_n$ converges absolutely.

(b) If $\sum\limits_{n=1}^{\infty} |a_n|$ diverges, then $\sum\limits_{n=1}^{\infty} a_n$ converges conditionally.

For example, the series $\sum\limits_{n=1}^{\infty} (-1)^{n+1} \dfrac{1}{n^2}$ converges absolutely because $\sum\limits_{n=1}^{\infty} \dfrac{1}{n^2}$ converges. And the series $\sum\limits_{n=1}^{\infty} (-1)^{n+1} \dfrac{1}{n}$ converges conditionally because $\sum\limits_{n=1}^{\infty} \dfrac{1}{n}$ diverges.

Exercises

Determine whether each alternating series converges or diverges. If it converges, find whether it converges absolutely or converges conditionally.

1. $\sum\limits_{n=1}^{\infty} (-1)^{n+1} \dfrac{1}{2n+1}$

2. $\sum\limits_{n=1}^{\infty} (-1)^{n+1} \dfrac{1}{2^n}$

3. $\sum\limits_{n=1}^{\infty} (-1)^{n} \dfrac{1}{(2n)^2}$

4. $\sum\limits_{n=1}^{\infty} (-1)^{n} \dfrac{n}{2n-1}$

5. $\sum\limits_{n=1}^{\infty} (-1)^{n+1} \dfrac{2n}{2n-1}$

6. $\sum\limits_{n=1}^{\infty} (-1)^{n-1} \dfrac{1}{n^2}$

7. $\sum\limits_{n=2}^{\infty} \dfrac{(-1)^{n-1}}{\ln n}$

8. $\sum\limits_{n=1}^{\infty} (-1)^{n+1} \dfrac{1}{\sqrt{n}}$

9. $\sum\limits_{n=1}^{\infty} (-1)^{n} \dfrac{n^2}{2^n}$

10. $\sum\limits_{n=1}^{\infty} (-1)^{n} \dfrac{1}{n!}$

11. $\sum\limits_{n=1}^{\infty} (-1)^{n+1} \dfrac{n^2}{n^2+1}$

12. $\sum\limits_{n=1}^{\infty} (-1)^{n+1} \dfrac{n}{e^n}$

13. $\sum\limits_{n=1}^{\infty} (-1)^{n+1} \dfrac{n!}{3^n}$

14. $\sum\limits_{n=1}^{\infty} (-1)^{n+1} \dfrac{n}{2^n}$

15. $\sum\limits_{n=1}^{\infty} (-1)^{n} \dfrac{2n+1}{n^2}$

16. $\sum\limits_{n=1}^{\infty} (-1)^{n} \dfrac{1}{\sqrt{n^2+1}}$

17. $\sum\limits_{n=2}^{\infty} (-1)^{n+1} \dfrac{n}{\ln n}$

18. $\sum\limits_{n=1}^{\infty} (-1)^{n+1} \dfrac{\sin \pi n}{n}$

19. $\sum\limits_{n=1}^{\infty} (-1)^{n} \dfrac{\cos n}{n^2}$

20. $\sum\limits_{n=1}^{\infty} (-1)^{n} \dfrac{n+1}{n\sqrt{n}}$

10.5 POWER SERIES

A more general example of an infinite series is the power series. A *power series* is an infinite series in the form

$$\sum_{n=0}^{\infty} a_n x^n = a_0 + a_1 x + a_2 x^2 + \cdots + a_n x^n + \cdots$$

Certain functions can be written as a power series. For example, the function $f(x) = \dfrac{1}{1+x}$ can be expressed as a power series as shown below:

$$\begin{array}{r} 1 - x + x^2 - x^3 + \cdots \\ 1 + x \overline{)1 } \\ \underline{1 + x} \\ - x \\ \underline{- x - x^2} \\ x^2 \\ \underline{x^2 + x^3} \\ - x^3 \\ \cdot \\ \cdot \\ \cdot \end{array}$$

Thus

$$f(x) = \frac{1}{1 + x} = 1 - x + x^2 - x^3 + \cdots$$

Note however that the infinite division process as indicated is valid only for $|x| < 1$, because the right-hand side is a geometric series with $a = 1$ and $r = -x$ which has sum $\dfrac{1}{1 - (-x)}$ only for $|x| < 1$. Thus, the equality is not valid for $|x| \geqslant 1$.

A more general series in the form

$$\sum_{n=0}^{\infty} a_n(x - a)^n = a_0 + a_1(x - a) + a_2(x - a)^2 + a_3(x - a)^3$$

$$+ \cdots + a_n(x - a)^n + \cdots$$

is called a *power series centered at a*.

For each particular value of x (x is the variable in the power series), we have an infinite series of constants, which either converges or diverges. In many power series, the series converges for some values of x and diverges for other values of x. The ratio test (when it works) is usually the simplest test to use on a power series.

Let

$$\lim_{n \to \infty} \left| \frac{u_{n+1}}{u_n} \right| = r(x)$$

where u_{n+1} is the $(n + 1)$st term and u_n is the nth term of a power series. Note that the ratio will most often be a function of x, or $r(x)$. Then the series will

1. converge absolutely for $r(x) < 1$ and
2. diverge for $r(x) > 1$.

Recall that the ratio test is not valid for $r = 1$ or $r(x) = 1$. These values of x have to be checked individually.

A power series always converges on an interval, which may vary from a single value of x to all real numbers x—the entire number line. This interval is called the *interval of convergence*. The interval of convergence may include both endpoints, only one endpoint, or neither endpoint. In general the interval of convergence of the power series

$$\sum_{n=0}^{\infty} a_n(x - a)^n$$

is always centered at $x = a$. The *radius of convergence* is the distance from the point $x = a$ to either endpoint of the interval. Thus the radius of convergence is one-half the length of the interval of convergence.

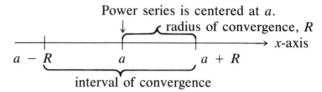

Example 1

For what values of x does the series $\sum_{n=0}^{\infty} 5(x - 3)^n$ converge? That is, find the interval of convergence.

$$r(x) = \lim_{n \to \infty} \left| \frac{u_{n+1}}{u_n} \right| = \lim_{n \to \infty} \left| \frac{5(x - 3)^{n+1}}{5(x - 3)^n} \right| = \lim_{n \to \infty} |x - 3| = |x - 3|$$

This series converges for

$$|x - 3| < 1$$
$$-1 < x - 3 < 1$$
$$2 < x < 4$$

Next, check the endpoints. For $x = 2$, the series is

$$\sum_{n=0}^{\infty} (-1)^n(5) \qquad \text{which diverges}$$

For $x = 4$, the series is

$$\sum_{n=0}^{\infty} 1^n(5) \qquad \text{which also diverges}$$

Thus the interval of convergence is $2 < x < 4$, whose graph is

$$\begin{array}{ccc} 2 & 3 & 4 \end{array} \longrightarrow x$$

Note: the interval of convergence is centered at $x = 3$ and the length of the radius of convergence is 1.

Example 2

Find the interval of convergence of the series $\sum_{n=0}^{\infty} \dfrac{x^n}{n!}$.

$$r(x) = \lim_{n \to \infty} \left| \frac{u_{n+1}}{u_n} \right| = \lim_{n \to \infty} \left| \frac{\dfrac{x^{n+1}}{(n + 1)!}}{\dfrac{x^n}{n!}} \right| = \lim_{n \to \infty} \left| \frac{x}{n + 1} \right| = 0$$

Since $r(x) < 1$ for all values of x, this series converges for all real numbers or for all values of x, which may also be written

$$-\infty < x < \infty$$

Example 3

Find the interval of convergence of the series $\sum\limits_{n=0}^{\infty} n! \, x^n$.

$$r(x) = \lim_{n\to\infty} \left| \frac{u_{n+1}}{u_n} \right| = \lim_{n\to\infty} \left| \frac{(n+1)! \, x^{n+1}}{n! \, x^n} \right|$$

$$= \lim_{n\to\infty} |(n+1)x| = \begin{cases} 0 & \text{if } x = 0 \\ \infty & \text{if } x \neq 0 \end{cases}$$

This series converges only for $x = 0$. This interval of convergence consists of one point.

Example 4

Find the interval of convergence of the series $\sum\limits_{n=0}^{\infty} \dfrac{n^2}{2^n}(x - 1)^n$.

$$r(x) = \lim_{n\to\infty} \left| \frac{u_{n+1}}{u_n} \right| = \lim_{n\to\infty} \left| \frac{\dfrac{(n+1)^2}{2^{n+1}}(x-1)^{n+1}}{\dfrac{n^2}{2^n}(x-1)^n} \right| = \lim_{n\to\infty} \left| \frac{(n+1)^2(x-1)}{2n^2} \right|$$

$$= \lim_{n\to\infty} \left| \frac{(n^2 + 2n + 1)(x-1)}{2n^2} \right| = \lim_{n\to\infty} \left| \frac{\left(1 + \dfrac{2}{n} + \dfrac{1}{n^2}\right)}{2}(x-1) \right| = \left| \frac{x-1}{2} \right|$$

This series converges for

$$\left| \frac{x-1}{2} \right| < 1$$

$$-1 < \frac{x-1}{2} < 1$$

$$-2 < x - 1 < 2$$

$$-1 < x < 3$$

Check the endpoints. For $x = -1$, the series is

$$\sum_{n=0}^{\infty} \frac{n^2(-2)^n}{2^n} = \sum_{n=0}^{\infty} (-1)^n n^2 \qquad \text{which diverges}$$

For $x = 3$, the series is

$$\sum_{n=0}^{\infty} \frac{n^2 \, 2^n}{2^n} = \sum_{n=0}^{\infty} n^2 \qquad \text{which also diverges}$$

Thus the interval of convergence is $-1 < x < 3$.

Example 5

Find the interval of convergence of the series $\sum\limits_{n=0}^{\infty} \dfrac{nx^n}{(n+1)^2}$.

$$r(x) = \lim_{n \to \infty} \left| \frac{u_{n+1}}{u_n} \right| = \lim_{n \to \infty} \left| \frac{\dfrac{(n + 1)x^{n+1}}{(n + 2)^2}}{\dfrac{nx^n}{(n + 1)^2}} \right| = \lim_{n \to \infty} \left| \frac{(n + 1)^3 \, x}{n(n + 2)^2} \right| = |x|$$

This series converges for $|x| < 1$ or $-1 < x < 1$. Check the endpoints. For $x = -1$, the series is

$$\sum_{n=0}^{\infty} \frac{n(-1)^n}{(n + 1)^2} \qquad \text{which converges conditionally}$$

For $x = 1$, the series is

$$\sum_{n=0}^{\infty} \frac{n \, 1^n}{(n + 1)^2} \qquad \text{which diverges}$$

Then the interval of convergence is $-1 \leqslant x < 1$.

Exercises

Find the interval of convergence of each series.

1. $\displaystyle\sum_{n=0}^{\infty} \left(\frac{x}{2}\right)^n$

2. $\displaystyle\sum_{n=1}^{\infty} \frac{(-1)^n x^n}{n}$

3. $\displaystyle\sum_{n=0}^{\infty} (4n)! \left(\frac{x}{2}\right)^n$

4. $\displaystyle\sum_{n=0}^{\infty} nx^n$

5. $\displaystyle\sum_{n=1}^{\infty} \frac{(4x)^n}{(2n)!}$

6. $\displaystyle\sum_{n=1}^{\infty} \frac{x^n}{n}$

7. $\displaystyle\sum_{n=1}^{\infty} \frac{(-1)^{n+1} x^n}{(n + 1)(n + 2)}$

8. $\displaystyle\sum_{n=0}^{\infty} \frac{(-1)^{n+1} x^n}{4^n}$

9. $\displaystyle\sum_{n=0}^{\infty} \frac{nx^n}{(n + 1)^2}$

10. $\displaystyle\sum_{n=0}^{\infty} 3^n x^n$

11. $\displaystyle\sum_{n=0}^{\infty} \frac{2^n x^n}{3^n}$

12. $\displaystyle\sum_{n=1}^{\infty} \frac{(-1)^{n+1}(x - 1)^{n+1}}{n + 1}$

13. $\displaystyle\sum_{n=1}^{\infty} \frac{(-1)^{n+1} x^n}{n \cdot 2^n}$

14. $\displaystyle\sum_{n=1}^{\infty} (-2)^n (x + 1)^n$

15. $\displaystyle\sum_{n=1}^{\infty} \frac{(-1)^n (x - 2)^n}{\sqrt{n}}$

16. $\displaystyle\sum_{n=1}^{\infty} \frac{(x + 3)^n}{n^2 \cdot 2^n}$

17. $\displaystyle\sum_{n=1}^{\infty} \frac{x^n}{n^2}$

18. $\displaystyle\sum_{n=1}^{\infty} \frac{x^n}{\sqrt{n} \, 3^n}$

19. $\displaystyle\sum_{n=1}^{\infty} \frac{(-1)^n x^{2n}}{n!}$

20. $\displaystyle\sum_{n=1}^{\infty} \frac{n! \, x^n}{(2n)!}$

21. $\displaystyle\sum_{n=1}^{\infty} \frac{2^n x^{n+1}}{n(3^{n+1})}$

22. $\displaystyle\sum_{n=1}^{\infty} \frac{(-1)^n x^n}{n^n}$

23. $\displaystyle\sum_{n=1}^{\infty} \frac{(2x - 5)^n}{n^2}$

24. $\displaystyle\sum_{n=1}^{\infty} \frac{\sin^n x}{n!}$

10.6 MACLAURIN SERIES

The Maclaurin series expansion of a function is a power series developed by differentiation. If a function can be differentiated repeatedly at $x = 0$, then it will have a Maclaurin series expansion. Thus we can write

$$f(x) = a_0 + a_1 x + a_2 x^2 + a_3 x^3 + a_4 x^4 + \cdots + a_n x^n + \cdots$$
$$f'(x) = a_1 + 2a_2 x + 3a_3 x^2 + 4a_4 x^3 + \cdots + na_n x^{n-1} + \cdots$$
$$f''(x) = 2a_2 + 2 \cdot 3a_3 x + 3 \cdot 4a_4 x^2 + \cdots + (n - 1)(n)a_n x^{n-2} + \cdots$$

$$f'''(x) = 2 \cdot 3a_3 + 2 \cdot 3 \cdot 4a_4 x + \cdots + (n-2)(n-1)(n)a_n x^{n-3} + \cdots$$

$$\vdots \qquad\qquad \vdots$$

If we let $x = 0$, then

$$f(0) = a_0, \quad f'(0) = a_1, \quad f''(0) = 2a_2 \quad \text{and} \quad f'''(0) = 2 \cdot 3a_3$$

If we continue differentiating, the nth derivative at $x = 0$ is

$$f^{(n)}(0) = (1) \cdots (n-2)(n-1)(n)a_n$$

or

$$f^{(n)}(0) = n!a_n$$

Rewriting, we have

$$a_0 = f(0)$$

$$a_1 = f'(0)$$

$$a_2 = \frac{f''(0)}{2!}$$

$$a_3 = \frac{f'''(0)}{3!}$$

$$\vdots$$

$$a_n = \frac{f^n(0)}{n!}$$

Replace the coefficients of the powers of x in the power series $f(x) = a_0 + a_1 x + a_2 x^2 + a_3 x^3 + \cdots + a_n x^n + \cdots$ by the equivalent expressions above to obtain

---

Maclaurin Series Expansion

$$f(x) = f(0) + f'(0)x + \frac{f''(0)}{2!}x^2 + \frac{f'''(0)}{3!}x^3 + \cdots + \frac{f^{(n)}(0)}{n!}x^n + \cdots$$

The expansion is valid for all values of x for which the power series converges and for which the function $f(x)$ is repeatedly differentiable.

Example 1

Find the first four terms of the Maclaurin expansion for $f(x) = \dfrac{1}{1 + x}$.

$$f(x) = \frac{1}{x + 1}, \qquad f(0) = 1$$

$$f'(x) = \frac{-1}{(1 + x)^2}, \qquad f'(0) = -1$$

$$f''(x) = \frac{2}{(1 + x)^3}, \qquad f''(0) = 2$$

$$f'''(x) = \frac{-6}{(1+x)^4}, \qquad f'''(0) = -6$$

So

$$f(x) = 1 - x + \frac{2}{2!}x^2 + \frac{-6}{3!}x^3 + \cdots$$

$$= 1 - x + x^2 - x^3 + \cdots$$

which is the same power series we obtained earlier by division.

Example 2

Find the first five terms of the Maclaurin series expansion for $f(x) = \cos 3x$.

$$
\begin{aligned}
f(x) &= \cos 3x, & f(0) &= 1 \\
f'(x) &= -3\sin 3x, & f'(0) &= 0 \\
f''(x) &= -9\cos 3x, & f''(0) &= -9 \\
f'''(x) &= 27\sin 3x, & f'''(0) &= 0 \\
f^{(iv)}(x) &= 81\cos 3x, & f^{(iv)}(0) &= 81
\end{aligned}
$$

Thus

$$f(x) = \cos 3x = 1 + 0 + \frac{-9}{2!}x^2 + 0 + \frac{81}{4!}x^4 + \cdots$$

$$= 1 - \tfrac{9}{2}x^2 + \tfrac{27}{8}x^4 - \cdots$$

Example 3

Find the Maclaurin series expansion for $f(x) = e^x$.

Since $\dfrac{d}{dx}(e^x) = e^x$, we have $f^{(n)}(x) = e^x$ for all n. So

$$f(0) = f'(0) = f''(0) = f'''(0) = \cdots = f^{(n)}(0) = e^0 = 1$$

Thus

$$f(x) = e^x = 1 + x + \frac{1}{2!}x^2 + \frac{1}{3!}x^3 + \frac{1}{4!}x^4 + \cdots + \frac{1}{n!}x^n + \cdots$$

$$= 1 + x + \frac{x^2}{2!} + \frac{x^3}{3!} + \frac{x^4}{4!} + \cdots + \frac{x^n}{n!} + \cdots$$

Note: One can show that this series converges (has a sum) and is a valid representation for $f(x) = e^x$ for all values of x.

Example 4

Find the first four nonzero terms of the Maclaurin expansion for $f(x) = e^x \cos x$.

$$
\begin{aligned}
f(x) &= e^x \cos x & f(0) &= 1 \\
f'(x) &= e^x(-\sin x) + e^x \cos x & \\
&= e^x(\cos x - \sin x) & f'(0) &= 1 \\
f''(x) &= e^x(-\sin x - \cos x) + e^x(\cos x - \sin x) & \\
&= -2e^x \sin x & f''(0) &= 0 \\
f'''(x) &= -2e^x \cos x - 2e^x \sin x & \\
&= -2e^x(\cos x + \sin x) & f'''(0) &= -2
\end{aligned}
$$

$$f^{iv}(x) = -2e^x(-\sin x + \cos x) - 2e^x(\cos x + \sin x)$$
$$= -4e^x \cos x \qquad\qquad f^{iv}(0) = -4$$

Thus

$$f(x) = 1 + x + 0x^2 - \frac{2}{3!}x^3 - \frac{4}{4!}x^4 + \cdots$$

$$= 1 + x - \frac{1}{3}x^3 - \frac{1}{6}x^4 + \cdots$$

Exercises

Find a Maclaurin series expansion for each function.

1. $f(x) = \sin x$ 2. $f(x) = \cos x$ 3. $f(x) = e^{-x}$

4. $f(x) = \dfrac{1}{1 - x}$ 5. $f(x) = \ln(1 + x)$ 6. $f(x) = e^{3x}$

7. $f(x) = \cos 2x$ 8. $f(x) = \sin 4x$ 9. $f(x) = xe^x$

10. $f(x) = x \sin x$ 11. $f(x) = \sqrt{4 - x}$ 12. $f(x) = \dfrac{1}{\sqrt{9 + x}}$

13. $f(x) = \sin\left(x - \dfrac{\pi}{2}\right)$ 14. $f(x) = \dfrac{1}{(1 + x)^2}$ 15. $f(x) = \dfrac{1}{(1 - x)^2}$

16. $f(x) = (1 + x)^{3/2}$ 17. $f(x) = (1 + x)^5$ 18. $f(x) = (2x - 1)^4$

19. $f(x) = e^{-x} \sin x$

20. Show that $(1 + x)^n = 1 + nx + \dfrac{n(n-1)}{2!}x^2 + \dfrac{n(n-1)(n-2)}{3!}x^3 + \cdots$ by using

the Maclaurin expansion. This series is called the binomial series, which is valid for all values of n for $|x| < 1$. (See also Appendix A.)

10.7 OPERATIONS WITH SERIES

We now summarize four important series which we have developed previously.

A. $e^x = 1 + x + \dfrac{x^2}{2!} + \dfrac{x^3}{3!} + \dfrac{x^4}{4!} + \cdots + \dfrac{x^n}{n!} + \cdots$

B. $\sin x = x - \dfrac{x^3}{3!} + \dfrac{x^5}{5!} - \cdots$

C. $\cos x = 1 - \dfrac{x^2}{2!} + \dfrac{x^4}{4!} - \cdots$

D. $\ln(1 + x) = x - \dfrac{x^2}{2} + \dfrac{x^3}{3} - \dfrac{x^4}{4} + \cdots$

From these and similar basic power series expansions we can often obtain power series of other functions.

Example 1

Find the Maclaurin series expansion for $f(x) = \cos 3x$.

Substituting $3x$ for x in equation C we obtain

$$\cos 3x = 1 - \frac{(3x)^2}{2!} + \frac{(3x)^4}{4!} - \cdots$$

$$= 1 - \frac{9x^2}{2!} + \frac{81}{4!}x^4 - \cdots$$

$$= 1 - \tfrac{9}{2}x^2 + \tfrac{27}{8}x^4 - \cdots$$

Compare this result with Example 2, Section 10.6.

Example 2

Find a power series expansion for $\dfrac{\ln(1+x)}{x}$.

Dividing each side of equation D by x, we have

$$\frac{\ln(1+x)}{x} = \frac{x}{x} - \frac{x^2}{2x} + \frac{x^3}{3x} - \frac{x^4}{4x} + \cdots$$

$$= 1 - \frac{x}{2} + \frac{x^2}{3} - \frac{x^3}{4} + \cdots$$

Example 3

Find the Maclaurin series expansion for $f(x) = e^{-3x}$.

Substituting $-3x$ for x in equation A, we have

$$e^{-3x} = 1 + (-3x) + \frac{(-3x)^2}{2!} + \frac{(-3x)^3}{3!} + \frac{(-3x)^4}{4!} + \cdots$$

$$= 1 - 3x + \frac{9x^2}{2} - \frac{9x^3}{2} + \frac{27x^4}{8} - \cdots$$

Example 4

Evaluate $\displaystyle\int_0^1 \frac{\sin x}{x}\,dx$.

In Chapter 6 we used the function $\dfrac{\sin x}{x}$ to find the derivative of $y = \sin x$.

We have no trouble differentiating this function, but none of the techniques of integration introduced in the preceding chapters lead to a solution of the integral $\displaystyle\int_0^1 \frac{\sin x}{x}\,dx$. However, if we divide each side of equation B by x, we have

$$\frac{\sin x}{x} = \frac{x}{x} - \frac{x^3}{3!x} + \frac{x^5}{5!x} - \cdots$$

$$= 1 - \frac{x^2}{3!} + \frac{x^4}{5!} - \cdots$$

Then

$$\int_0^1 \frac{\sin x}{x}\,dx = \int_0^1 \left(1 - \frac{x^2}{3!} + \frac{x^4}{5!} - \cdots\right) dx$$

$$= \left(x - \frac{x^3}{3!3} + \frac{x^5}{5!5} - \cdots\right)\Bigg|_0^1$$

$$= (1 - \tfrac{1}{18} + \tfrac{1}{600} - \cdots) - [0]$$

$$= 0.946111 \quad \text{(sum of the first three terms)}$$

Example 5

Evaluate $\displaystyle\int_0^1 \frac{e^{-x} - 1}{x}\, dx$.

Using equation A, we have

$$\frac{e^{-x}}{x} = \frac{1}{x} + \frac{-x}{x} + \frac{(-x)^2}{2!x} + \frac{(-x)^3}{3!x} + \cdots$$

$$= \frac{1}{x} - 1 + \frac{x}{2!} - \frac{x^2}{3!} + \cdots$$

So

$$\frac{e^{-x} - 1}{x} = \left(\frac{1}{x} - 1 + \frac{x}{2!} - \frac{x^2}{3!} + \cdots\right) - \frac{1}{x}$$

$$= -1 + \frac{x}{2!} - \frac{x^2}{3!} + \cdots$$

$$\int_0^1 \frac{(e^{-x} - 1)}{x}\, dx = \int_0^1 \left(-1 + \frac{x}{2!} - \frac{x^2}{3!} + \cdots\right) dx$$

$$= \left(-x + \frac{x^2}{4} - \frac{x^3}{18} + \cdots\right)\Bigg|_0^1$$

$$= (-1 + \tfrac{1}{4} - \tfrac{1}{18} + \cdots) - (0)$$

$$= -0.8056 \quad \text{(sum of first three terms)}$$

The exponential form of a complex number is based on the expression $e^{j\theta} = \cos\theta + j\sin\theta$, where $j = \sqrt{-1}$. We will now show that this is a valid identity. We have seen that

$$e^x = 1 + x + \frac{x^2}{2!} + \frac{x^3}{3!} + \frac{x^4}{4!} + \frac{x^5}{5!} + \cdots \tag{1}$$

$$\cos x = 1 - \frac{x^2}{2!} + \frac{x^4}{4!} - \cdots \tag{2}$$

$$\sin x = x - \frac{x^3}{3!} + \frac{x^5}{5!} - \cdots \tag{3}$$

If we let $x = j\theta$ in equation (1) and $x = \theta$ in equations (2) and (3), then we have equations (4), (5), and (6), respectively:

$$e^{j\theta} = 1 + j\theta + \frac{(j\theta)^2}{2!} + \frac{(j\theta)^3}{3!} + \frac{(j\theta)^4}{4!} + \frac{(j\theta)^5}{5!} + \cdots$$

$$= 1 + j\theta - \frac{\theta^2}{2!} - j\frac{\theta^3}{3!} + \frac{\theta^4}{4!} + j\frac{\theta^5}{5!} - \cdots \tag{4}$$

$$\cos\theta = 1 - \frac{\theta^2}{2!} + \frac{\theta^4}{4!} - \cdots \tag{5}$$

$$j \sin \theta = j\left(\theta - \frac{\theta^3}{3!} + \frac{\theta^5}{5!} - \cdots\right)$$

$$= j\theta - j\frac{\theta^3}{3!} + j\frac{\theta^5}{5!} - \cdots \tag{6}$$

Adding equations (5) and (6), we have

$$\cos \theta + j \sin \theta = \left(1 - \frac{\theta^2}{2!} + \frac{\theta^4}{4!} - \cdots\right) + \left(j\theta - j\frac{\theta^3}{3!} + j\frac{\theta^5}{5!} - \cdots\right)$$

$$= 1 + j\theta - \frac{\theta^2}{2!} - j\frac{\theta^3}{3!} + \frac{\theta^4}{4!} + j\frac{\theta^5}{5!} - \cdots$$

$$= e^{j\theta}$$

Exercises

Find a Maclaurin series expansion for each function.

1. $f(x) = e^{-x}$ **2.** $f(x) = \cos \sqrt{x}$ **3.** $f(x) = e^{x^2}$

4. $f(x) = \sin x^2$ **5.** $f(x) = \ln (1 - x)$ **6.** $f(x) = e^{-4x}$

7. $f(x) = \cos 5x^2$ **8.** $f(x) = \ln (1 + 3x)$ **9.** $f(x) = \sin x^3$

10. $f(x) = e^{-2x^2}$ **11.** $f(x) = xe^x$ **12.** $f(x) = x^2 \sin x$

13. $\dfrac{\cos x - 1}{x}$. **14.** $\dfrac{e^x}{x - 1}$.

15. Evaluate $\displaystyle\int_0^1 e^{-x^2}\, dx$. (Use first four nonzero terms.)

16. Evaluate $\displaystyle\int_0^1 \frac{\cos x - 1}{x}\, dx$. (Use first three nonzero terms.)

17. Evaluate $\displaystyle\int_2^3 \frac{e^{x-1}}{x - 1}\, dx$. (Use first three nonzero terms.)

18. Evaluate $\displaystyle\int_0^1 \cos x^2\, dx$. (Use first four nonzero terms.)

19. Evaluate $\displaystyle\int_0^1 \sin \sqrt{x}\, dx$. (Use first three nonzero terms.)

20. Evaluate $\displaystyle\int_0^{\pi/2} \sqrt{x} \cos x\, dx$. (Use first six nonzero terms.)

21. The hyperbolic sine function is defined by $\sinh x = \frac{1}{2}(e^x - e^{-x})$. Find its Maclaurin series.

22. The hyperbolic cosine function is defined by $\cosh x = \frac{1}{2}(e^x + e^{-x})$. Find its Maclaurin series.

23. If $i = \sin t^2$ amperes, find the amount of charge q (in coulombs) transmitted by this current from $t = 0$ to $t = 0.5$ s $\left(q = \displaystyle\int i\, dt\right)$.

24. The current supplied to a capacitor is given by $i = \dfrac{1 - \cos t}{t}$ amperes. Find the voltage V across the capacitor after 0.1 s, where the capacitance $C = 1 \times 10^{-6}$ farad $\left(V = \dfrac{1}{C}\displaystyle\int i\, dt\right)$.

Using power series for $\sin x$, $\cos x$, and e^x, show that

25. $\sin x = \dfrac{e^{jx} - e^{-jx}}{2j}$

26. $\cos x = \dfrac{e^{jx} + e^{-jx}}{2}$

10.8 TAYLOR SERIES

When a function $f(x)$ is repeatedly differentiable at a number a and at x as well as all numbers between a and x, then the function usually has a Taylor series expansion which is a valid representation of the given function at x.

> **A _Taylor series expansion_ of a function $f(x)$ is a power series in the form**
>
> $$f(x) = f(a) + f'(a)(x - a) + \frac{f''(a)}{2!}(x - a)^2 + \frac{f'''(a)}{3!}(x - a)^3 + \cdots$$
>
> $$+ \frac{f^{(n)}(a)}{n!}(x - a)^n + \cdots$$

Note that a Maclaurin series is a special case of a Taylor series with $a = 0$.

Example 1

Find the Taylor series expansion for $f(x) = \ln x$ with $a = 2$.

$$f(x) = \ln x, \qquad f(2) = \ln 2$$
$$f'(x) = \frac{1}{x}, \qquad f'(2) = \frac{1}{2}$$
$$f''(x) = -\frac{1}{x^2}, \qquad f''(2) = -\frac{1}{4}$$
$$f'''(x) = \frac{2}{x^3}, \qquad f'''(2) = \frac{2}{8} = \frac{1}{4}$$
$$f^{(iv)}(x) = -\frac{6}{x^4}, \qquad f^{(iv)}(2) = -\frac{6}{16} = -\frac{3}{8}$$

So

$$f(x) = \ln x = \ln 2 + \frac{1}{2}(x - 2) + \frac{(-\frac{1}{4})}{2!}(x - 2)^2 + \frac{1}{3!}(x - 2)^3$$
$$+ \frac{(-\frac{3}{8})}{4!}(x - 2)^4 + \cdots$$
$$= \ln 2 + \tfrac{1}{2}(x - 2) - \tfrac{1}{8}(x - 2)^2 + \tfrac{1}{24}(x - 2)^3$$
$$- \tfrac{1}{64}(x - 2)^4 + \cdots$$

Example 2

Find the Taylor series expansion for $f(x) = e^x$ with $a = 1$.

Since $\dfrac{d}{dx}(e^x) = e^x$, we have

$$f(1) = f'(1) = f''(1) = f'''(1) = \cdots = f^{(n)}(1) = e^1 = e$$

So,

$$f(x) = e^x = e + e(x - 1) + \frac{e}{2!}(x - 1)^2 + \frac{e}{3!}(x - 1)^3 + \cdots$$

$$+ \frac{e}{n!}(x - 1)^n + \cdots$$

$$= e\left[1 + (x - 1) + \frac{1}{2!}(x - 1)^2 + \frac{1}{3!}(x - 1)^3 + \cdots \right.$$

$$\left. + \frac{1}{n!}(x - 1)^n + \cdots \right]$$

Example 3

Find the Taylor series expansion for $f(x) = \sin x$ at $a = \pi/2$.

$$f(x) = \sin x, \qquad f\left(\frac{\pi}{2}\right) = 1$$

$$f'(x) = \cos x, \qquad f'\left(\frac{\pi}{2}\right) = 0$$

$$f''(x) = -\sin x, \qquad f''\left(\frac{\pi}{2}\right) = -1$$

$$f'''(x) = -\cos x, \qquad f'''\left(\frac{\pi}{2}\right) = 0$$

$$f^{(iv)}(x) = \sin x, \qquad f^{(iv)}\left(\frac{\pi}{2}\right) = 1$$

So,

$$f(x) = \sin x = 1 - \frac{\left(x - \frac{\pi}{2}\right)^2}{2!} + \frac{\left(x - \frac{\pi}{2}\right)^4}{4!} - \cdots$$

Exercises

Find the Taylor series expansions for each function for the given value of a.

1. $f(x) = \cos x$, $a = \dfrac{\pi}{2}$ **2.** $f(x) = \sin x$, $a = \dfrac{\pi}{4}$ **3.** $f(x) = e^x$, $a = 2$

4. $f(x) = \sqrt{x}$, $a = 4$ **5.** $f(x) = \sqrt{x}$, $a = 9$ **6.** $f(x) = \tan x$, $a = \dfrac{\pi}{4}$

7. $f(x) = \dfrac{1}{x}$, $a = 2$ **8.** $f(x) = e^{-x}$, $a = 1$ **9.** $f(x) = \ln x$, $a = 1$

10. $f(x) = x \ln x$, $a = 1$ **11.** $f(x) = \dfrac{1}{\sqrt{x}}$, $a = 1$ **12.** $f(x) = \dfrac{1}{1 + 2x}$, $a = 1$

13. $f(x) = \dfrac{1}{x^2}$, $a = 1$ **14.** $f(x) = \cos x$, $a = \dfrac{\pi}{3}$ **15.** $f(x) = \cos x$, $a = \pi$

16. $f(x) = e^{-x}$, $a = -3$

10.9 COMPUTATIONAL APPROXIMATIONS

One of the important uses of power series expansions is to compute the numerical values of transcendental functions.

Example 1

Calculate ln 1.1.

From Exercise 9 in Section 10.8 we found that

$$\ln x = (x - 1) - \frac{(x-1)^2}{2} + \frac{(x-1)^3}{3} - \frac{(x-1)^4}{4} + \cdots$$

then

$$\ln 1.1 = (1.1 - 1) - \frac{(1.1-1)^2}{2} + \frac{(1.1-1)^3}{3} - \frac{(1.1-1)^4}{4} + \cdots$$

$$= 0.1 - \frac{(0.1)^2}{2} + \frac{(0.1)^3}{3} - \frac{(0.1)^4}{4} + \cdots$$

$$= 0.095308 \qquad \text{(sum of the first four terms)}$$

Unlike the geometric series, it is difficult to compute the sum of a power series. Usually we must settle for an approximate value by simply evaluating only the first several terms of the series.

Example 2

Calculate $e^{-0.2}$.

From

$$e^x = 1 + x + \frac{x^2}{2!} + \frac{x^3}{3!} + \cdots,$$

we find

$$e^{-0.2} = 1 - 0.2 + \frac{(-0.2)^2}{2!} + \frac{(-0.2)^3}{3!} + \cdots$$

$$= 0.81867 \qquad \text{(sum of the first four terms)}$$

Example 3

Calculate sin 3°.

From

$$\sin x = x - \frac{x^3}{3!} + \frac{x^5}{5!} - \cdots$$

and the first two terms, we have

$$\sin 3° = \sin \frac{\pi}{60} = \frac{\pi}{60} - \frac{\left(\frac{\pi}{60}\right)^3}{3!} + \cdots$$

$$= 0.05236 - 0.00002$$

$$= 0.05234$$

Example 4

Calculate cos 32°.

The Taylor series expansion for $f(x) = \cos x$ at $a = \pi/6$ is found as follows:

$$f(x) = \cos x \qquad f\left(\frac{\pi}{6}\right) = \frac{\sqrt{3}}{2}$$

$$f'(x) = -\sin x \qquad f'\left(\frac{\pi}{6}\right) = -\frac{1}{2}$$

$$f''(x) = -\cos x \qquad f''\left(\frac{\pi}{6}\right) = -\frac{\sqrt{3}}{2}$$

$$f'''(x) = \sin x \qquad f'''\left(\frac{\pi}{6}\right) = \frac{1}{2}$$

So

$$\cos x = \frac{\sqrt{3}}{2} - \frac{1}{2}\left(x - \frac{\pi}{6}\right) - \frac{\sqrt{3}}{2}\frac{(x - \pi/6)^2}{2!} + \frac{1}{2}\frac{(x - \pi/6)^3}{3!} - \cdots$$

Note: We need to write $x = 32° = 30° + 2° = \dfrac{\pi}{6} + \dfrac{\pi}{90}$.

Then

$$x - a = x - \frac{\pi}{6} = \left(\frac{\pi}{6} + \frac{\pi}{90}\right) - \frac{\pi}{6} = \frac{\pi}{90}.$$

and

$$\cos x = \frac{\sqrt{3}}{2} - \frac{1}{2}\left(\frac{\pi}{90}\right) - \frac{\sqrt{3}}{2}\frac{(\pi/90)^2}{2!} + \frac{1}{2}\frac{(\pi/90)^3}{3!} - \cdots$$

$$= 0.848048 \qquad \text{(using the first four nonzero terms)}$$

In more advanced texts it is possible to determine how many terms need to be used in approximating the value of a function to a desired accuracy. In the exercises that follow, the number of terms to be used will be specified.

If one desires to evaluate $e^{1.1}$, then it is better to use the Taylor expansion with $a = 1$:

$$e^x = e\left[1 + (x - 1) + \frac{1}{2!}(x - 1)^2 + \frac{1}{3!}(x - 1)^3 + \cdots\right]$$

rather than the Maclaurin series

$$e^x = 1 + x + \frac{x^2}{2!} + \frac{x^3}{3!} + \cdots$$

because the powers of $x - 1$ become smaller faster than do the powers of x when $x = 1.1$. Thus, an accurate approximation can be obtained with the Taylor expansion using fewer terms. This observation illustrates the importance of the Taylor series expansion.

Exercises

Calculate the value of each expression.

1. $e^{0.1}$ (Use first four nonzero terms.)
2. $e^{-0.3}$ (Use first four nonzero terms.)
3. $\cos 1°$ (Use first two nonzero terms.)
4. $\sin 2°$ (Use first two nonzero terms.)
5. $\ln 0.5$ (Use first four nonzero terms.)
6. $\ln 1.5$ (Use first four nonzero terms.)

7. $\sqrt{1.1}$ (Use first four nonzero terms.) 8. $\sqrt{0.9}$ (Use first four nonzero terms.)

9. $e^{1.3}$ (Use first four nonzero terms.) 10. $\sqrt{3.9}$ (Use first four nonzero terms.)

11. $\sin 29°$ (Use first three nonzero terms.)

12. $e^{0.9}$ (Use first four nonzero terms.)

13. Find the value of a current $i = \sin \omega t$ when $\omega t = 0.03$ rad.

14. Find the value of a current $i = 3e^{t^2}$ when $t = 0.1$ s.

10.10 FOURIER SERIES

One of the difficulties with the Taylor series expansions is that, in general, they can be used to represent a given function only for values of x close to a [when expanded in powers of $(x - a)$]. A Fourier series expansion is often used when it is necessary to approximate a function over a larger interval of values of x.

The following expression is called a Fourier series expansion representing $f(x)$.

Fourier Series

$$f(x) = a_0 + a_1 \cos x + a_2 \cos 2x + \cdots + a_n \cos nx + \cdots$$
$$+ b_1 \sin x + b_2 \sin 2x + \cdots + b_n \sin nx + \cdots$$

The coefficients are determined as follows:

$$a_0 = \frac{1}{2\pi} \int_0^{2\pi} f(x) \, dx$$

$$a_n = \frac{1}{\pi} \int_0^{2\pi} f(x) \cos nx \, dx$$

$$b_n = \frac{1}{\pi} \int_0^{2\pi} f(x) \sin nx \, dx$$

Note that n is a positive integer and that the Fourier series expansion is periodic with period 2π.

Four of the basic periodic waves that commonly occur in the analysis of electrical and mechanical systems are shown in Fig. 10.1.

Example 1

Find the Fourier series which represents the wave function $f(x) = x$ $(0 \le x < 2\pi)$ with period 2π shown in Fig. 10.2.

Finding a_0: $a_0 = \frac{1}{2\pi} \int_0^{2\pi} x \, dx = \frac{1}{2\pi} \frac{x^2}{2} \Big|_0^{2\pi} = \pi$

Finding a_n: $a_n = \frac{1}{\pi} \int_0^{2\pi} x \cos nx \, dx$

From Appendix B, Table 10, Formula 81, we have

$$\int u \cos u \, du = \cos u + u \sin u + C$$

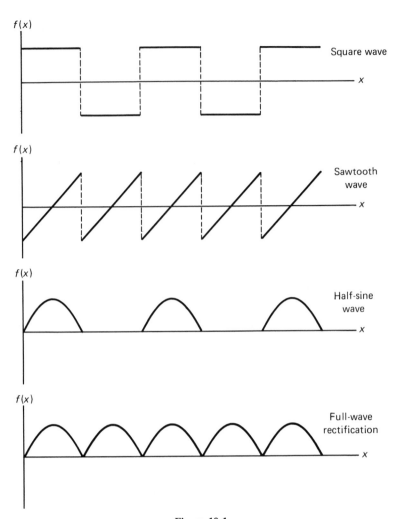

Figure 10.1

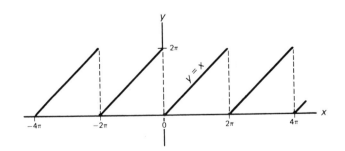

Figure 10.2

$$\int x \cos nx \, dx = \int \frac{u}{n} \cos u \, \frac{du}{n}$$

$u = nx$
$du = n \, dx$

$$= \frac{1}{n^2} \int u \cos u \, du$$

$$= \frac{1}{n^2}(\cos u + u \sin u) + C$$

$$= \frac{1}{n^2}(\cos nx + nx \sin nx) + C$$

Thus

$$a_n = \frac{1}{\pi n^2}(\cos nx + nx \sin nx)\Big|_0^{2\pi}$$

$$= \frac{1}{\pi n^2}[(\cos 2n\pi + 2n\pi \sin 2n\pi) - (\cos 0 + n(0)\sin 0)]$$

$$= \frac{1}{\pi n^2}(1 + 0 - 1 - 0) = 0 \qquad \text{(recall that } n \text{ is a positive integer.)}$$

Finding b_n: $\qquad b_n = \frac{1}{\pi}\int_0^{2\pi} x \sin nx \, dx$

From Table 10, Formula 80, we have

$$\int u \sin u \, du = \sin u - u \cos u + C$$

So

$$\int x \sin nx \, dx = \int \frac{u}{n} \sin u \frac{du}{n}$$

$u = nx$
$du = n \, dx$

$$= \frac{1}{n^2} \int u \sin u \, du$$

$$= \frac{1}{n^2}(\sin u - u \cos u) + C$$

$$= \frac{1}{n^2}(\sin nx - nx \cos nx) + C$$

Thus

$$b_n = \frac{1}{\pi n^2}(\sin nx - nx \cos nx)\Big|_0^{2\pi}$$

$$= \frac{1}{\pi n^2}\{(\sin 2n\pi - 2n\pi \cos 2n\pi) - [\sin 0 - n(0)\cos 0]\}$$

$$= \frac{1}{\pi n^2}(0 - 2n\pi - 0 + 0)$$

$$= -\frac{2}{n}$$

That is, $b_1 = -2$, $b_2 = -1$, $b_3 = -\frac{2}{3}, \ldots$ And the Fourier series is

$$f(x) = \pi - 2 \sin x - \sin 2x - \tfrac{2}{3} \sin 3x - \cdots$$

In this example there are no terms involving the cosine because $a_1 = a_2 = a_3 = \ldots = a_n = 0$. The graph of the sum of the first few terms is sketched in Fig. 10.3.

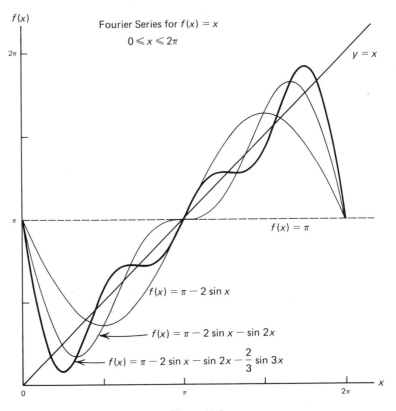

Figure 10.3

Example 2

In Fig. 10.4, find the Fourier series for the wave function given by

$$f(x) = \begin{cases} \pi, & 0 \leqslant x < \pi \\ 2\pi - x, & \pi \leqslant x < 2\pi \end{cases}$$

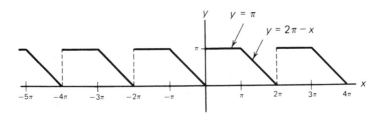

Figure 10.4

Finding a_0: $\quad a_0 = \dfrac{1}{2\pi}\displaystyle\int_0^\pi \pi\, dx + \dfrac{1}{2\pi}\int_\pi^{2\pi} (2\pi - x)\, dx$

$$= \dfrac{1}{2\pi}(\pi x)\Big|_0^\pi + \dfrac{1}{2\pi}\left(2\pi x - \dfrac{x^2}{2}\right)\Big|_\pi^{2\pi}$$

$$= \dfrac{\pi}{2} + \dfrac{\pi}{4} = \dfrac{3\pi}{4}$$

Note: Two separate integrals must be used to determine the coefficients. This is because the function is defined differently on the two intervals $0 \leq x < \pi$ and $\pi \leq x < 2\pi$.

Finding a_n: $\quad a_n = \dfrac{1}{\pi}\displaystyle\int_0^\pi \pi \cos nx\, dx + \dfrac{1}{\pi}\int_\pi^{2\pi} (2\pi - x) \cos nx\, dx$

$$= \int_0^\pi \cos nx\, dx + 2\int_\pi^{2\pi} \cos nx\, dx - \dfrac{1}{\pi}\int_\pi^{2\pi} x \cos nx\, dx$$

$$= \dfrac{1}{n}(\sin nx)\Big|_0^\pi + \dfrac{2}{n}(\sin nx)\Big|_\pi^{2\pi}$$

$$\quad - \dfrac{1}{\pi n^2}(\cos nx + nx \sin nx)\Big|_\pi^{2\pi}$$

$$= 0 + 0 - \dfrac{2}{\pi n^2} = -\dfrac{2}{\pi n^2} \qquad \text{(for } n \text{ odd)}$$

$$= 0 \qquad \text{(for } n \text{ even)}$$

Finding b_n: $\quad b_n = \dfrac{1}{\pi}\displaystyle\int_0^\pi \pi \sin nx\, dx + \dfrac{1}{\pi}\int_\pi^{2\pi} (2\pi - x) \sin nx\, dx$

$$= \int_0^\pi \sin nx\, dx + 2\int_\pi^{2\pi} \sin nx\, dx - \dfrac{1}{\pi}\int_\pi^{2\pi} x \sin nx\, dx$$

$$= \left(-\dfrac{1}{n}\right)(\cos nx)\Big|_0^\pi + \left(-\dfrac{2}{n}\right)(\cos nx)\Big|_\pi^{2\pi}$$

$$\quad - \dfrac{1}{\pi n^2}(\sin nx - nx \cos nx)\Big|_\pi^{2\pi}$$

$$= \dfrac{2}{n} - \dfrac{4}{n} + \dfrac{3}{n} = \dfrac{1}{n} \qquad \text{(for } n \text{ odd)}$$

$$= 0 \qquad \text{(for } n \text{ even)}$$

We thus obtain the Fourier series

$$f(x) = \dfrac{3\pi}{4} - \dfrac{2}{\pi}\left(\cos x + \dfrac{1}{9}\cos 3x + \dfrac{1}{25}\cos 5x + \cdots\right)$$

$$+ \left(\sin x + \dfrac{1}{3}\sin 3x + \dfrac{1}{5}\sin 5x + \cdots\right)$$

We will now show how the coefficients a_0, a_n, and b_n were obtained. Note that if we integrate each side of the equation

$$f(x) = a_0 + a_1 \cos x + a_2 \cos 2x + a_3 \cos 3x + \cdots$$

$$+ b_1 \sin x + b_2 \sin 2x + b_3 \sin 3x + \cdots$$

then the integrals should be equal. That is,

$$\int_0^{2\pi} f(x) \, dx = \int_0^{2\pi} a_0 \, dx + \int_0^{2\pi} a_1 \cos x \, dx + \int_0^{2\pi} a_2 \cos 2x \, dx + \cdots$$

$$+ \int_0^{2\pi} b_1 \sin x \, dx + \int_0^{2\pi} b_2 \sin 2x \, dx + \cdots$$

All terms on the right-hand side are zero except for $\int_0^{2\pi} a_0 \, dx = 2\pi a_0$.

So

$$\int_0^{2\pi} f(x) \, dx = 2\pi a_0$$

$$a_0 = \frac{1}{2\pi} \int_0^{2\pi} f(x) \, dx$$

Multiply each side of the integral series equation above by $\cos nx$. Then

$$\int_0^{2\pi} f(x) \cos nx \, dx = \int_0^{2\pi} a_0 \cos nx \, dx + \int_0^{2\pi} a_1 (\cos nx) \cos x \, dx$$

$$+ \int_0^{2\pi} a_2 (\cos nx) \cos 2x \, dx + \cdots$$

$$+ \int_0^{2\pi} b_1 (\cos nx) \sin x \, dx$$

$$+ \int_0^{2\pi} b_2 (\cos nx) \sin 2x \, dx + \cdots$$

All terms on the right-hand side are found to be zero except the term

$$\int_0^{2\pi} a_n (\cos nx)(\cos nx) \, dx = \pi a_n$$

So

$$\int_0^{2\pi} f(x) \cos nx \, dx = \pi a_n$$

$$a_n = \frac{1}{\pi} \int_0^{2\pi} f(x) \cos nx \, dx$$

In a similar manner (multiplying each side of the Fourier series equation by $\sin nx$) we can show that

$$b_n = \frac{1}{\pi} \int_0^{2\pi} f(x) \sin nx \, dx.$$

Note: If the function to be analyzed ranges periodically from $-\pi$ to π, then the coefficients become

$$a_0 = \frac{1}{2\pi} \int_{-\pi}^{\pi} f(x) \, dx$$

$$a_n = \frac{1}{\pi} \int_{-\pi}^{\pi} f(x) \cos nx \, dx$$

$$b_n = \frac{1}{\pi} \int_{-\pi}^{\pi} f(x) \sin nx \, dx$$

Exercises

Find the Fourier series for each function.

1. $f(x) = -x, \ 0 \le x < 2\pi$

2. $f(x) = 2x, \ 0 \le x < 2\pi$

3. $f(x) = \frac{1}{3} x, \ 0 \le x < 2\pi$

4. $f(x) = 2x, \ -\pi \le x < \pi$

5. $f(x) = \begin{cases} 0, & 0 \le x < \pi \\ 1, & \pi \le x < 2\pi \end{cases}$

6. $f(x) = \begin{cases} \pi, & 0 \le x < \pi \\ 0, & \pi \le x < 2\pi \end{cases}$

7. $f(x) = \begin{cases} 1, & 0 \le x < \pi \\ -1, & \pi \le x < 2\pi \end{cases}$

8. $f(x) = \begin{cases} x, & 0 \le x < \pi \\ \pi, & \pi \le x < 2\pi \end{cases}$

9. $f(x) = \begin{cases} x, & 0 \le x < \pi \\ 2\pi - x, & \pi \le x < 2\pi \end{cases}$

10. $f(x) = \begin{cases} 0, & 0 \le x < \pi \\ x, & \pi \le x < 2\pi \end{cases}$

11. $f(x) = e^x, \ 0 \le x < 2\pi$

12. $f(x) = e^{-2x}, \ 0 \le x < 2\pi$

13. $f(x) = \begin{cases} \sin x, & 0 \le x < \pi \\ 0, & \pi \le x < 2\pi \end{cases}$

CHAPTER SUMMARY

1. An arithmetic progression is a sequence of terms where each term differs from the immediately preceding term by a fixed number, d, which is called the common difference. The general form of an arithmetic progression is written

$$a, a + d, a + 2d, a + 3d, \ldots, a + (n - 1)d$$

(a) The nth or last term l of such a finite arithmetic progression is given by

$$l = a + (n - 1)d$$

(b) The sum of the first n terms of a finite arithmetic progression is given by

$$S = \frac{n}{2}(a + l)$$

2. A geometric progression is a sequence of terms each of which can be obtained by multiplying the preceding term by a fixed number, r, which is called the common ratio. The general form of a geometric progression is given by

$$a, ar, ar^2, ar^3, \ldots, ar^{n-1}$$

(a) The nth or last term l of such a finite geometric progression is given by

$$l = ar^{n-1}$$

(b) The sum of the first n terms of a finite geometric progression is given by

$$S = \frac{a(1 - r^n)}{1 - r}$$

(c) The sum of the infinite geometric series, where $|r| < 1$, is

$$S = \frac{a}{1 - r}$$

3. *Definition of convergence and divergence*

$$\sum_{n=1}^{\infty} a_n = a_1 + a_2 + a_3 + \cdots$$

Then

$$S_1 = a_1$$
$$S_2 = a_1 + a_2$$
$$S_3 = a_1 + a_2 + a_3$$
$$\cdot$$
$$\cdot$$
$$\cdot$$
$$S_n = a_1 + a_2 + a_3 + \cdots + a_n$$
$$\cdot$$
$$\cdot$$
$$\cdot$$

where S_1 is the sum of the first term, S_2 is the sum of the first two terms, S_3 is the sum of the first three terms, ..., S_n is the sum of the first n terms (or sometimes called the nth partial sum).

(a) If $\lim_{n \to \infty} S_n = S$ (where S is finite), the series $\sum_{n=1}^{\infty} a_n$ *converges* and S is the *sum of the infinite series.*

(b) If $\lim_{n \to \infty} S_n$ does not exist, the series $\sum_{n=1}^{\infty} a_n$ *diverges.*

4. *Test for divergence of a series:* If $\lim_{n \to \infty} a_n \neq 0$, then the series *diverges.*

5. *nth Term test for convergence of a series:* If the series $\sum_{n=1}^{\infty} a_n$ *converges,* then $\lim_{n \to \infty} a_n = 0$.

6. *Convergence and divergence of a p-series:* Any series in the form $$\sum_{n=1}^{\infty} \frac{1}{n^p} = \frac{1}{1^p} + \frac{1}{2^p} + \frac{1}{3^p} + \cdots,$$ where p is a real number, is called a *p*-series. The *p*-series
(a) converges for $p > 1$, and
(b) diverges for $p \leq 1$.

7. *Comparison test for convergence and divergence:* Let N be a positive integer, $\sum_{n=1}^{\infty} a_n$ and $\sum_{n=1}^{\infty} b_n$ be series of positive terms, and $0 \leqslant a_n \leqslant b_n$ for all $n > N$, then:

(a) If $\sum_{n=1}^{\infty} b_n$ converges, then $\sum_{n=1}^{\infty} a_n$ also converges.

(b) If $\sum_{n=1}^{\infty} a_n$ diverges, then $\sum_{n=1}^{\infty} b_n$ also diverges.

   In other words, the comparison test says:
(a) A series of positive terms that is term by term smaller than a known convergent series must also converge.
(b) A series of positive terms that is term by term larger than a known divergent series must also diverge.

8. Let $\sum_{n=1}^{\infty} a_n$ and $\sum_{n=1}^{\infty} b_n$ be two series of positive terms.

(a) Then $\sum_{n=1}^{\infty} a_n$ and $\sum_{n=1}^{\infty} b_n$ have the *same order of magnitude* if $\lim_{n\to\infty} \dfrac{a_n}{b_n} = L$, where L is a real number and $L > 0$.

(b) The series $\sum_{n=1}^{\infty} a_n$ has a *lesser order of magnitude* than $\sum_{n=1}^{\infty} b_n$ if
$$\lim_{n\to\infty} \frac{a_n}{b_n} = 0.$$

(c) The series $\sum_{n=1}^{\infty} a_n$ has a *greater order of magnitude* than $\sum_{n=1}^{\infty} b_n$ if
$$\lim_{n\to\infty} \frac{a_n}{b_n} = \infty.$$

9. *Limit comparison test:* Let $\sum_{n=1}^{\infty} a_n$ and $\sum_{n=1}^{\infty} b_n$ be series of positive terms.

(a) If both series have the same order of magnitude, then either both series converge or both series diverge.

(b) If the series $\sum_{n=1}^{\infty} a_n$ has a lesser order of magnitude than $\sum_{n=1}^{\infty} b_n$ and $\sum_{n=1}^{\infty} b_n$ is known to converge, then $\sum_{n=1}^{\infty} a_n$ also converges.

(c) If $\sum_{n=1}^{\infty} a_n$ has a greater order of magnitude than $\sum_{n=1}^{\infty} b_n$ and $\sum_{n=1}^{\infty} b_n$ is known to diverge, then $\sum_{n=1}^{\infty} a_n$ also diverges.

10. *Ratio test for convergence and divergence:* Let $\displaystyle\sum_{n=1}^{\infty} a_n$ be a series of positive terms and

$$r = \lim_{n\to\infty} \frac{a_{n+1}}{a_n}$$

(a) If $r < 1$, the series converges.
(b) If $r > 1$ (including $r = \infty$), the series diverges.
(c) If $r = 1$, the test fails. Some other test must be used.

11. *Integral test for convergence and divergence:* Let $\displaystyle\sum_{n=1}^{\infty} a_n$ be a series of positive terms, $f(x)$ be a continuous, decreasing function for $x \geqslant 1$ such that $f(n) = a_n$ for all positive integers n. Then either $\displaystyle\sum_{n=1}^{\infty} a_n$ and $\displaystyle\int_1^{\infty} f(x)\,dx$ both converge or both diverge.

12. *Alternating series test:* The alternating series

$$\sum_{n=1}^{\infty} (-1)^{n+1} a_n = a_1 - a_2 + a_3 - a_4 + \cdots \qquad (a_n > 0 \text{ for each } n)$$

converges provided that both conditions are fulfilled:
(a) $0 < a_{n+1} \leqslant a_n$ for $n \geqslant 1$ and
(b) $\displaystyle\lim_{n\to\infty} a_n = 0$

In addition, if S is the sum of the infinite series and S_n is the nth partial sum, then

$$|S - S_n| \leqslant a_{n+1}$$

13. *Absolute and conditional convergence:* Suppose that $\displaystyle\sum_{n=1}^{\infty} a_n$ converges.

(a) If $\displaystyle\sum_{n=1}^{\infty} |a_n|$ converges, then $\displaystyle\sum_{n=1}^{\infty} a_n$ *converges absolutely.*

(b) If $\displaystyle\sum_{n=1}^{\infty} |a_n|$ diverges, then $\displaystyle\sum_{n=1}^{\infty} a_n$ *converges conditionally.*

14. A *power series* is an infinite series in the form

$$\sum_{n=0}^{\infty} a_n x^n = a_0 + a_1 x + a_2 x^2 + \cdots + a_n x^n + \cdots$$

A series in the form

$$\sum_{n=0}^{\infty} a_n(x - a)^n = a_0 + a_1(x - a) + a_2(x - a)^2 + a_3(x - a)^3$$

$$+ \cdots + a_n(x - a)^n + \cdots$$

is called a *power series centered at a.* Let

$$\lim_{n \to \infty} \left| \frac{u_{n+1}}{u_n} \right| = r(x)$$

where u_{n+1} is the $(n + 1)$st term and u_n is the nth term of a power series. Note that the ratio will most often be a function of x, or $r(x)$. Then the series

(a) converges absolutely for $r(x) < 1$ and

(b) diverges for $r(x) > 1$.

Recall that the ratio test is not valid for $r = 1$ or $r(x) = 1$. These values of x have to be checked individually.

15. A power series converges on an interval, called the *interval of convergence*. The interval of convergence may include both endpoints, only one endpoint, or neither endpoint. The interval of convergence of the power series

$$\sum_{n=0}^{\infty} a_n (x - a)^n$$

is centered at $x = a$. The *radius of convergence* is the distance from the point $x = a$ to either endpoint of the interval. Thus the radius of convergence is one-half the length of the interval of convergence.

Power series is centered at a.
radius of convergence, R
x-axis

$a - R$ a $a + R$

interval of convergence

16. The Maclaurin series expansion of the function $f(x)$ is

$$f(x) = f(0) + f'(0)x + \frac{f''(0)}{2!} x^2 + \frac{f'''(0)}{3!} x^3 + \cdots + \frac{f^{(n)}(0)}{n!} x^n + \cdots$$

The expansion is valid for all values of x for which the power series converges and for which the function $f(x)$ is repeatedly differentiable.

17. A Taylor series expansion of $f(x)$ is a power series in the form

$$f(x) = f(a) + f'(a)(x - a) + \frac{f''(a)}{2!}(x - a)^2 + \frac{f'''(a)}{3!}(x - a)^3 + \cdots$$

$$+ \frac{f^{(n)}(a)}{n!}(x - a)^n + \cdots$$

Note that a Maclaurin series is a special case of a Taylor series with $a = 0$.

18. The Fourier series expansion of the function $f(x)$ is

$$f(x) = a_0 + a_1 \cos x + a_2 \cos 2x + \cdots + a_n \cos nx + \cdots$$

$$+ b_1 \sin x + b_2 \sin 2x + \cdots + b_n \sin nx + \cdots$$

The coefficients are determined as follows:

$$a_0 = \frac{1}{2\pi} \int_0^{2\pi} f(x)\, dx$$

$$a_n = \frac{1}{\pi} \int_0^{2\pi} f(x) \cos nx\, dx$$

$$b_n = \frac{1}{\pi} \int_0^{2\pi} f(x) \sin nx\, dx$$

Note that n is a positive integer and that the Fourier series expansion is periodic with period 2π.

CHAPTER REVIEW

Find the nth term of each progression.

1. $3, 7, 11, 15, \ldots, n = 12$

2. $4, 2, 1, \frac{1}{2}, \ldots, n = 7$

3. $\sqrt{3}, -3, 3\sqrt{3}, -9, \ldots, n = 8$

4. $4, -2, -8, -14, \ldots, n = 12$

5. $6, 2, \frac{2}{3}, \frac{2}{9}, \ldots, n = 6$

6. $5, 15, 25, 35, \ldots, n = 10$

7–12. Find the sum of the first n terms of the progressions in Exercises 1–6.

13. Find the sum of the first 1000 even positive integers.

14. If \$500 is deposited annually in a savings account at 6% interest compounded annually, find the total amount in the account after five years.

Find the sum, when possible, for each infinite geometric series.

15. $3 + 6 + 12 + \cdots$

16. $5 + \frac{5}{7} + \frac{5}{49} + \cdots$

17. $2 - \frac{2}{3} + \frac{2}{9} - \cdots$

18. $3 + \frac{9}{2} + \frac{27}{4} + \cdots$

19. Find the fraction equivalent to $0.454545\ldots$.

20. Find the fraction equivalent to $0.9212121\ldots$.

Write the expanded form of each series.

21. $\displaystyle\sum_{n=1}^{6} (1 - 3n)$

22. $\displaystyle\sum_{n=1}^{n} \frac{n+1}{n}$

Write each sum using sigma notation.

23. $\dfrac{1}{3} + \dfrac{1}{9} + \dfrac{1}{27} + \cdots + \dfrac{1}{2187}$

24. $\dfrac{1}{4} + \dfrac{2}{5} + \dfrac{3}{6} + \dfrac{4}{7} + \cdots + \dfrac{10}{13}$

Determine whether each series converges or diverges.

25. $\displaystyle\sum_{n=1}^{\infty} \frac{1}{n^3}$

26. $\displaystyle\sum_{n=1}^{\infty} \frac{1}{\sqrt[4]{n}}$

27. $\displaystyle\sum_{n=1}^{\infty} \frac{1}{6n^2 + 2}$

28. $\displaystyle\sum_{n=2}^{\infty} \frac{\sqrt{n}}{n^2 - 1}$

29. $\displaystyle\sum_{n=2}^{\infty} \frac{n}{\ln n}$

30. $\displaystyle\sum_{n=1}^{\infty} \frac{5n + 2}{(3n + 1) 4^n}$

31. $\displaystyle\sum_{n=1}^{\infty} \frac{n^3}{2^n}$

32. $\displaystyle\sum_{n=1}^{\infty} \frac{3n + 1}{4n - 5}$

33. $\displaystyle\sum_{n=1}^{\infty} \frac{n+1}{n^2+4n}$

34. $\displaystyle\sum_{n=1}^{\infty} \frac{\sin n}{n^2+1}$

Determine whether each alternating series converges or diverges. If it converges, find whether it converges absolutely or converges conditionally.

35. $\displaystyle\sum_{n=1}^{\infty} \frac{(-1)^{n+1}\, 3^n}{n!}$

36. $\displaystyle\sum_{n=1}^{\infty} \frac{(-1)^{n+1}\, 2^n}{5^n\,(n+1)}$

37. $\displaystyle\sum_{n=2}^{\infty} (-1)^n \frac{n+1}{n-1}$

38. $\displaystyle\sum_{n=1}^{\infty} (-1)^{n+1} \frac{n+2}{n(n+3)}$

Find the interval of convergence of each series.

39. $\displaystyle\sum_{n=0}^{\infty} \frac{n}{2}(x-2)^n$

40. $\displaystyle\sum_{n=0}^{\infty} n^2(x-3)^n$

41. $\displaystyle\sum_{n=1}^{\infty} \frac{(x-1)^n}{n!}$

42. $\displaystyle\sum_{n=1}^{\infty} \frac{3^n(x-4)^n}{n^2}$

Find a Maclaurin series expansion for each function.

43. $f(x) = \dfrac{1}{1-x}$

44. $f(x) = \sqrt{x+1}$

45. $f(x) = \sin x + \cos x$

46. $f(x) = e^x \sin x$

47. $f(x) = \dfrac{1-e^x}{x}$

48. $f(x) = \cos x^2$

49. $f(x) = \sin 3x$

50. $f(x) = e^{\sin x}$

51. Evaluate $\displaystyle\int_0^{0.1} \frac{\ln(x+1)}{x}\,dx$. (Use first four nonzero terms.)

52. If $i = \dfrac{\sin t}{t}$ amp, find the amount of charge q (in coulombs) transmitted by this current from $t = 0$ to $t = 0.1$ s. $\left(q = \displaystyle\int i\,dt. \right)$

Find the Taylor series expansions for each function for the given value of a.

53. $f(x) = \cos 2x,\ a = 0$

54. $f(x) = \ln x,\ a = 4$

55. $f(x) = e^{x^2},\ a = 1$

56. $f(x) = \sin x,\ a = \dfrac{3\pi}{2}$

Calculate the value of each expression.

57. $\sin 31°$ (Use first three nonzero terms.)

58. $e^{1.2}$ (Use first four nonzero terms.)

59. $\ln 1.2$ (Use first four nonzero terms.)

60. $\sqrt{4.1}$ (Use first four nonzero terms.)

Find the Fourier series for each function.

61. $f(x) = \begin{cases} 0 & 0 \leqslant x < \pi \\ -1 & \pi \leqslant x < 2\pi \end{cases}$

62. $f(x) = \begin{cases} x^2 & 0 \leqslant x < \pi \\ 0 & \pi \leqslant x < 2\pi \end{cases}$

11

FIRST-ORDER DIFFERENTIAL EQUATIONS

11.1 SOLVING DIFFERENTIAL EQUATIONS

Often in physics, engineering, and other technical areas there is a need to search for an unknown function. In many cases, this search leads to an equation involving derivatives (or differentials) of the unknown function. Such equations involving derivatives (or differentials) are called *differential equations*.

In Chapters 11 and 12 we will present methods of solving differential equations. That is, we will find ways in which we can successfully use differential equations to determine an unknown function. The following are examples of differential equations.

A. $\dfrac{dy}{dx} = x^2 y$

B. $\dfrac{dy}{dx} = \sin x$

C. $\dfrac{d^2 y}{dx^2} + x \dfrac{dy}{dx} + y = 0$

D. $x^2 \dfrac{d^3 y}{dx^3} + 2y \dfrac{d^2 y}{dx^2} - \left(\dfrac{dy}{dx}\right)^4 + 3 = 0$

E. $e^x \, dy - x^2 y \, dx = 2$

The *order of a differential equation* is the order of the derivative of highest order in the equation.

Example 1

Determine the order of the differential equation

$$\frac{d^2y}{dx^2} + 2\left(\frac{dy}{dx}\right)^3 + 5 = 0$$

The order is 2 since the second derivative $\frac{d^2y}{dx^2}$ (order 2 derivative) is the highest-order derivative appearing in the equation.

The *degree of a differential equation* is the highest power of the derivative of highest order.

Example 2

Determine the degree and order of each differential equation:

(a)
$$\frac{d^2y}{dx^2} - 7\frac{dy}{dx} + \left(\frac{dy}{dx}\right)^3 = 0$$

(b)
$$\left(\frac{dy}{dx}\right)^2 - 3\frac{dy}{dx} + y = 0$$

Equation (a) is a first-degree differential equation of order 2 since $\frac{d^2y}{dx^2}$ is the highest-order derivative in the equation and is raised to the first power. Note that the third power of $\frac{dy}{dx}$ has no effect on the degree of equation (a) because $\frac{dy}{dx}$ is of lesser order than $\frac{d^2y}{dx^2}$.

Equation (b) is a second-degree, first-order differential equation. $\frac{dy}{dx}$ is the highest-order derivative (order 1) and 2 is the highest power of $\frac{dy}{dx}$ appearing in the equation.

A *solution* of a differential equation is a function $y = f(x)$ which together with its derivatives satisfies the given differential equation.

Example 3

Verify that $y = x^2 + 5x$ is a solution of the second-order, first-degree differential equation $x\frac{d^2y}{dx^2} - \frac{dy}{dx} + 5 = 0$.

First, find $\frac{dy}{dx} = 2x + 5$ and $\frac{d^2y}{dx^2} = 2$. Then substitute in the differential equation:

$$x(2) - (2x + 5) + 5 = 0$$
$$2x - 2x - 5 + 5 = 0$$
$$0 = 0$$

Example 4

Verify that $y = \dfrac{1}{x^2 + C}$ is a solution of the first-order, first-degree differential equation $\dfrac{dy}{dx} = -2xy^2$.

First, from the given equation, $\dfrac{dy}{dx} = -\dfrac{2x}{(x^2 + C)^2}$. Substitute this result in the given differential equation:

$$\frac{-2x}{(x^2 + C)^2} = -2x\left(\frac{1}{x^2 + C}\right)^2$$

$$\frac{-2x}{(x^2 + C)^2} = \frac{-2x}{(x^2 + C)^2}$$

The solution $y = x^2 + 5x$ in Example 3 is an example of a *particular solution* of a differential equation. One can verify that $y = x^2 + 5x - 7$ is also a particular solution of the differential equation in Example 3. Thus a differential equation can have more than one particular solution.

A solution $y = f(x)$ of a differential equation of order n containing n arbitrary constants is called a *general solution*. Thus the solution $y = \dfrac{1}{x^2 + C}$ in Example 4 or $y = x^2 + 5x + C$ in Example 3 is an example of a general solution.

A complete study of differential equations would include a study of differential equations of all degrees. However, we will be considering only first-degree equations. Differential equations which do not contain partial derivatives are called *ordinary differential equations*. We restrict our considerations to first-degree ordinary differential equations.

Recall the use of other notations for derivatives:

$$y' = \frac{dy}{dx},\ y'' = \frac{d^2y}{dx^2},\ y''' = \frac{d^3y}{dx^3},\ \text{etc.}$$

Example 5

Verify that $y = C_1 + C_2x + C_3e^x$ is a general solution of the differential equation $y''' = y''$.

First, find the first three derivatives of the given function:

$$y' = C_2 + C_3e^x$$

$$y'' = C_3e^x$$

$$y''' = C_3e^x$$

Substituting in the differential equation, we have

$$y''' = y''$$

$$C_3e^x = C_3e^x$$

Therefore, $y = C_1 + C_2x + C_3e^x$ is a general solution of the given differential equation with three distinct arbitrary constants.

Exercises

State the order and degree of each differential equation.

1. $\dfrac{dy}{dx} = x^2 - y^2$

2. $\left(\dfrac{dy}{dx}\right)^2 - 3x\dfrac{dy}{dx} + 2 = 0$

3. $\dfrac{d^2y}{dx^2} + 5xy\dfrac{dy}{dx} = x^2y$

4. $x^2\dfrac{dy}{dx} + y\left(\dfrac{dy}{dx}\right)^2 = 0$

5. $y''' - 4y'' + xy = 0$

6. $y' + x\cos x = 0$

7. $(y'')^3 - xy' + y'' = 0$ **8.** $y'' + e^x y = 2$

Verify that each function $y = f(x)$ is a solution of the differential equation.

9. $\dfrac{dy}{dx} = 3;\ y = 3x - 7$

10. $\dfrac{dy}{dx} + y + 2x + 4 = x^2;\ y = x^2 - 4x$

11. $x\dfrac{dy}{dx} - 2y = 4x;\ y = x^2 - 4x$

12. $\dfrac{d^2y}{dx^2} + y = 0;\ y = 2\sin x + 3\cos x$

13. $\dfrac{dy}{dx} + y = e^{-x};\ y = (x + 2)e^{-x}$ **14.** $x\dfrac{dy}{dx} = x^2 + y;\ y = x^2 + Cx$

15. $\dfrac{d^2y}{dx^2} + 16y = 0;\ y = C_1 \sin 4x + C_2 \cos 4x$

16. $\dfrac{d^2y}{dx^2} = 20x^3;\ y = x^5 + 3x - 2$

17. $\dfrac{dy}{dx} + y - 2\cos x = 0;\ y = \sin x + \cos x - e^{-x}$

18. $\dfrac{d^2y}{dx^2} - y + x^2 = 2;\ y = e^{-x} + x^2$

19. $\left(\dfrac{d^2y}{dx^2}\right)^2 + 4\left(\dfrac{dy}{dx}\right)^2 = 4;\ y = \sin x \cos x$

20. $\dfrac{d^2y}{dx^2} = 9y;\ y = e^{3x}$

21. $\dfrac{d^2y}{dx^2} - 5\left(\dfrac{dy}{dx}\right) + 4y = 0;\ y = e^{4x}$ **22.** $\dfrac{d^2y}{dx^2} + 2\left(\dfrac{dy}{dx}\right) + y = 0;\ y = e^{-x}$

23. $\dfrac{d^2y}{dx^2} + 2\left(\dfrac{dy}{dx}\right) + y = 0;\ y = xe^{-x}$

24. $\dfrac{d^2y}{dx^2} - 2\left(\dfrac{dy}{dx}\right) + y = -\dfrac{e^{-x}}{x^2}$

11.2 SEPARATION OF VARIABLES

There are numerous methods developed for solving ordinary differential equations. We present a few of these methods. Certain first-order differential equations can be solved most easily by using the method of separation of variables.

A first-order differential equation is a relation involving the first derivative. That is, it can be written in the form

$$N(x, y)\frac{dy}{dx} + M(x, y) = 0 \tag{1}$$

or (by multiplying each side by the differential dx, where $dx \neq 0$).

$$M(x, y)\, dx + N(x, y)\, dy = 0 \tag{2}$$

where $M(x, y)$ and $N(x, y)$ are functions involving the variables x and y.

Example 1

Rewrite the first degree differential equation $x^2y' - e^{xy} = 0$ in the form of equation (2).

$$x^2y' - e^{xy} = 0$$

$$x^2 \frac{dy}{dx} - e^{xy} = 0$$

$$x^2\, dy - e^{xy}\, dx = 0 \quad \text{(Multiply each side by } dx.\text{)}$$

$$-e^{xy}\, dx + x^2\, dy = 0$$

In this example, $M(x, y) = -e^{xy}$ and $N(x, y) = x^2$.

Some first-degree equations in the form $M(x, y)\, dx + N(x, y)\, dy = 0$ can be rewritten in the form

$$f(x)\, dx + g(y)\, dy = 0 \tag{3}$$

where $f(x)$ is an expression only in variable x and $g(y)$ is an expression only in variable y.

Example 2

Rewrite the first-degree differential equation $x^2yy' - 2xy^3 = 0$ in the form of equation (3).

$$x^2yy' - 2xy^3 = 0$$

$$x^2y \frac{dy}{dx} - 2xy^3 = 0$$

$$x^2y\, dy - 2xy^3\, dx = 0 \qquad \text{(Multiply each side by } dx.\text{)}$$

$$\left(\frac{1}{x^2y^3}\right)(x^2y\, dy - 2xy^3\, dx) = (0)\left(\frac{1}{x^2y^3}\right) \qquad \text{(Divide each side by } x^2y^3.\text{)}$$

$$\frac{1}{y^2}\, dy - \frac{2}{x}\, dx = 0$$

or

$$-\frac{2}{x}\, dx + \frac{1}{y^2}\, dy = 0$$

The process demonstrated in Example 2 is called "separating the variables." By appropriate multiplications and divisions separate the equation into terms where each term involves only one variable and its differential. Because of the expression e^{xy} in Example 1, it is impossible to separate the variables.

When a first-order differential equation can be separated so that we can collect all y terms with dy and all x terms with dx, then the general solution can be obtained by integrating each term. If we separate the variables to each side of the equation, the general solution of a differential equation in the form

$$f(x)\, dx = g(y)\, dy$$

is

$$\int f(x)\, dx = \int g(y)\, dy$$

$$F(x) = G(y) + C$$

where $F(x)$ is the integral of $f(x)$, $G(y)$ is the integral of $g(y)$, and C is the constant of integration.

Example 3

Find the general solution of the differential equation $x^2yy' - 2xy^3 = 0$.

In Example 2 we wrote $x^2yy' - 2xy^3 = 0$ as

$$-\frac{2}{x}\,dx + \frac{1}{y^2}\,dy = 0$$

$$\frac{1}{y^2}\,dy = \frac{2}{x}\,dx$$

Integrating each side of the equation, we have

$$\int \frac{1}{y^2}\,dy = \int \frac{2}{x}\,dx$$

$$-\frac{1}{y} = 2 \ln x + C$$

or

$$1 + 2y \ln x + Cy = 0$$

Note: When the solution of a differential equation involves integrating a term in the form $\dfrac{du}{u}$, we will now write $\displaystyle\int \frac{du}{u} = \ln u + C$ rather than $\displaystyle\int \frac{du}{u} = \ln |u| + C$. We now assume that the solution is valid only when u is positive. Remember also to include the constant of integration C.

Example 4

Solve the differential equation $y' = \dfrac{y}{x^2 + 1}$.

Rewriting we have

$$\frac{dy}{dx} = \frac{y}{x^2 + 1}$$

$$dy = \frac{y}{x^2 + 1}\,dx \qquad \text{(Multiply each side by } dx.\text{)}$$

$$\frac{dy}{y} = \frac{dx}{x^2 + 1} \qquad \left(\text{Multiply each side by } \frac{1}{y}.\right)$$

$$\int \frac{dy}{y} = \int \frac{dx}{x^2 + 1} \qquad \text{(Integrate each side.)}$$

$$\ln y = \text{Arctan } x + C$$

Example 5

Solve the differential equation $x(1 + y^2) - y(1 + x^2)y' = 0$.

Rewriting, we have

$$x(1 + y^2) - y(1 + x^2)\frac{dy}{dx} = 0$$

$$x(1 + y^2)\,dx - y(1 + x^2)\,dy = 0$$

$$\frac{x}{1+x^2}\,dx - \frac{y}{1+y^2}\,dy = 0$$

$$\frac{x}{1+x^2}\,dx = \frac{y}{1+y^2}\,dy$$

$$\int \frac{x}{1+x^2}\,dx = \int \frac{y}{1+y^2}\,dy$$

$$\tfrac{1}{2}\ln(1+x^2) = \tfrac{1}{2}\ln(1+y^2) + C$$

$$\tfrac{1}{2}\ln(1+x^2) - \tfrac{1}{2}\ln(1+y^2) = C$$

$$\frac{1}{2}\ln\left(\frac{1+x^2}{1+y^2}\right) = C$$

Since C is an arbitrary constant, we could rewrite this constant as $C = \tfrac{1}{2}\ln k$ where $k > 0$. Then we have

$$\frac{1}{2}\ln\left(\frac{1+x^2}{1+y^2}\right) = \frac{1}{2}\ln k$$

$$\frac{1+x^2}{1+y^2} = k$$

$$1+x^2 = k + ky^2$$

$$x^2 - ky^2 + 1 - k = 0$$

This last equation is easier to work with since it no longer involves natural logarithms. The equations $\tfrac{1}{2}\ln\left(\dfrac{1+x^2}{1+y^2}\right) = C$ and $x^2 - ky^2 + 1 - k = 0$ are equivalent. They differ only in the form of the constant of integration. By working the exercises you will gain experience in choosing the most appropriate form for this arbitrary constant.

Example 6

Solve the differential equation $y' - 2x = 0$.

Rewriting, we have

$$\frac{dy}{dx} = 2x$$

$$dy = 2x\,dx$$

$$\int dy = \int 2x\,dx$$

$$y = x^2 + C$$

This general solution represents a family of functions where each function $y = f(x)$ is a particular solution of the differential equation $y' - 2x = 0$. In this case, the solution is a family which consists of parabolas some of which are sketched in Fig. 11.1.

A particular solution can be obtained if certain initial conditions are given. An *initial condition* of a differential equation is a condition which specifies a particular value of y, y_0, corresponding to a particular value of x, x_0. That is, if $y = f(x)$ is a solution of the differential equation, then the function must satisfy

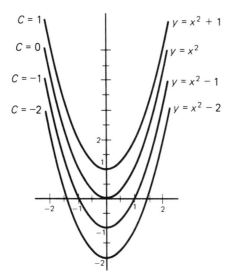

$C = 1$ $y = x^2 + 1$
$C = 0$ $y = x^2$
$C = -1$ $y = x^2 - 1$
$C = -2$ $y = x^2 - 2$

Figure 11.1

the condition: $y_0 = f(x_0)$. A differential equation with initial conditions is called an *initial value problem*.

Example 7

Solve the differential equation $y' - 2x = 0$ subject to the initial condition that $y = 1$ when $x = 2$.

From Example 6 the general solution is

$$y = x^2 + C$$

Substituting $y = 1$ and $x = 2$, we have

$$1 = (2)^2 + C$$
$$1 = 4 + C$$
$$-3 = C$$

So the particular solution is

$$y = x^2 - 3$$

Example 8

Solve the differential equation $y + xy' = 0$ subject to the initial condition that $y = 2$ when $x = 3$, which may also be written $f(3) = 2$.

Rewriting, we have

$$y + x\frac{dy}{dx} = 0$$
$$y\,dx + x\,dy = 0$$
$$\frac{dx}{x} + \frac{dy}{y} = 0$$
$$\frac{dx}{x} = -\frac{dy}{y}$$
$$\int \frac{dx}{x} = -\int \frac{dy}{y}$$

$$\ln x = -\ln y + C$$
$$\ln x + \ln y = C$$
$$\ln xy = C$$

or

$$\ln xy = \ln k, \qquad \text{where } C = \ln k$$
$$xy = k$$

Substituting $y = 2$ when $x = 3$, we have
$$(3)(2) = k$$
$$6 = k$$

So the required particular solution is
$$xy = 6$$

Exercises

Solve each differential equation.

1. $x\,dy - y^2\,dx = 0$ 2. $3x^3y^2\,dx - xy\,dy = 0$

3. $x\,dy + y\,dx = 0$ 4. $\sec x\,dy + \csc y\,dx = 0$

5. $\dfrac{dy}{dx} = y^{3/2}$ 6. $(y^2 - 4)\dfrac{dy}{dx} = 1$ 7. $\dfrac{dy}{dx} = x^2 + x^2y^2$

8. $y\dfrac{dy}{dx} = \sin x$ 9. $x\dfrac{dy}{dx} + y = 3$ 10. $\dfrac{dy}{dx} = 1 - y$

11. $\dfrac{dy}{dx} = \dfrac{x^2}{y}$ 12. $\dfrac{dy}{dx} = \dfrac{xy}{x^2 + 3}$

13. $\dfrac{dy}{dx} + y^3 \cos x = 0$ 14. $\dfrac{dy}{dx} - \dfrac{e^x}{y^2} = 0$

15. $e^{3x}\dfrac{dy}{dx} - e^x = 0$ 16. $x\sqrt{1 - y^2}\,dx - 3\,dy = 0$

17. $(1 + x^2)\,dy - dx = 0$ 18. $(1 + x^2)\,dy + x\,dx = 0$

19. $\dfrac{dy}{dx} = 1 + x^2 + y^2 + x^2y^2$ 20. $\dfrac{dy}{dx} = e^{x+y}$

21. $y' = e^{x-y}$ 22. $y' = xe^{x^2+2y}$

23. $(x + 1)y' = y^2 + 4$ 24. $\sqrt{1 - 16y^2} = (4x^2 + 9)y'$

25. $(4xy + 12x)\,dx = (5x^2 + 5)\,dy$ 26. $8x^3(1 - y^2) = 3y(1 + x^4)y'$

Find the particular solution of each differential equation subject to the given conditions.

27. $\dfrac{dy}{dx} = x^2y^4$; $y = 1$ when $x = 1$.

28. $ye^{-x}\dfrac{dy}{dx} + 2 = 0$; $y = 2$ when $x = 0$

29. $\dfrac{dy}{dx} = \dfrac{2x}{y + x^2y}$; $y = 4$ when $x = 0$ 30. $x^2\,dy = y\,dx$; $y = 1$ when $x = 1$

31. $y\dfrac{dy}{dx} = e^x$; $f(0) = 6$ 32. $y^2x\dfrac{dy}{dx} - 2x + 4 = 0$; $f(1) = 3$

                 First-Order Differential Equations   Chap. 11

33. $\sqrt{x} + \sqrt{y}\,\dfrac{dy}{dx} = 0;\ f(1) = 4$

34. $e^{-2y}\,\dfrac{dy}{dx} = x - 2;\ f(1) = 0$

35. $xy\,\dfrac{dy}{dx} = \ln x;\ f(1) = 0$

36. $\dfrac{dy}{dx} = e^{x-y};\ f(0) = 1$

11.3 USE OF INTEGRATING FACTORS

Not all differential equations can be solved by separating the variables. There are some differential equations, however, that can be solved by recognizing a combination of differentials which can be integrated. For example, the left-hand side of the differential equation

$$\frac{x\,dy - y\,dx}{x^2} = 3\,dx$$

we recognize as

$$d\left(\frac{y}{x}\right) = \frac{x\,dy - y\,dx}{x^2}$$

so that we can integrate as follows

$$\int \frac{x\,dy - y\,dx}{x^2} = \int 3\,dx$$

$$\int d\left(\frac{y}{x}\right) = \int 3\,dx$$

$$\frac{y}{x} = 3x + C$$

$$y = 3x^2 + xC$$

is the general solution.

Some easily recognizable differentials are expressed below:

A.	$d\left(\dfrac{x}{y}\right) = \dfrac{y\,dx - x\,dy}{y^2}$
B.	$d\left(\dfrac{y}{x}\right) = \dfrac{x\,dy - y\,dx}{x^2} = -\left(\dfrac{y\,dx - x\,dy}{x^2}\right)$
C.	$d\left(\text{Arctan}\,\dfrac{y}{x}\right) = \dfrac{x\,dy - y\,dx}{x^2 + y^2}$
D.	$d\left(\text{Arctan}\,\dfrac{x}{y}\right) = \dfrac{y\,dx - x\,dy}{x^2 + y^2}$
E.	$d(xy) = x\,dy + y\,dx$
F.	$d(\ln \sqrt{x^2 + y^2}) = \dfrac{x\,dx + y\,dy}{x^2 + y^2}$
G.	$d(x^2 + y^2) = 2(x\,dx + y\,dy)$

Thus, be on the lookout for the terms $x \, dy - y \, dx$, $x \, dy + y \, dx$, and $x \, dx + y \, dy$ appearing in a differential equation. It may be possible to integrate the differential equation after multiplying or dividing the equation by an appropriate term, called the *integrating factor*.

Example 1

Solve the differential equation

$$x \, dy - y \, dx = (x^2 + y^2) \, dy$$

By dividing each side of the equation by $(x^2 + y^2)$, we obtain

$$\frac{x \, dy - y \, dx}{x^2 + y^2} = dy$$

$$d \left(\text{Arctan} \frac{y}{x} \right) = dy$$

$$\int d \left(\text{Arctan} \frac{y}{x} \right) = \int dy$$

Then $\text{Arctan} \dfrac{y}{x} = y + C$ is the general solution.

Example 2

Solve the differential equation

$$x \, dy + y \, dx = 3xy \, dx$$

By dividing each side of the equation by xy, we obtain

$$\frac{x \, dy + y \, dx}{xy} = 3 \, dx$$

$$\frac{d(xy)}{xy} = 3 \, dx \qquad [d(xy) = x \, dy + y \, dx]$$

$$\int \frac{d(xy)}{xy} = \int 3 \, dx$$

Then $\ln xy = 3x + C$ is the general solution.

Example 3

Find the particular solution of the differential equation

$$y \, dx - x \, dy + dx = 4x^4 \, dx$$

subject to the initial condition that $y = \frac{1}{3}$ when $x = 1$.

Rewriting, we have

$$y \, dx - x \, dy = 4x^4 \, dx - dx$$

$$\frac{y \, dx - x \, dy}{x^2} = 4x^2 \, dx - \frac{dx}{x^2} \qquad \text{(Divide each side by } x^2.)$$

$$-d \left(\frac{y}{x} \right) = 4x^2 \, dx - \frac{dx}{x^2}$$

$$-\int d \left(\frac{y}{x} \right) = \int 4x^2 \, dx - \int \frac{dx}{x^2}$$

$$-\frac{y}{x} = \frac{4x^3}{3} + \frac{1}{x} + C$$

and $y = -\frac{4}{3}x^4 - 1 - Cx$ is the general solution. Substituting $y = \frac{1}{3}$ when $x = 1$, we have

$$(\tfrac{1}{3}) = -\tfrac{4}{3}(1)^4 - 1 - C(1)$$
$$\tfrac{1}{3} = -\tfrac{4}{3} - 1 - C$$
$$-\tfrac{8}{3} = C$$

So

$$y = -\tfrac{4}{3}x^4 + \tfrac{8}{3}x - 1$$

is the desired particular solution.

Exercises

Solve each differential equation.

1. $x\,dy + y\,dx = y^2\,dy$
2. $x\,dx + y\,dy = (x^2 + y^2)\,dx$
3. $x\,dy - y\,dx = 5x^2\,dy$
4. $x\,dx + y\,dy = x^3\,dx$
5. $y\,dx - x\,dy + y^2\,dx = 3\,dy$
6. $x\,dy - y\,dx = x^4\,dx + x^2y^2\,dx$
7. $x\sqrt{x^2 + y^2}\,dx - 2x\,dx = 2y\,dy$
8. $x\,dy - y\,dx + x^4y^2\,dx = 0$
9. $x\,dx + y\,dy = (x^3 + xy^2)\,dy + (x^2y + y^3)\,dx$
10. $(y + x)\,dx + (y - x)\,dy = (x^2 + y^2)\,dx$

Find the particular solution of each differential equation for the given initial conditions.

11. $x\,dy + y\,dx = 2x\,dx + 2y\,dy$ for $y = 1$ when $x = 0$
12. $y\,dx - x\,dy = y^2\,dx$ for $y = 3$ when $x = 1$
13. $x\,dy - 4\,dx = (x^3 + y^2x)\,dy + (x^2y + y^3)\,dx$ for $y = 2$ when $x = 2$
14. $x\,dy - y\,dx = x^5\,dx$ for $y = 4$ when $x = 1$

11.4 LINEAR EQUATIONS OF FIRST ORDER

A differential equation where both the unknown function $y = f(x)$ and its derivative are of first power is called a *linear differential equation* (of the first order). The equation for a linear differential equation of first order can be written in the form

$$\frac{dy}{dx} + P(x)y = Q(x) \tag{1}$$

or

$$dy + P(x)y\,dx = Q(x)\,dx \tag{2}$$

where $P(x)$ and $Q(x)$ are functions of x.

A method has been devised for solving linear equations of first order. An *integrating factor* is an expression which, when multiplied on both sides of a differential equation, gives a differential equation that can be integrated in order to find its solution. We will show that $e^{\int P(x)dx}$ becomes an integrating factor for these equations. Multiplying each side of Equation (2) by $e^{\int P(x)dx}$, we have

$$e^{\int P(x)dx}\,dy + e^{\int P(x)dx}P(x)y\,dx = e^{\int P(x)dx}Q(x)\,dx$$

First, observe that

$$d(ye^{\int P(x)dx}) = y\,d(e^{\int P(x)dx}) + e^{\int P(x)dx}\,dy$$

$$= ye^{\int P(x)dx}\, d\left[\int P(x)\, dx\right] + e^{\int P(x)dx}\, dy$$

$$= ye^{\int P(x)dx}P(x)\, dx + e^{\int P(x)dx}\, dy$$

$$= e^{\int P(x)dx}[yP(x)\, dx + dy]$$

$$= e^{\int P(x)dx}Q(x)\, dx \qquad \text{[from equation (2)]}$$

Thus

$$d(ye^{\int P(x)dx}) = [Q(x)e^{\int P(x)dx}]\, dx$$

So $e^{\int P(x)dx}$ becomes an integrating factor as the general solution of the first-order linear differential equation

$$\frac{dy}{dx} + P(x)y = Q(x)$$

is

$$\boxed{\; ye^{\int P(x)dx} = \int Q(x)e^{\int P(x)dx}\, dx \;}$$

Example 1

Solve $\dfrac{dy}{dx} + 2yx = 5x.$

Here $P(x) = 2x$, $Q(x) = 5x$ and $\int P(x)\, dx = \int 2x\, dx = x^2$. We do not write a constant of integration for $\int P(x)\, dx$ since we are merely obtaining an integrating factor. The solution is

$$ye^{\int P(x)dx} = \int Q(x)e^{\int P(x)dx}\, dx$$

$$ye^{x^2} = \int 5xe^{x^2}\, dx$$

$$ye^{x^2} = \int \frac{5}{2}e^{u}\, du$$

$$ye^{x^2} = \tfrac{5}{2}e^{u} + C$$

$$ye^{x^2} = \tfrac{5}{2}e^{x^2} + C$$

> let $u = x^2$
> $du = 2x\, dx$

or

$$y = \tfrac{5}{2} + Ce^{-x^2}$$

Example 2

Solve $y' + y = e^{-x}\cos x.$

Here $P(x) = 1$, $Q(x) = e^{-x}\cos x$, and $\int P(x)\, dx = \int dx = x$. The solution is

$$ye^{\int P(x)dx} = \int Q(x)e^{\int P(x)dx}\, dx$$

$$ye^{x} = \int (e^{-x}\cos x)(e^{x})\, dx$$

$$ye^x = \int \cos x \, dx$$
$$ye^x = \sin x + C$$

Example 3

Solve $(2y - 6x^2) \, dx + x \, dy = 0$

Rewriting, this equation becomes

$$dy + \frac{2y}{x} \, dx = 6x \, dx$$

where $P(x) = \frac{2}{x}$, $\dot{Q}(x) = 6x$ and

$$\int P(x) \, dx = \int \frac{2}{x} \, dx = 2 \ln x = \ln x^2$$

The solution is then

$$ye^{\int P(x)dx} = \int Q(x)e^{\int P(x)dx} \, dx$$
$$ye^{\ln x^2} = \int 6xe^{\ln x^2} \, dx$$
$$yx^2 = \int 6x^3 \, dx \qquad \text{(since } e^{\ln x^2} = x^2\text{)}$$
$$yx^2 = \tfrac{3}{2}x^4 + C$$

Exercises

Solve each differential equation.

1. $\dfrac{dy}{dx} - 5y = e^{3x}$

2. $\dfrac{dy}{dx} + 3y = e^{-2x}$

3. $\dfrac{dy}{dx} + \dfrac{3y}{x} = x^3 - 2$

4. $\dfrac{dy}{dx} - \dfrac{2y}{x} = x^2 + 5$

5. $\dfrac{dy}{dx} + 2xy = e^{3x}(3 + 2x)$

6. $\dfrac{dy}{dx} - 3x^2y = e^x(3x^2 - 1)$

7. $dy - 4y \, dx = x^2 e^{4x} \, dx$

8. $dy - 3x^2y \, dx = x^2 \, dx$

9. $x \, dy - 5y \, dx = (x^6 + 4x) \, dx$

10. $x \dfrac{dy}{dx} + 2y = (x^2 + 4)^3$

11. $(1 + x^2) \, dy + 2xy \, dx = 3x^2 \, dx$

12. $\dfrac{dy}{dx} - y \tan x = \sin x$

13. $x^2 \dfrac{dy}{dx} + 2xy = x^4 - 7$

14. $x^2 \dfrac{dy}{dx} - 2xy = x^3 + 5$

15. $\dfrac{dy}{dx} + 2y = e^{-x}$

16. $(x + 1)\dfrac{dy}{dx} + 5y = 10$

17. $x \dfrac{dy}{dx} - y = 3x^2$

18. $\dfrac{dy}{dx} + \dfrac{y}{3x - 1} = 8$

19. $\dfrac{dy}{dx} + y \cos x = \cos x$

20. $\dfrac{dy}{dx} + y \sec^2 x = \sec^2 x$

Find a particular solution of each differential equation subject to the given initial conditions.

21. $\dfrac{dy}{dx} - 3y = e^{2x}$; $y = 2$ when $x = 0$

22. $\dfrac{dy}{dx} - \dfrac{y}{x} = x^2 + 3;\ y = 3$ when $x = 1$

23. $\dfrac{dy}{dx} = \csc x - y \cot x;\ f\left(\dfrac{\pi}{2}\right) = \dfrac{3\pi}{2}$ **24.** $dy = (x - 3y)dx;\ f(0) = 1$

25. $\dfrac{dy}{dx} = e^x - y;\ f(0) = \dfrac{3}{2}$ **26.** $\dfrac{dy}{dx} + 8x^2y = 4x^2;\ f(0) = 2$

27. $x \dfrac{dy}{dx} + y = 3;\ f(1) = -2$ **28.** $\dfrac{dy}{dx} - \dfrac{3y}{x - 4} = (x - 4)^2;\ f(5) = 3$

29. $x \dfrac{dy}{dx} + y = 4x^3;\ f(2) = 3$ **30.** $\dfrac{dy}{dx} + y \sin x = \sin x;\ f(\pi/2) = 3$

11.5 APPLICATIONS OF FIRST-ORDER DIFFERENTIAL EQUATIONS

Many technical problems involve first-order, first-degree differential equations. In the applications that follow observe that certain phenomena involve a rate of change (a derivative). The presence of a derivative will often lead to a differential equation as we attempt to describe the physical situation in mathematical terms. Our goal is then to solve the differential equation. Finally, the mathematical solution is given a physical interpretation that will provide an answer to the original technical problem.

As with all mathematical applications, we create a mathematical model of certain physical phenomena. The mathematical model can only be an approximation. The model is only as accurate as the interpretation given the physical phenomena.

Example 1 Radioactive Decay

Radioactive material has been observed to decay at a rate proportional to the amount present. For a certain radioactive substance approximately 10% of the original quantity decomposes in 25 years. Find the half-life of this radioactive material. That is, find the time that elapses for the quantity of material to decay to one-half of its original quantity.

Step 1: Set up the mathematical model.

Let Q_0 represent the original quantity present and Q the amount present at any time t (in years). The mathematical model describing the observed rate of decay process then becomes

$$\frac{dQ}{dt} = kQ$$

where k is the constant of proportionality of decay.

Step 2: Solve the differential equation.

Separate the variables:

$$dQ = kQ\, dt$$

$$\frac{dQ}{Q} = k\, dt$$

then

$$\int \frac{dQ}{Q} = \int k \, dt$$

$$\ln Q = kt + \ln C$$

We have the initial condition that $Q = Q_0$ when $t = 0$. So

$$\ln Q_0 = k(0) + \ln C$$

$$\ln Q_0 = \ln C$$

$$C = Q_0$$

We then have

$$\ln Q = kt + \ln Q_0$$

$$\ln Q - \ln Q_0 = kt$$

$$\ln \frac{Q}{Q_0} = kt$$

$$\frac{Q}{Q_0} = e^{kt} \quad \text{(Rewrite each side as power of } e.)$$

$$Q = Q_0 e^{kt}$$

which is an expression for the amount of radioactive material present after t years with k, the constant of proportionality of decay.

Step 3: Solve the particular problem in question.

To find the value of t which represents the material's half-life we must first find k. We are given that after 25 yr, 10% of the material has decayed. That is, at $t = 25$ years $Q = (1 - 0.1)Q_0 = 0.9Q_0$. Thus we can write

$$0.9Q_0 = Q_0 e^{25k}$$

$$0.9 = e^{25k}$$

$$\ln (0.9) = 25k$$

$$\frac{\ln (0.9)}{25} = k$$

$$-0.00421 = k$$

Now determine the half-life. Since the half-life represents the time that must elapse for the material to decay to one-half its original quantity (that is, $Q = Q_0/2$), the half-life can be determined by substituting $Q = Q_0/2$ in the equation $Q = Q_0 e^{kt}$:

$$\frac{Q_0}{2} = Q_0 e^{kt}$$

$$\tfrac{1}{2} = e^{kt}$$

$$\ln \tfrac{1}{2} = kt \qquad \text{(Take the ln of each side.)}$$

$$-\ln 2 = kt \qquad (\ln \tfrac{1}{2} = \ln 1 - \ln 2 = 0 - \ln 2)$$

$$\frac{-\ln 2}{k} = t \qquad \textbf{(half-life formula)}$$

$$\frac{-\ln 2}{-0.00421} = t$$

$$t = 165 \text{ years}$$

Example 2   Electrical Circuit

The current i in a series circuit with constant inductance L, constant resistance R, and a constant voltage V applied is described by the differential equation

$$L\frac{di}{dt} + Ri = V$$

Find an equation for the current i as a function of time t.

Rewrite the equation as

$$\frac{di}{dt} + \frac{R}{L}i = \frac{V}{L}$$

and apply the method described in Section 11.4:

$$P(t) = \frac{R}{L} \qquad Q(t) = \frac{V}{L}$$

and

$$\int P(t)\, dt = \int \frac{R}{L}\, dt = \frac{R}{L}t$$

and

$$ie^{(R/L)t} = \int \frac{V}{L} e^{(R/L)t}\, dt$$

$$= \frac{V}{L} \int e^{(R/L)t}\, dt$$

$$= \frac{V}{L} \cdot \frac{L}{R} e^{(R/L)t} + C$$

$$= \frac{V}{R} e^{(R/L)t} + C$$

If there is no initial current, we assume that $i = 0$ when $t = 0$. So we have

$$0 \cdot e^{(R/L)(0)} = \frac{V}{R} e^{(R/L)(0)} + C$$

$$0 = \frac{V}{R} + C$$

$$C = -\frac{V}{R}$$

Finally, we solve for i:

$$ie^{(R/L)t} = \frac{V}{R} e^{(R/L)t} - \frac{V}{R}$$

$$i = \frac{\frac{V}{R}(e^{(R/L)t} - 1)}{e^{(R/L)t}}$$

$$i = \frac{V}{R}(1 - e^{-(R/L)t})$$

This shows that V/R is a limiting value for the current i. As $t \to \infty$,

$$e^{-(R/L)t} = \frac{1}{e^{(R/L)t}}$$

approaches 0. Thus as $t \to \infty$ (see Fig. 11.2),

$$i = \frac{V}{R}\left(1 - \frac{1}{e^{(R/L)t}}\right) \to \frac{V}{R}(1 - 0) = \frac{V}{R}$$

The equation for i involves two terms, V/R and $(V/R)e^{-(R/L)t}$. The term V/R is called the steady state solution and represents the solution when no inductance is present ($Ri = V$); the other term, $(V/R)e^{-(R/L)t}$, represents the effect of inductance. The addition of this term gives what is called the transient solution.

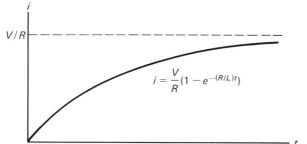

$$i = \frac{V}{R}(1 - e^{-(R/L)t})$$

Figure 11.2

Thus the study of this type of differential equation enables a consideration of the effect of inductance in an electric circuit.

Example 3 Mixtures

Salt is being dissolved in a tank filled with 100 gallons (gal) of water. Originally 25 lb of salt was dissolved in the tank. Salt water containing 1 lb of salt per gallon is poured in at the rate of 3 gal/min; the solution is kept well stirred and the mixture is poured out at the same rate of 3 gal/min. Find an expression for the amount of salt Q in the tank at time t. How much salt remains after 1 h?

The rate of change of salt $\frac{dQ}{dt}$ is equal to the rate at which salt enters the tank less the rate at which it leaves. That is,

$$\frac{dQ}{dt} = \text{rate of gain} - \text{rate of loss}$$

$$\text{rate of gain} = (1\ \text{lb/gal})(3\ \text{gal/min}) = 3\ \text{lb/min}$$

$$\text{rate of loss} = \left(\frac{Q\ \text{lb}}{100\ \text{gal}}\right)(3\ \text{gal/min}) = \frac{3Q}{100}\ \text{lb/min}$$

so

$$\frac{dQ}{dt} = 3 - \frac{3Q}{100}$$

$$\frac{dQ}{dt} = 3\left(1 - \frac{Q}{100}\right)$$

$$\frac{dQ}{dt} = 3\left(\frac{100 - Q}{100}\right)$$

Separating the variables, we have

$$\frac{dQ}{100 - Q} = \frac{3}{100}\,dt$$

Sec. 11.5   Applications of First-Order Differential Equations

453

or rewriting,

$$\int \frac{dQ}{Q - 100} = \int -\frac{3}{100} dt$$

$$\ln(Q - 100) = -\frac{3}{100}t + \ln C$$

$$\ln(Q - 100) - \ln C = -\frac{3}{100}t$$

$$\ln\left(\frac{Q - 100}{C}\right) = -\frac{3}{100}t$$

$$\frac{Q - 100}{C} = e^{-(3/100)t}$$

$$Q = Ce^{-(3/100)t} + 100$$

At $t = 0$, $Q = 25$, so

$$25 = Ce^{-(3/100)(0)} + 100$$

$$25 = C + 100$$

$$C = -75$$

So

$$Q = -75e^{-(3/100)t} + 100$$

After $1\ h = 60$ min we have

$$Q = -75e^{-(3/100)(60)} + 100$$
$$= -75e^{-1.8} + 100$$
$$= -12.4 + 100$$
$$= 87.6\ lb$$

Example 4   Temperature Change

Under certain conditions the temperature of an object changes at a rate proportional to the difference between the temperature outside the object and the temperature of the object (Newton's law of cooling). A thermometer registering 75°F is taken outside where the temperature is 40°F. After 5 min, the thermometer registers 60°F. Find the temperature reading on the thermometer 7 min after having been outside.

Let T represent the temperature reading of the thermometer at any time t

$$\frac{dT}{dt} = k(T - T_{\text{outside}})$$

$$\frac{dT}{dt} = k(T - 40)$$

The equation is subject to the conditions $T = 75$ when $t = 0$ and $T = 60$ when $t = 5$.

Separating the variables, we obtain

$$\frac{dT}{T - 40} = k\ dt$$

$$\int \frac{dT}{T - 40} = \int k\ dt$$

$$\ln(T - 40) = kt + \ln C$$

First-Order Differential Equations   Chap. 11

or
$$T = Ce^{kt} + 40$$

Since $T = 75$ when $t = 0$, we can find C:
$$75 = Ce^{k(0)} + 40$$
$$75 - 40 = C$$
$$35 = C$$

So
$$T = 35e^{kt} + 40$$

Now determine k: Since $T = 60$ when $t = 5$, we have
$$60 = 35e^{k(5)} + 40$$
$$\tfrac{20}{35} = e^{5k}$$
$$\frac{\ln \tfrac{4}{7}}{5} = k$$
$$k = -0.11$$

So
$$T = 35e^{-0.11t} + 40$$

After 7 min we have
$$T = 35e^{(-0.11)(7)} + 40$$
$$= 35e^{-0.77} + 40$$
$$= 56.2°F$$

Exercises

1. The rate of change of velocity with respect to time of a body moving along a straight line with acceleration $a = \dfrac{dv}{dt}$. A body is moving along a straight line with constant acceleration of $a = 5$ m/s². If the body had an initial velocity of $v = 10$ m/s (when $t = 0$), find the equation for velocity. Find also the velocity at $t = 3$ s.

2. Find the equation for the velocity of an object moving along a straight line with acceleration $a = t^2$ m/s², if the object started moving from rest. Find also the velocity at $t = 2$ s. (See Exercise 1.)

3. Find the equation of the curve passing through the point $(1, 1)$ and having slope $\dfrac{x^2 - y}{x}$.

4. Find the equation of the curve passing through the point $(-1, 0)$ and having slope $\dfrac{x^2}{y^3}$.

5. Find the equation for the current i in a series circuit with inductance $L = 0.1$ H, resistance $R = 80\ \Omega$, and voltage $V = 120$ V if the initial current $i_0 = 2$ A when $t = 0$. (See Example 2.)

6. Find the equation relating charge q in terms of time t in a series circuit containing only a resistance and a capacitance if the initial charge $q_0 = 3$ C when $t = 0$. This circuit is described by the differential equation $Ri + \dfrac{q}{C} = 0$ where $i = \dfrac{dq}{dt}$.

7. Find the equation for the current i in a series circuit with inductance $L = 0.1$ H, resistance $R = t$ ohms, and voltage $V = t$ volts. The initial current $i_0 = 0$ when $t = 0$.

8. Find the equation for the current i in a series circuit with inductance $L = 0.1$ H, resistance $R = t^2$ ohms, and voltage $V = t^2$ volts. The initial current $i_0 = 2$ A when $t = 0$.

9. The isotope ^{238}U of uranium has a half-life of 4.5×10^9 years. Determine the amount of ^{238}U left after t years if the initial quantity is 1 g. ^{238}U decays at a rate proportional to the amount present. (See Example 1.)

10. The isotope ^{234}U of uranium has a half-life of 2.7×10^5 years. Determine the amount of ^{234}U left after t years if the initial quantity is 2 g. ^{234}U decays at a rate proportional to the amount present.

11. A block of a certain radioactive material with a mass of 5 g decays 10% after 36 h. The material decays at a rate proportional to the amount present. Find an equation expressing the amount of material present at any given time t. Find the half-life of this material.

12. A block of a certain radioactive material with an original mass of 10 g has a mass of 8 g after 50 years. Find an expression for the amount of material present at any time t. Find the half-life of this material.

13. Fifty gallons of brine originally containing 10 lb of salt is in a tank. Salt water containing 2 lb of salt per gallon is poured in at the rate of 2 gal/min; the solution is kept well stirred; and the mixture is poured out at the same rate of 2 gal/min. Find the amount of salt remaining after 30 min. (See Example 3.)

14. One hundred gallons of brine originally containing 20 lb of salt is in a tank. Pure water is poured in at the rate of 1 gal/min; the solution is kept well stirred; and the mixture is poured out at the same rate of 1 gal/min. Find the amount of salt remaining after 1 h.

15. An object at 90°C is cooled in air, which is at 10°C. If the object is 70°C after 5 min, find the temperature of the object after 30 min. (See Example 4.)

16. A thermometer registering 80°F is taken outside where the temperature is 35°F. After 3 min, the thermometer registers 50°F. Find the temperature reading on the thermometer 5 min after having been outside.

17. Many populations tend to have a growth rate that is proportional to the present population. That is, $y' = ky$, where y is the present population. Suppose that the population of a country has doubled in the last 20 years. Find the expected population in 80 years if the current population is 2,000,000. (*Hint:* First solve the differential equation $y' = ky$ before determining k.)

18. An amount of money earning compound interest continuously also satisfies the growth equation $y' = ky$ (see Exercise 17), where k is the rate of compound interest. Find the value of a deposit of $500 after ten years if it is earning 9% interest compounded continuously.

19. A gas undergoing an adiabatic change has a rate of change of pressure with respect to volume that is directly proportional to the pressure and inversely proportional to the volume. Solve the resulting differential equation and express the pressure in terms of the volume.

20. A lake contains 5×10^7 litres (L) of water. Industrial waste in the form of a chemical compound begins to be dumped into the lake at the rate of 5 L/h. A freshwater stream feeds the lake at a rate of 345 L/h. Determine how long it will take for the

lake in Fig. 11.3 to become polluted if the lake is considered polluted when the water contains 0.2% of the chemical compound. (*Hint:* $Q' = 5 - 7 \times 10^{-6}Q$, where Q is the amount of the chemical compound present in the water at any time t.)

Figure 11.3

CHAPTER SUMMARY

1. The order of a differential equation is the order of the derivative of highest order in the equation.

2. The degree of a differential equation is the highest power of the derivative of highest order.

3. A solution of a differential equation is a function $y = f(x)$ which together with its derivatives satisfies the given differential equation.

4. *Separation of variables method:* By appropriate multiplications and divisions, separate the differential equation into terms where each term involves only one variable and its differential.

5. Some easily recognizable differentials:

(a) $\quad d\left(\dfrac{x}{y}\right) = \dfrac{y\,dx - x\,dy}{y^2}$

(b) $\quad d\left(\dfrac{y}{x}\right) = \dfrac{x\,dy - y\,dx}{x^2} = -\left(\dfrac{y\,dx - x\,dy}{x^2}\right)$

(c) $\quad d\left(\text{Arctan }\dfrac{y}{x}\right) = \dfrac{x\,dy - y\,dx}{x^2 + y^2}$

(d) $d\left(\text{Arctan }\dfrac{x}{y}\right) = \dfrac{y\,dx - x\,dy}{x^2 + y^2}$

(e) $d(xy) = x\,dy + y\,dx$

(f) $d(\ln \sqrt{x^2 + y^2}) = \dfrac{x\,dx + y\,dy}{x^2 + y^2}$

(g) $d(x^2 + y^2) = 2(x\,dx + y\,dy)$

6. *Linear differential equation of the first order:* a differential equation where both the unknown function $y = f(x)$ and its derivative are of first power. A method for solving such equations involves using the integrating factor, $e^{\int P(x)dx}$. The general solution of $\dfrac{dy}{dx} + P(x)\,y = Q(x)$ is

$$ye^{\int P(x)dx} = \int Q(x)e^{\int P(x)dx}\,dx$$

CHAPTER REVIEW

State the order and degree of each differential equation.

1. $\dfrac{d^2y}{dx^2} - 3x\dfrac{dy}{dx} + x^2y^3 = 0$

2. $\left(\dfrac{dy}{dx}\right)^2 - xy\dfrac{dy}{dx} = 3$

3. $(y'')^3 - 3y' + x^2y^5 = 2$

4. $\left(\dfrac{dy}{dx}\right)^2 + 3y^3 = 7$

Verify that each function $y = f(x)$ is a solution of the given differential equation.

5. $\dfrac{d^2y}{dx^2} + 3y = 3x^3;\ y = x^3 - 2x$

6. $y'' - 2y' + y = 4e^{-x};\ y = 2e^x - 3xe^x + e^{-x}$

7. $y'' + y = x^2 + 2;\ y = \sin x + x^2$

8. $\dfrac{dy}{dx} - 2y = 5x^4e^{2x};\ y = e^{2x}(x^5 - 1)$

Solve each differential equation.

9. $x^3\dfrac{dy}{dx} - y = 0$

10. $e^{2x}\dfrac{dy}{dx} - 3y = 0$

11. $(9 + x^2)y^2\,dy + x\,dx = 0$

12. $\dfrac{dy}{dx} = e^y\sec^2x$

13. $x\,dy - y\,dx = 3x^4\,dx$

14. $\dfrac{dy}{dx} + 6x^2y = 12x^2$

15. $\dfrac{dy}{dx} - \dfrac{y}{x^2} = \dfrac{5}{x^2}$

16. $x\,dx + y\,dy = (x^2y + y^3)\,dy$

17. $x\,dy - y\,dx = (x^4 + x^2y^2)\,dx$

18. $x\,dy + y\,dx = 14x^5\,dx$

19. $dy + 3y\,dx = e^{-2x}\,dx$

20. $\dfrac{dy}{dx} - \dfrac{5y}{x} = x^3 + 7$

21. $x\,dy - 3y\,dx = (4x^3 - x^2)\,dx$

22. $\dfrac{dy}{dx} + y\cot x = \cos x$

Find a particular solution of each differential equation subject to the given initial conditions.

23. $3x^2y^3\ dx - xy\ dy = 0;\ y = -1$ when $x = -2$

24. $ye^{2x}\ dy + 3\ dx = 0;\ y = -2$ when $x = 0$

25. $x\ dy - y\ dx = x^5\ dx;\ f(1) = 1$ **26.** $x\ dy + y\ dx = x\ \ln x\ dx;\ f(1) = \frac{3}{4}$

27. $\dfrac{dy}{dx} - y = e^{5x};\ f(0) = -3$ **28.** $\dfrac{dy}{dx} - \dfrac{2y}{x} = x^3 - 5x;\ f(1) = 3$

29. Find the equation for the current i in a series circuit with inductance $L = 0.2$ H, resistance $R = 60\ \Omega$, and voltage $V = 120$ V if the initial circuit $i = 1$ amp when $t = 0$.

30. The isotope ^{235}U of uranium has a half-life of 8.8×10^8 years. Determine the amount of ^{235}U left after t years if the initial quantity is 1 g. ^{235}U decays at a rate proportional to the amount present.

31. Two hundred litres of brine originally containing 5 kg of salt is in a tank. Pure water is poured in at the rate of 1 L/min; the solution is kept well stirred; and the mixture is poured out at the same rate of 1 L/min. Find the amount of salt remaining after 20 min.

32. A block of material registers 30°C. The block is cooled in air which is at 12°C. If the block is at 25°C after 20 min, find the temperature of the material after 45 min.

33. Find the equation of the curve passing through the point (0, 2) and having slope equal to x^2y.

34. Find the equation for the velocity of an object moving along a straight line with acceleration $a = 4t^3$ m/s², if the object started moving from rest. Find also the velocity after $t = 3$ s.

12

SECOND-ORDER DIFFERENTIAL EQUATIONS

12.1 HIGHER-ORDER HOMOGENEOUS DIFFERENTIAL EQUATIONS

We now consider higher-order linear differential equations. That is, consider equations which have derivatives of an unknown function with order higher than first order. These equations are called linear because they contain no powers of the unknown function and its derivatives higher than the first power. Thus a linear differential equation of order n is represented by the form

$$a_0 \frac{d^n y}{dx^n} + a_1 \frac{d^{n-1}y}{dx^{n-1}} + \cdots + a_{n-1} \frac{dy}{dx} + a_n y = b$$

where $a_1, a_2, \ldots, a_{n-1}, a_n$, and b can be either functions of x or constants. If $b = 0$, the linear differential equation is called *homogeneous*. If $b \neq 0$, the equation is called *nonhomogeneous*.

We have already seen nonhomogeneous linear differential equations where the coefficients of y and $\frac{dy}{dx}$ were not constants. However, in this chapter we consider only equations where b and the coefficients a_i are constants. The following are examples of linear differential equations with constant coefficients and with b constant.

A. $3\dfrac{d^4 y}{dx^4} - 2\dfrac{d^2 y}{dx^2} + 5\dfrac{dy}{dx} + 3y = 7$

B. $y''' + 2y'' - 5y = 2$

$$\text{C. } 4\frac{d^5y}{dx^5} - \frac{dy}{dx} + y = 0$$

D. $y'' - 3y' + 2y = 0$

Equations A and B are nonhomogeneous while equations C and D are homogeneous.

We now present a method of solving homogeneous linear differential equations with constant coefficients. These equations arise in many practical situations. While this method (and others to be presented) is applicable to equations with order higher than the second, we restrict our attention in the remaining sections to second-order equations.

We introduce a *differential operator D*, which operates on a function $y = f(x)$ as follows:

$$Dy = \frac{dy}{dx}, \qquad D^2y = \frac{d^2y}{dx^2}, \qquad D^3y = \frac{d^3y}{dx^3}, \qquad \text{etc.}$$

For example, if $y = 2x^3 + 5 \ln x$, then

$$Dy = D(2x^3 + 5 \ln x) = 6x^2 + \frac{5}{x}$$

and

$$D^2y = D(Dy) = D\left(6x^2 + \frac{5}{x}\right) = 12x - \frac{5}{x^2}$$

Operator D is an example of a linear operator. An algebraic expression in the form $ax + by$ where a and b are constants is called a *linear combination*. A *linear operator* is any process which after operating on a linear combination results in another linear combination. We thus observe that for the operator D, we have

$$D[a f(x) + b g(x)] = a D[f(x)] + b D[g(x)]$$

We can apply algebraic operations to linear operators. In particular, we can factor expressions involving the differential operator D. For example, $D^2 - 2D - 3 = (D - 3)(D + 1)$. That is, $(D^2 - 2D - 3)y = (D - 3)(D + 1)y$. This is used to solve differential equations. The method is explained in Example 1.

Example 1

Solve the differential equation

$$y'' - 2y' - 3y = 0$$

Step 1: First rewrite this equation using the differential operator D.

$$y'' - 2y' - 3y = D^2y - 2 Dy - 3y$$
$$= (D^2 - 2D - 3)y = 0$$

Factor:

$$= (D - 3)(D + 1)y = 0$$

Step 2: Next, let $z = (D + 1)y$. Note that z is a function of x and is a linear combination of the functions $\frac{dy}{dx}$ and y: that is,

$$z = \frac{dy}{dx} + y \quad .$$

Step 3: Then $(D - 3)(D + 1)y = (D - 3)z = 0$. $(D - 3)z = 0$ is a linear differential equation of first order which can be solved by the method of separation of variables:

$$(D - 3)z = 0$$

$$\frac{dz}{dx} - 3z = 0$$

$$dz = 3z\,dx$$

$$\frac{dz}{z} = 3\,dx$$

$$\int \frac{dz}{z} = \int 3\,dx$$

$$\ln z = 3x + \ln C_1 \qquad \text{(We use } \ln C_1 \text{ instead of } C_1$$
$$\ln z - \ln C_1 = 3x \qquad\qquad \text{as our constant of integration}$$
$$\qquad\qquad\qquad\qquad\qquad \text{in order to obtain a simpler}$$
$$\ln \frac{z}{C_1} = 3x \qquad\qquad \text{expression for } z.)$$

$$\frac{z}{C_1} = e^{3x}$$

$$z = C_1 e^{3x}$$

Step 4: Now replace z by $(D + 1)y$ and obtain

$$(D + 1)y = C_1 e^{3x}$$

$$\frac{dy}{dx} + y = C_1 e^{3x}$$

which is another differential equation of first order.

Step 5: Next, solve this differential equation by the method shown in Section 11.4.

$$P(x) = 1, \qquad Q(x) = C_1 e^{3x}, \qquad \int P(x)dx = \int dx = x$$

Then

$$ye^x = \int (C_1 e^{3x})e^x\,dx$$

$$= \int C_1 e^{4x}\,dx$$

$$= \frac{C_1}{4}e^{4x} + C_2$$

$$y = \frac{C_1}{4}e^{3x} + C_2 e^{-x}$$

or

$$y = k_1 e^{3x} + k_2 e^{-x}$$

where $k_1 = C_1/4$ and $k_2 = C_2$.

Second-Order Differential Equations   Chap. 12

A simpler method can now be shown. The general solution $y = k_1e^{3x} + k_2e^{-x}$ of the differential equation $D^2y - 2Dy - 3y = 0$ in Example 1 suggests that $y = ke^{mx}$ would be a particular solution of a second-order linear differential equation

$$a_0 D^2y + a_1 Dy + a_2y = 0.$$

Since $Dy = D(ke^{mx}) = mke^{mx}$ and $D^2y = D(D(ke^{mx})) = D(mke^{mx}) = m^2ke^{mx}$, we can write

$$a_0 D^2y + a_1 Dy + a_2y = 0$$

$$a_0(m^2ke^{mx}) + a_1(mke^{mx}) + a_2ke^{mx} = 0$$

$$(a_0m^2 + a_1m + a_2)ke^{mx} = 0$$

Note that ke^{mx} is not zero (e^{mx} is never zero and k cannot be zero if y is not zero). Thus $a_0m^2 + a_1m + a_2 = 0$ if $y = ke^{mx}$ is a solution.

The equation $a_0m^2 + a_1m + a_2 = 0$ is called the *auxiliary equation (characteristic equation)* for the differential equation

$$a_0 D^2y + a_1 Dy + a_2y = 0$$

The auxiliary equation is found by substituting m for Dy and m^2 for D^2y in the differential equation.

Observe that $y = ke^{mx}$ is a solution if m satisfies the auxiliary equation $a_0m^2 + a_1m + a_2 = 0$. To find a solution of the differential equation, find the two roots, m_1 and m_2, of the auxiliary equation.

Example 2

Solve the differential equation $y'' - 2y' - 3y = 0$ by solving its auxiliary equation.

The auxiliary equation is $m^2 - 2m - 3 = 0$. Since this factors as $(m - 3)(m + 1) = 0$, the roots are $m_1 = 3$ and $m_2 = -1$. Thus

$$y = k_1e^{m_1x} = k_1e^{3x}$$

and

$$y = k_2e^{m_2x} = k_2e^{-x}$$

are both solutions of the differential equation. Since the sum of the solutions of a linear differential equation can also be shown to be a solution, $y = k_1e^{3x} + k_2e^{-x}$ is the general solution. Observe that the same solution is obtained in Example 2 as in Example 1, as should be the case.

General Solution of Homogeneous Linear Differential Equation

1. Write the auxiliary equation for a given differential equation $a_0 D^2y + a_1 Dy + a_2y = 0$ using the substitutions $m = Dy$ and $m^2 = D^2y$:

$$a_0m^2 + a_1m + a_2 = 0$$

2. Determine the two roots, m_1 and m_2 of the auxiliary equation.
3. Write the general solution as

$$y = k_1e^{m_1x} + k_2e^{m_2x}$$

Example 3

Solve the differential equation $y'' + 3y' = 0$.

Step 1: Since $D^2y + 3\,Dy = 0$ the auxiliary equation is $m^2 + 3m = 0$.

Step 2: Solving for m, we have

$$m^2 + 3m = 0$$

$$m(m + 3) = 0$$

$$m_1 = 0 \quad \text{and} \quad m_2 = -3$$

Step 3: The general solution is

$$y = k_1 e^{m_1 x} + k_2 e^{m_2 x}$$

$$y = k_1 e^0 + k_2 e^{-3x}$$

$$y = k_1 + k_2 e^{-3x} \qquad (\text{since } e^0 = 1)$$

Example 4

Solve the differential equation $\dfrac{d^2y}{dx^2} - 5\dfrac{dy}{dx} + 4y = 0$ subject to the initial conditions

that $y = -2$ and $\dfrac{dy}{dx} = 1$ when $x = 0$.

Step 1: $D^2y - 5\,Dy + 4y = 0$

$$m^2 - 5m + 4 = 0$$

Step 2: $(m - 4)(m - 1) = 0$

$$m_1 = 4 \quad \text{and} \quad m_2 = 1$$

Step 3: The general solution is

$$y = k_1 e^{m_1 x} + k_2 e^{m_2 x}$$

$$y = k_1 e^{4x} + k_2 e^x$$

Now find the particular solution subject to the conditions that $y = -2$ and $\dfrac{dy}{dx} = 1$ when $x = 0$. First find $\dfrac{dy}{dx}$.

$$\frac{dy}{dx} = D(k_1 e^{4x} + k_2 e^x)$$

$$= 4k_1 e^{4x} + k_2 e^x$$

Next substitute $y = -2$, $\dfrac{dy}{dx} = 1$ and $x = 0$ into the equations for y and $\dfrac{dy}{dx}$:

$$y = k_1 e^{4x} + k_2 e^x$$

$$-2 = k_1 e^{4(0)} + k_2 e^0$$

$$-2 = k_1 + k_2 \qquad (\text{since } e^0 = 1)$$

and

$$\frac{dy}{dx} = 4k_1 e^{4x} + k_2 e^x$$

$$1 = 4k_1 e^{4(0)} + k_2 e^0$$

$$1 = 4k_1 + k_2$$

Now solve the resulting system of equations for k_1 and k_2:

$$-2 = k_1 + k_2$$
$$1 = 4k_1 + k_2$$

The solution of this system is $k_1 = 1$ and $k_2 = -3$. The particular solution is then

$$y = e^{4x} - 3e^x$$

Exercises

State whether each differential equation is homogeneous or nonhomogeneous. Also state the order of each equation.

1. $3y^{(iv)} - 7y''' + 2y = 0$
2. $y'' - 7y' + 3y = 2$
3. $y''' - 5y' + 2y + 6 = 0$
4. $y^{(v)} - 6y''' - y = 0$
5. $4y'' - y' + 3y = 0$
6. $y''' - y' + 2 = 0$
7. $y''' - y'' - 5y = 3$
8. $y'' - 2y' = 3y$

Solve each differential equation.

9. $\dfrac{d^2y}{dx^2} - 5\dfrac{dy}{dx} - 14y = 0$
10. $\dfrac{d^2y}{dx^2} + 4\dfrac{dy}{dx} - 5y = 0$
11. $y'' - 2y' - 8y = 0$

12. $y'' - y' - 6y = 0$
13. $y'' - y = 0$
14. $y'' - 9y = 0$

15. $\dfrac{d^2y}{dx^2} - 3\dfrac{dy}{dx} = 0$
16. $\dfrac{d^2y}{dx^2} + 2\dfrac{dy}{dx} = 0$

17. $2D^2y - 13Dy + 15y = 0$
18. $3D^2y + 2Dy - 5y = 0$

19. $3\dfrac{d^2y}{dx^2} - 7\dfrac{dy}{dx} + 2y = 0$
20. $2y'' + y' - 15y = 0$

Find the particular solution of each differential equation subject to the given conditions.

21. $y'' - 4y' = 0$; $y = 3$ and $y' = 4$ when $x = 0$
22. $y'' + 3y' = 0$; $y = -1$ and $y' = 6$ when $x = 0$
23. $y'' - y' - 2y = 0$; $y = 2$ and $y' = 1$ when $x = 0$
24. $y'' + 3y' - 10y = 0$; $y = 7$ and $y' = 0$ when $x = 0$
25. $y'' - 8y' + 15y = 0$; $y = 4$ and $y' = 2$ when $x = 0$
26. $y'' + y' = 0$; $y = 2$ and $y' = 1$ when $x = 1$

12.2 REPEATED ROOTS AND COMPLEX ROOTS

When the auxiliary equation of a differential equation results in a repeated root ($m_1 = m_2 = m$), then the general solution is in the form

---
Repeated Roots

$$y = k_1 e^{mx} + k_2 x e^{mx}$$

Note carefully that the second term includes the factor x.

Example 1

Solve the differential equation

$$y'' - 2y' + y = 0$$

Since $D^2y - 2\,Dy + y = 0$, the auxiliary equation is

$$m^2 - 2m + 1 = 0$$
$$(m - 1)^2 = 0$$
$$m_1 = m_2 = 1$$

The general solution is

$$y = k_1e^{mx} + k_2xe^{mx}$$
$$y = k_1e^x + k_2xe^x$$
$$y = e^x(k_1 + k_2x)$$

Example 2

Solve the differential equation $y'' + 4y' + 4y = 0$ subject to the conditions that $y = 2$ and $y' = 1$ when $x = 0$.

$$D^2y + 4\,Dy + 4y = 0$$
$$m^2 + 4m + 4 = 0$$
$$(m + 2)^2 = 0$$
$$m = -2$$

The general solution is

$$y = k_1e^{-2x} + k_2xe^{-2x}$$

Next,

$$y' = -2k_1e^{-2x} - 2k_2xe^{-2x} + k_2e^{-2x}$$

Substitute $y = 2$, $y' = 1$ and $x = 0$ in the equations for y and y':

$$y = k_1e^{-2x} + k_2xe^{-2x}$$
$$2 = k_1e^{-2(0)} + k_2(0)e^{-2(0)}$$
$$2 = k_1$$
$$y' = -2k_1e^{-2x} - 2k_2xe^{-2x} + k_2e^{-2x}$$
$$1 = -2k_1e^{-2(0)} - 2k_2(0)e^{-2(0)} + k_2e^{-2(0)}$$
$$1 = -2k_1 + k_2$$

But since $k_1 = 2$, we have

$$1 = -4 + k_2 \quad \text{or} \quad k_2 = 5$$

The desired particular solution is then

$$y = 2e^{-2x} + 5xe^{-2x} \quad \text{or} \quad y = e^{-2x}(2 + 5x)$$

When the roots of the auxiliary equation of a differential equation are complex numbers, the method used in Section 12.1 still applies.

Example 3

Solve the differential equation $y'' - y' + 2y = 0$.

The auxiliary equation is $m^2 - m + 2 = 0$, which is solved using the quadratic formula,

$$m = \frac{-(-1) \pm \sqrt{(-1)^2 - 4(1)(2)}}{2(1)}$$

$$= \frac{1 \pm \sqrt{1 - 8}}{2} = \frac{1 \pm \sqrt{-7}}{2} = \frac{1 \pm j\sqrt{7}}{2}$$

So the general solution is

$$y = k_1 e^{[(1/2) + (j\sqrt{7}/2)]x} + k_2 e^{[(1/2) - (j\sqrt{7}/2)]x}$$

However, we can simplify this expression

$$y = k_1 e^{(x/2)} e^{j(\sqrt{7}/2)x} + k_2 e^{(x/2)} e^{-j\sqrt{7}/2)x}$$

$$y = e^{(x/2)} (k_1 e^{j(\sqrt{7}/2)x} + k_2 e^{-j(\sqrt{7}/2)x})$$

One can show that $z = re^{j\theta}$ represents the trigonometric form of a complex number. That is,

$$z = re^{j\theta} = r(\cos\theta + j\sin\theta) \qquad \text{(see Section 10.7)}$$

$$\text{or, } e^{j\theta} = \cos\theta + j\sin\theta$$

For $\theta = \dfrac{\sqrt{7}}{2}x$,

$$e^{j(\sqrt{7}/2)x} = \cos\frac{\sqrt{7}}{2}x + j\sin\frac{\sqrt{7}}{2}x$$

and for $\theta = -\dfrac{\sqrt{7}}{2}x$,

$$e^{-j(\sqrt{7}/2)x} = \cos\left(-\frac{\sqrt{7}}{2}x\right) + j\sin\left(-\frac{\sqrt{7}}{2}x\right)$$

$$= \cos\frac{\sqrt{7}}{2}x - j\sin\frac{\sqrt{7}}{2}x$$

But then

$$k_1 e^{j(\sqrt{7}/2)x} = k_1 \cos\frac{\sqrt{7}}{2}x + jk_1 \sin\frac{\sqrt{7}}{2}x \qquad (1)$$

$$k_2 e^{-j(\sqrt{7}/2)x} = k_2 \cos\frac{\sqrt{7}}{2}x - jk_2 \sin\frac{\sqrt{7}}{2}x \qquad (2)$$

Adding equations (1) and (2), we have

$$k_1 e^{j(\sqrt{7}/2)x} + k_2 e^{-j(\sqrt{7}/2)x} = (k_1 + k_2)\cos\frac{\sqrt{7}}{2}x$$

$$+ j(k_1 - k_2)\sin\frac{\sqrt{7}}{2}x$$

$$= C_1 \cos\frac{\sqrt{7}}{2}x + C_2 \sin\frac{\sqrt{7}}{2}x$$

where $C_1 = k_1 + k_2$ and $C_2 = j(k_1 - k_2)$.
The general solution can now be written as

$$y = e^{(x/2)}\left[C_1 \cos\left(\frac{\sqrt{7}}{2}x\right) + C_2 \sin\left(\frac{\sqrt{7}}{2}x\right)\right]$$

In general, if the auxiliary equation of a differential equation has complex roots $m = a \pm bj$, the solution of the differential equation is:

Complex Roots

$$y = e^{ax}(k_1 \sin bx + k_2 \cos bx)$$
---

where k_1 and k_2 are arbitrary constants.

Example 4

Solve the differential equation $y'' - 2y' + 3y = 0$.
    The auxiliary equation is $m^2 - 2m + 3 = 0$. So

$$m = \frac{2 \pm \sqrt{-8}}{2} = 1 \pm j\sqrt{2}$$

Then $a = 1$ and $b = \sqrt{2}$ and the general solution is

$$y = e^x(k_1 \sin \sqrt{2}\, x + k_2 \cos \sqrt{2}x)$$

Example 5

Solve the differential equation $y'' + 9y = 0$.
    The auxiliary equation is $m^2 + 9 = 0$. So $m = \pm 3j$. Since $a = 0$ and $b = 3$, the general solution is

$$y = e^{0x}(k_1 \sin 3x + k_2 \cos 3x)$$

$$y = k_1 \sin 3x + k_2 \cos 3x$$

Exercises

Solve each differential equation.

1. $\dfrac{d^2y}{dx^2} - 4\dfrac{dy}{dx} + 4y = 0$ 2. $\dfrac{d^2y}{dx^2} + 6\dfrac{dy}{dx} + 9y = 0$ 3. $y'' - 4y' + 5y = 0$

4. $y'' - 2y' + 4y = 0$ 5. $4y'' - 4y' + y = 0$ 6. $y'' - 8y' + 16y = 0$

7. $\dfrac{d^2y}{dx^2} - 4\dfrac{dy}{dx} + 13y = 0$ 8. $\dfrac{d^2y}{dx^2} + 2\dfrac{dy}{dx} + y = 0$

9. $D^2y - 10\,Dy + 25y = 0$ 10. $D^2y + 12\,Dy + 36y = 0$

11. $D^2y + 9y = 0$ 12. $D^2y + 16y = 0$

13. $\dfrac{d^2y}{dx^2} = 0$ 14. $9\dfrac{d^2y}{dx^2} + 16y = 0$

Find the particular solution of each differential equation satisfying the given conditions.

15. $y'' - 6y' + 9y = 0$; $y = 2$ and $y' = 4$ when $x = 0$

16. $D^2y + 10\,Dy + 25y = 0$; $y = 0$ and $y' = 3$ when $x = 0$

17. $y'' + 25y = 0$; $y = 2$ and $y' = 0$ when $x = 0$

18. $y'' + 16y = 0$; $y = 1$ and $y' = -4$ when $x = 0$

19. $D^2y - 12\,Dy + 36y = 0$; $y = 1$ and $y' = 0$ when $x = 0$

20. $y'' - 6y' + 25y = 0$; $y = 0$ and $y' = 8$ when $x = 0$

12.3 NONHOMOGENEOUS EQUATIONS

Now consider the solutions of nonhomogeneous second-order linear differential equations with constant coefficients. We represent such an equation in the form

$$a_0 D^2 y + a_1 Dy + a_2 y = b$$

where $b = g(x)$ is a function of x (b may be a constant).

Obtaining a solution of a nonhomogeneous differential equation involves two basic steps. First, obtain a solution, called the *complementary solution* (denoted y_c), that is, the solution of the homogeneous equation obtained by substituting 0 for b; that is,

$$a_0 D^2 y + a_1 Dy + a_2 y = 0$$

Next obtain a *particular solution* (denoted y_p) for the given nonhomogeneous equation. The general solution of the nonhomogeneous equation will then be

$$y = y_c + y_p$$

Since we have shown methods of obtaining the complementary solution y_c in Sections 12.1 and 12.2, our only problem is to determine a particular solution of a given nonhomogeneous equation. We find y_p using the *method of undetermined coefficients*.

The method relies on inspecting the expression $b = g(x)$ and determining a combination of all possible functions which after differentiating twice would result in at least one of the terms in the expression $g(x)$. In most applications the functions that most often occur are where $g(x)$ involve polynomials, exponentials, sines, and cosines. For these functions, we offer the following table for finding y_p.

FINDING y_p

If $g(x)$ =	Try y_p =
$b_n x^n + b_{n-1}x^{n-1} + \cdots + b_2 x^2 +$ $b_1 x + b_0$ (polynomial of degree n)	$A_n x^n + A_{n-1}x^{n-1} + \cdots + A_2 x^2 + A_1 x + A_0$
be^{nx}	Ae^{nx}
$b \sin nx$ $b \cos nx$ $b \sin nx + c \cos nx$	$A \sin nx + B \cos nx$

Note 1: If $g(x)$ is the sum of two or more types in the left column, try the corresponding sum in the right column.

Note 2: If a term of $g(x)$ is a solution of the homogeneous equation, try multiplying y_p by x or some higher power of x.

To show how to use the table above, six examples of a differential equation and its corresponding trial solution for y_p using the table are shown below.

Differential equation	$g(x) =$	Try $y_p =$
(a) $y'' - 4y' - 5y = 6x^2 - 1$	$6x^2 - 1$	$Ax^2 + Bx + C$
(b) $y'' - y = 3e^{-4x}$	$3e^{-4x}$	Ae^{-4x}
(c) $y'' + 16y = 3 \sin 5x$	$3 \sin 5x$	$A \sin 5x + B \cos 5x$
(d) $y'' - 9y = 2e^{3x}$ Note: $2e^{3x}$ is a solution of the homogeneous equation.	$2e^{3x}$	Axe^{3x}
(e) $y'' + y' = 6x + 4$ Note: 4 is a solution of the homogeneous equation.	$6x + 4$	$Ax^2 + Bx$
(f) $y'' + 9y = \cos 3x$ Note: $\cos 3x$ is a solution of the homogeneous equation.	$\cos 3x$	$Ax \sin 3x + Bx \cos 3x$

This method is illustrated in the examples that follow.

Example 1

Find a particular solution of the differential equation

$$y'' - 2y' - 3y = e^x$$

Try $y_p = Ae^x$. We need to find the value of A. First, find $y_p' = Ae^x$ and $y_p'' = Ae^x$. If these are solutions, they must satisfy the given differential equation.

$$y_p'' - 2y_p' - 3y_p = e^x$$
$$Ae^x - 2Ae^x - 3Ae^x = e^x$$
$$-4Ae^x = e^x$$
$$-4A = 1$$
$$A = -\tfrac{1}{4}$$

Then

$$y_p = -\tfrac{1}{4}e^x$$

Example 2

Find a particular solution of the differential equation

$$y'' - 2y' - 3y = x^2 + e^{-2x}$$

Try $y_p = A + Bx + Cx^2 + De^{-2x}$, which is the sum of a second-degree polynomial and an exponential that appear in $g(x)$. To find the values of A, B, C, and D, we first find

$$y_p' = B + 2Cx - 2De^{-2x}$$

and

$$y_p'' = 2C + 4De^{-2x}$$

Then substitute into the given differential equation.

$$y_p'' - 2y_p' - 3y_p = x^2 + e^{-2x}$$

Then

$$(2C + 4De^{-2x}) - 2(B + 2Cx - 2De^{-2x})$$
$$- 3(A + Bx + Cx^2 + De^{-2x}) = x^2 + e^{-2x}$$

or
$$(2C - 2B - 3A) + (-4C - 3B)x$$
$$+ (-3C)x^2 + (4D + 4D - 3D)e^{-2x} = x^2 + e^{-2x}$$

From this last equation, we equate the coefficients on each side of the equation.

$$2C - 2B - 3A = 0 \tag{1}$$
$$-4C - 3B = 0 \tag{2}$$
$$-3C = 1 \tag{3}$$
$$5D = 1 \tag{4}$$

Solving this system of four equations, we find that

$$D = \frac{1}{5}, \quad C = -\frac{1}{3}, \quad B = \frac{4}{9}, \quad \text{and } A = -\frac{14}{27}$$

The particular equation is then

$$y_p = -\frac{14}{27} + \frac{4}{9}x - \frac{1}{3}x^2 + \frac{1}{5}e^{-2x}$$

General Solution of Nonhomogeneous Differential Equation

Finding the general solution of the nonhomogeneous equation $a_0 D^2y + a_1 Dy + a_2y = b$ involves the following steps:

1. Determine the complementary solution y_c by solving the homogeneous equation $a_0 D^2y + a_1 Dy + a_2y = 0$ using the methods developed in Sections 12.1 and 12.2.
2. Determine a particular solution y_p by using the method of undetermined coefficients described in Examples 1 and 2.
3. Determine the general solution y by adding the complementary solution y_c from step 1 and the particular solution y_p from step 2:
$$y = y_c + y_p$$

Example 3

Find the general solution of
$$y'' - 2y' - 3y = e^x$$

Step 1: Determine the complementary solution y_c which is the solution of the homogeneous equation $y'' - 2y' - 3y = 0$. We solved this equation in Example 2, Section 12.1:
$$y_c = k_1e^{3x} + k_2e^{-x}$$

Step 2: Determine a particular solution y_p of the given equation. We obtained a particular solution in Example 1 of this section:
$$y_p = -\tfrac{1}{4}e^x$$

Step 3: The desired general solution is
$$y = y_c + y_p$$
$$y = k_1e^{3x} + k_2e^{-x} - \tfrac{1}{4}e^x$$

Example 4

Find the general solution of $y'' - 2y' - 3y = x^2 + e^{-2x}$.

Step 1: From Example 3, the complementary solution is
$$y_c = k_1 e^{3x} + k_2 e^{-x}$$

Step 2: From Example 2, a particular solution is
$$y_p = -\tfrac{14}{27} + \tfrac{4}{9}x - \tfrac{1}{3}x^2 + \tfrac{1}{5}e^{-2x}$$

Step 3: The general solution is
$$y = y_c + y_p$$
$$y = k_1 e^{3x} + k_2 e^{-x} - \tfrac{14}{27} + \tfrac{4}{9}x - \tfrac{1}{3}x^2 + \tfrac{1}{5}e^{-2x}$$

Example 5

Solve the differential equation
$$y'' + 6y' + 9y = x + \sin x$$

Step 1: Find y_c:
$$y'' + 6y' + 9y = 0$$
$$m^2 + 6m + 9 = 0$$
$$(m + 3)^2 = 0$$
$$m = -3 \qquad \text{a repeated root}$$
$$y_c = e^{mx}(k_1 + k_2 x)$$
$$= e^{-3x}(k_1 + k_2 x)$$

Step 2: Find y_p. Try $y_p = A + Bx + C \sin x + D \cos x$. Differentiating y_p, we have
$$y_p' = B + C \cos x - D \sin x$$
$$y_p'' = -C \sin x - D \cos x$$

Substitute y_p, y_p' and y_p'' in the given differential equation.
$$y'' + 6y' + 9y = x + \sin x$$
$$(-C \sin x - D \cos x) + 6(B + C \cos x - D \sin x)$$
$$+ 9(A + Bx + C \sin x + D \cos x) = x + \sin x$$
$$(6B + 9A) + 9Bx + (-C - 6D + 9C) \sin x$$
$$+ (-D + 6C + 9D) \cos x = x + \sin x$$

Equating coefficients, we have
$$6B + 9A = 0$$
$$9B = 1$$
$$8C - 6D = 1$$
$$6C + 8D = 0$$

Solving for A, B, C, and D, we have
$$A = -\tfrac{2}{27}, \qquad B = \tfrac{1}{9}, \qquad C = \tfrac{2}{25}, \qquad \text{and } D = -\tfrac{3}{50}$$

The particular solution is then
$$y_p = -\tfrac{2}{27} + \tfrac{1}{9}x + \tfrac{2}{25} \sin x - \tfrac{3}{50} \cos x$$

Step 3: The general solution is then

$$y = y_c + y_p$$
$$y = e^{-3x}(k_1 + k_2x) - \tfrac{2}{27} + \tfrac{1}{6}x + \tfrac{2}{25} \sin x - \tfrac{3}{50} \cos x$$

Example 6

Solve the differential equation
$$y'' - 3y' - 4y = e^{4x}$$

Step 1: Find y_c:
$$m^2 - 3m - 4 = 0$$
$$(m - 4)(m + 1) = 0$$
$$m_1 = 4 \quad \text{and} \quad m_2 = -1$$
$$y_c = k_1e^{4x} + k_2e^{-x}$$

Step 2: Find y_p: First, note that $g(x) = e^{4x}$ is a solution of the homogeneous equation
$$y'' - 3y' - 4y = 0$$
$$16e^{4x} - 3(4e^{4x}) - 4(e^{4x}) = 0$$

Then, following Note 2 after the Table for finding y_p, we try
$$y_p = Axe^{4x}$$
$$y_p' = 4Axe^{4x} + Ae^{4x}$$
$$y_p'' = 16Axe^{4x} + 4Ae^{4x} + 4Ae^{4x} = 16Axe^{4x} + 8Ae^{4x}$$

Substituting into the differential equation, we have
$$y'' - 3y' - 4y = e^{4x}$$
$$(16Axe^{4x} + 8Ae^{4x}) - 3(4Axe^{4x} + Ae^{4x}) - 4(Axe^{4x}) = e^{4x}$$
$$16Axe^{4x} + 8Ae^{4x} - 12Axe^{4x} - 3Ae^{4x} - 4Axe^{4x} = e^{4x}$$
$$5Ae^{4x} = e^{4x}$$
$$5A = 1$$
$$A = \tfrac{1}{5}$$

Step 3:
$$y = y_c + y_p$$
$$y = k_1e^{4x} + k_2e^{-x} + \tfrac{1}{5}xe^{4x}$$

Exercises

Solve each differential equation.

1. $y'' + y' = \sin x$
2. $y'' + 4y = 3e^{-x}$
3. $y'' - y' - 2y = 4x$
4. $y'' - 4y' + 4y = e^x$
5. $y'' - 10y' + 25y = x$
6. $y'' - 2y' - 3y = 3x^2$
7. $y'' - y = x^2$
8. $y'' + y = e^x$
9. $y'' + 4y = e^x - 2$
10. $y'' + 4y = e^x - 2x$
11. $y'' - 3y' - 4y = 6e^x$
12. $y'' - 4y' + 3y = 20 \cos x$
13. $y'' + y = 5 + \sin 3x$
14. $y'' - y' - 6y = x + \cos x$
15. $y'' - y = e^x$
16. $y'' + 6y' = 18x^2 - 6x + 3$
17. $y'' + 4y = \cos 2x$
18. $y'' - 4y = e^{-2x}$

Find the particular solution of each differential equation satisfying the given conditions.

19. $y'' + y = 10e^{2x}$; $y = 0$ and $y' = 0$ when $x = 0$

20. $y'' + y' = \sin x$; $y = 0$ and $y' = 0$ when $x = 0$

21. $y'' + y = e^x$; $y = 0$ and $y' = 3$ when $x = 0$

22. $y'' - 4y = 2 - 8x$; $y = 0$ and $y' = 5$ when $x = 0$

12.4 APPLICATIONS OF SECOND-ORDER DIFFERENTIAL EQUATIONS

Mathematical models of numerous technical applications result in second-order linear differential equations. We will focus our attention on three basic applications: mechanical vibrations, buoyancy, and electric circuits.

We saw in Section 5.6 that the force required to stretch or compress a spring obeys Hooke's law. That is, the required force $F = f(x)$ is directly proportional to the distance x that the spring is stretched or compressed:

$$f(x) = kx$$

where k is called the spring constant.

Another basic mechanical principle, Newton's law, states that the force F acting on a mass m at any time t is equal to the product of the mass m and the acceleration a of the mass:

$$F = ma$$

We now relate these two mechanical principles and investigate the motion of a vibrating spring. Consider a spring hanging downward with a natural length l. If a weight W is attached to the spring, the force of gravity acting on the weight will stretch the spring downward a distance s until coming to rest. (See Fig. 12.1.)

By Hooke's law the force used to stretch the spring a distance s to this rest position is

$$F = ks$$

and by Newton's law this same force is

$$F = mg$$

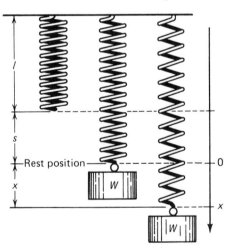

Figure 12.1

where m is the mass of the weight and g is the acceleration of the weight as a result of gravity (approximately 32 ft/s^2). Then,

$$mg = ks$$

If the spring is now stretched beyond the rest length $l + s$ and then released, a vibrating motion will result. When in motion, at any time t, the force F acting on the weight is

$$F = mg - k(s + x)$$

where x is the distance measured from the rest position at time t. The term mg represents the force of gravity acting downward on the weight. The term $-k(s + x)$ represents the spring tension, which is a force acting to restore the system to the rest position.

Again, by Newton's law this same force acting on the weight is given by

$$F = ma$$

$$F = m\frac{d^2x}{dt^2}$$

since $\dfrac{d^2x}{dt^2}$ is the acceleration of the weight at any time t. Then

$$m\frac{d^2x}{dt^2} = mg - k(s + x)$$

$$= mg - ks - kx$$

$$= mg - mg - kx \qquad (\text{since } mg = ks)$$

$$= -kx$$

We then have

$$m\frac{d^2x}{dt^2} + kx = 0$$

$$\frac{d^2x}{dt^2} + \frac{k}{m}x = 0$$

This is a second-order linear differential equation whose solution $x = f(t)$ describes the vibrating motion of the weight and spring. The motion described by this differential equation is called *simple harmonic motion*. Another visual example of simple harmonic motion is the motion of a simple pendulum like the pendulum of a clock.

The solution of this differential equation is obtained by using the methods of Section 12.2. The solution becomes

$$x = C_1 \sin \sqrt{\frac{k}{m}}t + C_2 \cos \sqrt{\frac{k}{m}}t$$

The period (time for one complete oscillation) of this simple harmonic motion is given by

$$P = \frac{2\pi}{\sqrt{\dfrac{k}{m}}}$$

since the period for $y = A \sin Bt$ and $y = A \cos Bt$ is $\dfrac{2\pi}{|B|}$.

Example 1

Find the equation expressing the simple harmonic motion of a weight attached to a spring. A 32-lb weight stretches the spring 6 in. to a rest position. The motion is started by stretching the spring an additional 3 in. and then releasing the spring.

We have

$$m = \frac{W}{g} = \frac{32 \text{ lb}}{32 \text{ ft/s}^2} = 1 \text{ lb-s}^2/\text{ft}$$

$$k = \frac{F}{s} = \frac{32 \text{ lb}}{6 \text{ in.}} = \frac{32 \text{ lb}}{\frac{1}{2} \text{ ft}} = 64 \text{ lb/ft}$$

Then

$$x = C_1 \sin \sqrt{\frac{k}{m}}t + C_2 \cos \sqrt{\frac{k}{m}}t$$

$$= C_1 \sin \sqrt{64}t + C_2 \cos \sqrt{64}t$$

$$= C_1 \sin 8t + C_2 \cos 8t$$

To determine constants C_1 and C_2, differentiate the above equation.

$$\frac{dx}{dt} = 8C_1 \cos 8t - 8C_2 \sin 8t$$

Substitute $x = 3$ in. $= 0.25$ ft and $\dfrac{dx}{dt} = 0$ when $t = 0$.

$$0.25 = C_1 \sin 0 + C_2 \cos 0$$

$$0 = 8C_1 \cos 0 - 8C_2 \sin 0$$

We find that $C_1 = 0$ and $C_2 = 0.25$. The solution is

$$x = 0.25 \cos 8t$$

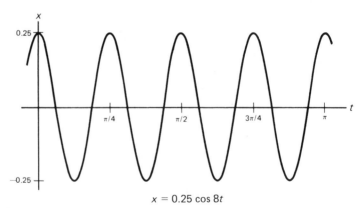

$$x = 0.25 \cos 8t$$

Figure 12.2

However, in reality, the vibrations do not continue forever. There exist resistant forces to the motion of vibration. The effect of these forces is called the *damping effect*. These resistant forces, such as friction, will cause the

476   Second-Order Differential Equations   Chap. 12

vibration to decrease. In many situations this force of resistance is directly proportional to the velocity of the vibrating system. That is, the damping force equals $p\dfrac{dx}{dt}$, where p is a constant.

The differential equation expressing the total forces acting on the system becomes

$$m\frac{d^2x}{dt^2} = -kx - p\frac{dx}{dt}$$

$$\frac{d^2x}{dt^2} + \frac{p}{m}\frac{dx}{dt} + \frac{k}{m}x = 0$$

The auxiliary equation is

$$\overline{m}^2 + \frac{p}{m}\overline{m} + \frac{k}{m}x = 0$$

(We use $\overline{m}$ in place of m in the auxiliary equation so as not to be confused with the mass m.) Solving for $\overline{m}$, we have

$$\overline{m} = \frac{-\dfrac{p}{m} \pm \sqrt{\left(\dfrac{p}{m}\right)^2 - \dfrac{4k}{m}}}{2}$$

$$= -\frac{p}{2m} \pm \sqrt{\left(\frac{p}{2m}\right)^2 - \frac{k}{m}}$$

Now let

$$d = \left(\frac{p}{2m}\right)^2 - \frac{k}{m}$$

Case 1 $d > 0$: There are two real roots of the auxiliary equation:

$$\overline{m}_1 = -\frac{p}{2m} + \sqrt{d}$$

$$\overline{m}_2 = -\frac{p}{2m} - \sqrt{d}$$

The solution of the differential equation is

$$x = C_1 e^{\overline{m}_1 t} + C_2 e^{\overline{m}_2 t}$$

This case is called *overdamped*.

Case 2 $d = 0$: There is a repeated real root of the auxiliary equation. The solution is

$$x = (C_1 + C_2 t)e^{\overline{m}t}, \qquad \text{where } \overline{m} = -\frac{p}{2m}$$

This case is called *critically damped*.

Case 3 $d < 0$: There are two complex roots of the auxiliary equation.

The solution is:
$$x = e^{-[p/(2m)]t} (C_1 \sin \omega t + C_2 \cos \omega t), \qquad \text{where } \omega = \sqrt{|d|}$$
This case is called *underdamped*. (See Fig. 12.3.)

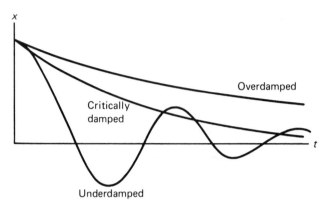

Critically damped

Overdamped

Underdamped

Figure 12.3

Example 2

Find the general solution for the motion of the vibrating spring of Example 1 if there is a resistant force that exerts a force of 0.04 lb with a velocity of 2 in./s.

Since $F_{\text{resistant}} = p\dfrac{dx}{dt}$

we have:

$$p = \frac{F_{\text{resistant}}}{\dfrac{dx}{dt}} = \frac{0.04 \text{ lb}}{2 \text{ in./s}} = \frac{0.04 \text{ lb-s}}{\frac{1}{6} \text{ ft}}$$

$$= 0.24 \text{ lb-s/ft}$$

$$d = \left(\frac{p}{2m}\right)^2 - \frac{k}{m}$$

$$= \left(\frac{0.24}{2(1)}\right)^2 - 64$$

$$= -63.99 \quad \text{and} \quad \omega = \sqrt{|d|} = \sqrt{63.99} = 8$$

Case 3 applies and the general solution is:

$$x = e^{[-p/(2m)]t} (C_1 \sin \omega t + C_2 \cot \omega t)$$

$$= e^{-0.12t} (C_1 \sin 8t + C_2 \cos 8t)$$

Example 3

A cylindrical buoy with radius 1 ft is floating in water with the axis of the buoy vertical as in Fig. 12.4. If the buoy is pushed a small distance into the water and released, the period of vibration is found to be 4 s. Find the weight of the buoy given that the density of water is 62.4 lb/ft³.

Let the origin be at the intersection of the axis of the buoy and the water when the buoy is in equilibrium. Let's choose the downward direction to be positive.

Second-Order Differential Equations   Chap. 12

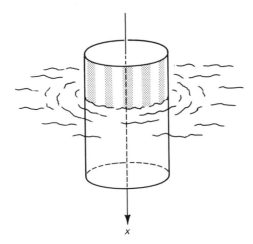

Figure 12.4

A body partly or wholly submerged in a fluid is buoyed up by a force equal to the weight of the fluid it displaces (Archimedes' principle). If the downward displacement is x, the change in submerged volume is $\pi(1)^2 x$ and the change in the buoyant force is $62.4\,\pi(1)^2 x$. If we let W (in pounds) be the weight of the buoy, then its mass is W/g, where $g = 32.2$ ft/s^2. The equation $F = ma$ thus becomes

$$-62.4\pi x = \frac{W}{g}\frac{d^2 x}{dt^2}$$

$$\frac{d^2 x}{dt^2} + \frac{2009\pi}{W} x = 0$$

By the methods of Section 12.2, the solution of this differential equation is

$$x = C_1 \sin\sqrt{\frac{2009\pi}{W}}\, t + C_2 \cos\sqrt{\frac{2009\pi}{W}}\, t$$

Recall that the period is 4 s, so

$$P = \frac{2\pi}{\sqrt{\dfrac{2009\pi}{W}}} = 4$$

or

$$\frac{\pi}{2} = \sqrt{\frac{2009\pi}{W}}$$

Squaring, we have

$$\frac{\pi^2}{4} = \frac{2009\pi}{W}$$

$$W = 2560 \text{ lb}$$

Electric circuits provide another example of the use of second-order linear differential equations. Kirchhoff's law states that the voltage drops across the elements of an electric circuit equal the voltage source V. If a circuit (see Fig. 12.5) contains a capacitor C (in farads), an inductance L (in henrys), and a resistance R (in ohms), then Kirchhoff's law results in the differential equation:

$$L\frac{di}{dt} + Ri + \frac{1}{C}\int_0^t i\, dt = V$$

If the voltage source V is constant, then differentiating we obtain:

$$L\frac{d^2i}{dt^2} + R\frac{di}{dt} + \frac{1}{C}i = \frac{dV}{dt} = 0$$

$$\frac{d^2i}{dt^2} + \frac{R}{L}\frac{di}{dt} + \frac{1}{CL}i = 0$$

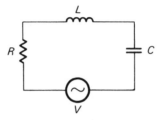

Figure 12.5

The auxiliary equation is $m^2 + \frac{R}{L}m + \frac{1}{CL} = 0$, whose roots are:

$$m = \frac{-\dfrac{R}{L} \pm \sqrt{\left(\dfrac{R}{L}\right)^2 - \dfrac{4}{CL}}}{2}$$

$$= -\frac{R}{2L} \pm \sqrt{\frac{R^2}{4L^2} - \frac{1}{CL}}$$

The form of the solution depends on the value of $d = \dfrac{R^2}{4L^2} - \dfrac{1}{CL}$

    Case 1 $d > 0$:   (overdamped)

$$i = k_1 e^{m_1 t} + k_2 e^{m_2 t}$$

where

$$m_1 = -\frac{R}{2L} + \sqrt{d}$$

and

$$m_2 = -\frac{R}{2L} - \sqrt{d}$$

    Case 2 $d = 0$:   (critically damped)

$$i = (k_1 + k_2 t)e^{-[R/(2L)]t}$$

    Case 3 $d < 0$:   (underdamped)

$$i = e^{-[R/(2L)]t}(k_1 \sin \omega t + k_2 \cos \omega t), \qquad \text{where } \omega = \sqrt{|d|}$$

Example 4

Find an expression for the current in an electric circuit containing an inductor of 0.2 H, a capacitor of 10^{-5} F and a resistor of 300 Ω. The voltage source is 12 V. Assume that $i = 0$ at $t = 0$.

$$d = \frac{R^2}{4L^2} - \frac{1}{CL}$$

$$= \frac{(300)^2}{4(0.2)^2} - \frac{1}{0.2 \times 10^{-5}}$$

$$= 6.25 \times 10^4$$

Then by case 1, we have

$$-\frac{R}{2L} = -\frac{300}{2(0.2)} = -7.5 \times 10^2$$

$$m_1 = -7.5 \times 10^2 + \sqrt{6.25 \times 10^4} = -500$$

$$m_2 = -7.5 \times 10^2 - \sqrt{6.25 \times 10^4} = -1000$$

The solution is then

$$i = k_1 e^{-500t} + k_2 e^{-1000t}$$

Differentiating, we have

$$\frac{di}{dt} = -500k_1 e^{-500t} - 1000k_2 e^{-1000t}$$

Using

$$L\frac{di}{dt} + Ri + \frac{1}{C}\int_0^t i\, dt = V$$

and the condition $i = 0$ at $t = 0$, we have

$$(0.2)\frac{di}{dt} + 300(0) + \left(\frac{1}{10^5}\right)(0) = 12$$

$$\frac{di}{dt} = 60$$

Then, substituting $t = 0$ in the equations for i and $\dfrac{di}{dt}$ above, we have

$$0 = k_1 + k_2$$

$$60 = -500k_1 - 1000k_2$$

Then

$$k_1 = 0.12$$

$$k_2 = -0.12$$

The particular solution is

$$i = 0.12e^{-500t} - 0.12e^{-1000t}$$

If the voltage source V in the electric circuit is not constant, then $\dfrac{dV}{dt}$ is not zero, and we obtain a nonhomogeneous differential equation:

$$\frac{d^2i}{dt^2} + \frac{R}{L}\frac{di}{dt} + \frac{1}{CL}i = \frac{1}{L}\frac{dV}{dt}$$

Similarly, if a force $f(t)$ external to a vibrating system of simple harmonic motion is applied (such as a periodic push or pull on the weight), then we also obtain a nonhomogeneous equation:

$$\frac{d^2x}{dt^2} + \frac{p}{m}\frac{dx}{dt} + \frac{k}{m}x = f(t)$$

We will not consider the nonhomogeneous cases in this text.

Exercises

1. A spring is stretched $\frac{1}{4}$ ft by a weight of 4 lb. If the weight is displaced a distance of 2 in. from the rest position and then released, find the equation of motion.

2. A spring is stretched 6 in. by a weight of 10 lb. If the weight is then displaced a distance of 6 in. from the rest position and then released, find the equation of motion.

3. A spring is stretched 6 in. by a weight of 20 lb. A damping force exerts a force of 5 lb for a velocity of 4 in./s. If the weight is displaced from the rest position and then released, find the general equation of motion.

4. A spring is stretched 2 in. by a weight of 8 lb. A damping force exerts a force of 4 lb for a velocity of 6 in./s. If the weight is displaced from the rest position and then released, find the general equation of motion.

5. A cylindrical buoy with a radius of 1.5 ft is floating in water with the axis of the buoy vertical. When the buoy is pushed a small distance into the water and released, the period of vibration is 6 s. Find the weight of the buoy.

6. A cylindrical buoy with a radius of 2 ft is floating in water with the axis of the buoy vertical. When the buoy is pushed a small distance into the water and released, the period of vibration is 1.6 s. Find the weight of the buoy.

7. A cylindrical buoy weighing 1000 lb is floating in water with the axis of the buoy vertical. When the buoy is pushed a small distance into the water and released, the period of vibration is 2 s. Find the radius of the buoy.

8. A cylindrical buoy weighing 1600 lb is floating in water with the axis of the buoy vertical. When the buoy is pushed a small distance into the water and released, the period of vibration is 1.5 s. Find the radius of the buoy.

9. An electric circuit has an inductance L of 0.1 H, a resistance R of 50 Ω, and a capacitance C of 2×10^{-4} F. Find the equation for the current i.

10. An electric circuit has an inductance L of 0.2 H, a resistance R of 400 Ω, and a capacitance C of 5×10^{-4} F. Find the equation for the current i.

11. An electric circuit has an inductance L of 0.4 H, a resistance R of 200 Ω, and a capacitance C of 5×10^{-5} F. Find the equation for the current i if the voltage source is 12 V.

12. An electric circuit has an inductance L of 0.5 H, a resistance R of 1000 Ω, and a capacitance C of 5.6×10^{-6} F. Find the equation for the current i if the voltage source is 12 V.

12.5 THE LAPLACE TRANSFORM

We have already used the linear differential operator D. We now look at another linear operator $\mathcal{L}$ called the *Laplace transform*. The use of this operator is very useful in solving differential equations with given initial conditions.

The Laplace transform $\mathcal{L}$, like the differential operator D, operates on functions. The Laplace transform operates on a function $f(t)$ and transforms it into another function $F(s)$. That is, $\mathcal{L}[f(t)] = F(s)$. The transform is defined by means of integration:

$$F(s) = \mathcal{L}[f(t)] = \int_0^\infty e^{-st} f(t)\, dt$$

Note that this equation involves an integral which indicates an infinite upper limit (∞). Such an integral is called an *improper integral*, whose value is found by a special limit process. In general, an integral of the form

$$\int_0^\infty g(t)\, dt$$

is evaluated by considering the limit of the integral $\int_0^t g(t)\, dt$ as $t \to \infty$ (as t takes on positive values without bound). That is,

$$\int_0^\infty g(t)\, dt = \lim_{t\to\infty} \int_0^t g(t)\, dt$$

And if $G(t)$ is an antiderivative of $g(t)$, then

$$\int_0^\infty g(t)\, dt = \lim_{t\to\infty} \int_0^t g(t)\, dt$$

$$= \lim_{t\to\infty} [G(t) - G(0)]$$

We illustrate this method of integration by using several examples of Laplace transforms.

Example 1

Find the Laplace transform of the function $f(t) = e^{at}$,

$$\mathcal{L}(e^{at}) = \int_0^\infty e^{-st}(e^{at})\, dt \qquad \text{(assume that } s > a)$$

$$= \int_0^\infty e^{-st+at}\, dt$$

$$= \int_0^\infty e^{-(s-a)t}\, dt$$

$$= \lim_{t\to\infty} \int_0^t e^{-(s-a)t}\, dt$$

$$= \lim_{t\to\infty} \left[\frac{-e^{-(s-a)t}}{s-a} \Big|_0^t \right]$$

$$= \lim_{t\to\infty} \left[\frac{-e^{-(s-a)t}}{s-a} - \frac{-e^0}{s-a} \right]$$

$$= \lim_{t \to \infty} \left[\frac{1}{s-a} \left(\frac{-1}{e^{(s-a)t}} + 1 \right) \right]$$

$$= \frac{1}{s-a} (0 + 1) \left[\frac{-1}{e^{(s-a)t}} \text{ approaches } 0 \text{ when } t \to \infty \right]$$

$$\mathcal{L}(e^{at}) = \frac{1}{s-a} \quad (s > a)$$

Example 2

Find the Laplace transform of the function $f(t) = 1$.

This is a special case of Example 1. If we let $a = 0$, we have $f(t) = e^{(0)t} = e^0 = 1$. Then

$$\mathcal{L}(1) = \mathcal{L}(e^{0t})$$

$$= \frac{1}{s-0}$$

$$\mathcal{L}(1) = \frac{1}{s} \quad (s > 0)$$

Example 3

Find $\mathcal{L}(\sin kt)$.

$$\mathcal{L}(\sin kt) = \int_0^\infty e^{-st} \sin kt \, dt$$

$$= \lim_{t \to \infty} \int_0^t e^{-st} \sin kt \, dt$$

$$= \lim_{t \to \infty} \left[\frac{e^{-st}(-s \sin kt - k \cos kt)}{s^2 + k^2} \Big|_0^t \right]$$

$$\text{(use Formula 78, Table 10)}$$

$$= 0 - \frac{1(0-k)}{s^2 + k^2} \quad (\text{since } e^{-st} \to 0 \text{ as } t \to \infty)$$

$$\mathcal{L}(\sin kt) = \frac{k}{s^2 + k^2}$$

We now develop the expression for the Laplace transform of the derivative $f'(t)$ of a function $f(t)$:

$$\mathcal{L}[f'(t)] = \int_0^\infty e^{-st} f'(t) \, dt$$

Integrating by parts, we let $u = e^{-st}$ and $dv = f'(t) \, dt$. Then

$$\int u \, dv \qquad\qquad \int v \, du$$

$u = e^{-st}$		$du = -se^{-st} \, dt$
$dv = f'(t) \, dt$	$\longrightarrow$	$v = f(t)$

$$\int u \, dv = uv - \int v \, du$$

$$\mathcal{L}[f'(t)] = \int_0^\infty e^{-st}f'(t)\,dt = f(t)e^{-st}\Big|_0^\infty - \int_0^\infty - se^{-st}f(t)\,dt$$

$$= \lim_{t\to\infty}\left[\frac{f(t)}{e^{st}}\Big|_0^t\right] + s\int_0^\infty e^{-st}f(t)\,dt$$

$$= \lim_{t\to\infty}\left[\frac{f(t)}{e^{st}} - \frac{f(0)}{e^0}\right] + s\,\mathcal{L}[f(t)]$$

$$= s\,\mathcal{L}[f(t)] - f(0) \qquad \left(\lim_{t\to\infty}\frac{f(t)}{e^{st}} = 0\right)$$

One can show that the Laplace transform of the second derivative is given as follows.

$$\mathcal{L}[f''(t)] = s^2\,\mathcal{L}[f(t)] - sf(0) - f'(0).$$

One can also show that the Laplace transform is a linear operator. That is,

$$\mathcal{L}[a\,f(t) + b\,g(t)] = a\,\mathcal{L}[f(t)] + b\,\mathcal{L}[g(t)]$$

where $a\,f(t)$ represents a constant multiple of the function $f(t)$ and $b\,g(t)$ represents a constant multiple of the function $g(t)$.

Using this linear property and the Laplace transforms for derivatives, we can obtain a Laplace transform for any linear combination of a function $y = f(t)$ and its first and second derivatives. A table of Laplace transforms appears on the next page.

Example 4

Express the Laplace transform for $y'' - 3y' + 2y$ in terms of $\mathcal{L}(y)$ and s, where $y = 2$ and $y' = 1$ when $t = 0$. That is, $[y(0) = 2$ and $y'(0) = 1]$.

$$\mathcal{L}(y'' - 3y' + 2y) = \mathcal{L}(y'') - 3\,\mathcal{L}(y') + 2\,\mathcal{L}(y)$$
$$= [s^2\,\mathcal{L}(y) - s\,y(0) - y'(0)] - 3[s\,\mathcal{L}(y) - y(0)] + 2\,\mathcal{L}(y)$$
$$= [s^2\,\mathcal{L}(y) - s(2) - 1] - 3[s\,\mathcal{L}(y) - 2] + 2\,\mathcal{L}(y)$$
$$= (s^2 - 3s + 2)\,\mathcal{L}(y) - 2s + 5$$

Example 5

Find the Laplace transform of $f(t) = te^{-3t}$

Using Formula 15 in the table, we have $a = -3$ and

$$\mathcal{L}(te^{-3t}) = \frac{1}{(s + 3)^2}.$$

When using the method of Laplace transforms to solve differential equations, we usually need to invert a transform at the end of the process. That is, we need to find the inverse transform

$$\mathcal{L}^{-1}[F(s)] = f(t)$$

where the symbol $\mathcal{L}^{-1}$ denotes an inverse transform.

If $F(s)$ is a function of s and $f(t)$ is a function of t such that

$$\mathcal{L}[f(t)] = F(s)$$

then $\qquad$ $\mathscr{L}^{-1}[F(s)] = f(t)$ is the inverse transform.

The table below lists some commonly used Laplace transforms. A more complete table can be found in standard books of mathematical tables.

TABLE OF LAPLACE TRANSFORMS

$f(t) = \mathscr{L}^{-1}[F(s)]$	$\mathscr{L}[f(t)] = F(s)$
1. $a\,f(t) + b\,g(t)$	$a\,\mathscr{L}[f(t)] + b\,\mathscr{L}[g(t)]$
2. $f'(t)$	$s\,\mathscr{L}[f(t)] - f(0)$
3. $f''(t)$	$s^2\,\mathscr{L}[f(t)] - s\,f(0) - f'(0)$
4. 1	$\dfrac{1}{s}$
5. t	$\dfrac{1}{s^2}$
6. $\dfrac{t^{n-1}}{(n-1)!}$	$\dfrac{1}{s^n},\ n = 1, 2, 3 \ldots$
7. e^{at}	$\dfrac{1}{s-a}$
8. $\sin kt$	$\dfrac{k}{s^2 + k^2}$
9. $e^{at}\sin kt$	$\dfrac{k}{(s-a)^2 + k^2}$
10. $\cos kt$	$\dfrac{s}{s^2 + k^2}$
11. $e^{at}\cos kt$	$\dfrac{s-a}{(s-a)^2 + k^2}$
12. $e^{at} - e^{bt}$	$\dfrac{a-b}{(s-a)(s-b)}$
13. $1 - e^{at}$	$\dfrac{-a}{s(s-a)}$
14. $ae^{at} - be^{bt}$	$\dfrac{s(a-b)}{(s-a)(s-b)}$
15. te^{at}	$\dfrac{1}{(s-a)^2}$
16. $\dfrac{t^{n-1}e^{at}}{(n-1)!}$	$\dfrac{1}{(s-a)^n},\quad n = 1, 2, 3, \ldots$
17. $e^{at}(1 + at)$	$\dfrac{s}{(s-a)^2}$
18. $t\sin kt$	$\dfrac{2ks}{(s^2 + k^2)^2}$
19. $t\cos kt$	$\dfrac{s^2 - k^2}{(s^2 + k^2)^2}$
20. $\sin kt - kt\cos t$	$\dfrac{2k^3}{(s^2 + k^2)^2}$
21. $\sin kt + kt\cos t$	$\dfrac{2ks^2}{(s^2 + k^2)^2}$
22. $1 - \cos kt$	$\dfrac{k^2}{s(s^2 + k^2)}$
23. $kt - \sin kt$	$\dfrac{k^3}{s^2(s^2 + k^2)}$

Second-Order Differential Equations Chap. 12

Example 6

Find the inverse transform of $F(s) = \dfrac{5}{s(s + 5)}$.

Using Formula 13 in the table, we have $a = -5$ and

$$-\left[\mathscr{L}^{-1}\left(\frac{-5}{s(s + 5)} \right) \right] = -(1 - e^{-5t}) = -1 + e^{-5t}$$

Example 7

Find the inverse transform of $\mathscr{L}[f(t)]$, where

$$s(s^2 + 9) \, \mathscr{L}[f(t)] - 9 = 0.$$

Solving for $\mathscr{L}[f(t)]$, we have

$$\mathscr{L}[f(t)] = \frac{9}{s(s^2 + 9)} = F(s)$$

Applying Formula 22 with $k = 3$, we obtain

$$\mathscr{L}^{-1}[F(s)] = \mathscr{L}^{-1}\left(\frac{9}{s(s^2 + 9)} \right)$$

$$= 1 - \cos 3t$$

An inverse transform cannot always be found directly by use of the table. Sometimes the method of completing the square or partial fractions can be used to rewrite a function $F(s)$ in a form applicable to the table.

Example 8

Find the inverse transform of $F(s) = \dfrac{s + 1}{s^2 - 4s + 5}$.

Completing the square, we have

$$s^2 - 4s + 5 = (s^2 - 4s + 4) + 1 = (s - 2)^2 + 1$$

Then we can write

$$F(s) = \frac{s + 1}{s^2 - 4s + 5} = \frac{(s - 2) + 3}{(s - 2)^2 + 1}$$

$$= \frac{s - 2}{(s - 2)^2 + 1} + \frac{3}{(s - 2)^2 + 1}$$

and

$$\mathscr{L}^{-1}[F(s)] = \mathscr{L}^{-1}\left[\frac{s - 2}{(s - 2)^2 + 1} + \frac{3}{(s - 2)^2 + 1} \right]$$

$$= \mathscr{L}^{-1}\left[\frac{s - 2}{(s - 2)^2 + 1} \right] + \mathscr{L}^{-1}\left[\frac{3}{(s - 2)^2 + 1} \right]$$

$$= e^{2t} \cos t + 3(e^{2t} \sin t)$$

where the first term is found from Formula 11 using $a = 2$ and $k = 1$ and the second term is found from Formula 9 using $a = 2$ and $k = 1$.

Exercises

1. Compute the Laplace transform for $f(t) = t$.
2. Compute the Laplace transform for $f(t) = \cos kt$.

Using the Table of Laplace Transforms, find the Laplace transform of each function $f(t)$.

3. $f(t) = \sin 3t$ **4.** $f(t) = e^{5t}$ **5.** $f(t) = 1 - e^{4t}$

6. $f(t) = 1 - \cos 7t$ **7.** $f(t) = t^2$ **8.** $f(t) = te^{3t}$

9. $f(t) = \sin 2t - 2t \cos t$ **10.** $f(t) = e^{3t} \sin 5t$

11. $f(t) = t - e^{2t}$ **12.** $f(t) = t + \sin 3t$ **13.** $f(t) = t \sin 4t$

14. $f(t) = 6t - \sin 6t$ **15.** $f(t) = e^{-3t} \cos 5t$ **16.** $f(t) = t^5$

17. $f(t) = 8t + 4t^3$ **18.** $f(t) = t \sin 4t + t \cos 4t$

Express the Laplace transform of each expression with the given conditions in terms of $\mathscr{L}(y)$ and s.

19. $y'' - 3y'$; $y(0) = 0$ and $y'(0) = 0$

20. $y'' - y' + 2y$; $y(0) = 0$ and $y'(0) = 0$

21. $y'' + y' + y$; $y(0) = 0$ and $y'(0) = 1$

22. $y'' - 2y$; $y(0) = -1$ and $y'(0) = 2$

23. $y'' - 3y' + y$; $y(0) = 1$ and $y'(0) = 0$

24. $y'' + 4y$; $y(0) = 2$ and $y'(0) = -3$

25. $y'' + 8y' + 2y$; $y(0) = 4$ and $y'(0) = 6$

26. $y'' - 3y' + 6y$; $y(0) = 7$ and $y'(0) = -3$

27. $y'' - 6y'$; $y(0) = 3$ and $y'(0) = 7$

28. $y'' + 2y' + 3y$; $y(0) = -4$ and $y'(0) = 5$

29. $y'' + 8y' - 3y$; $y(0) = -6$ and $y'(0) = 2$

30. $y'' + 3y' - 2y$; $y(0) = -5$ and $y'(0) = -3$

Using the Table of Laplace Transforms, find the inverse transform of each function $F(s)$.

31. $F(s) = \dfrac{1}{s}$ **32.** $F(s) = \dfrac{4}{s^2 + 16}$ **33.** $F(s) = \dfrac{1}{(s - 5)^2}$

34. $F(s) = \dfrac{2}{s(s + 2)}$ **35.** $F(s) = \dfrac{s}{s^2 + 64}$ **36.** $F(s) = \dfrac{1}{s^4}$

37. $F(s) = \dfrac{4}{(s - 6)(s - 2)}$ **38.** $F(s) = \dfrac{1}{s^2} - \dfrac{3}{(s - 2)^2 + 9}$

39. $F(s) = \dfrac{2}{s^2 - 6s + 13}$ **40.** $F(s) = \dfrac{1}{s^2 + 4s + 13}$ **41.** $F(s) = \dfrac{2}{(s^2 + 1)^2}$

42. $F(s) = \dfrac{9s^2 + 3s + 62}{(s^2 + 9)(s + 5)}$ **43.** $F(s) = \dfrac{2}{(s - 1)(s^2 + 1)}$ **44.** $F(s) = \dfrac{11s - 3}{s^2 - s - 6}$

45. $F(s) = \dfrac{6s^2 + 42s + 54}{s^3 + 9s^2 + 18s}$ **46.** $F(s) = \dfrac{4s^2 - 22s + 38}{(s + 1)(s - 3)^2}$ **47.** $F(s) = \dfrac{s + 11}{s^2 + 10s + 29}$

48. $F(s) = \dfrac{3s - 4}{(s + 2)(s^2 + s + 3)}$

12.6 SOLUTIONS BY THE METHOD OF LAPLACE TRANSFORM

We find particular solutions y_p of linear differential equations using Laplace transforms as follows:

To solve $a_0 D^2 y + a_1 Dy + a_2 y = b$ subject to initial conditions $y(0) = c$ and $y'(0) = d$:

1. Take the Laplace transform of each side of the homogeneous differential equation $a_0 D^2 y + a_1 Dy + a_2 y = 0$, noting that $\mathcal{L}(0) = \int_0^\infty 0 \, dt = 0$. Substitute the initial values c for $y(0)$ and d for $y'(0)$ in this new equation.
2. Solve the resulting equation in step 1 for $\mathcal{L}(y)$ obtaining an equation in the form $\mathcal{L}(y) = F(s)$.
3. Find the inverse transform of $\mathcal{L}(y) = F(s)$ from step 2 obtaining the particular solution $y_p = \mathcal{L}^{-1}[F(s)]$.

Example 1

Solve the homogeneous differential equation $y'' + 4y' + 4y = 0$ subject to the initial conditions that $y = 2$ and $y' = 1$ when $x = 0$. [Write this as $y(0) = 2$ and $y'(0) = 1$.]

Step 1: Take the Laplace transform of each side of the equation.
$$\mathcal{L}(y'' + 4y' + 4y) = \mathcal{L}(0)$$
$$[s^2 \mathcal{L}(y) - s\, y(0) - y'(0)] + 4[s\, \mathcal{L}(y) - y(0)] + 4 \mathcal{L}(y) = 0$$
$$(s^2 + 4s + 4) \mathcal{L}(y) + (-s - 4)y(0) - y'(0) = 0$$
$$(s^2 + 4s + 4) \mathcal{L}(y) - (s + 4)(2) - 1 = 0$$
$$(s^2 + 4s + 4) \mathcal{L}(y) - 2s - 9 = 0$$

Step 2: Solve for $\mathcal{L}(y)$:
$$\mathcal{L}(y) = \frac{2s + 9}{s^2 + 4s + 4}$$

Step 3: Find the inverse transform of $\mathcal{L}(y) = F(s)$.

While $F(s) = \dfrac{2s + 9}{s^2 + 4s + 4}$ does not appear in the table, we can rewrite $F(s)$ as follows:
$$F(s) = \frac{2s + 9}{s^2 + 4s + 4} = \frac{2s}{(s + 2)^2} + \frac{9}{(s + 2)^2}$$

then
$$y = \mathcal{L}^{-1}[F(s)]$$
$$= \mathcal{L}^{-1}\left[\frac{2s}{(s + 2)^2} + \frac{9}{(s + 2)^2}\right]$$
$$= 2 \mathcal{L}^{-1}\left(\frac{s}{(s + 2)^2}\right) + 9 \mathcal{L}^{-1}\left(\frac{1}{(s + 2)^2}\right)$$
$$= 2 [e^{-2t}(1 - 2t)] + 9(te^{-2t})$$
$$= 2e^{-2t} - 4te^{-2t} + 9te^{-2t}$$

or
$$y = (2 + 5t)e^{-2t}$$

We used Formulas 15 and 17 with $a = -2$. Compare this result with the solution obtained in Example 2, Section 12.2.

Example 2

Solve the nonhomogeneous differential equation

$$y'' + y = \sin t \quad \text{if} \quad y(0) = 1 \quad \text{and} \quad y'(0) = -1$$

Step 1: Take the Laplace transform of each side of the equation.

$$\mathcal{L}(y'' + y) = \mathcal{L}(\sin t)$$

$$\mathcal{L}(y'') + \mathcal{L}(y) = \mathcal{L}(\sin t)$$

$$s^2 \mathcal{L}(y) - s\, y(0) - y'(0) + \mathcal{L}(y) = \frac{1}{s^2 + 1}$$

$$s^2 \mathcal{L}(y) - s(1) - (-1) + \mathcal{L}(y) = \frac{1}{s^2 + 1}$$

$$(s^2 + 1)\,\mathcal{L}(y) - s + 1 = \frac{1}{s^2 + 1}$$

Step 2: Solve for $\mathcal{L}(y)$.

$$(s^2 + 1)\,\mathcal{L}(y) = \frac{1}{s^2 + 1} + s - 1$$

$$\mathcal{L}(y) = \frac{1}{(s^2 + 1)(s^2 + 1)} + \frac{s - 1}{s^2 + 1}$$

$$\mathcal{L}(y) = \frac{1}{(s^2 + 1)^2} + \frac{s - 1}{s^2 + 1}$$

Step 3: Find the inverse Laplace transform of $\mathcal{L}(y) = F(s)$.

$$y = \mathcal{L}^{-1}[(F(s)]$$

$$= \mathcal{L}^{-1}\left[\frac{1}{(s^2 + 1)^2} + \left(\frac{s - 1}{s^2 + 1}\right)\right]$$

$$= \mathcal{L}^{-1}\left[\frac{1}{(s^2 + 1)^2}\right] + \mathcal{L}^{-1}\left(\frac{s - 1}{s^2 + 1}\right)$$

$$= \mathcal{L}^{-1}\left[\frac{1}{(s^2 + 1)^2}\right] + \mathcal{L}^{-1}\left(\frac{s}{s^2 + 1} - \frac{1}{s^2 + 1}\right)$$

$$= \mathcal{L}^{-1}\left[\frac{1}{(s^2 + 1)^2}\right] + \mathcal{L}^{-1}\left(\frac{s}{s^2 + 1}\right) - \mathcal{L}^{-1}\left(\frac{1}{s^2 + 1}\right)$$

Since

$$\mathcal{L}^{-1}\left[\frac{1}{(s^2 + 1)^2}\right] = \tfrac{1}{2}(\sin t - t\cos t) \qquad \text{(Formula 20 with } k = 1)$$

$$\mathcal{L}^{-1}\left(\frac{s}{s^2 + 1}\right) = \cos t \qquad \text{(Formula 10 with } k = 1)$$

and

$$\mathcal{L}^{-1}\left(\frac{1}{s^2 + 1}\right) = \sin t \qquad \text{(Formula 8 with } k = 1)$$

we have

$$y = \tfrac{1}{2}(\sin t - t \cos t) + \cos t - \sin t$$

$$y = -\frac{1}{2}\sin t + \left(1 - \frac{t}{2}\right)\cos t$$

Example 3

Solve the differential equation $y'' + 4y' - 5y = 0$ with $y(0) = 1$ and $y'(0) = -4$.

Step 1: Take the Laplace transform of each side of the equation.

$$\mathcal{L}(y'' + 4y' - 5y) = \mathcal{L}(0)$$

$$[s^2 \mathcal{L}(y) - s\, y(0) - y'(0)] + 4[s\, \mathcal{L}(y) - y(0)] - 5\, \mathcal{L}(y) = 0$$

$$(s^2 + 4s - 5)\, \mathcal{L}(y) + (-s - 4)y(0) - y'(0) = 0$$

$$(s^2 + 4s - 5)\, \mathcal{L}(y) - (s + 4)(1) - (-4) = 0$$

$$(s^2 + 4s - 5)\, \mathcal{L}(y) - (s + 4) + 4 = 0$$

Step 2: Solve for $\mathcal{L}(y)$.

$$(s^2 + 4s - 5)\, \mathcal{L}(y) = s$$

$$\mathcal{L}(y) = \frac{s}{s^2 + 4s - 5}$$

Step 3: Find the inverse Laplace transform of $\mathcal{L}(y) = F(s)$. $F(s) = \frac{s}{s^2 + 4s - 5}$ does not appear in the table. However, using partial fractions, we express $F(s)$ as a sum of two fractions:

$$\frac{s}{s^2 + 4s - 5} = \frac{s}{(s + 5)(s - 1)}$$

$$= \frac{A}{s + 5} + \frac{B}{s - 1}$$

$$= \frac{A(s - 1)}{(s + 5)(s - 1)} + \frac{B(s + 5)}{(s + 5)(s - 1)}$$

$$= \frac{As - A + Bs + 5B}{(s + 5)(s - 1)}$$

$$= \frac{(A + B)s + (-A + 5B)}{(s + 5)(s - 1)}$$

Setting

$$A + B = 1$$

$$-A + 5B = 0$$

we solve for A and B and find $A = \tfrac{5}{6}$ and $B = \tfrac{1}{6}$.
Then

$$y = \mathcal{L}^{-1}[F(s)]$$

$$= \mathcal{L}^{-1}\left(\frac{s}{s^2 + 4s - 5}\right)$$

$$= \mathscr{L}^{-1}\left[\left(\frac{\frac{5}{6}}{s+5}\right) + \left(\frac{\frac{1}{6}}{s-1}\right)\right]$$

$$= \frac{5}{6}\mathscr{L}^{-1}\left(\frac{1}{s+5}\right) + \frac{1}{6}\mathscr{L}^{-1}\left(\frac{1}{s-1}\right)$$

or

$$y = \tfrac{5}{6}e^{-5t} + \tfrac{1}{6}e^{t} \qquad \text{(Formula 7)}$$

Exercises

Solve each differential equation subject to the given conditions by using Laplace transforms.

1. $y' - y = 0$; $y(0) = 2$ **2.** $y' + 3y = 0$; $y(0) = -1$

3. $4y' + 3y = 0$; $y(0) = 1$ **4.** $y' - 2y = 0$; $y(0) = 5$

5. $y' - 7y = e^{t}$; $y(0) = 5$ **6.** $y' + 2y = 3$; $y(0) = 0$

7. $y'' + y = 0$; $y(0) = 1$ and $y'(0) = 0$ **8.** $y'' + 3y = 0$; $y(0) = 2$ and $y'(0) = 5$

9. $y'' - 2y' = 0$; $y(0) = 1$ and $y'(0) = -1$

10. $y'' + 3y' = 0$; $y(0) = 0$ and $y'(0) = 2$

11. $y'' + 2y' + y = 0$; $y(0) = 1$ and $y'(0) = 0$

12. $y'' - 6y' + 9y = 0$; $y(0) = 1$ and $y'(0) = 2$

13. $y'' - 4y' + 4y = te^{2t}$; $y(0) = 0$ and $y'(0) = 0$

14. $y'' + 10y' + 25y = e^{-5t}$; $y(0) = 0$ and $y'(0) = 0$

15. $y'' + 2y' + y = 3te^{-t}$; $y(0) = 4$ and $y'(0) = 2$

16. $y'' - 6y' + 9y = e^{3t}$; $y(0) = 1$ and $y'(0) = 2$

17. $y'' + 3y' - 4y = 0$; $y(0) = 1$ and $y'(0) = -2$

18. $y'' - y' - 2y = 0$; $y(0) = 2$ and $y'(0) = 3$

CHAPTER SUMMARY

1. A linear differential equation of order n is represented by the form

$$a_0\frac{d^n y}{dx^n} + a_1\frac{d^{n-1}y}{dx^{n-1}} + \cdots + a_{n-1}\frac{dy}{dx} + a_n y = b$$

where $a_1, a_2, \ldots, a_{n-1}, a_n$, and b can be either functions of x or constants. If $b = 0$, the linear differential equation is called *homogeneous*. If $b \neq 0$, the equation is called *nonhomogeneous*.

2. *Differential operator D operates on a function* $y = f(x)$ *as follows:*

$$Dy = \frac{dy}{dx}, \qquad D^2y = \frac{d^2y}{dx^2}, \qquad D^3y = \frac{d^3y}{dx^3}, \qquad \text{etc.}$$

Operator D is an example of a linear operator. An algebraic expression in the form $ax + by$ where a and b are constants is called a *linear combination*. A *linear operator* is any process which after operating on a linear combination results in another linear combination. We thus observe that for the operator D, we have

$$D[a\,f(x) + b\,g(x)] = a\,D[f(x)] + b\,D[g(x)]$$

Second-Order Differential Equations   Chap. 12

We are able to apply algebraic operations to linear operators.

3. The equation $a_0 m^2 + a_1 m + a_2 = 0$ is called the *auxiliary equation (characteristic equation)* for the differential equation

$$a_0 D^2 y + a_1 Dy + a_2 y = 0$$

The auxiliary equation is found by merely substituting m for Dy and m^2 for $D^2 y$ in the differential equation.

Observe that $y = ke^{mx}$ will be a solution if m satisfies the auxiliary equation $a_0 m^2 + a_1 m + a_2 = 0$. The problem of determining a solution of the differential equation is now a problem of finding the two roots, m_1 and m_2, of the auxiliary equation.

The method may be summarized as follows:

(a) Write the auxiliary equation for a given differential equation $a_0 D^2 y + a_1 Dy + a_2 y = 0$ using the substitution $m = Dy$ and $m^2 = D^2 y$:

$$a_0 m^2 + a_1 m + a_2 = 0$$

(b) Determine the two roots, m_1 and m_2 of the auxiliary equation.
(c) Write the general solution as

$$y = k_1 e^{m_1 x} + k_2 e^{m_2 x}$$

4. Some differential equations will have auxiliary equations which result in repeated roots or complex roots. When the auxiliary equation results in a repeated root ($m_1 = m_2 = m$), then the general solution is in the form

$$y = k_1 e^{mx} + k_2 x e^{mx}$$

Note carefully that the second term includes the factor x.

5. If the auxiliary equation of a differential equation has complex roots in the form $m = a \pm bj$, then the solution of the differential equation becomes

$$y = e^{ax}(k_1 \sin bx + k_2 \cos bx)$$

where k_1 and k_2 are arbitrary constants.

6. Finding the general solution of the nonhomogeneous equation $a_0 D^2 y + a_1 Dy + a_2 y = b$ involves the following steps:

(a) Determine the complementary solution y_c which is found by solving the homogeneous equation $a_0 D^2 y + a_1 Dy + a_2 y = 0$ using the methods developed in Sections 12.1 and 12.2.
(b) Determine a particular solution y_p by using the method of undetermined coefficients.
(c) Determine the general solution y by adding the complementary solution y_c from step (a) and the particular solution y_p from step (b):

$$y = y_c + y_p$$

7. The Laplace transform $\mathcal{L}$, like the differential operator D, operates on functions. The Laplace transform operates on a function $f(t)$ and transforms it into another function $F(s)$. That is, $\mathcal{L}[f(t)] = F(s)$. The transform is defined by means of integration:

$$F(s) = \mathcal{L}[f(t)] = \int_0^\infty e^{-st} f(t) \, dt$$

8. The particular solution y_p of a linear differential equation $a_0 D^2y + a_1 Dy + a_2y = b$ subject to given initial conditions $y(0) = c$ and $y'(0) = d$ can be found by using Laplace transforms. The method involves the following steps:

(a) Take the Laplace transform of each side of the homogeneous differential equation $a_0 D^2y + a_1 Dy + a_2y = 0$, noting that $\mathcal{L}(0) = \int_0^\infty 0\, dt = 0$.

Substitute the initial values c for $y(0)$ and d for $y'(0)$ in this new equation.

(b) Solve the resulting equation in step (a) for $\mathcal{L}(y)$ obtaining an equation in the form $\mathcal{L}(y) = F(s)$.

(c) Find the inverse transform of $\mathcal{L}(y) = F(s)$ from step (b) obtaining the particular solution $y_p = \mathcal{L}^{-1}[F(s)]$.

CHAPTER REVIEW

State whether each differential equation is homogeneous or nonhomogeneous. Also state the order of each equation.

1. $2y'' + y' - 7y = 0$

2. $\dfrac{d^2y}{dx^2} - 7y = 8$

3. $\dfrac{dy}{dx} - 3y - 5 = 0$

4. $y''' - y'' + y = 0$

Solve each differential equation.

5. $\dfrac{d^2y}{dx^2} + 4\dfrac{dy}{dx} - 5y = 0$

6. $\dfrac{d^2y}{dx^2} - 5\dfrac{dy}{dx} + 6y = 0$

7. $y'' - 6y' = 0$

8. $2y'' - y' - 3y = 0$

9. $y'' - 6y' + 9y = 0$

10. $D^2y + 10\, Dy + 25y = 0$

11. $D^2y - 2\, Dy + y = 0$

12. $9\, D^2y - 6\, Dy + y = 0$

13. $y'' + 16y = 0$

14. $4\, D^2y + 25y = 0$

15. $D^2y - 2\, Dy + 3y = 0$

16. $y'' - 3y' + 8y = 0$

17. $D^2y + Dy - 2y = x$

18. $y'' - 6y' + 9y = e^x$

19. $y'' + 4y = \cos x$

20. $y'' - 2y' + 3y = 6e^{2x}$

Find the particular solution of each differential equation satisfying the given conditions.

21. $\dfrac{d^2y}{dx^2} + 2\dfrac{dy}{dx} - 8y = 0$; $y = 6$ and $y' = 0$ when $x = 0$

22. $D^2y - 3\, Dy = 0$; $y = 3$ and $y' = 6$ when $x = 0$

23. $D^2y - 4\, Dy + 4y = 0$; $y = 0$ and $y' = 3$ when $x = 0$

24. $y'' + 6y' + 9y = 0$; $y = 8$ and $y' = 0$ when $x = 0$

25. $y'' + 4y = 0$; $y = 1$ and $y' = -4$ when $x = 0$

26. $D^2y - 8\, Dy + 25y = 0$; $y = 2$ and $y' = 11$ when $x = 0$

27. $y'' - y' = e^{2x}$; $y = 1$ and $y' = 0$ when $x = 0$

28. $D^2y + 4y = \sin x$; $y = 2$ and $y' = \frac{7}{3}$ when $x = 0$

29. A spring is stretched 4 in. by a weight of 16 lb. If the weight is displaced a distance of 6 in. from the rest position and then released, determine the equation of motion.

494   Second-Order Differential Equations Chap. 12

30. A spring is stretched 2 in. by a weight of 12 lb. A damping force exerts a force of 8 lb for a velocity of 4 in./s. If the weight is displaced from the rest position and then released, determine the general equation of motion.

31. An electric circuit has an inductance L of 2 H, a resistance R of 400 Ω, and a capacitance C of 10^{-5} F. Find the equation for the current i.

32. An electric circuit has an inductance L of 1 H, a resistance R of 2000 Ω, and a capacitance C of 4×10^{-6} F. Find the equation for the current i.

Using the Table of Laplace Transforms, find the Laplace transform for each function $f(t)$.

33. e^{6t} 　　　　34. $e^{-2t} \cos 3t$ 　　　35. $t^3 + \cos t$ 　　　36. $3t - e^{5t}$

Using the Table of Laplace Transforms, find the inverse transform for each function $F(s)$.

37. $\dfrac{1}{s^2}$ 　　　38. $\dfrac{3}{s(s-2)}$ 　　　39. $\dfrac{2}{(s-3)(s-4)}$ 　　　40. $\dfrac{3}{s^2+9}$

Solve each differential equation subject to the given conditions by using Laplace transforms.

41. $4y' - 5y = 0$; $y(0) = 2$

42. $y'' + 9y = 0$; $y(0) = 3$ and $y'(0) = 0$

43. $y'' + 5y' = 0$; $y(0) = 0$ and $y'(0) = 2$

44. $y'' + 4y' + 4y = e^{-2t}$; $y(0) = 0$ and $y'(0) = 0$

Appendix A

THE BINOMIAL THEOREM

The *binomial theorem* provides us with a convenient means of expressing any power of a binomial as a sum of terms. It is used in Chapter 2 to develop the derivative of certain functions.

For small nonnegative integers n, we can find $(a + b)^n$ by actual multiplication.

$$n = 0: \quad (a + b)^0 = \qquad\qquad\qquad 1$$
$$n = 1: \quad (a + b)^1 = \qquad\qquad\qquad a + b$$
$$n = 2: \quad (a + b)^2 = \qquad\qquad\quad a^2 + 2ab + b^2$$
$$n = 3: \quad (a + b)^3 = \qquad\quad\; a^3 + 3a^2b + 3ab^2 + b^3$$
$$n = 4: \quad (a + b)^4 = \qquad a^4 + 4a^3b + 6a^2b^2 + 4ab^3 + b^4$$
$$n = 5: \quad (a + b)^5 = a^5 + 5a^4b + 10a^3b^2 + 10a^2b^3 + 5ab^4 + b^5$$

We could continue this process, but the multiplications become more complicated for larger values of n.

No matter what positive integral value of n is chosen, the following results:

1. $(a + b)^n$ has $n + 1$ terms.
2. The first term is a^n.
3. The second term is $na^{n-1}b$.
4. The exponent of a decreases by 1 and the exponent of b increases by 1 for each successive term.
5. In each term, the sum of the exponents of a and b is n.
6. The last term is b^n.

The kth term is given by the formula

$$\frac{n(n - 1)(n - 2) \cdots (n - k + 2)}{(k - 1)!} a^{n-k+1} b^{k-1}$$

where $k!$ (k factorial) indicates the product of the first k positive integers. (For example $3! = 3 \cdot 2 \cdot 1 = 6$; $5! = 5 \cdot 4 \cdot 3 \cdot 2 \cdot 1 = 120$; and $6! = 6 \cdot 5 \cdot 4 \cdot 3 \cdot 2 \cdot 1 = 720$). The three dots in the numerator indicate that the multiplication of decreasing numbers is to continue until the number $n - k + 2$ is reached. For example, if $n = 8$ and $k = 4$, then the formula gives us

$$\frac{8 \cdot 7 \cdot 6}{1 \cdot 2 \cdot 3} = 56$$

Binomial Theorem:

$$(a + b)^n = a^n + na^{n-1}b + \frac{n(n - 1)}{2!} a^{n-2}b^2 + \frac{n(n - 1)(n - 2)}{3!} a^{n-3}b^3 + \cdots$$

$$+ \frac{n(n - 1)(n - 2) \cdots (n - k + 2)}{(k - 1)!} a^{n-k+1} b^{k-1} + \cdots + b^n$$

where the three dots indicate that you are to complete the process of calculating the terms. The expression for the kth term is also given.

Example 1

Expand $(x + 4y)^5$ by using the binomial theorem.

Let $a = x$, $b = 4y$, and $n = 5$.

$$(x + 4y)^5 = x^5 + 5x^{5-1}(4y) + \frac{5(5 - 1)}{2!} x^{5-2}(4y)^2$$

$$+ \frac{5(5 - 1)(5 - 2)}{3!} x^{5-3}(4y)^3$$

$$+ \frac{5(5 - 1)(5 - 2)(5 - 3)}{4!} x^{5-4}(4y)^4 + (4y)^5$$

$$= x^5 + 20x^4y + 160x^3y^2 + 640x^2y^3 + 1280xy^4 + 1024y^5$$

For small values of n it is possible to determine the coefficients of each term of the expansion by the use of *Pascal's triangle* as shown below:

$n = 0$:					1					
$n = 1$:				1		1				
$n = 2$:			1		2		1			
$n = 3$:		1		3		3		1		
$n = 4$:	1		4		6		4		1	
$n = 5$:	1	5		10		10		5		1

Observe the similarity of this triangle with the triangular format shown earlier when expanding $(a + b)^n$ for $n = 0, 1, 2, 3, 4,$ and 5. Each row gives the coefficients for all terms of the binomial expansion for a given integer n. Each successive row provides the coefficients for the next integer n. Each row is read from left to right. The first and last coefficients are always one as observed in the triangle. Beginning with the third row ($n = 2$), coefficients of terms other than the first and last are found by adding together the two nearest coefficients found in the row above. For example, the coefficient of the fourth term for the expansion with $n = 5$ is 10, which is the sum of 6 and 4. 6 and 4 appear just above 10 in Pascal's triangle.

We can enlarge Pascal's triangle to obtain a row for any desired integer n. However, this is not very practical for very large values of n.

Example 2

Using Pascal's triangle, find the coefficients of the terms of the binomial expansion for $n = 7$.

We need two more rows of the triangle:

$n = 6$:	1		6		15		20		15		6		1
$n = 7$:	1		7		21		35		35		21		7		1

The last row provides the desired coefficients.

Example 3

Expand $(2m + k)^7$ using the binomial theorem.

Let $a = 2m$, $b = k$, and $n = 7$.

$$(2m + k)^7 = 1(2m)^7 + 7(2m)^{7-1}k^1 + 21(2m)^{7-2}k^2 + 35(2m)^{7-3}k^3$$
$$+ 35(2m)^{7-4}k^4 + 21(2m)^{7-5}k^5 + 7(2m)^{7-6}k^6 + (1)k^7$$
$$= 128m^7 + 448m^6k + 672m^5k^2 + 560m^4k^3$$
$$+ 280m^3k^4 + 84m^2k^5 + 14mk^6 + k^7$$

Example 4

Find the seventh term of $(x^3 - 2y)^{10}$.

First, note that $k = 7$, $n = 10$, $a = x^2$, and $b = -2y$.

$$k\text{th term} = \frac{n(n - 1)(n - 2) \cdots (n - k + 2)}{(k - 1)!} a^{n-k+1}b^{k-1}$$

$$7\text{th term} = \frac{10 \cdot 9 \cdot 8 \cdot 7 \cdot 6 \cdot 5}{6!}(x^3)^4(-2y)^6$$

$$= 210(x^{12})(64y^6)$$
$$= 13,440x^{12}y^6$$

Exercises

Expand each binomial using the binomial theorem.

1. $(3x + y)^3$
2. $(2x - 3y)^4$
3. $(2a + 3b)^6$
4. $(a - 2b)^8$
5. $(\frac{2}{3}x - 2)^5$
6. $(\frac{3}{4}m + \frac{2}{3}k)^5$
7. $\left(\frac{x}{y} - \frac{2}{z}\right)^4$
8. $\left(\frac{2x}{y} + \frac{3}{z}\right)^6$

Find the indicated term of each binomial expansion.

9. $(x - y)^9$; 6th term
10. $(4x + 2y)^5$; 3rd term
11. $(2x + y^2)^7$; 5th term
12. $(x^2 - y^3)^8$; middle term

Appendix B

TABLES AND FORMULAS

TABLE 1 ENGLISH WEIGHTS AND MEASURES

Units of length	Units of Weight
Standard unit—inch (in. or ″)	Standard unit—pound (lb)
12 inches = 1 foot (ft or ′)	16 ounces (oz) = 1 pound
3 feet = 1 yard (yd)	2000 pounds = 1 ton (T)
$5\frac{1}{2}$ yards or $16\frac{1}{2}$ feet = 1 rod (rd)	
5280 feet = 1 mile (mi)	

Volume measure

Liquid

16 ounces (fl oz) = 1 pint (pt)
2 pints = 1 quart (qt)
4 quarts = 1 gallon (gal)

Dry

2 pints (pt) = 1 quart (qt)
8 quarts = 1 peck (pk)
4 pecks = 1 bushel (bu)

TABLE 2 METRIC SYSTEM PREFIXES

Multiple or submultiple[a] decimal form	Power of 10	Prefix[b]	Prefix symbol	Pronun- ciation	Meaning
1,000,000,000,000	10^{12}	tera	T	tĕr′ă	one trillion times
1,000,000,000	10^{9}	giga	G	jĭg′ă	one billion times
1,000,000	10^{6}	mega	M	mĕg′ă	one million times
1,000	10^{3}	kilo	k	kĭl′ō	one thousand times
100	10^{2}	hecto	h	hĕk′tō	one hundred times
10	10^{1}	deka	da	dĕk′ă	ten times
0.1	10^{-1}	deci	d	dĕs′ĭ	one tenth of
0.01	10^{-2}	centi	c	sĕnt′ĭ	one hundredth of
0.001	10^{-3}	milli	m	mĭl′ĭ	one thousandth of
0.000001	10^{-6}	micro	μ	mī′krō	one millionth of
0.000000001	10^{-9}	nano	n	năn′ō	one billionth of
0.000000000001	10^{-12}	pico	p	pē′kō	one trillionth of

[a] Factor by which the unit is multiplied.

[b] The same prefixes are used with all SI metric units.

TABLE 3 CONVERSION TABLES

Length

	cm	m	km	in.	ft	mi
1 centimetre	1	10^{-2}	10^{-5}	0.394	3.28×10^{-2}	6.21×10^{-6}
1 metre	100	1	10^{-3}	39.4	3.28	6.21×10^{-4}
1 kilometre	10^5	1000	1	3.94×10^4	3280	0.621
1 inch	2.54	2.54×10^{-2}	2.54×10^{-5}	1	8.33×10^{-2}	1.58×10^{-5}
1 foot	30.5	0.305	3.05×10^{-4}	12	1	1.89×10^{-4}
1 mile	1.61×10^5	1610	1.61	6.34×10^4	5280	1

Area

Metric	*English*
$1 \text{ m}^2 = 10{,}000 \text{ cm}^2$	$1 \text{ ft}^2 = 144 \text{ in}^2$
$= 1{,}000{,}000 \text{ mm}^2$	$1 \text{ yd}^2 = 9 \text{ ft}^2$
$1 \text{ cm}^2 = 100 \text{ mm}^2$	$1 \text{ rd}^2 = 30.25 \text{ yd}^2$
$= 0.0001 \text{ m}^2$	$1 \text{ acre} = 160 \text{ rd}^2$
$1 \text{ km}^2 = 1{,}000{,}000 \text{ m}^2$	$= 4840 \text{ yd}^2$
$1 \text{ ha} = 10{,}000 \text{ m}^2$	$= 43{,}560 \text{ ft}^2$
	$1 \text{ mi}^2 = 640 \text{ acres}$

	m^2	cm^2	ft^2	in^2
1 m^2	1	10^4	10.8	1550
1 cm^2	10^{-4}	1	1.08×10^{-3}	0.155
1 ft^2	9.29×10^{-2}	929	1	144
1 in^2	6.45×10^{-4}	6.45	6.94×10^{-3}	1

$1 \text{ mi}^2 = 2.79 \times 10^7 \text{ ft}^2 = 640 \text{ acres}$
$1 \text{ circular mil} = 5.07 \times 10^{-6} \text{ cm}^2 = 7.85 \times 10^{-7} \text{ in}^2$
$1 \text{ hectare} = 2.47 \text{ acres}$

TABLE 3—*Cont.* CONVERSION TABLES

Volume

	Metric		English	
	1 m³ = 10⁶ cm³		1 ft³ = 1728 in³	
	1 cm³ = 10⁻⁶ m³		1 yd³ = 27 ft³	
	= 10³ mm³			

	m³	cm³	L	ft³	in³
1 m³	1	10⁶	1000	35.3	6.10 × 10⁴
1 cm³	10⁻⁶	1	1.00 × 10⁻³	3.53 × 10⁻⁵	6.10 × 10⁻²
1 L	1.00 × 10⁻³	1000	1	3.53 × 10⁻²	61.0
1 ft³	2.83 × 10⁻²	2.83 × 10⁴	28.3	1	1728
1 in³	1.64 × 10⁻⁵	16.4	1.64 × 10⁻²	5.79 × 10⁻⁴	1

1 U. S. fluid gallon = 4 U. S. fluid quarts = 8 U. S. pints = 128 U. S. fluid ounces = 231 in³ = 0.134 ft³ = 3.79 litres.

1 L = 1000 cm³ = 1.06 qt

Other useful conversion factors

1 newton (N) = 0.225 lb
1 pound (lb) = 4.45 N
1 slug = 14.6 kg
1 joule (J) = 0.738 ft-lb
    = 2.39 × 10⁻⁴ kcal
1 calorie (cal) = 4.185 J
1 kilocalorie (kcal) = 4185 J
1 foot-pound (ft-lb) = 1.36 J
1 watt (W) = 1 J/s = 0.738 ft-lb/s
1 kilowatt (kW) = 1000 W
    = 1.34 hp
1 hp = 550 ft-lb/s = 746 W

1 atm = 101.32 kPa
    = 14.7 lb/in²
1 Btu = 0.252 kcal
1 kcal = 3.97 Btu
$F = \frac{9}{5}C + 32°$
$C = \frac{5}{9}(F - 32°)$
1 kg = 2.20 lb (on the earth's surface)
1 lb = 454 g
    = 16 oz
1 metric ton = 1000 kg
    = 2200 lb

TABLE 4 GEOMETRY FORMULAS

Area		Volume	
Rectangle	$A = lw$	Prism	$V = (\text{area of base})h$
Square	$A = s^2$	Cylinder	$V = \pi r^2 h$
Parallelogram	$A = bh$	Cone	$V = \frac{1}{3}\pi r^2 h$
Trapezoid	$A = \dfrac{h}{2}(b_1 + b_2)$	Sphere	$V = \frac{4}{3}\pi r^3$
Triangle	$A = \frac{1}{2}bh$		
Circle	$A = \pi r^2$		

Pythagorean theorem

$$c^2 = a^2 + b^2$$

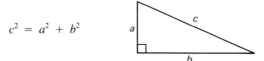

TABLE 5 ALGEBRA REVIEW

Laws of exponents

1. $a^m \cdot a^n = a^{m+n}$

2. $\dfrac{a^m}{a^n} = a^{m-n}$

3. $(a^m)^n = a^{mn}$

4. $(ab)^n = a^n b^n$

5. $\left(\dfrac{a}{b}\right)^n = \dfrac{a^n}{b^n}$ $\qquad (b \neq 0)$

6. $a^0 = 1$

Fractional exponents

$$a^{1/n} = \sqrt[n]{a}$$

where n is called the index and a is called the radicand.

$$a^{m/n} = \sqrt[n]{a^m} = (\sqrt[n]{a})^m$$

Note: m is an integer, n is a positive integer, and a is a real number. If n is even, $a \geq 0$.

Quadratic formula

The solution of $ax^2 + bx + c = 0$ is

$$x = \frac{-b \pm \sqrt{b^2 - 4ac}}{2a}$$

TABLE 5—*Cont.* ALGEBRA REVIEW

Operations with radical expressions

1. $\sqrt[n]{a} \cdot \sqrt[n]{b} = \sqrt[n]{ab}$

2. $\dfrac{\sqrt[n]{a}}{\sqrt[n]{b}} = \sqrt[n]{\dfrac{a}{b}}$ $(b \ne 0)$

3. $\sqrt[m]{\sqrt[n]{a}} = \sqrt[mn]{a}$

4. $\sqrt[cn]{a^{cm}} = \sqrt[n]{a^m}$

TABLE 6 COMMON TRIGONOMETRIC IDENTITIES

Basic identities

1. $\sin \theta = \dfrac{1}{\csc \theta}$

4. $\cot \theta = \dfrac{1}{\tan \theta}$

7. $\tan \theta = \dfrac{\sin \theta}{\cos \theta}$

2. $\cos \theta = \dfrac{1}{\sec \theta}$

5. $\sec \theta = \dfrac{1}{\cos \theta}$

8. $\cot \theta = \dfrac{\cos \theta}{\sin \theta}$

3. $\tan \theta = \dfrac{1}{\cot \theta}$

6. $\csc \theta = \dfrac{1}{\sin \theta}$

9. $\sin^2 \theta + \cos^2 \theta = 1$

12. $\sin (-\theta) = -\sin \theta$

15. $\cot (-\theta) = -\cot \theta$

10. $1 + \tan^2 \theta = \sec^2 \theta$

13. $\cos (-\theta) = \cos \theta$

16. $\sec (-\theta) = \sec \theta$

11. $\cot^2 \theta + 1 = \csc^2 \theta$

14. $\tan (-\theta) = -\tan \theta$

17. $\csc (-\theta) = -\csc \theta$

Sum and difference formulas

18. $\sin (\theta + \phi) = \sin \theta \cos \phi + \cos \theta \sin \phi$

19. $\sin (\theta - \phi) = \sin \theta \cos \phi - \cos \theta \sin \phi$

20. $\cos (\theta + \phi) = \cos \theta \cos \phi - \sin \theta \sin \phi$

21. $\cos (\theta - \phi) = \cos \theta \cos \phi + \sin \theta \sin \phi$

22. $\tan (\theta + \phi) = \dfrac{\tan \theta + \tan \phi}{1 - \tan \theta \tan \phi}$

23. $\tan (\theta - \phi) = \dfrac{\tan \theta - \tan \phi}{1 + \tan \theta \tan \phi}$

Double-angle formulas

24. $\sin 2\theta = 2 \sin \theta \cos \theta$

25a. $\cos 2\theta = \cos^2 \theta - \sin^2 \theta$
25b. $= 2 \cos^2 \theta - 1$
25c. $= 1 - 2 \sin^2 \theta$

26. $\tan 2\theta = \dfrac{2 \tan \theta}{1 - \tan^2 \theta}$

Half-angle formulas

27. $\sin \dfrac{\theta}{2} = \pm \sqrt{\dfrac{1 - \cos \theta}{2}}$

28. $\cos \dfrac{\theta}{2} = \pm \sqrt{\dfrac{1 + \cos \theta}{2}}$

29. $\tan \dfrac{\theta}{2} = \dfrac{1 - \cos \theta}{\sin \theta}$

Alternate half-angle formulas

30. $\sin^2 x = \dfrac{1}{2}(1 - \cos 2x)$

31. $\cos^2 x = \dfrac{1}{2}(1 + \cos 2x)$

32. $\sin \theta + \sin \phi = 2 \sin \left(\dfrac{\theta + \phi}{2}\right) \cos \left(\dfrac{\theta - \phi}{2}\right)$

33. $\sin\theta - \sin\phi = 2\cos\left(\dfrac{\theta + \phi}{2}\right)\sin\left(\dfrac{\theta - \phi}{2}\right)$

34. $\cos\theta + \cos\phi = 2\cos\left(\dfrac{\theta + \phi}{2}\right)\cos\left(\dfrac{\theta - \phi}{2}\right)$

35. $\cos\theta - \cos\phi = -2\sin\left(\dfrac{\theta + \phi}{2}\right)\sin\left(\dfrac{\theta - \phi}{2}\right)$

36. $\sin(\theta + \phi) + \sin(\theta - \phi) = 2\sin\theta\cos\phi$

37. $\sin(\theta + \phi) - \sin(\theta - \phi) = 2\cos\theta\sin\phi$

38. $\cos(\theta + \phi) + \cos(\theta - \phi) = 2\cos\theta\cos\phi$

39. $\cos(\theta + \phi) - \cos(\theta - \phi) = -2\sin\theta\sin\phi$

Law of sines

$$\frac{a}{\sin A} = \frac{b}{\sin B} = \frac{c}{\sin C}$$

Law of cosines

$$a^2 = b^2 + c^2 - 2bc\cos A$$
$$b^2 = a^2 + c^2 - 2ac\cos B$$
$$c^2 = a^2 + b^2 - 2ab\cos C$$

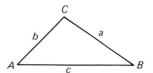

TABLE 7 PROPERTIES OF LOGARITHMS

M and N are positive real numbers, $a > 0$ and $a \neq 1$.

1. $\log_a(M \cdot N) = \log_a M + \log_a N$

2. $\log_a\left(\dfrac{M}{N}\right) = \log_a M - \log_a N$

3. $\log_a M^n = n\log_a M$

4. $\log_a\dfrac{1}{M} = -\log_a M$

5. $\log_a\sqrt[n]{M} = \dfrac{1}{n} \cdot \log_a M$

6. $\log_a 1 = 0$

7. $\log_a(a^x) = x$

8. $a^{(\log_a x)} = x$

TABLE 8 DIFFERENTIATION FORMULAS

1. $\dfrac{d}{dx}(c) = 0$, c is a constant

2. $\dfrac{d}{dx}(x) = 1$

3. $\dfrac{d}{dx}(x^n) = nx^{n-1}, \qquad x \neq 0$

4. $\dfrac{d}{dx}(cu) = c\dfrac{du}{dx}$, c is a constant

5. $\dfrac{d}{dx}(u + v) = \dfrac{du}{dx} + \dfrac{dv}{dx}$

6. $\dfrac{d}{dx}(u - v) = \dfrac{du}{dx} - \dfrac{dv}{dx}$

7. $\dfrac{d}{dx}(uv) = u\dfrac{dv}{dx} + v\dfrac{du}{dx}$

8. $\dfrac{d}{dx}\left(\dfrac{u}{v}\right) = \dfrac{v\dfrac{du}{dx} - u\dfrac{dv}{dx}}{v^2}, \qquad v \neq 0$

9. $\dfrac{d}{dx}(u^n) = nu^{n-1}\dfrac{du}{dx}, \qquad u \neq 0$

10. *chain rule.* $\dfrac{dy}{dx} = \dfrac{dy}{du} \cdot \dfrac{du}{dx}$

11. $\dfrac{d}{dx}(\ln u) = \dfrac{1}{u}\dfrac{du}{dx}$

12. $\dfrac{d}{dx}(e^u) = e^u\dfrac{du}{dx}$

13. $\dfrac{d}{dx}(b^u) = \dfrac{b^u}{\log_b e}\dfrac{du}{dx}$

14. $\dfrac{d}{dx}(\sin u) = \cos u\dfrac{du}{dx}$

15. $\dfrac{d}{dx}(\cos u) = -\sin u\dfrac{du}{dx}$

16. $\dfrac{d}{dx}(\tan u) = \sec^2 u\dfrac{du}{dx}$

17. $\dfrac{d}{dx}(\cot u) = -\csc^2 u\dfrac{du}{dx}$

18. $\dfrac{d}{dx}(\sec u) = \sec u \tan u\dfrac{du}{dx}$

19. $\dfrac{d}{dx}(\csc u) = -\csc u \cot u\dfrac{du}{dx}$

20. $\dfrac{d}{dx}(\text{Arcsin } u) = \dfrac{1}{\sqrt{1-u^2}}\dfrac{du}{dx}$

21. $\dfrac{d}{dx}(\text{Arccos } u) = -\dfrac{1}{\sqrt{1-u^2}}\dfrac{du}{dx}$

22. $\dfrac{d}{dx}(\text{Arctan } u) = \dfrac{1}{1+u^2}\dfrac{du}{dx}$

23. $\dfrac{d}{dx}(\text{Arccot } u) = -\dfrac{1}{1+u^2}\dfrac{du}{dx}$

24. $\dfrac{d}{dx}(\text{Arcsec } u) = \dfrac{1}{\sqrt{u^2(u^2-1)}}\dfrac{du}{dx}$

25. $\dfrac{d}{dx}(\text{Arccsc } u) = -\dfrac{1}{\sqrt{u^2(u^2-1)}}\dfrac{du}{dx}$

TABLE 9   INTEGRATION FORMULAS

1. $\displaystyle\int x^n\, dx = \dfrac{x^{n+1}}{n+1} + C, \qquad n \neq -1$

2. $\displaystyle\int k\, f(x)\, dx = k\int f(x)\, dx,\ k \text{ is a constant}$

3. $\displaystyle\int [f(x) \pm g(x)]\, dx = \int f(x)\, dx \pm \int g(x)\, dx$

4. $\displaystyle\int u^n\, du = \dfrac{u^{n+1}}{n+1} + C, \qquad n \neq -1$

5. $\displaystyle\int \dfrac{du}{u} = \ln|u| + C$

6. $\displaystyle\int e^u\, du = e^u + C$

7. $\displaystyle\int \ln u\, du = u\ln|u| - u + C$

8. $\displaystyle\int \sin u\, du = -\cos u + C$

9. $\displaystyle\int \cos u\, du = \sin u + C$

10. $\displaystyle\int \tan u\, du = -\ln|\cos u| + C$

11. $\displaystyle\int \cot u\, du = \ln|\sin u| + C$

12. $\displaystyle\int \sec u\, du = \ln|\sec u + \tan u| + C$

13. $\displaystyle\int \csc u\, du = \ln|\csc u - \cot u| + C$

14. $\displaystyle\int \sec^2 u\, du = \tan u + C$

15. $\displaystyle\int \csc^2 u\, du = -\cot u + C$

16. $\displaystyle\int \sec u \tan u\, du = \sec u + C$

17. $\displaystyle\int \csc u \cot u\, du = -\csc u + C$

18. $\displaystyle\int u\, dv = uv - \int v\, du \qquad \text{(integration by parts)}$

19. $\displaystyle\int \dfrac{du}{a^2+u^2} = \dfrac{1}{a}\,\text{Arctan}\,\dfrac{u}{a} + C$

20. $\displaystyle\int \dfrac{du}{\sqrt{a^2-u^2}} = \text{Arcsin}\,\dfrac{u}{a} + C$

TABLE 10 TABLE OF INTEGRALS

1. $\int u^n \, du = \dfrac{u^{n+1}}{n+1} + C, \; n \neq -1$

2. $\int \dfrac{du}{a+bu} = \dfrac{1}{b} \ln|a+bu| + C$

3. $\int \dfrac{u}{a+bu} \, du = \dfrac{1}{b^2}[(a+bu) - a \ln|a+bu|] + C$

4. $\int \dfrac{u^2 \, du}{a+bu} = \dfrac{1}{b^3} \left[\dfrac{1}{2}(a+bu)^2 - 2a(a+bu) + a^2 \ln|a+bu| \right] + C$

5. $\int \dfrac{du}{u(a+bu)} = \dfrac{1}{a} \ln\left| \dfrac{u}{a+bu} \right| + C$

6. $\int \dfrac{du}{u^2(a+bu)} = -\dfrac{1}{au} + \dfrac{b}{a^2} \ln\left| \dfrac{a+bu}{u} \right| + C$

7. $\int \dfrac{u \, du}{(a+bu)^2} = \dfrac{1}{b^2} \left(\ln|a+bu| + \dfrac{a}{a+bu} \right) + C$

8. $\int \dfrac{u^2 \, du}{(a+bu)^2} = \dfrac{1}{b^3} \left[a+bu - \dfrac{a^2}{a+bu} - 2a \ln|a+bu| \right] + C$

9. $\int \dfrac{du}{u(a+bu)^2} = \dfrac{1}{a(a+bu)} + \dfrac{1}{a^2} \ln\left| \dfrac{u}{a+bu} \right| + C$

10. $\int \dfrac{du}{u^2(a+bu)^2} = -\dfrac{a+2bu}{a^2u(a+bu)} + \dfrac{2b}{a^3} \ln\left| \dfrac{a+bu}{u} \right| + C$

Forms containing $\sqrt{a+bu}$

11. $\int u\sqrt{a+bu} \, du = -\dfrac{2(2a-3bu)(a+bu)^{3/2}}{15b^2} + C$

12. $\int u^2\sqrt{a+bu} \, du = \dfrac{2(8a^2 - 12abu + 15b^2u^2)(a+bu)^{3/2}}{105b^3} + C$

13. $\int \dfrac{u \, du}{\sqrt{a+bu}} = -\dfrac{2(2a-bu)\sqrt{a+bu}}{3b^2} + C$

14. $\int \dfrac{u^2 \, du}{\sqrt{a+bu}} = \dfrac{2(3b^2u^2 - 4abu + 8a^2)\sqrt{a+bu}}{15b^3} + C$

15. $\int \dfrac{du}{u\sqrt{a+bu}} = \dfrac{1}{\sqrt{a}} \ln\left| \dfrac{\sqrt{a+bu}-\sqrt{a}}{\sqrt{a+bu}+\sqrt{a}} \right| + C, \; a > 0$

16. $\int \dfrac{du}{u\sqrt{a+bu}} = \dfrac{2}{\sqrt{-a}} \arctan\sqrt{\dfrac{a+bu}{-a}} + C, \; a < 0$

17. $\int \dfrac{\sqrt{a+bu} \, du}{u} = 2\sqrt{a+bu} + a \int \dfrac{du}{u\sqrt{a+bu}} + C$

TABLE 10—*Cont.* TABLE OF INTEGRALS

Rational forms containing $a^2 \pm u^2$ and $u^2 \pm a^2$

18. $\displaystyle\int \frac{du}{a^2 + u^2} = \frac{1}{a} \arctan \frac{u}{a} + C$

19. $\displaystyle\int \frac{du}{a^2 - u^2} = \frac{1}{2a} \ln \left| \frac{a+u}{a-u} \right| C, \quad a^2 > u^2$

20. $\displaystyle\int \frac{du}{u^2 - a^2} = \frac{1}{2a} \ln \left| \frac{u-a}{u+a} \right| + C, \quad a^2 < u^2$

Irrational forms containing $\sqrt{a^2 - u^2}$

21. $\displaystyle\int (a^2 - u^2)^{1/2}\, du = \frac{u}{2}\sqrt{a^2 - u^2} + \frac{a^2}{2} \arcsin \frac{u}{a} + C$

22. $\displaystyle\int \frac{du}{(a^2 - u^2)^{1/2}} = \arcsin \frac{u}{a} + C, \quad a > 0$

23. $\displaystyle\int \frac{du}{(a^2 - u^2)^{3/2}} = \frac{u}{a^2\sqrt{a^2 - u^2}} + C$

24. $\displaystyle\int \frac{u^2\, du}{(a^2 - u^2)^{1/2}} = -\frac{u}{2}\sqrt{a^2 - u^2} + \frac{a^2}{2} \arcsin \frac{u}{a} + C$

25. $\displaystyle\int \frac{u^2\, du}{(a^2 - u^2)^{3/2}} = \frac{u}{\sqrt{a^2 - u^2}} - \arcsin \frac{u}{a} + C$

26. $\displaystyle\int \frac{du}{u(a^2 - u^2)^{1/2}} = -\frac{1}{a} \ln \left| \frac{a + \sqrt{a^2 - u^2}}{u} \right| + C$

27. $\displaystyle\int \frac{du}{u^2(a^2 - u^2)^{1/2}} = -\frac{\sqrt{a^2 - u^2}}{a^2 u} + C$

28. $\displaystyle\int \frac{(a^2 - u^2)^{1/2}\, du}{u} = \sqrt{a^2 - u^2} - a \ln \left| \frac{a + \sqrt{a^2 - u^2}}{u} \right| + C$

29. $\displaystyle\int \frac{(a^2 - u^2)^{1/2}\, du}{u^2} = -\frac{\sqrt{a^2 - u^2}}{u} - \arcsin \frac{u}{a} + C$

Irrational forms containing $\sqrt{u^2 \pm a^2}$

30. $\displaystyle\int \sqrt{u^2 \pm a^2}\, du = \tfrac{1}{2}[u\sqrt{u^2 \pm a^2} \pm a^2 \ln |u + \sqrt{u^2 \pm a^2}|] + C$

31. $\displaystyle\int u^2\sqrt{u^2 \pm a^2}\, du = \tfrac{1}{8}u(2u^2 \pm a^2)\sqrt{u^2 \pm a^2} - \tfrac{1}{8}a^4 \ln |u + \sqrt{u^2 \pm a^2}| + C$

32. $\displaystyle\int \frac{\sqrt{u^2 + a^2}}{u}\, du = \sqrt{u^2 + a^2} - a \ln \left| \frac{a + \sqrt{u^2 + a^2}}{u} \right| + C$

33. $\displaystyle\int \frac{\sqrt{u^2 - a^2}}{u}\, du = \sqrt{u^2 - a^2} - a \arccos \frac{a}{u} + C$

34. $\displaystyle\int \frac{\sqrt{u^2 \pm a^2}}{u^2}\, du = -\frac{\sqrt{u^2 \pm a^2}}{u} + \ln |u + \sqrt{u^2 \pm a^2}| + C$

TABLE 10—*Cont.*   TABLE OF INTEGRALS

35. $\displaystyle\int \frac{du}{\sqrt{u^2 \pm a^2}} = \ln|u + \sqrt{u^2 \pm a^2}| + C$

36. $\displaystyle\int \frac{du}{u\sqrt{u^2 - a^2}} = \frac{1}{a}\arccos\frac{a}{u} + C$

37. $\displaystyle\int \frac{du}{u\sqrt{u^2 + a^2}} = \frac{1}{a}\ln\left|\frac{u}{a + \sqrt{u^2 + a^2}}\right| + C$

38. $\displaystyle\int \frac{u^2\,du}{\sqrt{u^2 \pm a^2}} = \frac{1}{2}(u\sqrt{u^2 \pm a^2} \mp a^2\ln|u + \sqrt{u^2 \pm a^2}|) + C$

39. $\displaystyle\int \frac{du}{u^2\sqrt{u^2 \pm a^2}} = -\frac{(\pm\sqrt{u^2 \pm a^2})}{a^2 u} + C$

40. $\displaystyle\int \frac{du}{(u^2 \pm a^2)^{3/2}} = \frac{\pm u}{a^2\sqrt{u^2 \pm a^2}} + C$

41. $\displaystyle\int \frac{u^2\,du}{(u^2 \pm a^2)^{3/2}} = \frac{-u}{\sqrt{u^2 \pm a^2}} + \ln|u + \sqrt{u^2 \pm a^2}| + C$

Forms containing $a + bu \pm cu^2$ $(c > 0)$

42. $\displaystyle\int \frac{du}{a + bu + cu^2} = \frac{2}{\sqrt{4ac - b^2}}\arctan\frac{2cu + b}{\sqrt{4ac - b^2}} + C, \quad b^2 < 4ac$

43. $\displaystyle\int \frac{du}{a + bu + cu^2} = \frac{1}{\sqrt{b^2 - 4ac}}\ln\left|\frac{2cu + b - \sqrt{b^2 - 4ac}}{2cu + b + \sqrt{b^2 - 4ac}}\right| + C, \quad b^2 > 4ac$

44. $\displaystyle\int \frac{du}{a + bu - cu^2} = \frac{1}{\sqrt{b^2 + 4ac}}\ln\left|\frac{\sqrt{b^2 + 4ac} + 2cu - b}{\sqrt{b^2 + 4ac} - 2cu + b}\right| + C$

45. $\displaystyle\int \sqrt{a + bu + cu^2}\,du = \frac{2cu + b}{4c}\sqrt{a + bu + cu^2}$

$\displaystyle\qquad\qquad -\frac{b^2 - 4ac}{8c^{3/2}}\ln|2cu + b + 2\sqrt{c}\sqrt{a + bu + cu^2}| + C$

46. $\displaystyle\int \sqrt{a + bu - cu^2}\,du = \frac{2cu - b}{4c}\sqrt{a + bu - cu^2} + \frac{b^2 + 4ac}{8c^{3/2}}\arcsin\left(\frac{2cu - b}{\sqrt{b^2 + 4ac}}\right) + C$

47. $\displaystyle\int \frac{du}{\sqrt{a + bu + cu^2}} = \frac{1}{\sqrt{c}}\ln|2cu + b + 2\sqrt{c}\sqrt{a + bu + cu^2}| + C$

48. $\displaystyle\int \frac{du}{\sqrt{a + bu - cu^2}} = \frac{1}{\sqrt{c}}\arcsin\left(\frac{2cu - b}{\sqrt{b^2 + 4ac}}\right) + C$

49. $\displaystyle\int \frac{u\,du}{\sqrt{a + bu + cu^2}} = \frac{\sqrt{a + bu + cu^2}}{c} - \frac{b}{2c^{3/2}}\ln|2cu + b + 2\sqrt{c}\sqrt{a + bu + cu^2}| + C$

50. $\displaystyle\int \frac{u\,du}{\sqrt{a + bu - cu^2}} = -\frac{\sqrt{a + bu - cu^2}}{c} + \frac{b}{2c^{3/2}}\arcsin\left(\frac{2cu - b}{\sqrt{b^2 + 4ac}}\right) + C$

TABLE 10—*Cont.* TABLE OF INTEGRALS

Exponential and logarithmic forms

51. $\int e^u \, du = e^u + C$

52. $\int a^u \, du = \dfrac{a^u}{\ln a} + C, \quad a > 0, \, a \neq 1$

53. $\int u e^{au} \, du = \dfrac{e^{au}}{a^2}(au - 1) + C$

54. $\int u^n e^{au} \, du = \dfrac{u^n e^{au}}{a} - \dfrac{n}{a} \int u^{n-1} e^{au} \, du$

55. $\int \dfrac{e^{au}}{u^n} \, du = -\dfrac{e^{au}}{(n-1)u^{n-1}} + \dfrac{a}{n-1} \int \dfrac{e^{au} \, du}{u^{n-1}}$

56. $\int \ln u \, du = u \ln |u| - u + C$

57. $\int u^n \ln u \, du = \dfrac{u^{n+1} \ln |u|}{n+1} - \dfrac{u^{n+1}}{(n+1)^2} + C$

58. $\int \dfrac{du}{u \ln u} = \ln |\ln u| + C$

59. $\int e^{au} \sin nu \, du = \dfrac{e^{au}(a \sin nu - n \cos nu)}{a^2 + n^2} + C$

60. $\int e^{au} \cos nu \, du = \dfrac{e^{au}(n \sin nu + a \cos nu)}{a^2 + n^2} + C$

Trigonometric forms

61. $\int \sin u \, du = -\cos u + C$

62. $\int \cos u \, du = \sin u + C$

63. $\int \tan u \, du = -\ln |\cos u| + C = \ln |\sec u| + C$

64. $\int \cot u \, du = \ln |\sin u| + C$

65. $\int \sec u \, du = \ln |\sec u + \tan u| + C$

66. $\int \csc u \, du = \ln |\csc u - \cot u| + C$

67. $\int \sec^2 u \, du = \tan u + C$

68. $\int \csc^2 u \, du = -\cot u + C$

69. $\int \sec u \tan u \, du = \sec u + C$

510

TABLE 10—*Cont.* TABLE OF INTEGRALS

70. $\displaystyle\int \csc u \cot u \, du = -\csc u + C$

71. $\displaystyle\int \sin^2 u \, du = \tfrac{1}{2}u - \tfrac{1}{4}\sin 2u + C$

72. $\displaystyle\int \cos^2 u \, du = \tfrac{1}{2}u + \tfrac{1}{4}\sin 2u + C$

73. $\displaystyle\int \cos^n u \sin u \, du = -\frac{\cos^{n+1} u}{n+1} + C$

74. $\displaystyle\int \sin^n u \cos u \, du = \frac{\sin^{n+1} u}{n+1} + C$

75. $\displaystyle\int \sin mu \sin nu \, du = -\frac{\sin (m+n)u}{2(m+n)} + \frac{\sin (m-n)u}{2(m-n)} + C$

76. $\displaystyle\int \cos mu \cos nu \, du = \frac{\sin (m+n)u}{2(m+n)} + \frac{\sin (m-n)u}{2(m-n)} + C$

77. $\displaystyle\int \sin mu \cos nu \, du = -\frac{\cos (m+n)u}{2(m+n)} - \frac{\cos (m-n)u}{2(m-n)} + C$

78. $\displaystyle\int e^{au} \sin nu \, du = \frac{e^{au}(a \sin nu - n \cos nu)}{a^2 + n^2} + C$

79. $\displaystyle\int e^{au} \cos nu \, du = \frac{e^{au}(n \sin nu + a \cos nu)}{a^2 + n^2} + C$

80. $\displaystyle\int u \sin u \, du = \sin u - u \cos u + C$

81. $\displaystyle\int u \cos u \, du = \cos u + u \sin u + C$

82. $\displaystyle\int \sin^m u \cos^n u \, du = \frac{\sin^{m+1} u \cos^{n-1} u}{m+n} + \frac{n-1}{m+n} \int \sin^m u \cos^{n-2} u \, du$

83. $\displaystyle\int \sin^n u \, du = -\frac{1}{n}\sin^{n-1} u \cos u + \frac{n-1}{n} \int \sin^{n-2} u \, du$

84. $\displaystyle\int \cos^n u \, du = \frac{1}{n}\cos^{n-1} u \sin u + \frac{n-1}{n} \int \cos^{n-2} u \, du$

85. $\displaystyle\int \tan^n u \, du = \frac{\tan^{n-1} u}{n-1} - \int \tan^{n-2} u \, du$

86. $\displaystyle\int \cot^n u \, du = -\frac{\cot^{n-1} u}{n-1} - \int \cot^{n-2} u \, du$

87. $\displaystyle\int \sec^n u \, du = \frac{\sec^{n-2} u \tan u}{n-1} + \frac{n-2}{n+1} \int \sec^{n-2} u \, du$

88. $\displaystyle\int \csc^n u \, du = -\frac{\csc^{n-2} u \cot u}{n-1} + \frac{n-2}{n-1} \int \csc^{n-2} u \, du$

ANSWERS
TO ODD-NUMBERED
EXERCISES
AND CHAPTER REVIEWS

CHAPTER 1

Page 6

	Function	Domain	Range
1.	Yes	{2, 3, 9}	{2, 4, 7}
3.	No	{1, 2, 7}	{1, 3, 5}
5.	Yes	{−2, 2, 3, 5}	{2}
7.	Yes	Real numbers	Real numbers
9.	Yes	Real numbers	Real numbers where $y \geq 1$
11.	No	Real numbers where $x \geq -2$	Real numbers
13.	Yes	Real numbers where $x \geq -3$	Real numbers where $y \geq 0$

15.

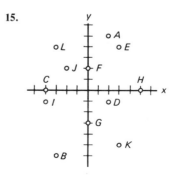

17.

$y = 2x + 1$

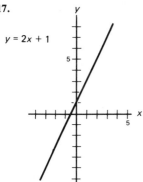

19.

$-2x - 3y = 6$

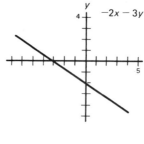

21.

$y = x^2 - 9$

(−4, 7)   (4, 7)

(−3, 0)   (3, 0)

(−2, −5)   (2, −5)

(−1, −8)   (1, −8)

(0, −9)

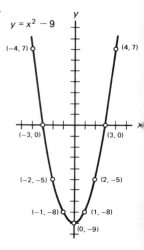

23.

$y = x^2 - 5x + 4$

(0, 4)   (5, 4)

(1, 0)   (4, 0)

(2, −2)   (3, −2)

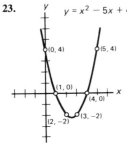

25.

$y = -2x^2 + 4x$

(1, 2)

(0, 0)   (2, 0)

(−1, −6)   (3, −6)

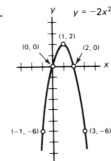

27.

(2, 8)

(−3, 3)   (1, 3)

(−2, 0)

(−1, −1)   (0, 0)

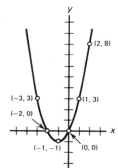

29. $y = x^3 - x^2 - 10x + 8$

(−1, 16)

(−2, 16)   (4, 16)

(−3, 2)

(1, −2)

(3, −4)

(2, −8)

(−4, −22)

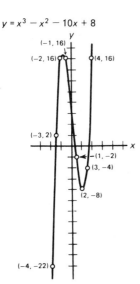

31. $y = \sqrt{x + 4}$

(0, 2)   (9, 3.6)

(5, 3)

(2, 2.4)

(−3, 1)

(−4, 0)

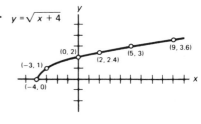

33. $y = \sqrt{12 - 6x}$

(−8, 7.7)

(−4, 6)

(−2, 4.9)

(−1, 4.2)   (0, 3.5)

(1, 2.4)

(2, 0)

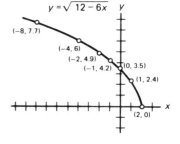

Answers to Odd-Numbered Exercises and Chapter Reviews

Pages 15–16

1. 1   **3.** −4   **5.** 0   **7.** $\frac{5}{8}$

9.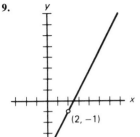

11.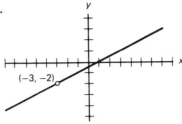

13.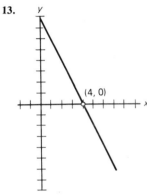

15.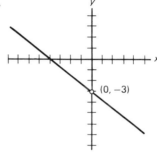

17. $3x + y − 2 = 0$ **19.** $x − 2y − 5 = 0$ **21.** $x + y − 5 = 0$ **23.** $x + 2y + 10 = 0$
25. $y = −5x − 2$ **27.** $y = 2x + 7$ **29.** $y = 5$ **31.** $x = −2$ **33.** $m = −\frac{1}{4},\ b = 3$
35. $m = 2,\ b = 7$ **37.** $m = 0,\ b = 6$

39.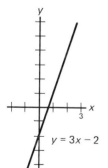

41.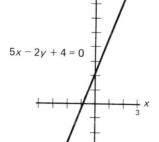

43.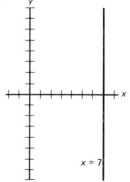

Answers to Odd-Numbered Exercises and Chapter Reviews   **515**

45.

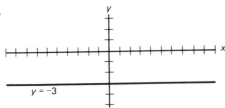

47. Perpendicular **49.** Neither **51.** Parallel **53.** $2x - y + 7 = 0$ **55.** $5x + y + 31 = 0$
57. $3x - 4y = 0$ **59.** $3x - 2y = 18$ **61.** $y = 8$ **63.** $x = 7$
65. (a) Yes, slopes of opposite sides are equal. (b) No, slopes of adjacent sides are not negative reciprocals.

Page 19

1. 15 **3.** 7 **5.** $4\sqrt{2}$ **7.** 7 **9.** $(3\frac{1}{2}, 5)$ **11.** $(1\frac{1}{2}, -1)$ **13.** $(0, -2\frac{1}{2})$
15. (a) 24 (b) Yes (c) No (d) 24 **17.** $4\sqrt{2}$ **19.** $x - 2y = 10$

Page 22

1.

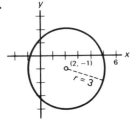

3.

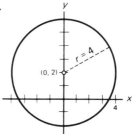

5. $(x - 1)^2 + (y + 1)^2 = 16$ **7.** $(x + 2)^2 + (y + 4)^2 = 34$ **9.** $x^2 + y^2 = 36$ **11.** $(0, 0); r = 4$
13. $(-3, 4); r = 8$ **15.** $(4, -6); r = 2\sqrt{15}$ **17.** $(6, 1); r = 7$ **19.** $(-\frac{1}{2}, -\frac{3}{2}); r = \sqrt{94}/2$
21. $x^2 + y^2 - 2y - 9 = 0; (0, 1); r = \sqrt{10}$ **23.** $x^2 + y^2 + 10x - 40y = 0; (-5, 20); r = 5\sqrt{17}$

Pages 27–28

1.

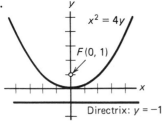

3.

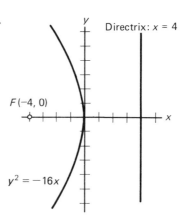

5.

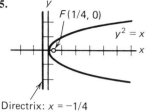

F(1/4, 0)

$y^2 = x$

Directrix: $x = -1/4$

7.

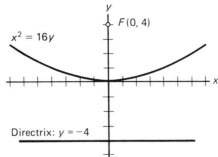

$x^2 = 16y$

F(0, 4)

Directrix: $y = -4$

9.

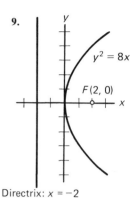

$y^2 = 8x$

F(2, 0)

Directrix: $x = -2$

11. $y^2 = 8x$ **13.** $y^2 = -32x$ **15.** $x^2 = 24y$ **17.** $y^2 = -16x$ **19.** $y^2 - 6y + 8x + 1 = 0$

21. 15 m, 7 m   **23.** $x^2 = 32y$

Page 32

Vertices	Foci	Major axis	Minor axis
1. (5, 0)(−5, 0)	(3, 0)(−3, 0)	10	8
3. (4, 0) (−4, 0)	($\sqrt{7}$, 0) (−$\sqrt{7}$, 0)	8	6
5. (0, 6)(0, −6)	(0, $\sqrt{35}$) (0, −$\sqrt{35}$)	12	2
7. (0, 4) (0, −4)	(0, $\sqrt{7}$) (0, −$\sqrt{7}$)	8	6

1.

(0, 4)

$\dfrac{x^2}{25} + \dfrac{y^2}{16} = 1$

(−5, 0)   (−3, 0)   (3, 0)   (5, 0)

(0, −4)

3.

(0, 3)

$9x^2 + 16y^2 = 144$

(−4, 0) (−$\sqrt{7}$, 0)   ($\sqrt{7}$, 0) (4, 0)

(0, −3)

5.

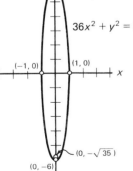

(0, 6)

(0, $\sqrt{35}$)

$36x^2 + y^2 = 36$

(−1, 0)   (1, 0)

(0, −$\sqrt{35}$)

(0, −6)

7.

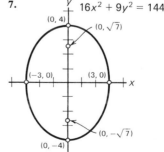

$16x^2 + 9y^2 = 144$

(0, 4)

(0, $\sqrt{7}$)

(−3, 0)   (3, 0)

(0, −$\sqrt{7}$)

(0, −4)

9. $\dfrac{x^2}{16} + \dfrac{y^2}{12} = 1$ or $3x^2 + 4y^2 = 48$ **11.** $\dfrac{x^2}{45} + \dfrac{y^2}{81} = 1$ or $9x^2 + 5y^2 = 405$

13. $\dfrac{x^2}{36} + \dfrac{y^2}{25} = 1$ or $25x^2 + 36y^2 = 900$ **15.** $\dfrac{x^2}{39} + \dfrac{y^2}{64} = 1$ or $64x^2 + 39y^2 = 2496$

17. $\dfrac{x^2}{5600^2} + \dfrac{y^2}{5000^2} = 1$ or $625x^2 + 784y^2 = 1.96 \times 10^{10}$

Pages 38–39

	Vertices	Foci	Transverse axis	Conjugate axis	Asymptotes
1.	(5, 0) (−5, 0)	(13, 0) (−13, 0)	10	24	$y = \pm\dfrac{12}{5}x$
3.	(0, 3) (0, −3)	(0, 5) (0, −5)	6	8	$y = \pm\dfrac{3}{4}x$
5.	$(\sqrt{2}, 0)$ $(-\sqrt{2}, 0)$	$(\sqrt{7}, 0)$ $(-\sqrt{7}, 0)$	$2\sqrt{2}$	$2\sqrt{5}$	$y = \pm\sqrt{\dfrac{5}{2}}x$
7.	(0, 1) (0, −1)	$(0, \sqrt{5})$ $(0, -\sqrt{5})$	2	4	$y = \pm\dfrac{1}{2}x$

1.

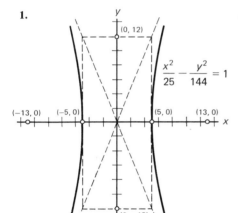

$$\dfrac{x^2}{25} - \dfrac{y^2}{144} = 1$$

3.

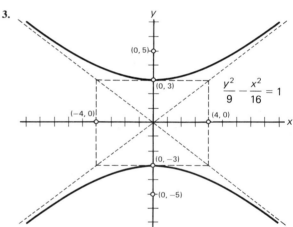

$$\dfrac{y^2}{9} - \dfrac{x^2}{16} = 1$$

5.

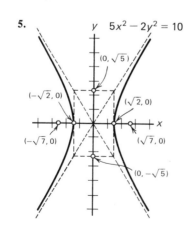

$5x^2 - 2y^2 = 10$

7.

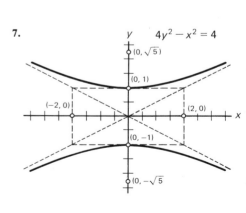

$4y^2 - x^2 = 4$

9. $\dfrac{x^2}{16} - \dfrac{y^2}{20} = 1$ or $5x^2 - 4y^2 = 80$ **11.** $\dfrac{y^2}{36} - \dfrac{x^2}{28} = 1$ or $7y^2 - 9x^2 = 252$

13. $\dfrac{x^2}{9} - \dfrac{y^2}{25} = 1$ or $25x^2 - 9y^2 = 225$ **15.** $\dfrac{x^2}{25} - \dfrac{y^2}{11} = 1$ or $11x^2 - 25y^2 = 275$

17.

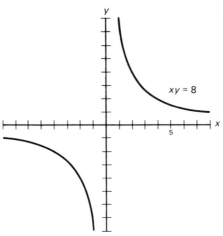

$xy = 8$

Pages 43–44

1. $\dfrac{(x - 1)^2}{16} + \dfrac{(y + 1)^2}{12} = 1$ **3.** $\dfrac{(y - 1)^2}{36} - \dfrac{(x - 1)^2}{28} = 1$ **5.** $(y + 1)^2 = 8(x - 3)$

7. Parabola; vertex: $(2, -3)$

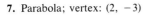
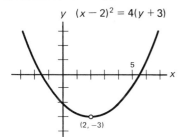

$y \quad (x - 2)^2 = 4(y + 3)$

$(2, -3)$

9. Hyperbola; center: $(-2, 0)$

$\dfrac{y^2}{9} - \dfrac{(x + 2)^2}{16} = 1$

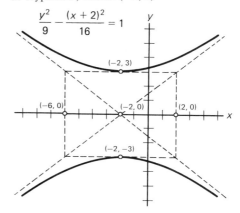

$(-2, 3)$

$(-6, 0)$ $(-2, 0)$ $(2, 0)$

$(-2, -3)$

11. Ellipse; center: $(2, 0)$

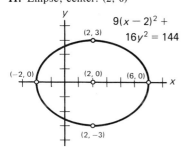

$9(x - 2)^2 + 16y^2 = 144$

$(2, 3)$

$(-2, 0)$ $(2, 0)$ $(6, 0)$

$(2, -3)$

13. Ellipse; center: (3, 1)

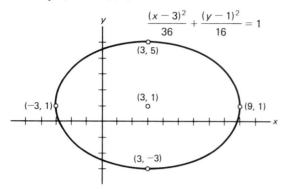

$$\frac{(x-3)^2}{36} + \frac{(y-1)^2}{16} = 1$$

15. Parabola; vertex: (1, −3)

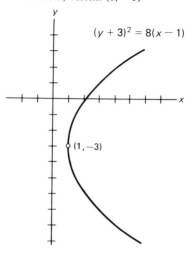

$(y + 3)^2 = 8(x - 1)$

17. Hyperbola; center: (−1, −1)

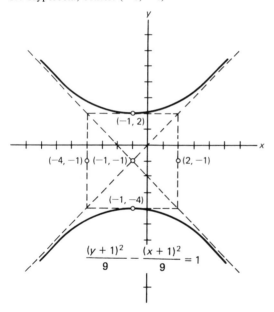

$$\frac{(y+1)^2}{9} - \frac{(x+1)^2}{9} = 1$$

19. Parabola; vertex: (2, −1)

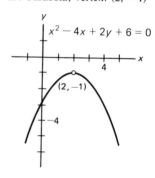

$x^2 - 4x + 2y + 6 = 0$

23. Hyperbola; center: (1, 1)

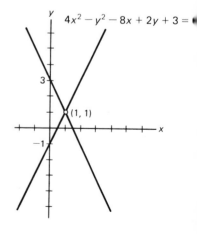

$4x^2 - y^2 - 8x + 2y + 3 = $

21. Ellipse; center: (−2, 1)

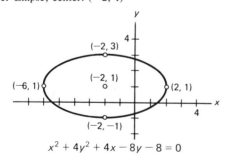

$x^2 + 4y^2 + 4x - 8y - 8 = 0$

25. Hyperbola; center: $(-3, 3)$

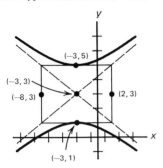

$$\frac{(y-3)^2}{4} - \frac{(x+3)^2}{25} = 1$$

27. Parabola; vertex: $(-8, -2)$

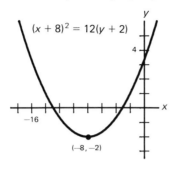

$(x + 8)^2 = 12(y + 2)$

29. Ellipse; center: $(-6, -2)$

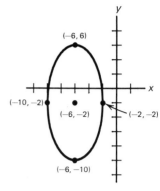

$$\frac{(x+6)^2}{16} + \frac{(y+2)^2}{64} = 1$$

Pages 45–46

1. Ellipse   **3.** Parabola   **5.** Hyperbola   **7.** Circle   **9.** Circle   **11.** Ellipse   **13.** Hyperbola
15. Parabola

Pages 49–50
(Chapter Review)

	Function	Domain	Range
1.	Yes	$\{2, 3, 4, 5\}$	$\{3, 4, 5, 6\}$
2.	No	$\{2, 4, 6\}$	$\{1, 3, 4, 6\}$
3.	Yes	Real numbers	Real numbers
4.	Yes	Real numbers	Real numbers where $y \geq -5$
5.	No	Real numbers where $x \geq 4$	Real numbers
6.	Yes	Real numbers where $x \leq \frac{1}{2}$	Real numbers where $y \geq 0$

7.

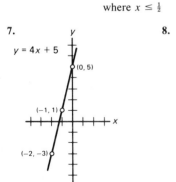

$y = 4x + 5$

8. $y = x^2 + 4$

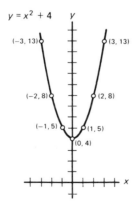

9. $y = x^2 + 2x - 8$

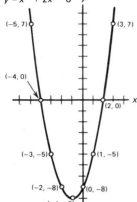

10. $y = 2x^2 + x - 6$

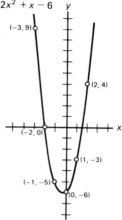

11. $-\frac{2}{9}$ **12.** $\sqrt{85}$ **13.** $(-\frac{3}{2}, -3)$ **14.** $11x + 2y - 58 = 0$ **15.** $2x - 3y + 9 = 0$

16. $x + 3y + 9 = 0$ **17.** $x = -3$ **18.** $m = \frac{3}{2}; b = -3$

19.

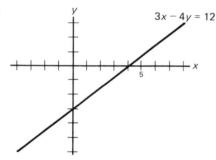

20. Perpendicular   **21.** Parallel   **22.** Neither   **23.** Perpendicular   **24.** Parallel

25. $2x - y - 8 = 0$ **26.** $5x - 3y + 20 = 0$

27. $(x - 5)^2 + (y + 7)^2 = 36$ or $x^2 + y^2 - 10x + 14y + 38 = 0$ **28.** $(4, -3); 7$

29. $(0, \frac{3}{2}); y = -\frac{3}{2}$

30. $y^2 = -16x$ **31.** $(y - 3)^2 = 8(x - 2)$ or $y^2 - 6y - 8x + 25 = 0$

32. $V(7, 0), (-7, 0); F(3\sqrt{5}, 0), (-3\sqrt{5}, 0)$

29.

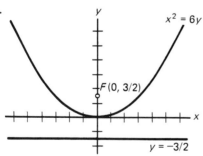

32.

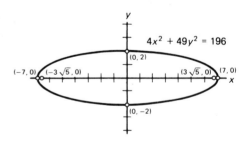

33. $\dfrac{x^2}{4} + \dfrac{y^2}{16} = 1$ or $4x^2 + y^2 = 16$ **34.** $V(6, 0), (-6, 0); F(2\sqrt{13}, 0), (-2\sqrt{13}, 0)$

35. $\dfrac{y^2}{25} - \dfrac{x^2}{16} = 1$ or $16y^2 - 25x^2 = 400$ **36.** $\dfrac{(x-3)^2}{9} + \dfrac{(y+4)^2}{25} = 1$ **37.** $\dfrac{(x+7)^2}{81} - \dfrac{(y-4)^2}{9} = 1$

38. Hyperbola

33.

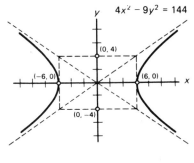

$4x^2 - 9y^2 = 144$

38.

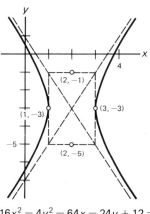

$16x^2 - 4y^2 - 64x - 24y + 12 = 0$

CHAPTER 2

Pages 57–58

1. 9   **3.** -16 **5.** $\frac{1}{7}$ **7.** $2\sqrt{7}$ **9.** $3h + 7$ **11.** $3(\Delta t)^2 + 14(\Delta t) + 11$ **13.** $3(\Delta t); 3$
15. $2(\Delta t)^2 + 4t(\Delta t); 2(\Delta t) + 4t$ **17.** $(\Delta t)^2 + 2t(\Delta t) - 2(\Delta t); \Delta t + 2t - 2$ **19.** 50 m/s
21. 17 m/s   **23.** 6×10^5 m/s   **25.** 100 μA   **27.** 6 m/s   **29.** 10 m/s   **31.** $-\frac{1}{18}$ m/s
33. $-\frac{1}{4}$ m/s   **35.** 64 ft/s

Pages 63–65

1. 4   **3.** 3   **5.** 1   **7.** -6 **9.** 1   **11.** 2   **13.** -6 **15.** 1   **17.** No limit   **19.** 0
21. $\frac{3}{4}$ **23.** 13   **25.** 11   **27.** 6   **29.** 102   **31.** -12 **33.** -11 **35.** $\frac{12}{5}$
37. -14 **39.** 10   **41.** 152   **43.** $2x$ **45.** $-1/x^2$ **47.** $1/(2\sqrt{a})$ **49.** Does not exist
51. b **53.** Does not exist   **55.** b **57.** No   **59.** No   **61.** No   **63.** Yes
65. Does not exist   **67.** 0   **69.** b **71.** Does not exist

Page 68

1. 6   **3.** -6 **5.** 9   **7.** -5 **9.** 12   **11.** $y = -4x - 4$ **13.** $y = -8x - 11$
15. $y = -13x - 3$ **17.** $y = -13x + 29$ **19.** $y = 4x - 3$ **21.** $(-\frac{1}{6}, \frac{1}{36})$
23. $(1, 2), (-1, -2)$

Pages 72–74

1. 3   **3.** -2 **5.** $6x$ **7.** $2x - 2$ **9.** $6x - 4$ **11.** $-12x$ **13.** $3x^2 + 4$ **15.** $-1/x^2$
17. $-2/(x-3)^2$ **19.** $-2/x^3$ **21.** $2x/(4 - x^2)^2$ **23.** $1/(2\sqrt{x+1})$ **25.** $-1/\sqrt{1-2x}$
27. $-1/(2(x-1)^{3/2})$ **29.** 6   **31.** 3/2   **33.** $y = x - 4$ **35.** $x - 4y = 3$ **37.** $(4, 1), (2, -1)$
39. $(-3, 1)$

Pages 79–80

1. 0 **3.** $5x^4$ **5.** 4 **7.** -3 **9.** $10x$ **11.** $2x - 3$ **13.** $8x - 3$ **15.** $-16x$
17. $9x^2 + 4x - 6$ **19.** $20x^4 - 6x^2 + 1$ **21.** $20x^7 - 6x^4 + 30x^3 - 3x^2$
23. $4\sqrt{7}x^3 - 3\sqrt{5}x^2 - \sqrt{3}$ **25.** -4 **27.** 92 **29.** 26 **31.** 2 **33.** 7107

35. $y = -5x + 2$ **37.** 120 W/A **39.** 0.4 V/Ω **41.** $\frac{3}{2}x^{1/2}$ or $\frac{3\sqrt{x}}{2}$ **43.** $-4x^{-5}$ or $\frac{-4}{x^5}$

45. $120x^{19}$ **47.** $-112x^{-9}$ or $\frac{-112}{x^9}$ **49.** $\frac{-5}{3}x^{-4/3}$ or $\frac{-5}{3x\sqrt[3]{x}}$ **51.** 0.5 V/Ω

Pages 82–83

1. $6x^2 + 2x$ **3.** $24x^2 + 12x - 10$ **5.** $20x + 7$ **7.** $12x^2 + 14x - 4$
9. $(x^2 + 3x + 4)(3x^2 - 4) + (x^3 - 4x)(2x + 3)$ or $5x^4 + 12x^3 - 24x - 16$
11. $(x^4 - 3x^2 - x)(6x^2 - 4) + (2x^3 - 4x)(4x^3 - 6x - 1)$ or $14x^6 - 50x^4 - 8x^3 + 36x^2 + 8x$
13. $\dfrac{5}{(2x + 5)^2}$ **15.** $\dfrac{-2x - 1}{(x^2 + x)^2}$ **17.** $\dfrac{7}{2(x + 2)^2}$ **19.** $\dfrac{2x^2 + 2x}{(2x + 1)^2}$ **21.** $\dfrac{-x^2 + 2x + 2}{(x^2 + x + 1)^2}$
23. $\dfrac{-12x^3 - 81x + 72}{x^3(3x - 4)^2}$ **25.** -7 **27.** 10 **29.** $-13/64$ **31.** $y = -5x + 21$ **33.** 10.4 V/s

Page 87

1. $160(4x + 3)^{39}$ **3.** $5(3x^2 - 7x + 4)^4(6x - 7)$ **5.** $\dfrac{-12x^2}{(x^3 + 3)^5}$ **7.** $\dfrac{10x - 7}{2\sqrt{5x^2 - 7x + 2}}$

9. $\dfrac{16x^2 + 2}{\sqrt[3]{8x^3 + 3x}}$ **11.** $-\frac{3}{2}(2x + 3)^{-7/4}$ **13.** $\dfrac{27x^3 - 5x^2 + 9x - 1}{\sqrt{9x^2 - 2x}}$ **15.** $\dfrac{33x^2 + 32x + 18}{(3x + 4)^{1/4}}$

17. $\dfrac{(x^3 + 2)^3(40x^4 - 33x^3 - 16x + 6)}{(4x^2 - 3x)^2}$ **19.** $\dfrac{3(3x + 2)^4(4x - 9)}{(2x - 1)^4}$ **21.** $\dfrac{2x + 8}{(4x + 3)^{3/2}(3x - 1)^{1/3}}$

23. 0.177 m/s

Pages 89–90

1. $-4/3$ **3.** x/y **5.** $\dfrac{-x}{y + 2}$ **7.** $\dfrac{2x - y}{y^2 + x}$ **9.** $\dfrac{y^2 - 2x}{4y^3 - 2xy}$ **11.** $\dfrac{2xy^2 - 3x}{2y^3 - 2x^2y}$

13. $\dfrac{(9x^2 - 12)(x^3 - 4x)^2}{4y^3 + 8y}$ **15.** $\dfrac{2x - 2y + 8 - 3x^2 - 6xy - 3y^2}{3x^2 + 6xy + 3y^2 + 2x - 2y + 8}$ **17.** $\dfrac{y}{x - y(x - y)^2}$

19. $\frac{4}{5}$ **21.** $\frac{1}{3}$ **23.** $y = 4x + 9$ **25.** $12x + 5y = -8$

Pages 95–96

1. $y' = 5x^4 + 6x$, $y'' = 20x^3 + 6$, $y''' = 60x^2$, $y^{iv} = 120x$
3. $y' = 25x^4 + 6x^2 - 8$, $y'' = 100x^3 + 12x$, $y''' = 300x^2 + 12$, $y^{iv} = 600x$ **5.** $-6/x^4$

7. 162 **9.** $-\frac{1215}{16}(3x + 2)^{-7/2}$ **11.** $\dfrac{6x^2 - 2}{(x^2 + 1)^3}$ **13.** $\dfrac{4}{(x - 1)^3}$ **15.** $y' = -x/y$; $y'' = -1/y^3$

17. $y' = \dfrac{2x - y}{x - 2y}$; $y'' = \dfrac{6}{(x - 2y)^3}$ **19.** $y' = -y^{1/2}/x^{1/2}$; $y'' = \dfrac{1}{2x^{3/2}}$ **21.** $y' = y^2/x^2$; $y'' = \dfrac{2y^3 - 2xy^2}{x^4}$

23. $y' = \dfrac{1}{2(1 + y)}$; $y'' = \dfrac{-1}{4(1 + y)^3}$ **25.** $a = 6t^2 - 36t - 8$ **27.** $a = \dfrac{-9}{(6t - 4)^{3/2}}$

29. $x - 5y = -14$; $4x - 5y = 14$

Answers to Odd-Numbered Exercises and Chapter Reviews

1. $6t(\Delta t) + 3(\Delta t)^2$; $6t + 3(\Delta t)$ **2.** $10t(\Delta t) + 5(\Delta t)^2$; $10t + 5(\Delta t)$
3. $2t(\Delta t) + (\Delta t)^2 - 3(\Delta t)$; $2t + \Delta t - 3$ **4.** $6t(\Delta t) + 3(\Delta t)^2 - 6(\Delta t)$; $6t + 3(\Delta t) - 6$ **5.** 27 m/s
6. 20 m/s   **7.** 10 m/s   **8.** 33 m/s   **9.** 12 m/s   **10.** 10 m/s   **11.** 4 m/s   **12.** 17 m/s
13. 4   **14.** −11   **15.** −4   **16.** 10   **17.** No limit   **18.** $\sqrt{15}$ **19.** $\frac{22}{9}$ **20.** $-\frac{37}{24}$
21. 168   **22.** 112   **23.** $-\frac{1}{4}$ **24.** $\frac{1}{2}$ **25.** $\frac{5}{2}$ **26.** $\frac{7}{10}$ **27.** Does not exist   **28.** c **29.** c
30. Does not exist   **31.** No   **32.** Yes   **33.** -10; $y = -10x + 2$ **34.** -1; $y = -x - 16$
35. 14; $y = 14x - 11$ **36.** -24; $y = -24x - 13$ **37.** -0.048 cm/s   **38.** -64 ft/s
39. $20x^3 - 9x^2 + 4x + 5$ **40.** $100x^{99} + 400x^4$ **41.** $6x^5 - 4x^3 + 15x^2 - 4$
42. $21x^6 - 25x^4 + 12x^3 - 36x^2 - 10x + 20$ **43.** $\dfrac{3x^2 - 8x - 3}{(3x - 4)^2}$ **44.** $\dfrac{6x^5 - 18x^4 - 4x + 4}{(3x^4 + 2)^2}$

45. $30x(3x^2 - 8)^4$ **46.** $\frac{3}{4}(x^4 + 2x^3 + 7)^{-1/4}(4x^3 + 6x^2)$ **47.** $\dfrac{-12}{(3x + 5)^5}$

48. $\dfrac{(7x^2 + 21x)(7x^2 - 5)^{-1/2} - 2\sqrt{7x^2 - 5}}{(x + 3)^3}$ or $\dfrac{-7x^2 + 21x + 10}{(x + 3)^3\sqrt{7x^2 - 5}}$ **49.** $\dfrac{-3x^2 - 45x + 20}{2(x + 5)^2\sqrt{2 - 3x}}$

50. $\dfrac{2y^3 - x}{y - 6xy^2}$ **51.** $\dfrac{y}{2y^3 - y - x}$ **52.** $\dfrac{4x}{3y(y^2 + 1)^2}$ **53.** $\dfrac{9x^2(2x^3 - 3)^2}{2(y + 2)^3}$ **54.** 648   **55.** 0

56. -7 **57.** $\dfrac{4}{9\sqrt{2}}$ or $\dfrac{2\sqrt{2}}{9}$ **58.** $11x + y + 14 = 0$ **59.** $26x - y + 62 = 0$

60. $x - y + 10 = 0$ **61.** $13x + 4y - 55 = 0$ **62.** 4 m/s   **63.** 48 m/s   **64.** 2.75 m/s

65. 0.0788 m/s   **66.** $2x + y + 18 = 0$ **67.** 50.3 A   **68.** $\dfrac{dc}{dT} = 0.5 + 0.000012T$

69. $-\dfrac{1}{4\pi C\sqrt{LC}}$

70. $y' = 24x^5 - 32x^3 + 27x^2 - 6$, $y'' = 120x^4 - 96x^2 + 54x$, $y''' = 480x^3 - 192x + 54$, $y^{iv} = 1440x^2 - 192$
71. $\dfrac{-1}{(2x - 3)^{3/2}}$ **72.** $\dfrac{72x^2 - 12}{(2x^2 + 1)^3}$ **73.** $y' = \dfrac{-y}{x + y}$, $y'' = \dfrac{4}{(x + y)^3}$ **74.** $y' = \dfrac{y^{3/2}}{x^{3/2}}$, $y'' = \dfrac{3y^2 - 3y^{3/2}x^{1/2}}{2x^3}$
75. $a = -\frac{3}{4}(2t + 3)^{-7/4}$

CHAPTER 3

Pages 110–111

1.

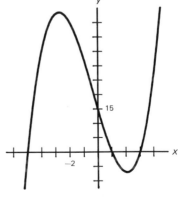

$y = 2x(x + 1)(x - 4)$

3.

$y = (x - 1)(x - 3)(x + 5)$

5.

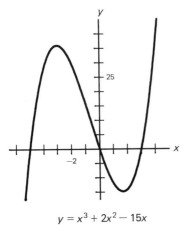

$$y = x^3 + 2x^2 - 15x$$

7.

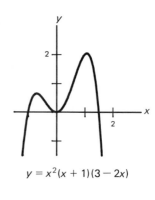

$$y = x^2(x + 1)(3 - 2x)$$

9.

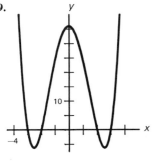

$$y = x^4 - 13x^2 + 36$$

11.

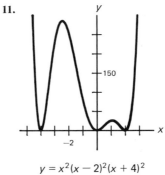

$$y = x^2(x - 2)^2(x + 4)^2$$

13.

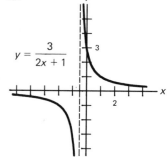

$$y = \frac{3}{2x + 1}$$

15.

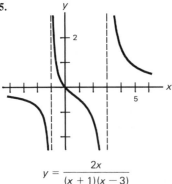

$$y = \frac{2x}{(x + 1)(x - 3)}$$

17.

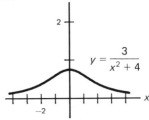

$$y = \frac{3}{x^2 + 4}$$

Answers to Odd-Numbered Exercises and Chapter Reviews

19.

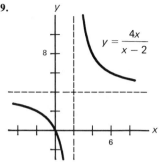

$$y = \frac{4x}{x-2}$$

8

6

21.

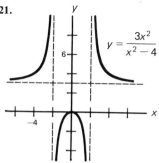

$$y = \frac{3x^2}{x^2-4}$$

6

−4

23.

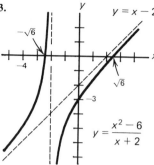

$y = x - 2$

$-\sqrt{6}$

−4

$\sqrt{6}$

−3

$$y = \frac{x^2-6}{x+2}$$

25.

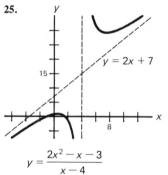

$y = 2x + 7$

15

8

$$y = \frac{2x^2-x-3}{x-4}$$

27.

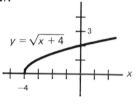

$y = \sqrt{x+4}$

3

−4

29.

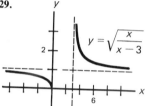

$$y = \sqrt{\frac{x}{x-3}}$$

2

6

31.

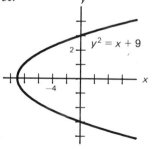

$y^2 = x + 9$

2

−4

33.

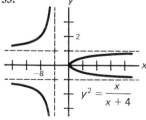

2

−8

$$y^2 = \frac{x}{x+4}$$

35.

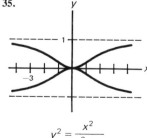

1

−3

$$y^2 = \frac{x^2}{x^2+4}$$

1.

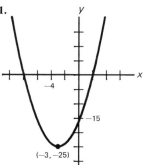

$$y = x^2 + 6x - 16$$

3.

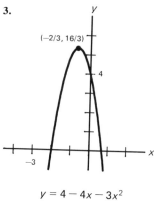

$$y = 4 - 4x - 3x^2$$

5.

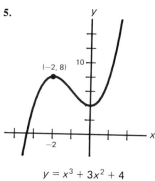

$$y = x^3 + 3x^2 + 4$$

7.

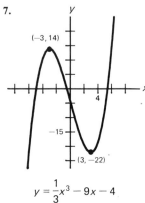

$$y = \frac{1}{3}x^3 - 9x - 4$$

9.

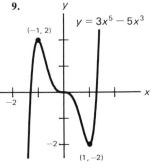

$$y = 3x^5 - 5x^3$$

11.

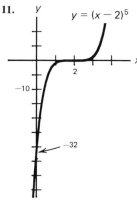

$$y = (x - 2)^5$$

13.

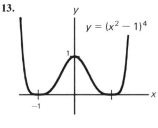

$$y = (x^2 - 1)^4$$

15.

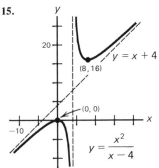

$$y = x + 4$$

$$y = \frac{x^2}{x - 4}$$

17.

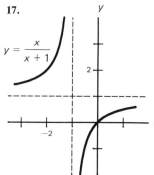

$$y = \frac{x}{x + 1}$$

19.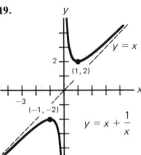

21.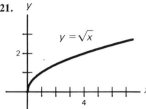

23.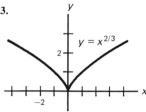

Pages 121–122

1. (a) Increasing for $x > 2$, decreasing for $x < 2$; (b) Relative minimum at $(2, -4)$; (c) Concave upward for all values of x; (d) No points of inflection; (e)

3. (a) Increasing for $x > 0$, decreasing for $x < 0$; (b) Relative minimum at $(0, 0)$; (c) Concave upward for all values of x; (d) No points of inflection; (e)

1e.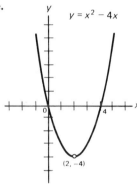

3e.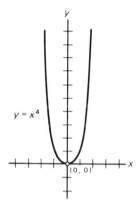

5. (a) Increasing for $-1 < x < 1$, decreasing for $x < -1$ and $x > 1$; (b) Relative minimum at $(-1, -2)$, relative maximum at $(1, 2)$; (c) Concave upward for $x < 0$, concave downward for $x > 0$; (d) Point of inflection at $(0, 0)$; (e)

7. (a) Increasing for $x < 2$, decreasing for $x > 2$; (b) Relative maximum at $(2, 3)$; (c) Concave downward for all values of x; (d) No point of inflection; (e)

5e.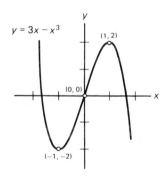

7e.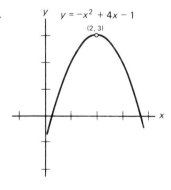

Answers to Odd-Numbered Exercises and Chapter Reviews

9. (a) Increasing for $-2 < x < 0$ and $x > 2$, decreasing for $x < -2$ and $0 < x < 2$; (b) Relative maximum at $(0, 5)$, relative minimums at $(2, -11)$ and $(-2, -11)$; (c) Concave upward for $x < -\dfrac{2}{\sqrt{3}}$ and $x > \dfrac{2}{\sqrt{3}}$, concave downward for $-\dfrac{2}{\sqrt{3}} < x < \dfrac{2}{\sqrt{3}}$; (d) Points of inflection at $\left(\dfrac{-2}{\sqrt{3}}, \dfrac{-35}{9}\right)$ and $\left(\dfrac{2}{\sqrt{3}}, \dfrac{-35}{9}\right)$; (e)

11. (a) Increasing for $x < 0$, decreasing for $x > 0$; (b) No relative maximum or minimum; (c) Concave upward for $x > 0$ and $x < 0$; (d) No point of inflection; (e)

9e.

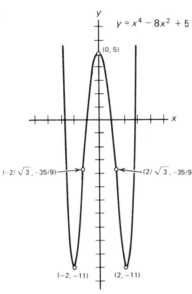

11e.

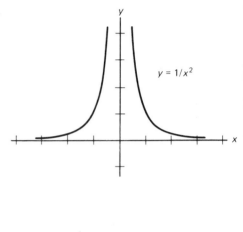

13. (a) Increasing for $x < -2$ and $x > -2$; (b) No relative maximum or minimum; (c) Concave upward for $x < -2$, concave downward for $x > -2$; (d) No points of inflection; (e)

15. (a) Increasing for $x < -1$ and $x > -1$; (b) No relative maximum or minimum; (c) Concave upward for $x < -1$, concave downward for $x > -1$; (d) No point of inflection; (e)

13e.

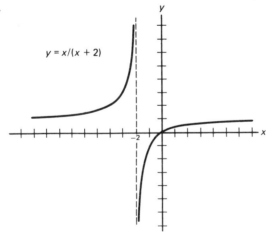

15e.

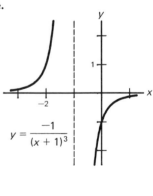

17. (a) Increasing for $-1 < x < 1$, decreasing for $x < -1$, $x > 1$; (b) Relative maximum at $(1, \frac{1}{2})$, relative minimum at $(-1, -\frac{1}{2})$; (c) Concave upward for $-\sqrt{3} < x < 0$, $x > \sqrt{3}$, concave downward for $x < -\sqrt{3}$, $0 < x < \sqrt{3}$; (d) Points of inflection at $(\sqrt{3}, \sqrt{3}/4)$, $(0, 0)$, $(-\sqrt{3}, -\sqrt{3}/4)$; (e)

19. (a) Increasing for $x < -3$, $x > -3$; (b) No relative maximum or minimum; (c) Concave upward for $x < -3$, concave downward for $x > -3$; (d) No point of inflection; (e)

17e.

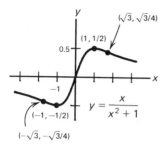

19e.

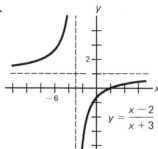

21. (a) Increasing for $x < 0$, decreasing for $x > 0$; (b) Relative maximum at $(0, 1)$; (c) Concave upward for $x < -2/\sqrt{3}$, $x > 2/\sqrt{3}$, concave downward for $-2/\sqrt{3} < x < 2/\sqrt{3}$; (d) Inflection points at $(2/\sqrt{3}, \frac{3}{4})$ and $(-2/\sqrt{3}, \frac{3}{4})$; (e)

23. (a) Increasing for $0 < x < 2$, decreasing for $x < 0$, $x > 2$; (b) Relative maximum at $(2, \frac{1}{4})$; (c) Concave upward for $x > 3$, concave downward for $x < 0$, $0 < x < 3$; (d) Point of inflection $(3, \frac{2}{9})$; (e)

21e.

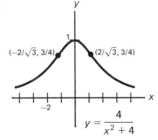

23e.

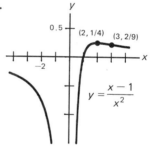

Pages 126–129

1. 28, 28   **3.** 2 cm × 2 cm × 0.5 cm   **5.** 200 m × 400 m   **7.** 81 cm²   **9.** 1600 m
11. $96\sqrt{3}$ **13.** $\frac{32}{3}$ **15.** 4 Ω   **19.** $m = \frac{3}{2}$ at $x = \frac{1}{2}$ **21.** $i = -16$ A at $t = 2$ s
23. 20 cm × 20 cm × 10 cm   **25.** 20   **27.** $r = 2\sqrt{2}$ cm, $h = 4\sqrt{2}$ cm
29. $w = 2r/\sqrt{3}$, $d = 2r\sqrt{6}/3$ **31.** 4 km

Pages 131–133

1. 0.06 Ω/s   **3.** 1.92 cm²/min   **5.** 43.2 cm³/min   **7.** −0.277 cm²/min   **9.** 3/(80π) m/s
11. −160 W/s   **13.** 0.509 cm/s   **15.** $\frac{1}{4}$ m/s   **17.** 12 lb/in²/min   **19.** −2.4 Ω/s   **21.** 240 ΩA
23. 8 V/s   **25.** (a) 8 m/min, (b) $\frac{8}{9}$ m/min

Answers to Odd-Numbered Exercises and Chapter Reviews

1. $dy = (10x - 24x^2)\, dx$ 3. $dy = \dfrac{-7\, dx}{(2x - 1)^2}$ 5. $dy = 16t(2t^2 + 1)^3\, dt$

7. $ds = -2(t^4 - t^{-2})^{-3}(4t^3 + 2t^{-3})\, dt$ or $ds = \dfrac{-4t^3(2t^6 + 1)\, dt}{(t^6 - 1)^3}$ 9. $dy = \dfrac{-x\, dx}{4y}$

11. $dy = \dfrac{x^{-1/2} - 6(x + y)^2}{6(x + y)^2 - y^{-1/2}}\, dx$ or $dy = \dfrac{y^{1/2} - 6x^{1/2}y^{1/2}(x + y)^2}{6x^{1/2}y^{1/2}(x + y)^2 - x^{1/2}}\, dx$ 13. 43.2   15. 1.2   17. 282.7

19. (a) 1.20 cm² (b) 1.2025 cm² (c) 0.833%   21. (a) 84.9 m³ (b) 662,000 kg   23. 0.3 hp

25. 4 V

Pages 140–141
(Chapter Review)

1.

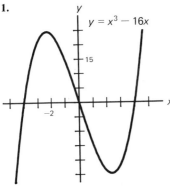

$y = x^3 - 16x$

2.

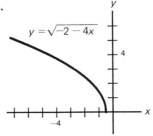

$y = \sqrt{-2 - 4x}$

3.

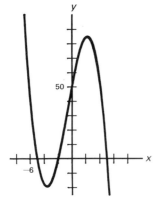

$y = (x + 2)(25 - x^2)$

4.

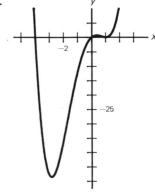

$y = (x^2 + 4x)(x - 1)^2$

5.

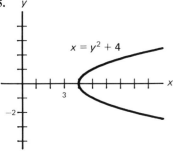

$x = y^2 + 4$

3

−2

6.

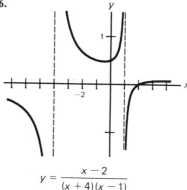

1

−2

$$y = \frac{x - 2}{(x + 4)(x - 1)}$$

7.

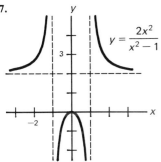

$$y = \frac{2x^2}{x^2 - 1}$$

3

−2

8.

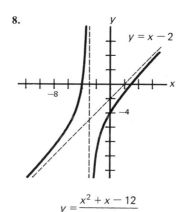

$y = x - 2$

−8

−4

$$y = \frac{x^2 + x - 12}{x + 3}$$

9.

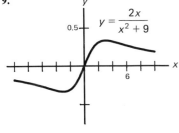

$$y = \frac{2x}{x^2 + 9}$$

0.5

6

10.

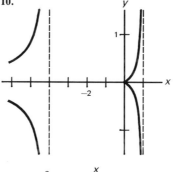

1

−2

$$y^2 = \frac{x}{(1 - x)(x + 4)}$$

Answers to Odd-Numbered Exercises and Chapter Reviews

11. (a) Increasing for $-\sqrt{2} < x < \sqrt{2}$, decreasing for $x < -\sqrt{2}$ and $x > \sqrt{2}$; (b) Relative maximum at $(\sqrt{2}, 4\sqrt{2})$, relative minimum at $(-\sqrt{2}, -4\sqrt{2})$; (c) Concave upward for $x < 0$, concave downward for $x > 0$; (d) Point of inflection at $(0, 0)$; (e)

12. (a) Increasing for $x > \frac{3}{2}$, decreasing for $x < \frac{3}{2}$; (b) Relative minimum at $\left(\frac{3}{2}, \frac{-25}{4}\right)$; (c) Concave upward for all values of x; (d) No points of inflection; (e)

13. (a) Increasing for all values of x; (b) No relative maximum or minimum; (c) Concave upward for $x > 0$, concave downward for $x < 0$; (d) Inflection point at $(0, -7)$; (e)

11e.

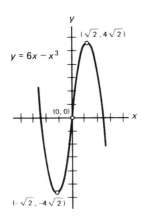

$y = 6x - x^3$
$(\sqrt{2}, 4\sqrt{2})$
$(0, 0)$
$(-\sqrt{2}, -4\sqrt{2})$

12e.

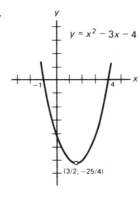

$y = x^2 - 3x - 4$
$(3/2, -25/4)$

13e.

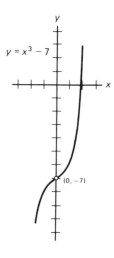

$y = x^3 - 7$
$(0, -7)$

14. (a) Increasing for $x < -1$, $x > 4$, decreasing for $-1 < x < 4$; (b) Relative maximum at $(-1, 11)$, relative minimum at $(4, -114)$; (c) Concave upward for $x > \frac{3}{2}$, concave downward for $x < \frac{3}{2}$; (d) Point of inflection $(1.5, -51.5)$; (e)

15. (a) Increasing for $x < -1$, decreasing for $x > -1$; (b) No relative maximum or minimum; (c) Concave upward for $x < -1$, $x > -1$; (d) No point of inflection; (e)

16. (a) Increasing for $x > 0$, decreasing for $x < 0$; (b) Relative minimum at $(0, -\frac{1}{4})$; (c) Concave upward for $-2/\sqrt{3} < x < 2/\sqrt{3}$, concave downward for $x < -2/\sqrt{3}$, $x > 2/\sqrt{3}$; (d) Points of inflection at $(2/\sqrt{3}, \frac{1}{16})$, $(-2/\sqrt{3}, \frac{1}{16})$; (e)

14e.

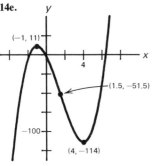

$(-1, 11)$
$(1.5, -51.5)$
-100
$(4, -114)$
$y = 2x^3 - 9x^2 - 24x - 2$

15e.

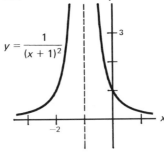

$y = \dfrac{1}{(x + 1)^2}$

16e.

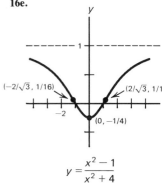

$(-2/\sqrt{3}, 1/16)$
$(2/\sqrt{3}, 1/16)$
$(0, -1/4)$
$y = \dfrac{x^2 - 1}{x^2 + 4}$

Answers to Odd-Numbered Exercises and Chapter Reviews

17. (a) Increasing for $x < 0$, decreasing for $x > 0$; (b) Relative maximum at $(0, 10)$; (c) Concave upward for $x < -1/\sqrt{3}$, $x > 1/\sqrt{3}$, concave downward for $-1/\sqrt{3} < x < 1/\sqrt{3}$; (d) Points of inflection at $(-1/\sqrt{3}, \frac{15}{2})$, $(1/\sqrt{3}, \frac{15}{2})$; (e)

18. (a) Increasing for $-2 < x < 0$, decreasing for $x < -2$, $x > 0$; (b) Relative minimum at $(-2, -\frac{1}{4})$; (c) Concave upward for $-3 < x < 0$, $x > 0$, concave downward for $x < -3$; (d) Point of inflection $(-3, -\frac{2}{9})$; (e)

17e.

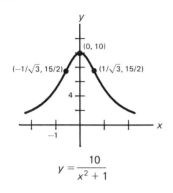

$$y = \frac{10}{x^2 + 1}$$

18e.

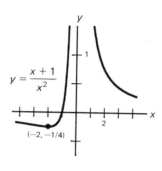

$$y = \frac{x + 1}{x^2}$$

19. 900 ft   **20.** $b = h = 10\sqrt{2}$ **21.** $1 \, \Omega$ **22.** 12 m²   **23.** $3\sqrt{2}$ cm   **24.** $(\frac{1}{2}, \sqrt{2}/2)$

25. $\dfrac{1}{288\pi}$ ft/s   **26.** 18 V/s   **27.** -0.0087 A/s   **28.** -0.8π cm²/min or -2.51 cm²/min

29. 0.255 km/day   **30.** 1.92 km/s   **31.** $dy = (20x^4 - 18x^2 + 2)\, dx$ **32.** $dy = -2(3x - 5)^{-5/3}\, dx$

33. $ds = \dfrac{(15t^2 + 6t + 20)\, dt}{(5t + 1)^2}$ **34.** $dy = \dfrac{2 - 4x(x^2 + y^2)}{4y(x^2 + y^2) - 1}\, dx$ **35.** 2.6   **36.** -0.000158

37. 45.2 in³   **38.** 0.3 m   **39.** 9.58 gal   **40.** 8.04 cm³   **41.** -74.1k N

CHAPTER 4

Pages 148–149

1. $\frac{1}{8}x^8 + C$ **3.** $\frac{1}{3}x^9 + C$ **5.** $4x + C$ **7.** $\frac{54}{11}x^{11/6} + C$ **9.** $-\dfrac{3}{x^2} + C$

11. $\dfrac{5x^3}{3} - 6x^2 + 8x + C$ **13.** $x^3 - \frac{1}{2}x^2 - \dfrac{5}{2x^2} + C$ **15.** $\frac{4}{5}x^5 - 4x^3 + 9x + C$

17. $\frac{1}{9}(6x + 2)^{3/2} + C$ **19.** $(x^2 + 3)^4 + C$ **21.** $\frac{3}{40}(5x^2 - 1)^{4/3} + C$ **23.** $\frac{1}{10}(x^2 - 1)^5 + C$

25. $2\sqrt{x^2 + 1} + C$ **27.** $\frac{1}{4}(x^3 + 2x)^4 + C$ **29.** $\dfrac{-1}{3(x^3 - 4)} + C$ **31.** $\frac{2}{3}(5x^2 - x)^{3/2} + C$

33. $2\sqrt{x^2 + x} + C$ **35.** $\frac{1}{6}(2x + 3)^3 + C$ **37.** $\frac{1}{10}(2x - 1)^5 + C$

39. $\dfrac{x^7}{7} + \dfrac{3x^5}{5} + x^3 + x + C$ **41.** $\frac{1}{2}(x^2 + 1)^4 + C$ **43.** $\frac{2}{5}(5x^3 + 1)^5 + C$ **45.** $4\sqrt{x^3 + 1} + C$

47. $3(x^3 + 3x)^{2/3} + C$ **49.** $-1/x + 1/(2x^2) + C$

Pages 152–154

1. $y = \frac{3}{2}x^2 + 1$ **3.** $y = x^3 + 3x + 6$ **5.** $y = \frac{1}{6}(x^2 - 3)^3 + 1$ **7.** $s = \frac{1}{3}t^3 + 16t + 50$

9. 64 ft (36 ft from the ground at $t = 2$); -80 ft/s   **11.** 240 ft/s   **13.** (a) 31.9 m   (b) 5.10 s

(c) 25.0 m/s   **15.** (a) $s = -16t^2 + 30t + 200$ (b) 4.59 s   **17.** 213 rev   **19.** 313 V

21. $q = \dfrac{(t^2 + 1)^{3/2} - 1}{3}$; $\dfrac{2\sqrt{2} - 1}{3}$

Answers to Odd-Numbered Exercises and Chapter Reviews

1. 2 **3.** $\frac{52}{3}$ **5.** 4 **7.** $\frac{3}{2}$ **9.** $\frac{14}{9}$ **11.** 4 **13.** $\frac{8}{5}$ **15.** $\frac{98}{3}$ **17.** $\frac{3}{4}$ **19.** $\frac{4}{5}$ **21.** $\frac{3}{7}$
23. 36 **25.** $\frac{4}{3}$ **27.** $\frac{1}{12}$ **29.** $\frac{4}{15}$

Page 164

1. $\frac{5}{2}$ **3.** $\frac{16}{3}$ **5.** -6 **7.** $\frac{14}{3}$ **9.** 40 **11.** $\frac{88}{3}$ **13.** 5546.2 **15.** $\frac{208}{3}$ **17.** $-\frac{15}{2}$
19. $-\frac{3}{10}$ **21.** $\dfrac{2\sqrt{2} - 1}{3}$ **23.** 12 **25.** $2\sqrt{10} - 2\sqrt{2}$

Pages 165–166
(Chapter Review)

1. $\frac{5}{3}x^3 - \frac{1}{2}x^2 + C$ **2.** $\frac{3}{8}x^8 + x^2 + 4x + C$ **3.** $\frac{4}{3}x^{9/2} + C$ **4.** $\frac{12}{5}x^{5/3} + C$
5. $\dfrac{-3}{4x^4} + C$ **6.** $\dfrac{-2}{\sqrt{x}} + C$ **7.** $\frac{1}{8}(3x^4 + 2x - 1)^4 + C$ **8.** $\frac{5}{16}(7x^2 + 8x + 2)^{8/5} + C$
9. $2\sqrt{x^2 + 5x} + C$ **10.** $3(5x^3 + 4x)^{1/3} + C$ **11.** $y = x^3 - 4$ **12.** $s = -16t^2 + 25t + 100$

13. 69.2 Ω **14.** 2.97 **15.** $\frac{14}{3}$ **16.** 6 **17.** $\frac{15}{64}$ **18.** 1 **19.** $\dfrac{144 - 4\sqrt{6}}{5}$ **20.** $\dfrac{16\sqrt{2} - 8}{3}$

21. $\frac{17}{12}$ **22.** $\frac{108}{5}$ **23.** $\frac{62}{3}$ **24.** $\frac{7}{18}$ **25.** $\dfrac{12\sqrt{6} - 4\sqrt{2}}{3}$ **26.** $\dfrac{4\sqrt{7} - 2}{3}$

27. -6 **28.** $\dfrac{6 - 3\sqrt{2}}{4}$

CHAPTER 5

Page 173

1. $\frac{1}{3}$ **3.** $\frac{1}{2}$ **5.** $\frac{9}{2}$ **7.** $\frac{32}{3}$ **9.** $\frac{1}{6}$ **11.** $\frac{9}{2}$ **13.** $\frac{1}{2}$ **15.** $\frac{9}{2}$ **17.** 4 **19.** $\frac{44}{15}$ **21.** $\frac{343}{6}$
23. $\frac{22}{5}$ **25.** $\frac{37}{12}$

Pages 181–182

1. $\dfrac{26\pi}{3}$ **3.** $\dfrac{178\pi}{15}$ **5.** $\dfrac{8\pi}{3}$ **7.** 2π **9.** $\dfrac{\pi}{3}$ **11.** $\dfrac{\pi}{3}$ **13.** 2π **15.** 8π **17.** $\dfrac{4\pi}{15}$ **19.** $\dfrac{5\pi}{6}$
21. $\dfrac{8\pi}{3}$ **23.** $\dfrac{32\pi}{3}$ **25.** 60π **27.** $\dfrac{100\pi\sqrt{5}}{9}$

Page 185

1. 2π **3.** 2π **5.** 4π **7.** $\dfrac{128\pi}{7}$ **9.** $\dfrac{8\pi}{3}$ **11.** $\dfrac{\pi}{3}$ **13.** $\dfrac{5\pi}{6}$ **15.** $\dfrac{4\pi}{21}$ **17.** $\dfrac{\pi}{2}$ **19.** $\dfrac{16\pi}{5}$

Pages 189–190

1. 15/7 **3.** -6.75 **5.** -6 **7.** $-\frac{11}{3}$ **9.** 15 **11.** 40 **13.** 11 mi north of Flatville
15. (20/7, 22/7) **17.** $(-3.8, -7.2)$ **19.** $(2.1, -8.4)$ **21.** 100
23. 1.50 mi east and 2.56 mi south of A

Pages 202–204

1. 10 cm **3.** $\frac{20}{3}$ cm **5.** 8.77 cm **7.** 4 cm from given end **9.** $(4\frac{2}{3}, 4\frac{2}{3})$ **11.** (8, 2)
13. (0, 2.95) **15.** $(\frac{27}{5}, \frac{9}{8})$ **17.** $(1, -0.4)$ **19.** (0, 1.6) **21.** $(\frac{8}{15}, \frac{8}{21})$ **23.** $\left(0, \dfrac{4}{3\pi}\right)$

25. $(\frac{7}{8}, 0)$ **27.** $(\frac{3}{4}, 0)$ **29.** $(0, \frac{2}{3})$

Pages 210–211

1. 516; 3.79 **3.** 1965; 5.98 **5.** 963; 5.17 **7.** 759; 4.87 **9.** $\frac{64}{3}$; 0.894 **11.** $\frac{1}{7}$; 0.463
13. $\frac{64}{9}$; 0.894 **15.** 2; 1.41 **17.** 576 π; 1.55 **19.** 2.29 × 10⁴; 8.43 **21.** 1458 π; 1.73
23. 4096 π; 2.53

Pages 221–222

1. $\frac{63}{4}$ **3.** 25 in.-lb **5.** 675 N · cm or 6.75 J **7.** 2.896 × 10⁻¹⁴ J
9. (a) 900 ft-lb (b) 1875 ft-lb (c) 2500 ft-lb **11.** 225,800 ft-lb **13.** 602,200 ft-lb
15. 58,810 ft-lb **17.** 25,000 lb **19.** 352,800 N **21.** 710,500 N **23.** 166.4 lb **25.** 54,660 lb
27. 21,542 lb **29.** $\frac{13}{3}$ **31.** $\frac{2}{5}$ **33.** 1.70 A

Pages 226–227
(Chapter Review)

1. $\frac{16}{3}$ **2.** $\frac{2}{3}$ **3.** $\frac{1}{12}$ **4.** 8 **5.** $\frac{128}{15}$ **6.** 36 **7.** 8π **8.** 8π
9. $\frac{\pi}{30}$ **10.** $\frac{8\pi}{3}$ **11.** $\frac{243\pi}{10}$ **12.** $\frac{99\pi}{2}$ **13.** $\frac{256\pi}{5}$ **14.** $\frac{63\pi}{2}$ **15.** 7.5
16. (−1, −6.8) **17.** (9.6, 5.3) **18.** (2$\frac{2}{3}$, 6$\frac{2}{3}$) **19.** ($\frac{3}{2}$, $\frac{27}{5}$) **20.** (3, −3.6)
21. (0, $\frac{3}{2}$) **22.** ($\frac{3}{8}$, 0) **23.** (−$\frac{8}{5}$, 0) **24.** 462; 4.39 **25.** 192; 2.83 **26.** 576; 4.90
27. $\frac{1}{8}$;0.354 **28.** $\frac{2\pi}{13}$;0.519 **29.** $\frac{8\pi}{7}$;0.845 **30.** 126π;2.65 **31.** 200 in.-lb **32.** 2.62 × 10⁻¹⁶ J
33. 130,000 ft-lb **34.** 49,920 lb **35.** 816,700 N **36.** 9.5 V **37.** 67.52 A **38.** 20 W

CHAPTER 6

Pages 235–236

37. sin 4θ **39.** cos θ **41.** tan 5θ **43.** 2 sin θ cos ϕ **45.** sin $\dfrac{x}{2}$ **47.** cos 6x **49.** cos $\dfrac{\theta}{8}$

51. cos $\dfrac{x}{3}$ **53.** 10 sin 8θ **55.** 4 cos 2θ

57.

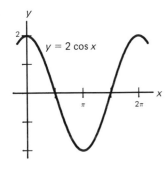

$y = 2 \cos x$

59.

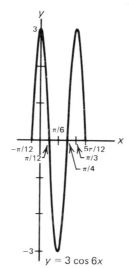

$y = 3 \cos 6x$

61.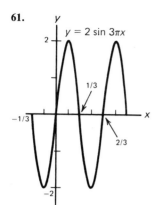

63.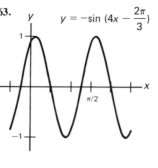

65.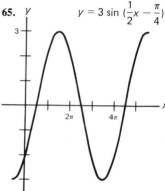

67.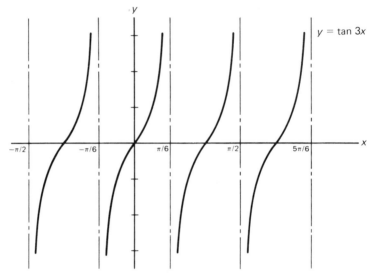

Pages 241–242

1. $7 \cos 7x$ **3.** $-10 \sin 5x$ **5.** $6x^2 \cos x^3$ **7.** $-24x \sin 4x^2$ **9.** $-4 \cos (1 - x)$

11. $6x \cos (x^2 + 4)$ **13.** $-4(10x + 1) \sin (5x^2 + x)$ **15.** $-4(x^3 - x) \sin (x^4 - 2x^2 + 3)$

17. $-6 \cos (3x - 1) \sin (3x - 1)$ **19.** $6 \sin^2(2x + 3) \cos (2x + 3)$ **21.** $4(2x - 5) \cos (2x - 5)^2$

23. $-12x^2(x^3 - 4)^3 \sin (x^3 - 4)^4$ **25.** $\cos 3x \cos x - 3 \sin x \sin 3x$

27. $-6 \sin 5x \sin 6x + 5 \cos 6x \cos 5x$ **29.** $-7 \cos 4x \sin 7x - 4 \cos 7x \sin 4x$

31. $2(x + 1) \cos x^3 \cos (x^2 + 2x) - 3x^2 \sin (x^2 + 2x) \sin x^3$

33. $5(x^2 + 3x) \cos (5x - 2) + (2x + 3) \sin (5x - 2)$

35. $\dfrac{5x \cos 5x - \sin 5x}{x^2}$ **37.** $\dfrac{2x \cos 3x + 3(x^2 - 1) \sin 3x}{\cos^2 3x}$ **39.** $5 \cos 5x - 6 \sin 6x$

41. $(2x - 3) \cos (x^2 - 3x) - 4 \sin 4x$ **43.** $\sec^2 x$ **45.** $-\cos x$ **47.** $-\cos x$

49. $-25 \sin 5x - 36 \cos 6x$ **51.** $6\sqrt{3}$ **53.** $y = -10x + \pi$

Pages 246–247

1. $3\sec^2 3x$ **3.** $7\sec 7x\tan 7x$ **5.** $-6x\csc^2(3x^2-7)$ **7.** $-9\csc(3x-4)\cot(3x-4)$

9. $10\tan(5x-2)\sec^2(5x-2)$ **11.** $-24\cot^2 2x\csc^2 2x$ **13.** $\dfrac{2x+1}{2\sqrt{x^2+x}}\sec\sqrt{x^2+x}\tan\sqrt{x^2+x}$

15. $-\dfrac{\csc x(x\cot x+1)}{3x^2}$ **17.** $3\sec^2 3x-2x\sec(x^2+1)\tan(x^2+1)$ **19.** $\sec x(2\sec^2 x-1)$ **21.** $\cos 2x$

23. $\sec x(x\tan x+1)$ **25.** $2x\sec^2 x(x\tan x+1)$ **27.** $-3\csc 3x(2\csc^2 3x-1)$ **29.** $-3\csc 3x\cot 3x$

31. $-2\cos 2x$ **33.** $4(x+\sec^2 3x)^3(1+6\sec^2 3x\tan 3x)$ **35.** $3\sec x(\sec x+\tan x)^3$

37. $\cos(\tan x)\sec^2 x$ **39.** $-\sin x\sec^2(\cos x)$ **41.** $-2\sin x\sin(\cos x)\cos(\cos x)$

43. $\dfrac{1+\sin^2 x}{\cos^3 x}$ **45.** $-2\sin x\cos x$ or $-\sin 2x$ **47.** $\dfrac{\cos x+\sin x-\sin x\sec^2 x}{(1+\tan x)^2}$

49. $18\sec^2 3x\tan 3x$ **51.** $2\csc^2 x(x\cot x-1)$ **53.** 2

Pages 255–256

1. $\dfrac{\pi}{3},\dfrac{2\pi}{3}$ **3.** $\dfrac{\pi}{4},\dfrac{5\pi}{4}$ **5.** $\dfrac{5\pi}{6},\dfrac{7\pi}{6}$ **7.** $\dfrac{2\pi}{3},\dfrac{4\pi}{3}$ **9.** $\dfrac{3\pi}{4},\dfrac{5\pi}{4}$ **11.** $\dfrac{\pi}{6},\dfrac{7\pi}{6}$ **13.** 0

15. $1.10,4.24$ **17.** $\dfrac{2\pi}{3}+2n\pi,\dfrac{5\pi}{3}+2n\pi$, for every integer n

19. $\dfrac{5\pi}{6}+2n\pi,\dfrac{7\pi}{6}+2n\pi$, for every integer n **21.** $\dfrac{\pi}{6}+2n\pi,\dfrac{5\pi}{6}+2n\pi$, for every integer n

23. $\pi+2n\pi$ or $(2n+1)\pi$, for every integer n **25.** $\dfrac{\pi}{6}+2n\pi,\dfrac{7\pi}{6}+2n\pi$, for every integer n

27. $\dfrac{\pi}{6}+2n\pi,\dfrac{11\pi}{6}+2n\pi$, for every integer n **29.** $2n\pi$, for every integer n

31. $4.17+2n\pi,5.25+2n\pi$, for every integer n **33.** $x=\frac13\arcsin y$ **35.** $x=\arccos\dfrac{y}{4}$

37. $x=2\arctan\dfrac{y}{5}$ **39.** $x=4\,\text{arccot}\dfrac{2y}{3}$ **41.** $x=1+\arcsin\dfrac{y}{3}$ **43.** $x=-\frac13+\frac13\arccos 2y$

45. $\dfrac{\pi}{3}$ **47.** $-\dfrac{\pi}{6}$ **49.** $\dfrac{5\pi}{6}$ **51.** $\dfrac{2\pi}{3}$ **53.** $\dfrac{\pi}{4}$ **55.** $\dfrac{\pi}{3}$

57. $\dfrac{\pi}{4}$ **59.** $-\dfrac{\pi}{3}$ **61.** $\frac12$ **63.** $\dfrac{1}{\sqrt{2}}$ **65.** 0 **67.** $\dfrac{\sqrt{3}}{2}$ **69.** 0.8

71. -0.1579 **73.** $\sqrt{1-x^2}$ **75.** $\dfrac{\sqrt{x^2-1}}{x}$ **77.** $\dfrac{1}{x}$ **79.** x **81.** $\sqrt{1-4x^2}$ **83.** $2x\sqrt{1-x^2}$

Pages 258–259

1. $\dfrac{5}{\sqrt{1-25x^2}}$ **3.** $\dfrac{3}{1+9x^2}$ **5.** $\dfrac{1}{|1-x|\sqrt{x^2-2x}}$ **7.** $\dfrac{-3}{\sqrt{2x-x^2}}$ **9.** $\dfrac{-12x}{1+9x^4}$

11. $\dfrac{15}{|x|\sqrt{x^6-1}}$ **13.** $\dfrac{3\,\text{Arcsin}^2 x}{\sqrt{1-x^2}}$ **15.** $\dfrac{-12\,\text{Arccos}3x}{\sqrt{1-9x^2}}$ **17.** $\dfrac{12\,\text{Arctan}^3\sqrt{x}}{\sqrt{x}(1+x)}$ **19.** 0 **21.** $\dfrac{1-x}{\sqrt{1-x^2}}$

23. $\dfrac{3x}{\sqrt{1-9x^2}}+\text{Arcsin}\,3x$ **25.** $\dfrac{x}{1+x^2}+\text{Arctan}\,x$ **27.** $\text{Arcsin}\,x$ **29.** $\dfrac{\sqrt{1-x^2}\,\text{Arcsin}\,x-x}{\sqrt{1-x^2}\,\text{Arcsin}^2 x}$

31. $2/\sqrt{3}$ **33.** $\dfrac{-\pi-2}{4}$

1.

3.

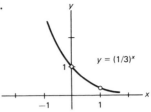

5.

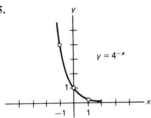

7.

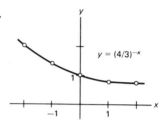

9. $\log_3 9 = 2$ **11.** $\log_5 125 = 3$ **13.** $\log_9 3 = \frac{1}{2}$ **15.** $\log_{10} 0.00001 = -5$ **17.** $5^2 = 25$
19. $25^{1/2} = 5$ **21.** $2^{-2} = \frac{1}{4}$ **23.** $10^{-2} = 0.01$

25.

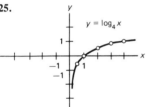

27.

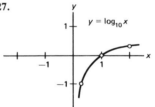

29.

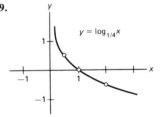

31. 64 **33.** $\frac{1}{2}$ **35.** 3 **37.** $\frac{1}{2}$ **39.** 5 **41.** 3 **43.** 144 **45.** 27 **47.** 4
49. $\log_2 5 + 3 \log_2 x + \log_2 y$ **51.** $3 \log_b y + \frac{1}{2} \log_b x - 2 \log_b z$ **53.** $\frac{2}{3} \log_b x - \frac{1}{3} \log_b y$
55. $\frac{1}{2} \log_2 y - \log_2 x - \frac{1}{2} \log_2 z$ **57.** $3 \log_b z + \frac{1}{2} \log_b x - \frac{1}{3} \log_b y$
59. $2 \log_b x + \log_b (x + 1) - \frac{1}{2} \log_b (x + 2)$ **61.** $\log_b xy^2$
63. $\log_b \dfrac{xy^2}{z^3}$ **65.** $\log_3 \dfrac{x\sqrt[3]{y}}{\sqrt{z}}$ **67.** $\log_{10} \dfrac{x^2}{(x + 1)\sqrt{x - 3}}$ **69.** $\log_b \dfrac{x^5\sqrt[3]{x - 1}}{x + 2}$
71. $\log_{10} \dfrac{x(x - 1)^2}{\sqrt[3]{(x + 2)(x - 5)}}$ **73.** 3 **75.** 2 **77.** 3 **79.** -2 **81.** -3 **83.** 0 **85.** 5

87. 36 **89.** $\frac{1}{25}$

Page 271

1. $\dfrac{4 \log e}{4x - 3}$ **3.** $\dfrac{\log_2 e}{x}$ **5.** $\dfrac{6x^2}{2x^3 - 3}$ **7.** $\dfrac{3 \sec^2 3x}{\tan 3x}$ or $3 \sec 3x \csc 3x$ **9.** $\dfrac{x \cos x + \sin x}{x \sin x}$
11. $\dfrac{3}{2(3x - 2)}$ **13.** $\dfrac{x^2 + 3}{x(x^2 + 1)}$ **15.** $\dfrac{\sec^2(\ln x)}{x}$ **17.** $\dfrac{1}{x \ln x}$ **19.** $\dfrac{2}{x(1 + \ln^2 x)}$
21. $\dfrac{-2}{(\text{Arccos } x)\sqrt{1 - x^2}}$ **23.** $(3x + 2)(6x - 1)^2(x - 4)\left[\dfrac{3}{3x + 2} + \dfrac{12}{6x - 1} + \dfrac{1}{x - 4}\right]$

Answers to Odd-Numbered Exercises and Chapter Reviews

25. $\dfrac{(x + 1)(2x + 1)}{(3x - 4)(1 - 8x)}\left[\dfrac{1}{x + 1} + \dfrac{2}{2x + 1} - \dfrac{3}{3x - 4} + \dfrac{8}{1 - 8x}\right]$ **27.** $x^x(1 + \ln x)$

29. $2x^{2/x}\left(\dfrac{1 - \ln x}{x^2}\right)$ **31.** $(\sin x)^x[x \cot x + \ln (\sin x)]$ **33.** $(1 + x)^{x^2}\left[\dfrac{x^2}{1 + x} + 2x \ln (1 + x)\right]$

35. $y = x - 1$ **37.** $y = \sqrt{3}(x - \pi/6) - \ln 2$

Pages 272–273

1. $5e^{5x}$ **3.** $12x^2e^{x^3}$ **5.** $\dfrac{3(10^{3x})}{\log_{10} e}$ **7.** $\dfrac{-6}{e^{6x}}$ **9.** $\dfrac{e^{\sqrt{x}}}{2\sqrt{x}}$ **11.** $(\cos x)e^{\sin x}$ **13.** $6e^{x^2-1}(2x^2 + 1)$

15. $e^{3x^2}(6x \cos x - \sin x)$ **17.** $-5e^{5x} \tan e^{5x}$ **19.** $e^x + e^{-x}$

21. $\dfrac{6xe^x(2 - x) - 2x^2}{(3e^x - x)^2}$ or $\dfrac{12xe^x - 6x^2e^x - 2x^2}{(3e^x - x)^2}$ **23.** $\dfrac{6e^{3x}}{1 + e^{6x}}$ **25.** $\dfrac{6e^{-2x} \text{Arccos}^2 e^{-2x}}{\sqrt{1 - e^{-4x}}}$

27. xe^x **29.** $2x$

Pages 275–276

1.

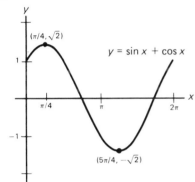

3.

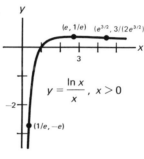

5.

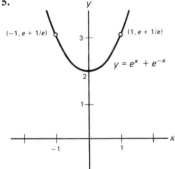

7.

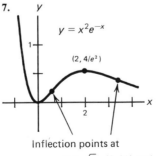

9.

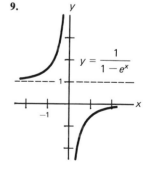

11. $\left(\dfrac{1}{2}, \dfrac{1}{2e}\right)$ maximum   **13.** $(1, e^{-2})$ maximum; $(0, 0)$ minimum   **15.** $\left(e^{1/2}, \dfrac{1}{2e}\right)$ maximum

17. $y = -2x + \pi$ **19.** $y = 3x - 2$ **21.** $-4e^x \sin x$ **23.** $V_L = -300 \sin 2t$ **25.** 33.2 V

27. 2.12 W/s   **29.** $-0.1e^{-0.02t}(\sin 2t + 0.01 \cos 2t)$ **31.** $v = 6e^{3t} - 15e^{-3t}$, $a = 18e^{3t} + 45e^{-3t}$

33. $v = \dfrac{\pi}{2e^2}$, $a = -\dfrac{\pi}{2e^2}$ **35.** -0.0614

Pages 279–281
(Chapter Review)

11. $\frac{1}{2} \sin 2\theta$ **12.** $\cos 6\theta$ **13.** $\cos^2 2\theta$ **14.** $\cos \dfrac{2\theta}{3}$ **15.** $\cos 5x$ **16.** $\sin x$

17.

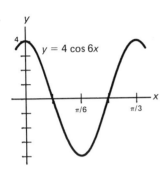

$y = 4 \cos 6x$

18.

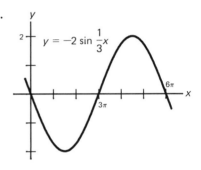

$y = -2 \sin \dfrac{1}{3}x$

19.

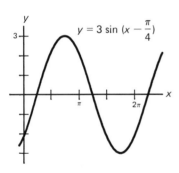

$y = 3 \sin \left(x - \dfrac{\pi}{4}\right)$

20.

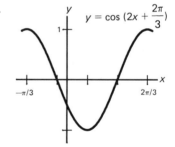

$y = \cos \left(2x + \dfrac{2\pi}{3}\right)$

23.

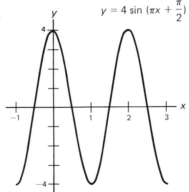

$y = 4 \sin \left(\pi x + \dfrac{\pi}{2}\right)$

21.

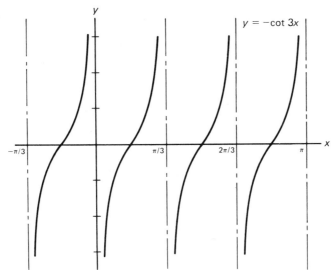

$y = -\cot 3x$

22.

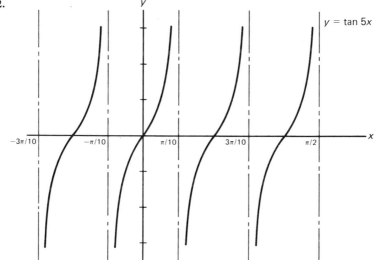

$y = \tan 5x$

Answers to Odd-Numbered Exercises and Chapter Reviews

24.

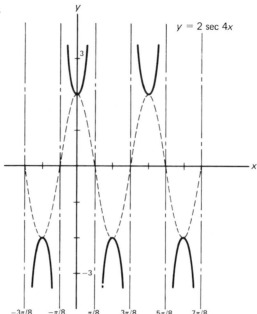

$y = 2 \sec 4x$

25. $2x \cos (x^2 + 3)$ **26.** $-8 \sin 8x$ **27.** $-15 \cos^2(5x - 1) \sin (5x - 1)$

28. $-2 \sin 2x \sin 3x + 3 \cos 2x \cos 3x$ **29.** $3 \sec^2(3x - 2)$ **30.** $4 \sec (4x + 3) \tan (4x + 3)$

31. $-12x \csc^2 6x^2$ **32.** $-2(16x + 1) \csc^2(8x^2 + x) \cot (8x^2 + x)$ **33.** $2 \sec^3 x - \sec x$

34. $2x + 2 \csc^2 x \cot x$ **35.** $\sec x \tan x \sec^2(\sec x)$ **36.** $\dfrac{-1}{1 + \sin x}$ **37.** $-3 \cos x (1 - \sin x)^2$

38. $8(1 + \sec 4x) \sec 4x \tan 4x$ **39.** $\dfrac{\pi}{6}, \dfrac{5\pi}{6}$ **40.** $\dfrac{5\pi}{6}, \dfrac{11\pi}{6}$ **41.** $\dfrac{\pi}{4}, \dfrac{7\pi}{4}$

42. $\dfrac{2\pi}{3}, \dfrac{4\pi}{3}$ **43.** $\dfrac{2\pi}{3} + 2n\pi, \dfrac{5\pi}{3} + 2n\pi$, for every integer n

44. $\dfrac{3\pi}{2} + 2n\pi$, for every integer n **45.** $x = \frac{1}{4} \arcsin 2y$ **46.** $x = \dfrac{1}{2}\left(1 - \arctan \dfrac{y}{5}\right)$ **47.** $\dfrac{\pi}{4}$

48. $-\dfrac{\pi}{6}$ **49.** π **50.** $\dfrac{2\pi}{3}$ **51.** $\dfrac{\sqrt{3}}{2}$ **52.** $\sqrt{3}$ **53.** $\dfrac{\sqrt{x^2 + 1}}{x^2 + 1}$ **54.** $\dfrac{3x^2}{\sqrt{1 - x^6}}$

55. $\dfrac{3}{1 + 9x^2}$ **56.** $\dfrac{3}{|x|\sqrt{4x^2 - 1}}$ **57.** $\dfrac{2}{|x|\sqrt{16x^2 - 1}}$ **58.** $\dfrac{3 \operatorname{Arcsin} 3\sqrt{x}}{\sqrt{x - 9x^2}}$ **59.** $\dfrac{x}{\sqrt{1 - x^2}} + \operatorname{Arcsin} x$

60. **61.**

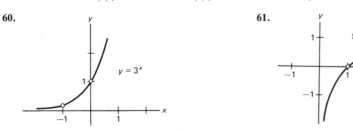

$y = 3^x$

$y = \log_3 x$

62. 81 **63.** 2 **64.** 5 **65.** $\log_4 6 + 2 \log_4 x + \log_4 y$ **66.** $\log_3 5 + \log_3 x + \frac{1}{2} \log_3 y - 3 \log_3 z$

67. $2 \log x + 3 \log (x + 1) - \frac{1}{2} \log (x - 4)$ **68.** $3 \ln x + 3 \ln (x - 1) - \frac{1}{2} \ln (x + 1)$ **69.** $\log_2 \dfrac{xy^3}{z^2}$

70. $\log \dfrac{\sqrt{x+1}}{(x-2)^3}$ **71.** $\ln \dfrac{x^4}{(x+1)^5(x+2)}$ **72.** $\ln \dfrac{\sqrt{x(x+2)}}{(x-5)^2}$ **73.** 3   **74.** x^2 **75.** 2   **76.** x

77. $\dfrac{3x^2}{x^3-2}$ **78.** $\dfrac{4\log_3 e}{4x+1}$ **79.** $\dfrac{6}{x^3+3x}$ **80.** $-\dfrac{1}{x}\sin(\ln x)$

81. $\dfrac{\sqrt{x+1}(3x-4)}{x^2(x+2)}\left[\dfrac{1}{2(x+1)}+\dfrac{3}{3x-4}-\dfrac{2}{x}-\dfrac{1}{x+2}\right]$ **82.** $x^{1-x}\left[\dfrac{1-x}{x}-\ln x\right]$ **83.** $2xe^{x^2+5}$

84. $\dfrac{3(8^{3x})}{\log_8 e}$ **85.** $-2e^{2x}\csc^2 e^{2x}$ **86.** $2x\cos x^2 e^{\sin x^2}$ **87.** $\dfrac{-4e^{-4x}}{\sqrt{1-e^{-8x}}}$ **88.** $x^2 e^{-4x}(3-4x)$

89.

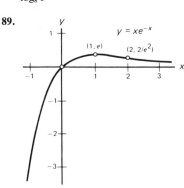

90.

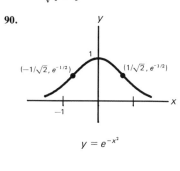

$y = e^{-x^2}$

91. $-1000\sin 5t$ **92.** $p = -360\cos 3t\sin 3t$ or $-180\sin 6t$ **93.** $y = 2x - 2$
94. $v = (\cos t)e^{\sin t}$ **95.** $-4e^{-1/2}$ or -2.43

CHAPTER 7

Pages 284–285

1. $\frac{2}{9}(3x+2)^{3/2}+C$ **3.** $2\sqrt{4+x}+C$ **5.** $\frac{2}{7}(x^2+4x)^{7/4}+C$ **7.** $-\frac{1}{4}\cos^4 x+C$
9. $\frac{1}{16}\tan^4 4x+C$ **11.** $-\frac{1}{8}(\cos 4x+1)^2+C$ **13.** $\frac{2}{3}(9+\sec x)^{3/2}+C$ **15.** $\frac{1}{3}(1+e^{2x})^{3/2}+C$
17. $\sqrt{1+e^{x^2}}+C$ **19.** $\frac{1}{6}\ln^2|3x-5|+C$ **21.** $-\dfrac{1}{\ln|x|}+C$ **23.** $\frac{1}{6}\text{Arcsin}^2 3x+C$
25. $\frac{1}{4}\sin^4 x+C$ **27.** $\frac{1}{3}\text{Arctan}^3 x+C$ **29.** $\frac{64}{3}$ **31.** $\frac{2}{3}(\sqrt{1+e^3}-\sqrt{2})$ **33.** $\frac{1}{4}\ln^2 3$ **35.** $\frac{1}{3}$

Pages 288–289

1. $\frac{1}{3}\ln|3x+2|+C$ **3.** $-\frac{1}{4}\ln|1-4x|+C$ **5.** $-2\ln|1-x^2|+C$ **7.** $\frac{1}{4}\ln|x^4-1|+C$
9. $-\ln|\cot x|+C$ or $\ln|\tan x|+C$ **11.** $\frac{1}{3}\ln|1+\tan 3x|+C$ **13.** $-\ln|1+\csc x|+C$
15. $\ln|1+\sin x|+C$ **17.** $\ln|\ln|x||+C$ **19.** $\frac{1}{2}e^{2x}+C$ **21.** $\dfrac{-1}{4e^{4x}}+C$ **23.** $\frac{1}{2}e^{x^2}+C$
25. $-\frac{1}{2}e^{-x^2-9}+C$ **27.** $-e^{\cos x}+C$ **29.** $\dfrac{e^2}{2}(e^4-1)$ **31.** $\dfrac{4^x}{\ln 4}+C$ **33.** $2\ln|e^x+4|+C$
35. $\frac{1}{2}\ln 2$ or 0.347 **37.** $2\ln 9$ or $\ln 81$ or 4.39 **39.** $2(e-1)$ or 3.44 **41.** $\frac{1}{3}(e-1)$ or 0.573
43. $\ln 2$ or 0.693 **45.** $\frac{1}{2}\ln 3$ or 0.549 **47.** $\frac{1}{2}(e^8-1)$ or 1490

Page 294

1. $-\frac{1}{5}\cos 5x+C$ **3.** $\frac{1}{3}\sin(3x-1)+C$ **5.** $-\frac{1}{2}\cos(x^2+5)+C$ **7.** $\sin(x^3-x^2)+C$
9. $-\frac{1}{5}\cot 5x+C$ **11.** $\frac{1}{3}\sec 3x+C$ **13.** $\frac{1}{4}\tan(4x+3)+C$ **15.** $-\frac{1}{2}\csc(2x-3)+C$
17. $\frac{1}{2}\tan(x^2+3)+C$ **19.** $-\frac{1}{3}\csc(x^3-1)+C$ **21.** $-\frac{1}{4}\ln|\cos 4x|+C$
23. $\frac{1}{5}\ln|\sec 5x+\tan 5x|+C$ **25.** $\ln|\sin e^x|+C$ **27.** $x+2\ln|\sec x+\tan x|+\tan x+C$

Answers to Odd-Numbered Exercises and Chapter Reviews

29. $5 \ln |\sec x + \tan x| - \ln |\cos x| + C$ **31.** $\frac{1}{2}$ **33.** 3 **35.** 1 **37.** $\frac{\sqrt{2} - 1}{2}$ **39.** 2

41. 1 **43.** $\frac{1}{2} \ln 2$ **45.** π

Pages 297–298

1. $\frac{1}{3} \cos^3 x - \cos x + C$ **3.** $\sin x - \frac{2}{3} \sin^3 x + \frac{1}{5} \sin^5 x + C$ **5.** $\frac{1}{3} \sin^3 x + C$ **7.** $\frac{-1}{2 \cos^2 x} + C$

9. $-\frac{1}{3} \cos^3 x + \frac{1}{5} \cos^5 x + C$ **11.** $\frac{x}{2} - \frac{1}{4} \sin 2x + C$ **13.** $\frac{3x}{8} + \frac{1}{12} \sin 6x + \frac{1}{96} \sin 12x + C$

15. $\frac{x}{8} - \frac{1}{32} \sin 4x + C$ **17.** $\frac{x}{16} - \frac{1}{64} \sin 4x + \frac{1}{48} \sin^3 2x + C$ **19.** $\frac{1}{2} \tan^2 x + \ln |\cos x| + C$

21. $x - \frac{1}{6} \cot^3 2x + \frac{1}{2} \cot 2x + C$ **23.** $\tan x + \frac{2}{3} \tan^3 x + \frac{1}{5} \tan^5 x + C$

25. $\frac{1}{6} \tan^3 2x - \frac{1}{2} \tan 2x + x + C$ **27.** $\frac{\pi}{2}$

Page 300

1. $\frac{1}{3} \text{Arcsin } 3x + C$ **3.** $\text{Arcsin } \frac{x}{3} + C$ **5.** $\frac{1}{5} \text{Arctan } \frac{x}{5} + C$ **7.** $\frac{1}{8} \text{Arctan } \frac{3x}{2} + C$

9. $\frac{1}{5} \text{Arcsin } \frac{5x}{6} + C$ **11.** $\frac{1}{2\sqrt{3}} \text{Arcsin } 2x + C$ **13.** $\frac{1}{2} \text{Arctan } \left(\frac{x - 1}{2} \right) + C$

15. $\frac{1}{4} \text{Arctan } \left(\frac{x + 3}{4} \right) + C$ **17.** $\text{Arcsin } e^x + C$ **19.** $-\text{Arctan } (\cos x) + C$

21. $\frac{\pi}{4}$ **23.** 0.215 **25.** 10.9 N

Page 307

1. $\frac{5}{x + 2} + \frac{3}{x - 7}$ **3.** $\frac{3}{2x + 3} - \frac{2}{x - 4}$ **5.** $\frac{7}{x} + \frac{2}{3x - 4} + \frac{5}{2x + 1}$ **7.** $\frac{1}{x + 1} + \frac{1}{(x + 3)^2}$

9. $\frac{3}{4x - 1} + \frac{1}{(4x - 1)^2} - \frac{7}{(4x - 1)^3}$ **11.** $\frac{3}{x} + \frac{8}{x - 1} - \frac{4}{(x - 1)^2}$

13. $\frac{x - 1}{x^2 + 1} - \frac{x}{x^2 - 3}$ **15.** $\frac{4x + 1}{x^2 + x + 1} - \frac{1}{x^2 - 5}$ **17.** $\frac{5x - 2}{x^2 + 5x + 3} + \frac{1}{x + 3} - \frac{2}{x - 3}$

19. $\frac{5}{x} + \frac{3x - 1}{x^2 + 1} - \frac{5}{(x^2 + 1)^2}$ **21.** $\frac{1}{x} - \frac{2}{x^2} - \frac{4x}{(x^2 + 2)^2}$ **23.** $\frac{2}{x + 3} + \frac{4}{x - 3} - \frac{6x}{x^2 + 9}$

25. $x + \frac{\frac{1}{2}}{x + 1} + \frac{\frac{1}{2}}{x - 1}$ **27.** $x - 1 + \frac{3}{x - 2} + \frac{1}{x + 2}$ **29.** $3x - 2 + \frac{5}{x} - \frac{8x}{x^2 + 1}$

Page 310

1. $\frac{1}{2} \ln \left| \frac{x + 1}{x - 1} \right| + C$ **3.** $\frac{1}{6} \ln \left| \frac{x - 2}{x + 4} \right| + C$ **5.** $2 \ln |x - 2| - \ln |x - 1| + C$ or $\ln \left| \frac{(x - 2)^2}{x - 1} \right| + C$

7. $\frac{2}{3} \ln |x + 5| + \frac{1}{3} \ln |x - 1| + C$ or $\frac{1}{3} \ln |(x + 5)^2 (x - 1)| + C$ **9.** $\ln \left| \frac{x}{x + 1} \right| + \frac{1}{x + 1} + C$

11. $2 \ln |x + 3| - \frac{1}{x} + C$ **13.** $\frac{x^2}{2} - 3x + \ln \left| \frac{(x + 2)^8}{x + 1} \right| + C$ **15.** $\frac{3}{2} \ln |x^2 + 1| - 2 \ln |x| + C$

17. $\frac{1}{2} \ln |x^2 + 9| + \text{Arctan } \frac{x}{3} + \frac{1}{x} + C$ **19.** $\frac{1}{2(x^2 + 1)} + \frac{1}{2} \ln |x^2 + 1| + C$ **21.** $\frac{3}{2} \ln \frac{2}{3}$

23. $\frac{11}{6} \ln 3 - \frac{5}{6} \ln 7$ **25.** $\ln 3 + 3 \ln \frac{7}{5}$

Page 315

1. $x \ln |x| - x + C$ **3.** $xe^x - e^x + C$ **5.** $\frac{2}{3}x^{3/2} \ln |x| - \frac{4}{9}x^{3/2} + C$
7. $x \ln x^2 - 2x + C$ **9.** $x \operatorname{Arccos} x - \sqrt{1 - x^2} + C$ **11.** $\frac{1}{2}e^x(\sin x + \cos x) + C$
13. $x^2 \sin x + 2x \cos x - 2 \sin x + C$ **15.** $x \tan x + \ln |\cos x| + C$
17. $x(\ln |x|)^2 - 2x \ln |x| + 2x + C$ **19.** $x \sec x - \ln |\sec x + \tan x| + C$ **21.** $\frac{1}{3}(2e^3 + 1)$
23. $\frac{18}{15}$ **25.** $3 \ln 3 - 2 \ln 2 - 1$ or $\ln \frac{27}{4} - 1$ **27.** $\ln 2 - \frac{1}{2}$ **29.** 2π

Pages 318–319

1. $\frac{1}{2} \ln |2x + \sqrt{9 + 4x^2}| + C$ **3.** $-\frac{x}{18}\sqrt{4 - 9x^2} + \frac{2}{27} \operatorname{Arcsin} \frac{3x}{2} + C$ **5.** $\operatorname{Arcsin} \frac{1}{3}$

7. $\frac{1}{4\sqrt{3}}$ **9.** $\frac{1}{2} \ln \left| \frac{\sqrt{x^2 + 4} - 2}{x} \right| + C$ **11.** $\ln |x + \sqrt{x^2 - 9}| - \frac{\sqrt{x^2 - 9}}{x} + C$

13. $\frac{-\sqrt{16 - x^2}}{16x} + C$ **15.** $\sqrt{9 + x^2} - 3 \ln \left| \frac{3 + \sqrt{9 + x^2}}{x} \right| + C$ **17.** $\frac{x}{25\sqrt{25 - x^2}} + C$

19. $\ln |x + \sqrt{x^2 + 9}| + C$ **21.** $\frac{9x^2 - 8}{243}\sqrt{9x^2 + 4} + C$ **23.** $\ln |x - 3 + \sqrt{x^2 - 6x + 8}| + C$

25. $-\frac{x + 4}{\sqrt{x^2 + 8x + 15}} + C$ **27.** $\ln (1 + \sqrt{2})$

Page 321

1. $\frac{1}{\sqrt{5}} \ln \left| \frac{\sqrt{x + 5} - \sqrt{5}}{\sqrt{x + 5} + \sqrt{5}} \right| + C; 15$ **3.** $\ln |x + \sqrt{x^2 - 4}| + C; 35$ **5.** $-\frac{(3 - x)\sqrt{2x + 3}}{3}; 13$

7. $\frac{1}{4}\left(\frac{\sin 4x}{2} - \frac{\sin 10x}{5} \right) + C; 75$ **9.** $-\frac{x}{2}\sqrt{9 - x^2} + \frac{9}{2} \operatorname{Arcsin} \frac{x}{3} + C; 24$

11. $\frac{1}{1 + 9x} + \ln \left| \frac{x}{1 + 9x} \right| + C; 9$ **13.** $\frac{1}{10} \ln \left| \frac{x - 5}{x + 5} \right| + C; 20$

15. $\frac{x}{2}\sqrt{x^2 + 4} + 2 \ln |x + \sqrt{x^2 + 4}| + C; 30$

17. $\frac{1}{16}\left(\frac{3}{3 + 4x} + \ln |3 + 4x| \right) + C; 7$ **19.** $\frac{1}{4} \operatorname{Arccos} \frac{4}{3x} + C; 36$

21. $\frac{e^{3x}}{25}(3 \sin 4x - 4 \cos 4x) + C; 59$ **23.** $\frac{1}{2} \sin (2x - 3) - \frac{(2x - 3)}{2} \cos (2x - 3) + C; 80$

25. $-\frac{1}{4} \sin^3 x \cos x + \frac{3x}{8} - \frac{3}{8} \sin x \cos x + C; 83$ **27.** $\sqrt{9x^2 - 16} - 4 \operatorname{Arccos} \frac{4}{3x} + C; 33$

Pages 326–330

1. 1.117   **3.** 0.783   **5.** 1.913   **7.** 7.395   **9.** 0.925   **11.** 219 ft-lb   **13.** 270
15. 1.443   **17.** 0.105   **19.** 1.464   **21.** 0.186   **23.** 1.910   **25.** (a) 6.889 (b) 6.998
27. (a) 1.023 (b) 1.000   **29.** (a) 1.006 (b) 1.006   **31.** (a) 373,650 ft² (b) 378,200 ft²

Pages 332–334
(Chapter Review)

1. $\frac{2}{3}\sqrt{2 + \sin 3x} + C$ **2.** $\frac{(5 + \tan 2x)^4}{8} + C$ **3.** $\frac{1}{3} \sin 3x + C$ **4.** $\frac{1}{2} \ln |x^2 - 5| + C$

5. $\frac{1}{6}e^{3x^2} + C$ **6.** $\frac{1}{3} \operatorname{Arctan} \frac{2x}{3} + C$ **7.** $\operatorname{Arcsin} \frac{x}{4} + C$ **8.** $\frac{1}{7} \tan (7x + 2) + C$

9. $\frac{1}{5} \ln |3 + 5 \tan x| + C$ **10.** $-\frac{1}{3} \cos (x^3 + 4) + C$ **11.** $\frac{1}{4}(\operatorname{Arctan} \frac{3}{4})$

12. $\frac{1}{\pi}$ **13.** $-\ln \frac{\sqrt{2}}{2}$ or $\ln \sqrt{2}$ **14.** $\operatorname{Arcsin} \frac{2}{3}$ **15.** $\frac{1}{3} \operatorname{Arccos} \frac{3}{4x} + C$

16. $-\frac{1}{12}\cos^4 3x + C$ **17.** $\frac{1}{6}\text{Arctan}^2 3x + C$ **18.** $\frac{1}{5}\ln|\sec 5x + \tan 5x| + C$

19. $-\frac{1}{2}\ln|\cos x^2| + C$ **20.** $-\dfrac{\cos^3 2x}{6} + \dfrac{1}{5}\cos^5 2x - \dfrac{\cos^7 2x}{14} + C$

21. $\dfrac{x}{16} - \dfrac{1}{192}\sin 12x + \dfrac{1}{144}\sin^3 6x + C$ **22.** $\dfrac{1}{4}\ln\left|\dfrac{x-1}{x+3}\right| + C$ **23.** $\dfrac{e^{3x}}{25}(3\cos 4x + 4\sin 4x)$

24. $\cos x(1 - \ln|\cos x|) + C$ **25.** $\dfrac{\tan^3 x}{3} - \tan x + x + C$ **26.** $\dfrac{x}{2} + \dfrac{\sin 10x}{20} + C$

27. $\frac{2}{3}\ln|3x+1| - \frac{1}{2}\ln|2x-1| + C$ **28.** $\frac{2}{3}x^{3/2}(\ln|x| - \frac{2}{3}) + C$ **29.** $\dfrac{\cos^3(e^{-x})}{3} + C$

30. $\ln\left|\dfrac{x-2}{x+2}\right| + C$ **31.** $-x^2\cos x + 2x\sin x + 2\cos x + C$

32. $\frac{1}{10}(\text{Arcsin } 5x)^2 + C$ **33.** Arcsec 3 − Arcsec 2 **34.** $\dfrac{3e^4 + 1}{16}$ **35.** ln 2 **36.** 2 ln 3 or ln 9

37. $\dfrac{-1}{(x-2)} + 3\ln|x| + C$ **38.** $\dfrac{1}{4}\ln\left|\dfrac{4 - \sqrt{16 - 9x^2}}{3x}\right| + C$ **39.** ln 3 **40.** $e^5 - e^3$ or $e^3(e^2 - 1)$

41. $\frac{1}{2}(e^2 - 1)$ **42.** $\dfrac{\pi}{4}$ **43.** $\dfrac{\pi}{6}$ **44.** $\frac{1}{2} + \ln 2$ **45.** 1 **46.** $\dfrac{\pi}{8}$ **47.** $\frac{1}{2}\ln 2$

48. $\pi(2\ln^2 2 - 4\ln 2 + 2)$ **49.** $\dfrac{\pi}{2}(e^2 - 1)$ **50.** $-\dfrac{4t}{3}\cos 3t + \dfrac{4}{9}\sin 3t + C$

51. $3\ln|t+2| + 2\ln|t-1| + C$ **52.** $\dfrac{e}{2}(e^3 - 1)$ **53.** 1.011 **54.** 0.155 **55.** 12.359

56. 6.489 **57.** 22.2 **58.** 0.96 **59.** 18.365 **60.** 5.608 **61.** 0.605 **62.** 72.632

CHAPTER 8
Pages 343–347

1.

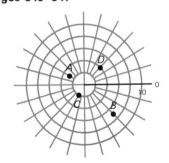

3.

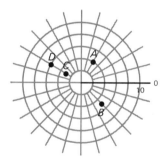

5. $(-3, 240°), (-3, -120°), (3, -300°)$ **7.** $(5, 135°), (-5, -45°), (5, -225°)$ **9.** $(-4, -315°),$ $(-4, 45°), (4, 225°)$ **11.** $(-3, 7\pi/6), (-3, -5\pi/6), (3, -11\pi/6)$ **13.** $(9, 5\pi/3), (9, -\pi/3),$ $(-9, -4\pi/3)$ **15.** $(4, -3\pi/4), (-4, \pi/4), (4, 5\pi/4)$

17.

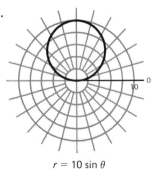

$r = 10\sin\theta$

19.

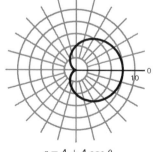

$r = 4 + 4\cos\theta$

21.

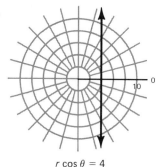

$r \cos \theta = 4$

23.

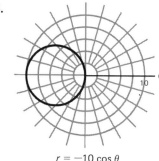

$r = -10 \cos \theta$

25.

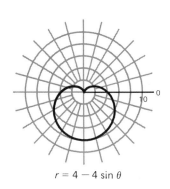

$r = 4 - 4 \sin \theta$

27.

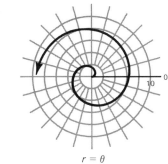

$r = \theta$

29. $(3\sqrt{3}/2, \, 3/2)$ **31.** $(1, \, \sqrt{3})$ **33.** $(2\sqrt{3}, \, -2)$ **35.** $(0, \, 6)$ **37.** $(2.5, \, -4.33)$ **39.** $(1.4, \, 1.4)$
41. $(7.1, \, 45°)$ **43.** $(4, \, 90°)$ **45.** $(4, \, 240°)$ **47.** $(4\sqrt{2}, \, 3\pi/4)$ **49.** $(2\sqrt{2}, \, 5\pi/6)$ **51.** $(4, \, 3\pi/2)$
53. $r \cos \theta = 3$ **55.** $r = 6$ **57.** $r + 2 \cos \theta + 5 \sin \theta = 0$ **59.** $r = 12/(4 \cos \theta - 3 \sin \theta)$
61. $r^2 = 36/(9 - 5 \sin^2 \theta)$ **63.** $r = 4 \sec \theta \tan^2 \theta$ **65.** $y = -3$ **67.** $x^2 + y^2 = 25$
69. $y = x$ **71.** $x^2 + y^2 - 5x = 0$ **73.** $x^2 + y^2 - 3x + 3\sqrt{3}y = 0$ **75.** $y^2 = 3x$
77. $xy = 1$ **79.** $x^4 + 2x^2y^2 + y^4 - 2xy = 0$ **81.** $y^2 = x^2(x^2 + y^2)$ **83.** $x^2 + 6y - 9 = 0$
85. $x^4 + 2x^2y^2 + y^4 + 4y^3 - 12x^2y = 0$ **87.** $x^4 + 2x^2y^2 + y^4 - 8x^2y - 8y^3 - 4x^2 + 12y^2 = 0$
89. $\sqrt{13}$ **91.** $d = \sqrt{r_1^2 + r_2^2 - 2r_1r_2 \cos (\theta_1 - \theta_2)}$

Pages 353–354

1.

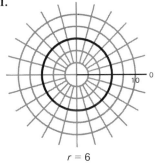

$r = 6$

3.

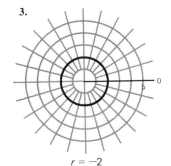

$r = -2$

5.

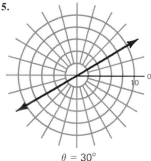

$\theta = 30°$

Answers to Odd-Numbered Exercises and Chapter Reviews

549

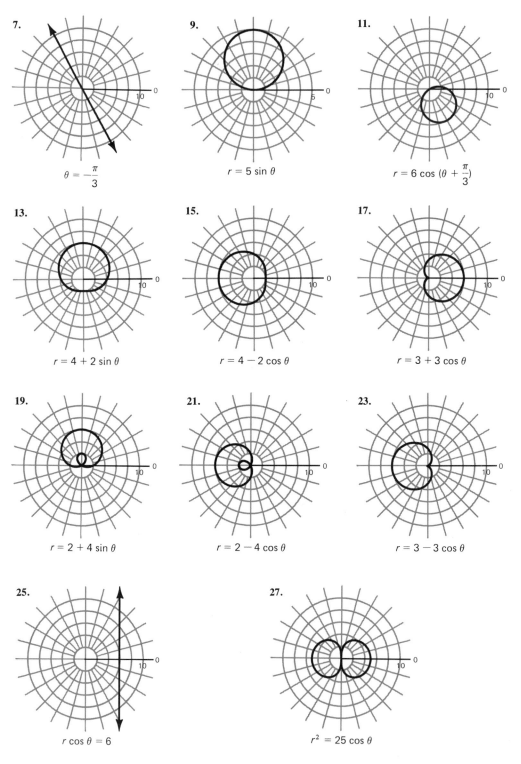

7.

$\theta = -\dfrac{\pi}{3}$

9.

$r = 5 \sin \theta$

11.

$r = 6 \cos \left(\theta + \dfrac{\pi}{3}\right)$

13.

$r = 4 + 2 \sin \theta$

15.

$r = 4 - 2 \cos \theta$

17.

$r = 3 + 3 \cos \theta$

19.

$r = 2 + 4 \sin \theta$

21.

$r = 2 - 4 \cos \theta$

23.

$r = 3 - 3 \cos \theta$

25.

$r \cos \theta = 6$

27.

$r^2 = 25 \cos \theta$

Answers to Odd-Numbered Exercises and Chapter Reviews

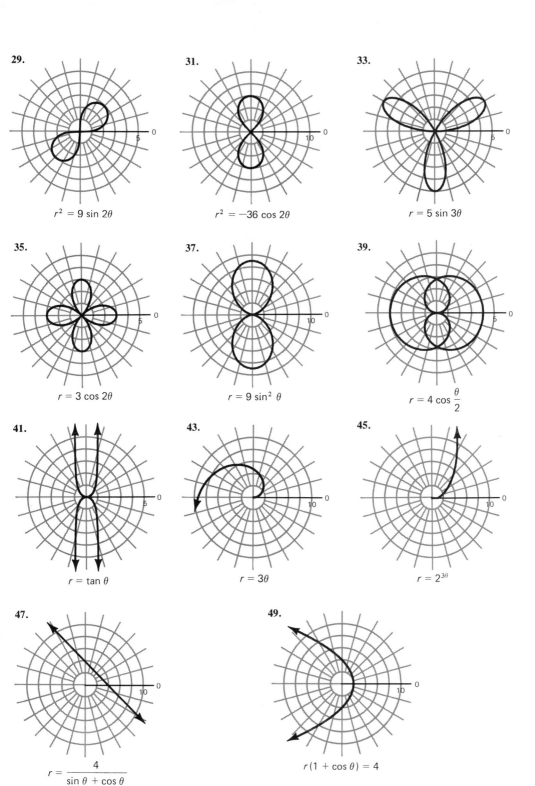

29.

$r^2 = 9 \sin 2\theta$

31.

$r^2 = -36 \cos 2\theta$

33.

$r = 5 \sin 3\theta$

35.

$r = 3 \cos 2\theta$

37.

$r = 9 \sin^2 \theta$

39.

$r = 4 \cos \dfrac{\theta}{2}$

41.

$r = \tan \theta$

43.

$r = 3\theta$

45.

$r = 2^{3\theta}$

47.

$r = \dfrac{4}{\sin \theta + \cos \theta}$

49.

$r(1 + \cos \theta) = 4$

Answers to Odd-Numbered Exercises and Chapter Reviews

551

Page 358

1. $\dfrac{4\pi}{3}$ **3.** $\dfrac{9\pi}{4}$ **5.** 9 **7.** $\dfrac{3\pi}{2}$ **9.** 8π **11.** 8π **13.** $\dfrac{41\pi}{2}$ **15.** $\frac{1}{4}(e^{2\pi} - 1)$

17. $\pi - \dfrac{3\sqrt{3}}{2}$ **19.** 2π **21.** $\dfrac{\pi}{4} - \dfrac{3\sqrt{3}}{16}$ **23.** $\dfrac{2\pi}{3} + \sqrt{3}$ **25.** $\dfrac{11\pi}{3} + \sqrt{3}$ **27.** $9\sqrt{2} + \dfrac{27\pi}{8} + \dfrac{9}{4}$

Pages 359–360
(Chapter Review)

1.

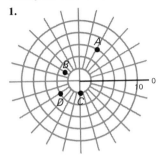

2.

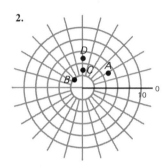

3. $(5, -225°), (-5, -45°), (-5, 315°)$ **4.** $(2, -11\pi/6), (-2, -5\pi/6), (2, \pi/6)$ **5.** $(-2.6, -1.5)$
6. $(-1, -1.7)$ **7.** $(-4.3, 2.5)$ **8.** $(0, 6)$ **9.** $(4.2, 135°)$ **10.** $(6, 270°)$ **11.** $(2, 120°)$
12. $(5, \pi)$ **13.** $(12, 5\pi/6)$ **14.** $(\sqrt{2}, 7\pi/4)$ **15.** $r = 7$ **16.** $r \sin^2 \theta = 9 \cos \theta$
17. $r = 8/(5 \cos \theta + 2 \sin \theta)$ **18.** $r^2 = 12/(1 - 5 \sin^2 \theta)$ **19.** $r = 6 \csc \theta \cot^2\theta$
20. $r = \cos \theta \cot \theta$ **21.** $x = 12$ **22.** $x^2 + y^2 = 81$ **23.** $y = -\sqrt{3}x$ **24.** $x^2 + y^2 - 8x = 0$
25. $y^2 = 5x$ **26.** $xy = 4$ **27.** $x^4 + 2x^2y^2 + y^4 + 4y^2 - 4x^2 = 0$ **28.** $y = 1$
29. $x^4 + y^4 + 2x^2y^2 - 2x^2y - 2y^3 - x^2 = 0$ **30.** $x^2 = 4(y + 1)$

31.

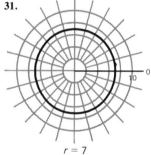

$r = 7$

32.

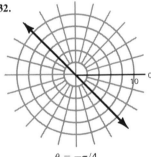

$\theta = -\pi/4$

33.

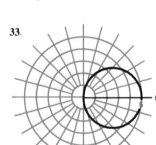

$r = 5 \cos \theta$

34.

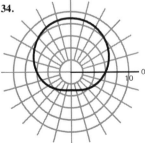

$r = 6 + 3 \sin \theta$

35.

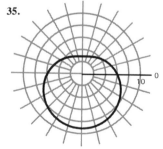

$r = 6 - 3 \sin \theta$

36.

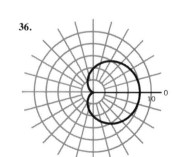

$r = 4 + 4 \cos \theta$

Answers to Odd-Numbered Exercises and Chapter Reviews

37.

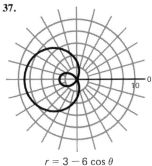

$r = 3 - 6 \cos \theta$

38.

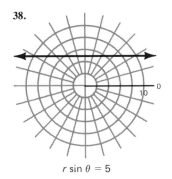

$r \sin \theta = 5$

39.

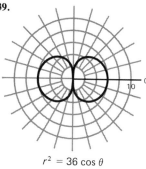

$r^2 = 36 \cos \theta$

40.

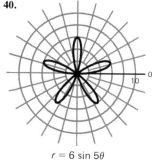

$r = 6 \sin 5\theta$

41.

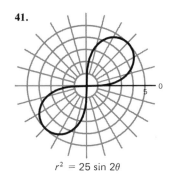

$r^2 = 25 \sin 2\theta$

42.

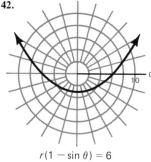

$r(1 - \sin \theta) = 6$

43. 4   **44.** $\dfrac{3\pi}{2}$ **45.** 11π **46.** $\dfrac{\pi}{16}$ **47.** $\dfrac{\pi^3}{8}$ **48.** $\dfrac{\sqrt{3}}{2} + \dfrac{\pi}{3}$ **49.** $4\pi - 6\sqrt{3}$

50. $2 - \pi/4$ **51.** $4\sqrt{3} - \dfrac{4\pi}{3}$ **52.** $1 - \sqrt{2}/2$

CHAPTER 9

Pages 368–370

1.

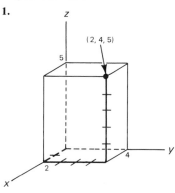

3.

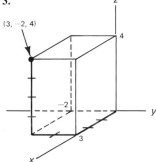

5.

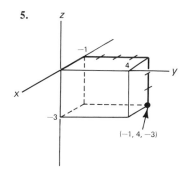

7. Plane

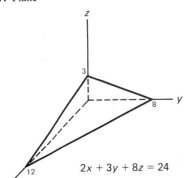

$2x + 3y + 8z = 24$

9. Plane

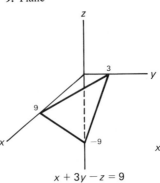

$x + 3y - z = 9$

11. Plane

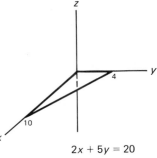

$2x + 5y = 20$

13. Plane

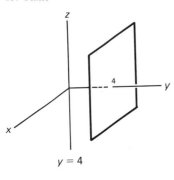

$y = 4$

15. Sphere

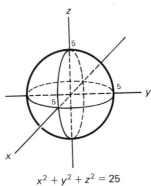

$x^2 + y^2 + z^2 = 25$

17. Sphere

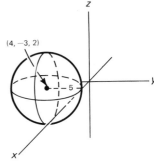

(4, −3, 2)

$x^2 + y^2 + z^2 - 8x + 6y - 4z + 4 = 0$

19. Cylindrical surface

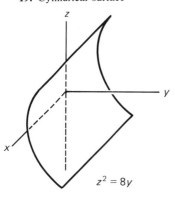

$z^2 = 8y$

21. Cylindrical surface

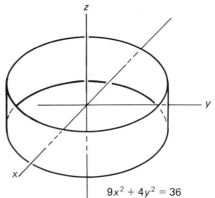

$9x^2 + 4y^2 = 36$

23. Ellipsoid

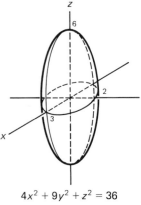

$4x^2 + 9y^2 + z^2 = 36$

Answers to Odd-Numbered Exercises and Chapter Reviews

25. Elliptic paraboloid

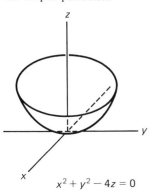

$x^2 + y^2 - 4z = 0$

27. Hyperboloid of two sheets

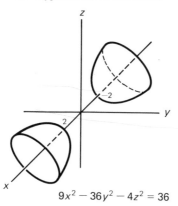

$9x^2 - 36y^2 - 4z^2 = 36$

29. Cylindrical surface

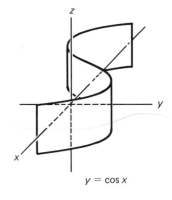

$y = \cos x$

31. Hyperbolic paraboloid

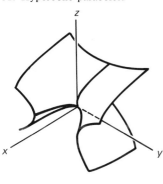

$y^2 - z^2 = 8x$

33. Elliptic cone

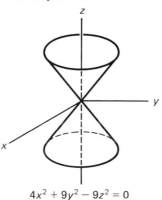

$4x^2 + 9y^2 - 9z^2 = 0$

35. Hyperboloid of one sheet

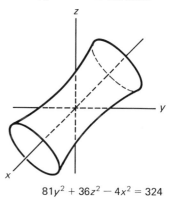

$81y^2 + 36z^2 - 4x^2 = 324$

37. $(x - 3)^2 + (y + 2)^2 + (z - 4)^2 = 36$ **39.** $(x - 3)^2 + (y + 2)^2 + (z - 4)^2 = 16$ **43.** 6

Pages 372–373

1. (a) $12x^2y^2$ (b) $8x^3y$ **3.** (a) $12xy^4 + 2y^2$ (b) $24x^2y^3 + 4xy$ **5.** (a) $\dfrac{x}{\sqrt{x^2 + y^2}}$ (b) $\dfrac{y}{\sqrt{x^2 + y^2}}$

7. (a) $\dfrac{x^2 + y^2}{2x^2y}$ (b) $\dfrac{-x^2 - y^2}{2xy^2}$ **9.** (a) ay (b) ax **11.** (a) $1/x$ (b) $-1/y$

13. (a) $\sec^2(x - y)$ (b) $-\sec^2(x - y)$ **15.** (a) $e^{3x}(y \cos xy + 3 \sin xy)$ (b) $xe^{3x} \cos xy$

17. (a) $\dfrac{e^{xy}(xy - 1)}{x^2 \sin y}$ (b) $\dfrac{e^{xy}(x \sin y - \cos y)}{x \sin^2 y}$ **19.** (a) $-2 \sin x \sin y$ (b) $2 \cos x \cos y$

21. (a) $xy^2 \sec^2xy + y \tan xy$ (b) $x^2y \sec^2xy + x \tan xy$ **23.** $2IR$ **25.** $\dfrac{-E}{(R + r)^2}$ **27.** $\dfrac{R}{\sqrt{R^2 + X_L^2}}$

29. $2\pi Ef \cos 2\pi ft$ **31.** $R_2 + R_3$ **33.** $\dfrac{-E}{R^2C}e^{-t/RC}$ **35.** $Ee^{-t/RC}\left(\dfrac{t}{RC} + 1\right)$ **37.** $\dfrac{-X_L}{R^2 + X_L^2}$

39. (a) 18 (b) -16 **41.** (a) $\frac{5}{3}$ (b) $-\frac{12}{5}$ **45.** 96π cm^3

Pages 378–379

1. $(6x + 4y)\,dx + (4x + 3y^2)\,dy$ **3.** $2x \cos y\,dx - x^2 \sin y\,dy$ **5.** $\dfrac{1}{x^2}dx - \dfrac{1}{y^2}dy$

7. $\dfrac{y}{2(1 + xy)} dx + \dfrac{x}{2(1 + xy)} dy$ **9.** 195 cm³ **11.** 17 Ω **13.** −64.1 cm³

15. Maximum, $(1, -2, 20)$ **17.** Minimum, $(1, 1, 3)$ **19.** Saddle point, $(1, -2, -1)$

21. Minimum, $(1, 1, 3)$; saddle point, $(0, 0, 4)$ **23.** Minimum, $(3, 1, -109)$; saddle point, $(-1, 1, 19)$

25. 10 cm × 10 cm × 5 cm **27.** 10, 10, 10

Page 384

1. 20 **3.** $\frac{17}{8}$ **5.** $\frac{14}{5}$ **7.** $\frac{2}{3}$ **9.** $\frac{1}{2}e^2 - e + \frac{1}{2}$ **11.** $\frac{244}{3}$ **13.** $1 - \pi/4$ **15.** 16 **17.** 2

19. 16

Chapter Review
Pages 386–387

1. Plane

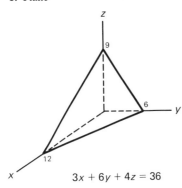

$3x + 6y + 4z = 36$

2. Elliptic paraboloid

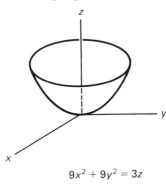

$9x^2 + 9y^2 = 3z$

3. Hyperboloid of one sheet

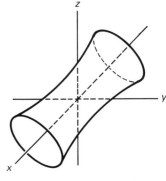

$36y^2 + 9z^2 - 16x^2 = 144$

4. Ellipsoid

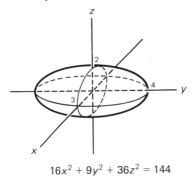

$16x^2 + 9y^2 + 36z^2 = 144$

5. Cylindrical surface

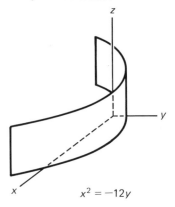

$x^2 = -12y$

Answers to Odd-Numbered Exercises and Chapter Reviews

7. Hyperboloid of two sheets   **8.** Elliptic cone

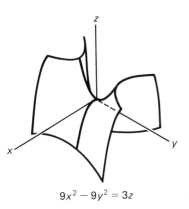

$$9x^2 - 9y^2 = 3z$$

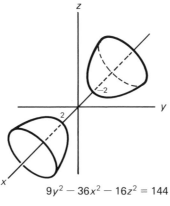

$$9y^2 - 36x^2 - 16z^2 = 144$$

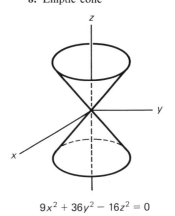

$$9x^2 + 36y^2 - 16z^2 = 0$$

9. $(x + 3)^2 + (y - 2)^2 + (z - 1)^2 = 9$ **10.** $5\sqrt{2}$ **11.** (a) $3x^2 + 6xy$ (b) $3x^2 + 4y$

12. (a) $6xe^{2y}$ (b) $6x^2e^{2y}$ **13.** (a) $2/x$ (b) $1/y$ **14.** (a) $3\cos 3x \sin 3y$ (b) $3\sin 3x \cos 3y$

15. (a) $\dfrac{ye^{x^2}(2x^2 - 1)}{x^2 \ln y}$ (b) $\dfrac{e^{x^2}(\ln y - 1)}{x \ln^2 y}$ **16.** (a) $\dfrac{e^x \sin y(\sin x - \cos x)}{y \sin^2 x}$ (b) $\dfrac{e^x(y \cos y - \sin y)}{y^2 \sin x}$

17. $\dfrac{-p^{1/2}}{2w^{3/2}}$ **18.** LI **19.** $\dfrac{X_C}{\sqrt{R^2 + X_C^2}}$ **20.** $\dfrac{rE}{(R + r)^2}$ **21.** (a) 8 (b) 2

22. $\dfrac{2x + y}{2(x^2 + xy)}\,dx + \dfrac{x}{2(x^2 + xy)}\,dy$ **23.** $\dfrac{2y}{(x + y)^2}\,dx - \dfrac{2x}{(x + y)^2}\,dy$ **24.** 452 L   **25.** 0.000666 A or

0.666 mA   **26.** Saddle point $(5, 2, -33)$ **27.** Saddle point, $(1, 2, -7)$ **28.** Maximum, $(0, -4, 25)$

29. 2 m × 2 m × 2 m   **30.** $\frac{1}{20}$ **31.** 516   **32.** $\frac{14}{3}$ **33.** $\ln \sqrt{2}$ or $\frac{1}{2} \ln 2$ **34.** $abc/6$ **35.** $\frac{2}{3}$

36. $\frac{4}{15}$

CHAPTER 10

Pages 394–395

1. 17   **3.** 24   **5.** -95 **7.** 57   **9.** 202.5   **11.** -546 **13.** $2, -1, -4, -7, -10$

15. $5, 5\frac{2}{3}, 6\frac{1}{3}, 7, 7\frac{2}{3}$ **17.** 4   **19.** 1,000,000   **21.** \$31,200   **23.** $\frac{20}{2187}$ **25.** 8   **27.** $-\frac{1}{64}$

29. $\dfrac{65,600}{2187}$ **31.** $7\sqrt{2} + 14$ **33.** $\frac{341}{64}$ **35.** $3, \frac{3}{2}, \frac{3}{4}, \frac{3}{8}, \frac{3}{16}$ **37.** $5, -\frac{5}{4}, \frac{5}{16}, -\frac{5}{64}, \frac{5}{256}$

39. $-4, -12, -36, -108, -324$ **41.** $\frac{1}{2}$ **43.** \$17,531   **45.** $\frac{3}{8}$ ft   **47.** 15.1°C   **49.** $\frac{14}{3}$

51. $\frac{8}{3}$ **53.** No sum   **55.** No sum   **57.** $\frac{1}{3}$ **59.** $\frac{2}{165}$ **61.** $\frac{13}{15}$

Pages 401–402

1. $5 + 9 + 13 + 17 + 21 + 25$ **3.** $10 + 17 + 26 + 37 + 50 + 65$

5. $\dfrac{1}{2} + \dfrac{4}{3} + \dfrac{9}{4} + \dfrac{16}{5} + \cdots + \dfrac{n^2}{n + 1}$ **7.** $-1 + \dfrac{1}{4} - \dfrac{1}{9} + \dfrac{1}{16} - \dfrac{1}{25} + \cdots$ **9.** $\displaystyle\sum_{n=1}^{12} n$

11. $\displaystyle\sum_{n=1}^{50} (2n)$ **13.** $\displaystyle\sum_{n=1}^{n} (2n - 1)$ **15.** $\displaystyle\sum_{n=3}^{n} (n^2 + 1)$ **17.** Diverges   **19.** Diverges

21. Diverges   **23.** Converges   **25.** Converges   **27.** Converges   **29.** Diverges   **31.** Converges

33. Diverges   **35.** Converges   **37.** Converges   **39.** Diverges   **41.** Converges   **43.** Diverges

Page 406

1. Converges   **3.** Converges   **5.** Converges   **7.** Diverges   **9.** Converges   **11.** Diverges

13. Diverges   **15.** Diverges   **17.** Diverges   **19.** Converges

Page 408

1. Converges conditionally   **3.** Converges absolutely   **5.** Diverges   **7.** Converges conditionally
9. Converges absolutely   **11.** Diverges   **13.** Diverges   **15.** Converges conditionally
17. Diverges   **19.** Converges absolutely

Page 412

1. $-2 < x < 2$ **3.** $x = 0$ **5.** $-\infty < x < \infty$ **7.** $-1 \le x \le 1$ **9.** $-1 \le x < 1$
11. $-\frac{3}{2} < x < \frac{3}{2}$ **13.** $-2 < x \le 2$ **15.** $1 < x \le 3$ **17.** $-1 \le x \le 1$ **19.** $-\infty < x < \infty$
21. $-\frac{3}{2} \le x < \frac{3}{2}$ **23.** $2 \le x \le 3$

Page 415

1. $x - \dfrac{x^3}{3!} + \dfrac{x^5}{5!} - \cdots$ **3.** $1 - x + \dfrac{x^2}{2!} - \dfrac{x^3}{3!} + \cdots$ **5.** $x - \dfrac{x^2}{2} + \dfrac{x^3}{3} - \dfrac{x^4}{4} + \cdots$

7. $1 - \dfrac{4x^2}{2!} + \dfrac{16x^4}{4!} - \cdots$ **9.** $x + x^2 + \dfrac{x^3}{2!} + \dfrac{x^4}{3!} + \cdots$ **11.** $2 - \dfrac{x}{4} - \dfrac{x^2}{(32)(2!)} - \dfrac{3x^3}{256(3!)} - \cdots$

13. $-1 + \dfrac{x^2}{2!} - \dfrac{x^4}{4!} + \cdots$ **15.** $1 + 2x + 3x^2 + 4x^3 + \cdots$

17. $1 + 5x + 10x^2 + 10x^3 + 5x^4 + x^5$ (Sum is finite.)   **19.** $x - x^2 + \frac{1}{3}x^3 - \frac{1}{30}x^5 + \cdots$

Pages 418–419

1. $1 - x + \dfrac{x^2}{2!} - \dfrac{x^3}{3!} + \dfrac{x^4}{4!} - \cdots$ **3.** $1 + x^2 + \dfrac{x^4}{2!} + \dfrac{x^6}{3!} + \cdots$ **5.** $-x - \dfrac{x^2}{2} - \dfrac{x^3}{3} - \dfrac{x^4}{4} - \cdots$

7. $1 - \dfrac{25x^4}{2!} + \dfrac{625x^8}{3!} - \cdots$ **9.** $x^3 - \dfrac{x^9}{3!} + \dfrac{x^{15}}{5!} - \cdots$ **11.** $x + x^2 + \dfrac{x^3}{2!} + \dfrac{x^4}{3!} + \dfrac{x^5}{4!} + \cdots$

13. $-\dfrac{x}{2!} + \dfrac{x^3}{4!} - \dfrac{x^5}{6!} + \cdots$ **15.** 0.743 **17.** $\ln 2 + \frac{7}{4}$ **19.** 0.602 **21.** $x + \dfrac{x^3}{3!} + \dfrac{x^5}{5!} + \cdots$

23. 0.041481 coulomb

Page 420

1. $-\left(x - \dfrac{\pi}{2}\right) + \dfrac{\left(x - \dfrac{\pi}{2}\right)^3}{3!} - \dfrac{\left(x - \dfrac{\pi}{2}\right)^5}{5!} + \cdots$ **3.** $e^2\left[1 + (x - 2) + \dfrac{(x - 2)^2}{2!} + \dfrac{(x - 2)^3}{3!} + \cdots\right]$

5. $3 + \dfrac{(x - 9)}{6} - \dfrac{(x - 9)^2}{108(2!)} + \dfrac{(x - 9)^3}{648(3!)} + \cdots$ **7.** $\dfrac{1}{2} - \dfrac{(x - 2)}{4} + \dfrac{(x - 2)^2}{4(2!)} - \dfrac{3(x - 2)^3}{8(3!)} + \cdots$

9. $(x - 1) - \dfrac{(x - 1)^2}{2!} + \dfrac{2(x - 1)^3}{3!} - \dfrac{6(x - 1)^4}{4!} + \cdots$ **11.** $1 - 2(x - 1) + \frac{3}{8}(x - 1)^2 - \frac{5}{16}(x - 1)^3 + \cdots$

13. $1 - 2(x - 1) + 3(x - 1)^2 - 4(x - 1)^3 + \cdots$ **15.** $-1 + \dfrac{(x - \pi)^2}{2!} - \dfrac{(x - \pi)^4}{4!} + \cdots$

Pages 422–423

1. 1.10517 **3.** 0.99985 **5.** -0.68229 **7.** 1.0488 **9.** 3.66832 **11.** 0.48481 **13.** 0.029996

Page 429

1. $-\pi + 2 \sin x + \sin 2x + \frac{2}{3} \sin 3x + \cdots$ **3.** $\dfrac{\pi}{3} - \dfrac{2}{3} \sin x - \dfrac{1}{3} \sin 2x - \dfrac{2}{9} \sin 3x - \cdots$

5. $\dfrac{1}{2} - \dfrac{2}{\pi} \sin x - \dfrac{2}{3\pi} \sin 3x - \dfrac{2}{5\pi} \sin 5x - \cdots$ **7.** $\dfrac{4}{\pi} \sin x + \dfrac{4}{3\pi} \sin 3x + \dfrac{4}{5\pi} \sin 5x + \cdots$

9. $\dfrac{\pi}{2} - \dfrac{4}{\pi} \cos x - \dfrac{4}{9\pi} \cos 3x - \dfrac{4}{25\pi} \cos 5x - \cdots$ **11.** $\dfrac{e^{2\pi} - 1}{2\pi} + \dfrac{1}{\pi} \cdot \dfrac{e^{2\pi} - 1}{2} \cos x + \dfrac{1}{\pi} \cdot \dfrac{e^{2\pi} - 1}{5} \cos 2x +$

$\dfrac{1}{\pi} \cdot \dfrac{e^{2\pi} - 1}{10} \cos 3x + \cdots + \dfrac{1}{\pi} \cdot \dfrac{1 - e^{2\pi}}{2} \sin x + \dfrac{1}{\pi} \cdot \dfrac{2 - e^{2\pi}}{5} \sin 2x + \cdots$

Answers to Odd-Numbered Exercises and Chapter Reviews

13. $\dfrac{1}{\pi} + \dfrac{1}{2}\sin x - \dfrac{2}{3\pi}\cos 2x - \dfrac{2}{15\pi}\cos 4x - \cdots$

Pages 434–435
(Chapter Review)

1. 47   **2.** $\frac{1}{16}$ **3.** -81 **4.** -62 **5.** $\frac{2}{81}$ **6.** 95   **7.** 300   **8.** $\frac{127}{16}$ **9.** $\dfrac{-80\sqrt{3}}{1 + \sqrt{3}}$

10. -348 **11.** $\frac{728}{81}$ **12.** 500   **13.** 1,001,000   **14.** \$2988 (approx.)   **15.** No sum

16. $\frac{35}{6}$ **17.** $\frac{3}{2}$ **18.** No sum   **19.** $\frac{5}{11}$ **20.** $\frac{152}{165}$ **21.** $-2 \ -5 \ -8 \ -11 \ -14 \ -17$

22. $2 + \dfrac{3}{2} + \dfrac{4}{3} + \dfrac{5}{4} + \cdots + \dfrac{n + 1}{n}$ **23.** $\displaystyle\sum_{n=1}^{7}\dfrac{1}{3^{n}}$ **24.** $\displaystyle\sum_{n=1}^{10}\dfrac{n}{n + 3}$ **25.** Converges   **26.** Diverges

27. Converges   **28.** Converges   **29.** Diverges   **30.** Converges   **31.** Converges   **32.** Diverges

33. Diverges   **34.** Converges   **35.** Converges absolutely   **36.** Converges absolutely

37. Diverges   **38.** Converges conditionally   **39.** $1 < x < 3$ **40.** $2 < x < 4$

41. $-\infty < x < \infty$ **42.** $\frac{11}{3} \le x \le \frac{13}{3}$ **43.** $1 + x + x^2 + x^3 + \cdots$ **44.** $1 + \dfrac{x}{2} - \dfrac{x^2}{8} + \dfrac{x^3}{16} - \cdots$

45. $1 + x - \dfrac{x^2}{2!} - \dfrac{x^3}{3!} + \dfrac{x^4}{4!} + \cdots$ **46.** $x + x^2 + \frac{1}{3}x^3 + \cdots$ **47.** $-1 - \dfrac{x}{2!} - \dfrac{x^2}{3!} - \cdots$

48. $1 - \dfrac{x^4}{2!} + \dfrac{x^8}{4!} - \dfrac{x^{12}}{6!} + \cdots$ **49.** $3x - \dfrac{9x^3}{2} + \dfrac{81x^5}{40} - \cdots$ **50.** $1 + \sin x + \dfrac{\sin^2 x}{2!} + \dfrac{\sin^3 x}{3!} + \cdots$

51. 0.09772   **52.** 0.09994   **53.** $1 - 2x^2 + \dfrac{2x^4}{3} - \cdots$

54. $\ln 4 + \dfrac{x - 4}{4} - \dfrac{(x - 4)^2}{32} + \dfrac{(x - 4)^3}{192} - \cdots$ **55.** $e\left[1 + 2(x - 1) + \dfrac{6(x - 1)^2}{2!} + \dfrac{20(x - 1)^3}{3!} + \cdots\right]$

56. $-1 + \dfrac{\left(x - \dfrac{3\pi}{2}\right)^2}{2!} - \dfrac{\left(x - \dfrac{3\pi}{2}\right)^4}{4!} + \cdots$ **57.** 0.5150   **58.** 3.208   **59.** 0.18227

60. 2.024846   **61.** $-\dfrac{1}{2} + \dfrac{2}{\pi}\sin x + \dfrac{2}{3\pi}\sin 3x + \dfrac{2}{5\pi}\sin 5x + \cdots$

62. $\dfrac{\pi^2}{6} - 2\cos x + \dfrac{1}{2}\cos 2x - \dfrac{2}{9}\cos 3x + \cdots + \dfrac{\pi^2 - 4}{\pi}\sin x - \dfrac{\pi}{2}\sin 2x + \dfrac{9\pi^2 - 4}{27\pi}\sin 3x - \cdots$

CHAPTER 11

Pages 438–439

1. Order 1; degree 1   **3.** Order 2; degree 1   **5.** Order 3; degree 1   **7.** Order 2; degree 3

Pages 444–445

1. $\dfrac{-1}{y} = \ln x - C$ or $y \ln x + 1 = Cy$ **3.** $y = \dfrac{C}{x}$ or $xy = C$ **5.** $-2y^{-1/2} = x + C$

7. $\operatorname{Arctan} y = \dfrac{x^3}{3} + C$ **9.** $-\ln(3 - y) = \ln x + C$ or $x(3 - y) = C$ **11.** $3y^2 = 2x^3 + C$ **13.** $1 = 2y^2(\sin x + C)$

15. $2y + e^{-2x} = C$ **17.** $y = \operatorname{Arctan} x + C$ **19.** $\operatorname{Arctan} y = x + \dfrac{x^3}{3} + C$ **21.** $e^y = e^x + C$

23. $\operatorname{Arctan} \dfrac{y}{2} = 2\ln(x + 1) + C$ **25.** $(x^2 + 1)^2 = C(y + 3)^5$ **27.** $1 = (2 - x^3)y^3$

29. $y^2 = 2\ln(x^2 + 1) + 16$ **31.** $y^2 = 2e^x + 34$ **33.** $x^{3/2} + y^{3/2} = 9$ **35.** $y^2 = \ln^2 x$

Page 447

1. $3xy = y^3 + C$ **3.** $y = 5xy + Cx$ **5.** $x + xy + Cy + 3 = 0$ **7.** $x^2 = 4\sqrt{x^2 + y^2} + C$

9. $\ln \sqrt{x^2 + y^2} = xy + C$ **11.** $xy = x^2 + y^2 - 1$ **13.** $\text{Arctan}\left(\dfrac{y}{x}\right) = xy + \dfrac{\pi - 16}{4}$

Pages 449–450

1. $2y + e^{3x} = Ce^{5x}$ **3.** $14x^3y = 2x^7 - 7x^4 + C$ **5.** $y = e^{3x} + Ce^{-x2}$ **7.** $3y = x^3e^{4x} + Ce^{4x}$
9. $y = x^6 - x + Cx^5$ **11.** $(1 + x^2)y = x^3 + C$ **13.** $5x^2y = x^5 - 35x + C$
15. $y = e^{-x} + Ce^{-2x}$ **17.** $y = 3x^2 + Cx$ **19.** $(y - 1)e^{\sin x} = C$ **21.** $y = e^{2x}(3e^x - 1)$

23. $y \sin x = x + \pi$ **25.** $y = \dfrac{e^x}{2} + e^{-x}$ **27.** $y = 3 - \dfrac{5}{x}$ **29.** $xy = x^4 - 10$

Pages 455–457

1. $v = 5t + 10;\ v = 25$ m/s   **3.** $3xy = x^3 + 2$ **5.** $i = \frac{1}{2}(3 + e^{-800t})$ **7.** $i = 1 - e^{-5t2}$
9. $Q = e^{-1.54 \times 10 - 10t}$ **11.** $Q = 5e^{-0.00293t};\ 237$ yr   **13.** 72.9 lb   **15.** 24.3°C   **17.** 3.2×10^7
19. $P = CV^k$

Pages 458–459
(Chapter Review)

1. Order 2; degree 1   **2.** Order 1; degree 2   **3.** Order 2; degree 3   **4.** Order 1; degree 2
9. $2x^2 \ln y = -1 + Cx^2$ **10.** $2 \ln y + 3e^{-2x} = C$ **11.** $2y^3 + 3 \ln (9 + x^2) = C$
12. $\tan x + e^{-y} = C$ **13.** $y = x^4 + Cx$ **14.** $(y - 2)e^{2x3} = C$ **15.** $(y + 5)e^{1/x} = C$

16. $\ln (x^2 + y^2) = y^2 + C$ **17.** $3 \text{ Arctan} \dfrac{y}{x} = x^3 + C$ **18.** $3xy = 7x^6 + C$ **19.** $ye^{3x} = e^x + C$

20. $4y = -4x^4 - 7x + Cx^5$ **21.** $y = 4x^3 \ln x - x^2 + Cx^3$ **22.** $2y \sin x = \sin^2 x + C$

23. $3x^2y - 10y + 2 = 0$ **24.** $y^2 = 3e^{-2x} + 1$ **25.** $4y = x^5 + 3x$ **26.** $xy = \dfrac{x^2}{4}(2 \ln x - 1) + 1$

27. $4y = e^{5x} - 13e^x$ **28.** $2y = x^4 - 10x^2 \ln x + 5x^2$ **29.** $i = 2 - e^{-300t}$ **30.** $q = e^{-7.88 \times 10 - 10t}$
31. 4.524 kg   **32.** 20.6°C   **33.** $y^3 = 8e^{x3}$ **34.** $v = t^4;\ 81$ m/s^2

CHAPTER 12

Page 465

1. Homogeneous, 4   **3.** Nonhomogeneous, 3   **5.** Homogeneous, 2   **7.** Nonhomogeneous, 3
9. $y = k_1e^{7x} + k_2e^{-2x}$ **11.** $y = k_1e^{4x} + k_2e^{-2x}$ **13.** $y = k_1e^x + k_2e^{-x}$ **15.** $y = k_1 + k_2e^{3x}$
17. $y = k_1e^{3x/2} + k_2e^{5x}$ **19.** $y = k_1e^{2x} + k_2e^{x/3}$ **21.** $y = 2 + e^{4x}$ **23.** $y = e^{2x} + e^{-x}$
25. $y = 9e^{3x} - 5e^{5x}$

Page 468

1. $y = e^{2x}(k_1 + k_2x)$ **3.** $y = e^{2x}(k_1 \sin x + k_2 \cos x)$ **5.** $y = e^{x/2}(k_1 + k_2x)$
7. $y = e^{2x}(k_1 \sin 3x + k_2 \cos 3x)$ **9.** $y = e^{5x}(k_1 + k_2x)$ **11.** $y = k_1 \sin 3x + k_2 \cos 3x$
13. $y = k_1 + k_2x$ **15.** $y = 2e^{3x}(1 - x)$ **17.** $y = 2 \cos 5x$ **19.** $y = e^{6x}(1 - 6x)$

Pages 473–474

1. $y = k_1 + k_2e^{-x} - \frac{1}{2} \sin x - \frac{1}{2} \cos x$ **3.** $y = k_1e^{2x} + k_2e^{-x} - 2x + 1$

5. $y = e^{5x}(k_1 + k_2x) + \dfrac{x}{25} + \dfrac{2}{125}$ **7.** $y = k_1e^x + k_2e^{-x} - 2 - x^2$

9. $y = k_1 \sin 2x + k_2 \cos 2x + \dfrac{e^x}{5} - \dfrac{1}{2}$ **11.** $y = k_1e^{4x} + k_2e^{-x} - e^x$

13. $y = k_1 \sin x + k_2 \cos x + 5 - \frac{1}{8} \sin 3x$ **15.** $y = k_1e^x + k_2e^{-x} + \frac{1}{2}xe^x$
17. $y = k_1 \sin 2x + k_2 \cos 2x + \frac{1}{4}x \sin 2x$ **19.** $y = 2(e^{2x} - \cos x - 2 \sin x)$
21. $y = \frac{1}{2}(5 \sin x - \cos x + e^x)$

Page 482

1. $x = 0.167 \cos 8\sqrt{2}t$ **3.** $x = c_1 e^{-3.06t} + c_2 e^{-20.9t}$ **5.** 13,000 lb   **7.** 1.25 ft
9. $i = k_1 e^{(-138t)} + k_2 e^{(-362t)}$ **11.** $i = 0.136 e^{-140t} - 0.136 e^{-360t}$

Pages 487–488

3. $\dfrac{3}{s^2 + 9}$ **5.** $\dfrac{-4}{s(s-4)}$ **7.** $\dfrac{2}{s^3}$ **9.** $\dfrac{16}{(s^2+4)^2}$ **11.** $\dfrac{-s^2 + s - 2}{(s-2)s^2}$

13. $\dfrac{8s}{(s^2+16)^2}$ **15.** $\dfrac{s+3}{s^2+6s+34}$ **17.** $\dfrac{8s^2+24}{s^4}$ **19.** $(s^2 - 3s)\mathscr{L}(y)$ **21.** $(s^2 + s + 1)\mathscr{L}(y) - 1$

23. $(s^2 - 3s + 1)\mathscr{L}(y) - s + 3$ **25.** $(s^2 + 8s + 2)\mathscr{L}(y) - 4s - 38$ **27.** $(s^2 - 6s)\mathscr{L}(y) - 3s + 11$

29. $(s^2 + 8s - 3)\mathscr{L}(y) + 6s + 46$ **31.** 1   **33.** te^{5t} **35.** $\cos 8t$ **37.** $e^{6t} - e^{2t}$ **39.** $e^{3t} \sin 2t$

41. $\sin t - t \cos t$ **43.** $e^t - \sin t - \cos t$ **45.** $2e^{-3t} + e^{-6t} + 3$ **47.** $e^{-5t}(3 \sin 2t + \cos 2t)$

Page 492

1. $y = 2e^t$ **3.** $y = e^{-3t/4}$ **5.** $6y = 31e^{7t} - e^t$ **7.** $y = \cos t$ **9.** $2y = 3 - e^{2t}$
11. $y = e^{-t}(1 + t)$ **13.** $6y = t^3 e^{2t}$ **15.** $y = (4 - 10t + \frac{1}{2}t^3)e^{-t}$ **17.** $5y = 3e^{-4t} + 2e^t$

Pages 494–495
(Chapter Review)

1. Homogeneous; order 2   **2.** Nonhomogeneous; order 2   **3.** Nonhomogeneous; order 1
4. Homogeneous; order 3   **5.** $y = k_1 e^{5x} + k_2 e^{-x}$ **6.** $y = k_1 e^{3x} + k_2 e^{2x}$ **7.** $y = k_1 + k_2 e^{6x}$
8. $y = k_1 e^{(3/2)x} + k_2 e^{-x}$ **9.** $y = k_1 e^{3x} + k_2 x e^{3x}$ **10.** $y = k_1 e^{-5x} + k_2 x e^{-5x}$ **11.** $y = k_1 e^x + k_2 x e^x$
12. $y = k_1 e^{(1/3)x} + k_2 x e^{(1/3)x}$ **13.** $y = k_1 \sin 4x + k_2 \cos 4x$ **14.** $y = k_1 \sin \frac{3}{2}x + k_2 \cos \frac{3}{2}x$

15. $y = e^x(k_1 \sin \sqrt{2}x + k_2 \cos\sqrt{2}x)$ **16.** $y = e^{3x/2}\left(k_1 \sin \dfrac{\sqrt{23}}{2}x + k_2 \cos \dfrac{\sqrt{23}}{2}x\right)$

17. $y = k_1 e^x + k_2 e^{-2x} - \frac{1}{4} - \frac{1}{2}x$ **18.** $y = k_1 e^{3x} + k_2 x e^{3x} + \frac{1}{4}e^x$
19. $y = k_1 \sin 2x + k_2 \cos 2x + \frac{1}{3}\cos x$ **20.** $y = e^x(k_1 \sin \sqrt{2}\,x + k_2 \cos \sqrt{2}\,x) + 2e^{2x}$
21. $y = 4e^{2x} + 2e^{-4x}$ **22.** $y = 1 + 2e^{3x}$ **23.** $y = 3xe^{2x}$ **24.** $y = 8e^{-3x}(1 + 3x)$
25. $y = -2 \sin 2x + \cos 2x$ **26.** $y = e^{4x}(\sin 3x + 2 \cos 3x)$ **27.** $2y = e^{2x} - 2e^x + 3$
28. $y = \sin 2x + 2 \cos 2x + \frac{1}{3}\sin x$ **29.** $x = \frac{1}{2}\cos 4\sqrt{6}t$ **30.** $x = k_1 e^{(-32+8\sqrt{13})t} + k_2 e^{(-32-8\sqrt{13})t}$
31. $i = e^{-100t}(k_1 \sin 200t + k_2 \cos 200t)$ **32.** $i = k_1 e^{(-1000+500\sqrt{3})t} + k_2 e^{(-1000-500\sqrt{3})t}$ **33.** $\dfrac{1}{s-6}$

34. $\dfrac{s+2}{(s+2)^2+9}$ **35.** $\dfrac{6}{s^4} + \dfrac{s}{s^2+1}$ **36.** $\dfrac{3}{s^2} - \dfrac{1}{s-5}$ **37.** t **38.** $\frac{3}{2}(e^{2t} - 1)$ **39.** $2(e^{4t} - e^{3t})$

40. $\sin 3t$ **41.** $y = 2e^{(5/4)t}$ **42.** $y = 3 \cos 3t$ **43.** $y = \frac{2}{5}(1 - e^{-2t})$ **44.** $y = \dfrac{t^2 e^{-2t}}{2}$

APPENDIX A

Page 498

1. $27x^3 + 27x^2y + 9xy^2 + y^3$ **2.** $16x^4 - 96x^3y + 216x^2y^2 - 216xy^3 + 81y^4$
3. $64a^6 + 576a^5b + 2160a^4b^2 + 4320a^3b^3 + 4860a^2b^4 + 2916ab^5 + 726b^6$
4. $a^8 - 16a^7b + 112a^6b^2 - 448a^5b^3 + 1120a^4b^4 - 1792a^3b^5 + 1792a^2b^6 - 1024ab^7 + 256b^8$
5. $\dfrac{32x^5}{243} - \dfrac{160x^4}{81} + \dfrac{320x^3}{27} - \dfrac{320x^2}{9} + \dfrac{160x}{3} - 32$
6. $\dfrac{243m^5}{1024} + \dfrac{135m^4k}{128} + \dfrac{15m^3k^2}{8} + \dfrac{5m^2k^3}{3} + \dfrac{20mk^4}{27} + \dfrac{32k^5}{243}$ **7.** $\dfrac{x^4}{y^4} - \dfrac{8x^3}{y^3z} + \dfrac{24x^2}{y^2z^2} - \dfrac{32x}{yz^3} + \dfrac{16}{z^4}$
8. $\dfrac{64x^6}{y^6} + \dfrac{576x^5}{y^5z} + \dfrac{2160x^4}{y^4z^2} + \dfrac{4320x^3}{y^3z^3} + \dfrac{4860x^2}{y^2z^4} + \dfrac{2916x}{yz^5} + \dfrac{729}{z^6}$
9. $-126x^4y^5$ **10.** $2560x^3y^2$ **11.** $280x^3y^8$ **12.** $70x^8y^{12}$

INDEX